Methods in Enzymology

Volume 104

ENZYME PURIFICATION AND RELATED TECHNIQUES

Part C

METHODS IN ENZYMOLOGY

EDITORS-IN-CHIEF

Sidney P. Colowick Nathan O. Kaplan

Methods in Enzymology

Volume 104

Enzyme Purification and Related Techniques

Part C

EDITED BY

William B. Jakoby
NATIONAL INSTITUTE OF ARTHRITIS, DIABETES,
AND DIGESTIVE AND KIDNEY DISEASES
NATIONAL INSTITUTES OF HEALTH
BETHESDA, MARYLAND

1984

ACADEMIC PRESS, INC.
(*Harcourt Brace Jovanovich, Publishers*)

Orlando San Diego San Francisco New York London
Toronto Montreal Sydney Tokyo São Paulo

ACADEMIC PRESS, INC.
Orlando, Florida 32887

United Kingdom Edition published by
ACADEMIC PRESS, INC. (LONDON) LTD.
24/28 Oval Road, London NW1 7DX

LIBRARY OF CONGRESS CATALOG CARD NUMBER: 54-9110
ISBN 0-12-182004-1

PRINTED IN THE UNITED STATES OF AMERICA

84 85 86 87 9 8 7 6 5 4 3 2 1

Table of Contents

Section I. Chromatography

Section II. Electrophoresis

Section III. Techniques for Membrane Proteins

Section IV. Other Separation Systems

Section V. Related Techniques

Contributors to Volume 104

Article numbers are in parentheses following the names of contributors.
Affiliations listed are current.

RÜDIGER V. BATTERSBY (15), *Abteilung Klinische Biochemie, Medizinische Hochschule Hannover, Hannover, Federal Republic of Germany*

PERRY J. BLACKSHEAR (12), *Diabetes Unit, Massachusetts General Hospital, Boston, Massachusetts 02114*

JOHN S. BLANCHARD (26), *Department of Biochemistry, Albert Einstein College of Medicine, Bronx, New York 10461*

HILARY BOLTON (17), *Department of Biology, The Open University, Walton Hall, Milton Keynes MK7 6AA, England*

WILLIAM M. BONNER (34), *Laboratory of Molecular Pharmacology, National Cancer Institute, National Institutes of Health, Bethesda, Maryland 20205*

ENRICO CABIB (27), *Laboratory of Biochemistry and Metabolism, National Institute of Arthritis, Diabetes, and Digestive and Kidney Diseases, National Institutes of Health, Bethesda, Maryland 20205*

GARY J. CALTON (24), *Purification Engineering, Inc., 9505 Berger Road, Columbia, Maryland 21046*

ANDREAS CHRAMBACH (16), *Section of Macromolecular Analysis, Laboratory of Theoretical and Physical Biology, National Institute of Child Health and Human Development, National Institutes of Health, Bethesda, Maryland 20205*

JOHN A. CUTTING (32), *Houston, Texas 77030*

DAVID R. DAVIES (23), *Laboratory of Molecular Biology, National Institute of Arthritis, Diabetes, and Digestive and Kidney Diseases, National Institutes of Health, Bethesda, Maryland 20205*

SUSANNE FLYGARE (22), *Department of Pure and Applied Biochemistry, Chemical Center, University of Lund, S-220 07 Lund, Sweden*

ANNA J. FURTH (17), *Department of Biology, The Open University, Walton Hall, Milton Keynes MK7 6AA, England*

OTHMAR GABRIEL (28), *Department of Biochemistry, Georgetown University Schools of Medicine and Dentistry, Washington, D.C. 20007*

JOHN E. GANDER (31), *Department of Microbiology and Cell Science, University of Florida, Gainesville, Florida 32611*

GARY C. GANZI (25), *Millipore Corporation, Ashby Road, Bedford, Massachusetts 01730*

GARY L. GILLILAND (23), *Department of Physical Chemistry, Hoffman La-Roche, Inc., Nutley, New Jersey 07110*

DAVID GOLDMAN (30), *Laboratory of Clinical Science, National Institute of Mental Health, National Institutes of Health, Bethesda, Maryland 20205*

MILTON T. W. HEARN (9), *St. Vincent's School of Medical Research, Fitzroy, Victoria 3065, Australia*

MARY J. HEEB (28), *Department of Immunology, Research Institute of Scripps Clinic, La Jolla, California 92037*

LEONARD M. HJELMELAND (16), *Laboratory of Vision Research, National Eye Institute, National Institutes of Health, Bethesda, Maryland 20205*

CHRISTOPHER J. HOLLOWAY (15), *Abteilung Klinische Biochemie, Medizinische Hochschule Hannover, Hannover, Federal Republic of Germany*

VÁCLAV HOŘEJŠÍ (14), *Institute of Molecular Genetics, Czechoslovak Academy of Sciences, 142 20 Praha 4, Czechoslovakia*

KENNETH C. INGHAM (20), *American Red Cross Blood Services, Plasma Derivatives Laboratory, 9312 Old Georgetown Road, Bethesda, Maryland 20814*

GÖTE JOHANSSON (21), *Department of Biochemistry, Chemical Center, University of Lund, S-220 07 Lund, Sweden*

CHRISTOPH KEMPF (18), *Institut für Hygiene und Medizinische Mikrobiologie, Universität Bern, Friedbuehlstrasse 59, Bern CH-3008, Switzerland*

RICHARD D. KLAUSNER (19), *Laboratory of Biochemistry and Metabolism, National Institute of Arthritis, Diabetes, and Digestive and Kidney Diseases, National Institutes of Health, Bethesda, Maryland 20205*

JOACHIM KOHN (1), *Department of Biophysics, The Weizmann Institute of Science, Rehovot 76100, Israel*

PER-OLOF LARSSON (10, 22), *Department of Pure and Applied Biochemistry, Chemical Center, University of Lund, S-220 07 Lund, Sweden*

CHRISTOPHER R. LOWE (4), *Department of Biochemistry, University of Southampton, Bassett Crescent East, Southampton S09 3TU, England*

CARL R. MERRIL (30), *Laboratory of General and Comparative Biochemistry, National Institute of Mental Health, National Institutes of Health, Bethesda, Maryland 20205*

TALIA MIRON (1), *Department of Biophysics, The Weizmann Institute of Science, Rehovot 76100, Israel*

KLAUS MOSBACH (2, 22), *Department of Pure and Applied Biochemistry, Chemical Center, University of Lund, S-220 07 Lund, Sweden*

KURT NILSSON (2), *Department of Pure and Applied Biochemistry, Chemical Center, University of Lund, S-220 07 Lund, Sweden*

JAMES C. PEARSON (4), *Department of Biochemistry, University of Southampton, Bassett Crescent East, Southampton S09 3TU, England*

ELBERT A. PETERSON (5), *National Cancer Institute, National Institutes of Health, Bethesda, Maryland 20205*

ITZHACK POLACHECK (27), *Department of Clinical Microbiology, Hadassah University Hospital, Ein Karem, Jerusalem, Israel*

JENNIFER POTTER (17), *Department of Biology, The Open University, Walton Hall, Milton Keynes MK7 6AA, England*

JOHN D. PRIDDLE (17), *Department of Gastroenterology, Radcliffe Infirmary, Oxford 0X2 6HE, England*

BERTOLD J. RADOLA (13), *Institut für Lebensmitteltechnologie und Analytische Chemie, Technische Universität München, D-8050 Freising-Weihenstephan, Federal Republic of Germany*

FRED E. REGNIER (8), *Department of Biochemistry, Purdue University, West Lafayette, Indiana 47907*

A. H. REISNER (29), *CSIRO, Division of Molecular Biology, North Ryde N.S.W. 2113, Australia*

JAIME RENART (33), *Instituto de Enzimología y Patología Molecular del C.S.I.C., Facultad de Medicina, Universidad Autónoma de Madrid, Madrid 34, Spain*

JAMES S. RICHEY (11), *Pharmacia Fine Chemicals, 800 Centennial Avenue, Piscataway, New Jersey 08854*

BENJAMIN RIVNAY (19), *Department of Membrane Research, The Weizmann Institute of Science, Rehovot 76100, Israel*

IGNACIO V. SANDOVAL (33), *Section on Enzymes and Cellular Biochemistry, Laboratory of Biochemistry and Metabolism, National Institute of Arthritis, Diabetes, and Digestive and Kidney Diseases, National Institutes of Health, Bethesda, Maryland 20205*

SHMUEL SHALTIEL (3), *Department of Chemical Immunology, The Weizmann Institute of Science, Rehovot 76100, Israel*

ANTHONY R. TORRES (5), *Department of Laboratory Medicine, Yale University School of Medicine, New Haven, Connecticut 06504*

KLAUS UNGER (7), *Institut für Anorganische Chemie und Analytische Chemie, Johannes-Gutenberg-Universität, 6500 Mainz, Federal Republic of Germany*

MARGARET L. VAN KEUREN (30), *The Eleanor Roosevelt Institute for Cancer Research, 4200 East Ninth Avenue, Denver, Colorado 80262*

JOS VAN RENSWOUDE (18, 19), *Laboratory of Biochemistry and Metabolism, National Institute of Arthritis, Diabetes, and Digestive and Kidney Diseases, National Institutes of Health, Bethesda, Maryland 20205*

C. TIMOTHY WEHR (6), *Varian Associates, Walnut Creek Instrument Division, Walnut Creek, California 94598*

MEIR WILCHEK (1), *Department of Biophysics, The Weizmann Institute of Science, Rehovot 76100, Israel*

Preface

A dozen years after publication of "Enzyme Purification and Related Techniques," Volume 22 of *Methods in Enzymology,* a number of novel purification methods had been introduced which, along with subsequent modifications of existing procedures, suggested the need for a supplement. It is not that Volume 22 is outdated; almost all of the procedures remain very much in use. Rather, it was felt that new, simpler, and possibly more productive approaches that have been developed in the interim should be made accessible in the Methods format.

Some of the subjects are entirely new. High-performance liquid chromatography, for example, is now being applied to proteins and, less frequently, to enzymes. One major purification technique that has grown steadily in importance since being introduced in Volume 22, affinity chromatography, merited a volume of its own (Volume 34); additional techniques and a resumé of existing ones are presented in detail here. The solubilization of those proteins that are attached to membranes remains difficult, but a variety of different approaches and detergents have been introduced and are recorded in this volume. And a large number of methods of somewhat less protean specificity but general utility are also presented with the intent of rounding out our present knowledge of the state of the art and science of protein purification.

William B. Jakoby

METHODS IN ENZYMOLOGY

EDITED BY

Sidney P. Colowick and Nathan O. Kaplan

VANDERBILT UNIVERSITY
SCHOOL OF MEDICINE
NASHVILLE, TENNESSEE

DEPARTMENT OF CHEMISTRY
UNIVERSITY OF CALIFORNIA
AT SAN DIEGO
LA JOLLA, CALIFORNIA

METHODS IN ENZYMOLOGY

EDITORS-IN-CHIEF

Sidney P. Colowick   Nathan O. Kaplan

VOLUME XXXIV. Affinity Techniques (Enzyme Purification: Part B)
Edited by WILLIAM B. JAKOBY AND MEIR WILCHEK

VOLUME XXXV. Lipids (Part B)
Edited by JOHN M. LOWENSTEIN

VOLUME XXXVI. Hormone Action (Part A: Steroid Hormones)
Edited by BERT W. O'MALLEY AND JOEL G. HARDMAN

VOLUME XXXVII. Hormone Action (Part B: Peptide Hormones)
Edited by BERT W. O'MALLEY AND JOEL G. HARDMAN

VOLUME XXXVIII. Hormone Action (Part C: Cyclic Nucleotides)
Edited by JOEL G. HARDMAN AND BERT W. O'MALLEY

VOLUME XXXIX. Hormone Action (Part D: Isolated Cells, Tissues, and Organ Systems)
Edited by JOEL G. HARDMAN AND BERT W. O'MALLEY

VOLUME XL. Hormone Action (Part E: Nuclear Structure and Function)
Edited by BERT W. O'MALLEY AND JOEL G. HARDMAN

VOLUME XLI. Carbohydrate Metabolism (Part B)
Edited by W. A. WOOD

VOLUME XLII. Carbohydrate Metabolism (Part C)
Edited by W. A. WOOD

VOLUME XLIII. Antibiotics
Edited by JOHN H. HASH

VOLUME XLIV. Immobilized Enzymes
Edited by KLAUS MOSBACH

VOLUME XLV. Proteolytic Enzymes (Part B)
Edited by LASZLO LORAND

VOLUME XLVI. Affinity Labeling
Edited by WILLIAM B. JAKOBY AND MEIR WILCHEK

VOLUME XLVII. Enzyme Structure (Part E)
Edited by C. H. W. HIRS AND SERGE N. TIMASHEFF

VOLUME XLVIII. Enzyme Structure (Part F)
Edited by C. H. W. HIRS AND SERGE N. TIMASHEFF

VOLUME XLIX. Enzyme Structure (Part G)
Edited by C. H. W. HIRS AND SERGE N. TIMASHEFF

VOLUME L. Complex Carbohydrates (Part C)
Edited by VICTOR GINSBURG

VOLUME LI. Purine and Pyrimidine Nucleotide Metabolism
Edited by PATRICIA A. HOFFEE AND MARY ELLEN JONES

VOLUME LII. Biomembranes (Part C: Biological Oxidations)
Edited by SIDNEY FLEISCHER AND LESTER PACKER

VOLUME LIII. Biomembranes (Part D: Biological Oxidations)
Edited by SIDNEY FLEISCHER AND LESTER PACKER

VOLUME LIV. Biomembranes (Part E: Biological Oxidations)
Edited by SIDNEY FLEISCHER AND LESTER PACKER

VOLUME LV. Biomembranes (Part F: Bioenergetics)
Edited by SIDNEY FLEISCHER AND LESTER PACKER

VOLUME LVI. Biomembranes (Part G: Bioenergetics)
Edited by SIDNEY FLEISCHER AND LESTER PACKER

VOLUME LVII. Bioluminescence and Chemiluminescence
Edited by MARLENE A. DELUCA

VOLUME LVIII. Cell Culture
Edited by WILLIAM B. JAKOBY AND IRA PASTAN

VOLUME LIX. Nucleic Acids and Protein Synthesis (Part G)
Edited by KIVIE MOLDAVE AND LAWRENCE GROSSMAN

VOLUME LX. Nucleic Acids and Protein Synthesis (Part H)
Edited by KIVIE MOLDAVE AND LAWRENCE GROSSMAN

VOLUME 61. Enzyme Structure (Part H)
Edited by C. H. W. HIRS AND SERGE N. TIMASHEFF

Volume 62. Vitamins and Coenzymes (Part D)
Edited by Donald B. McCormick and Lemuel D. Wright

Volume 63. Enzyme Kinetics and Mechanism (Part A: Initial Rate and Inhibitor Methods)
Edited by Daniel L. Purich

Volume 64. Enzyme Kinetics and Mechanism (Part B: Isotopic Probes and Complex Enzyme Systems)
Edited by Daniel L. Purich

Volume 65. Nucleic Acids (Part I)
Edited by Lawrence Grossman and Kivie Moldave

Volume 66. Vitamins and Coenzymes (Part E)
Edited by Donald B. McCormick and Lemuel D. Wright

Volume 67. Vitamins and Coenzymes (Part F)
Edited by Donald B. McCormick and Lemuel D. Wright

Volume 68. Recombinant DNA
Edited by Ray Wu

Volume 69. Photosynthesis and Nitrogen Fixation (Part C)
Edited by Anthony San Pietro

Volume 70. Immunochemical Techniques (Part A)
Edited by Helen Van Vunakis and John J. Langone

Volume 71. Lipids (Part C)
Edited by John M. Lowenstein

Volume 72. Lipids (Part D)
Edited by John M. Lowenstein

Volume 73. Immunochemical Techniques (Part B)
Edited by John J. Langone and Helen Van Vunakis

Volume 74. Immunochemical Techniques (Part C)
Edited by John J. Langone and Helen Van Vunakis

VOLUME 75. Cumulative Subject Index Volumes XXXI, XXXII, and XXXIV–LX
Edited by EDWARD A. DENNIS AND MARTHA G. DENNIS

VOLUME 76. Hemoglobins
Edited by ERALDO ANTONINI, LUIGI ROSSI-BERNARDI, AND EMILIA CHIANCONE

VOLUME 77. Detoxication and Drug Metabolism
Edited by WILLIAM B. JAKOBY

VOLUME 78. Interferons (Part A)
Edited by SIDNEY PESTKA

VOLUME 79. Interferons (Part B)
Edited by SIDNEY PESTKA

VOLUME 80. Proteolytic Enzymes (Part C)
Edited by LASZLO LORAND

VOLUME 81. Biomembranes (Part H: Visual Pigments and Purple Membranes, I)
Edited by LESTER PACKER

VOLUME 82. Structural and Contractile Proteins (Part A: Extracellular Matrix)
Edited by LEON W. CUNNINGHAM AND DIXIE W. FREDERIKSEN

VOLUME 83. Complex Carbohydrates (Part D)
Edited by VICTOR GINSBURG

VOLUME 84. Immunochemical Techniques (Part D: Selected Immunoassays)
Edited by JOHN J. LANGONE AND HELEN VAN VUNAKIS

VOLUME 85. Structural and Contractile Proteins (Part B: The Contractile Apparatus and the Cytoskeleton)
Edited by DIXIE W. FREDERIKSEN AND LEON W. CUNNINGHAM

VOLUME 86. Prostaglandins and Arachidonate Metabolites
Edited by WILLIAM E. M. LANDS AND WILLIAM L. SMITH

VOLUME 87. Enzyme Kinetics and Mechanism (Part C: Intermediates, Stereochemistry, and Rate Studies)
Edited by DANIEL L. PURICH

VOLUME 88. Biomembranes (Part I: Visual Pigments and Purple Membranes, II)
Edited by LESTER PACKER

VOLUME 89. Carbohydrate Metabolism (Part D)
Edited by WILLIS A. WOOD

VOLUME 90. Carbohydrate Metabolism (Part E)
Edited by WILLIS A. WOOD

VOLUME 91. Enzyme Structure (Part I)
Edited by C. H. W. HIRS AND SERGE N. TIMASHEFF

VOLUME 92. Immunochemical Techniques (Part E: Monoclonal Antibodies and General Immunoassay Methods)
Edited by JOHN J. LANGONE AND HELEN VAN VUNAKIS

VOLUME 93. Immunochemical Techniques (Part F: Conventional Antibodies, Fc Receptors, and Cytotoxicity)
Edited by JOHN J. LANGONE AND HELEN VAN VUNAKIS

VOLUME 94. Polyamines
Edited by HERBERT TABOR AND CELIA WHITE TABOR

VOLUME 95. Cumulative Subject Index Volumes 61–74 and 76–80
Edited by EDWARD A. DENNIS AND MARTHA G. DENNIS

VOLUME 96. Biomembranes [Part J: Membrane Biogenesis: Assembly and Targeting (General Methods; Eukaryotes)]
Edited by SIDNEY FLEISCHER AND BECCA FLEISCHER

VOLUME 97. Biomembranes [Part K: Membrane Biogenesis: Assembly and Targeting (Prokaryotes, Mitochondria, and Chloroplasts)]
Edited by SIDNEY FLEISCHER AND BECCA FLEISCHER

VOLUME 98. Biomembranes [Part L: Membrane Biogenesis (Processing and Recycling)]
Edited by SIDNEY FLEISCHER AND BECCA FLEISCHER

VOLUME 99. Hormone Action (Part F: Protein Kinases)
Edited by JACKIE D. CORBIN AND JOEL G. HARDMAN

VOLUME 100. Recombinant DNA (Part B)
Edited by RAY WU, LAWRENCE GROSSMAN, AND KIVIE MOLDAVE

VOLUME 101. Recombinant DNA (Part C)
Edited by RAY WU, LAWRENCE GROSSMAN, AND KIVIE MOLDAVE

VOLUME 102. Hormone Action (Part G: Calmodulin and Calcium-Binding Proteins)
Edited by ANTHONY R. MEANS AND BERT W. O'MALLEY

VOLUME 103. Hormone Action (Part H: Neuroendocrine Peptides)
Edited by P. MICHAEL CONN

VOLUME 104. Enzyme Purification and Related Techniques (Part C)
Edited by WILLIAM B. JAKOBY

VOLUME 105. Oxygen Radicals in Biological Systems
Edited by LESTER PACKER

VOLUME 106. Posttranslational Modifications (Part A) (in preparation)
Edited by FINN WOLD AND KIVIE MOLDAVE

VOLUME 107. Posttranslational Modifications (Part B) (in preparation)
Edited by FINN WOLD AND KIVIE MOLDAVE

Methods in Enzymology

Volume 104

ENZYME PURIFICATION AND RELATED TECHNIQUES

Part C

Section I

Chromatography

[1] Affinity Chromatography

By MEIR WILCHEK,[1] TALIA MIRON, and JOACHIM KOHN

Fifteen years after its introduction, affinity chromatography,[1a,2] a method of purification based on biological recognition, has become a major means for the purification of biologically active molecules. In 1974, a volume in this series explored affinity chromatography,[3] and at that time we were able to summarize the major literature in a table of fewer than 150 entries[4] and to present another 50 or so new examples.[3] The Medline service now lists 1800 papers, published during a 3-year period, 1980–1982, in which affinity chromatography is mentioned in the title.

Despite the enormous expansion of the field and the many materials that have been purified, the directions for the investigation of methods have not changed markedly. The same carriers, basically an agarose,[1a] and the same activating reagent, cyanogen bromide,[1a,5] remain the predominant means for applying the technique. It is not surprising, therefore, that the same concepts and difficulties remained. The problems reside mainly in the lack of good protocols for efficient elution, particularly with high-affinity interacting systems. Certain nonspecific interactions are inherent with cyanogen bromide systems due to ion-exchange and hydrophobic interactions,[6] both of which are obviously useful in their own right and are covered elsewhere in this volume. In addition, there is the matter of leakage of ligand from the carrier due to the unstable isourea linkage.[7] These two difficulties have been treated experimentally with some success, as discussed subsequently.

It should be clear that means other than cyanogen bromide are being made available, albeit at a slow pace. Indeed, other chapters in this volume treat such imaginative methods.

This chapter describes the chemistry of activation and coupling of ligands and presents newer and effective methods for the preparation and

[1] Prepared during tenure as a Fogarty Scholar-in-Residence at the Fogarty International Center of the National Institutes of Health, Bethesda, Maryland.

[1a] P. Cuatrecasas, M. Wilchek, and C. B. Anfinsen, *Proc. Natl. Acad. Sci. U.S.A.* **61,** 636 (1968).

[2] P. Cuatrecasas and M. Wilchek, *Biochem. Biophys. Res. Commun.* **33,** 735 (1968).

[3] This series, Vol. 34, "Affinity Techniques."

[4] M. Wilchek and W. B. Jakoby, this series, Vol. 34, p. 3.

[5] R. Axén, J. Porath, and S. Ernbäck, *Nature (London)* **214,** 1302 (1967).

[6] R. Jost, T. Miron, and M. Wilchek, *Biochim. Biophys. Acta* **362,** 75 (1974).

[7] M. Wilchek, T. Oka, and Y. J. Topper, *Proc. Natl. Acad. Sci. U.S.A.* **72,** 1055 (1975).

METHODS IN ENZYMOLOGY, VOL. 104

ISBN 0-12-182004-1

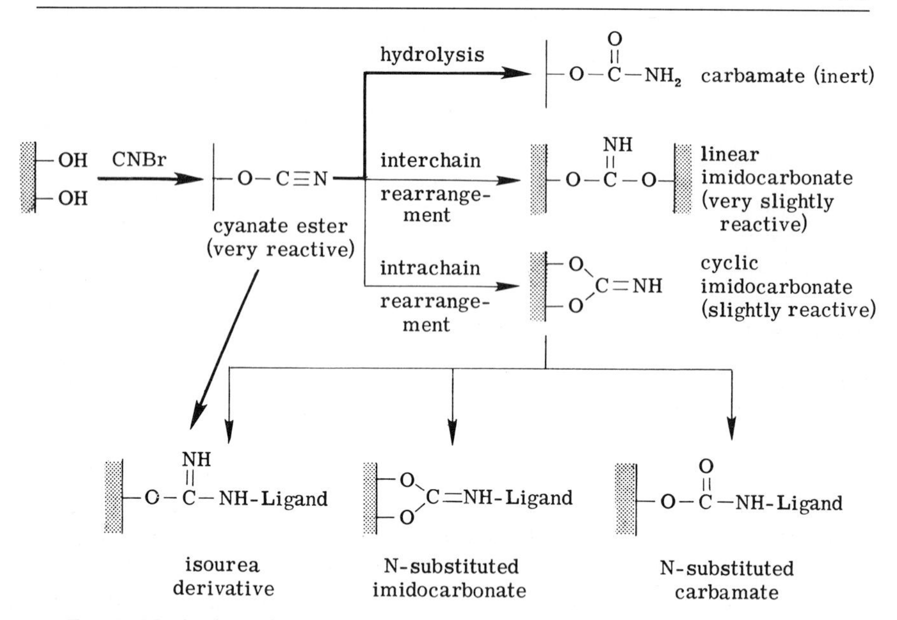

FIG. 1. Mechanism of activation of Sepharose by CNBr and subsequent coupling of ligand. Heavy lines indicate major reaction pathways.

assay of affinity matrices. Specific ligands that have been coupled to prepare affinity resins are so many and so varied that it has become unreasonable to present each one in detail. Nevertheless, knowledge of the scope and variety of what has been done is obviously useful, even if presented only in the form of representative examples. A selective list of biomaterial that has been purified, along with comments on methods, is appended.

CNBr Activation and Analysis

The preparation of an affinity chromatography column is a two-step process. Reactive groups are first introduced into the chemically inert polymeric resin, the process of activation. The second step is the covalent attachment of a ligand to the activated resin. There is universal agreement on the coupling condition: It must be performed in a mild base. On the other hand, owing to limited knowledge of the mechanism of activation of polysaccharides by CNBr, many different conditions for activation have been described on the basis of "trial and error."[3]

The study of the mechanism of activation of agarose by CNBr (Fig. 1) became possible after the development of analytical methods for the de-

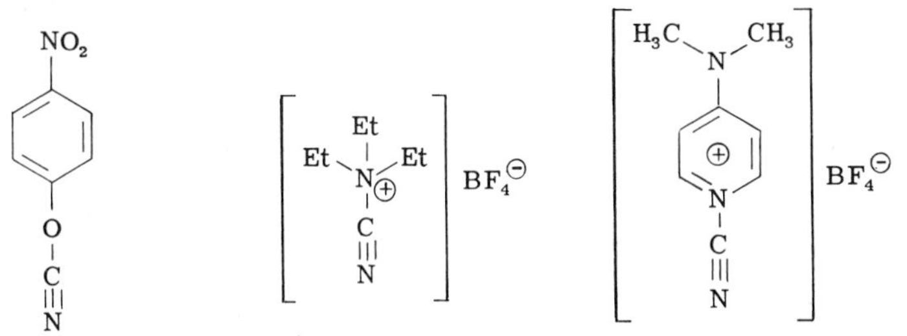

p-nitrophenyl cyanate | *N*-cyanotriethylammonium tetrafluoroborate | 1-cyano-4-dimethylamino pyridinium tetrafluoroborate

FIG. 2. Molecular structure of *p*-nitrophenyl cyanate (pNPC), *N*-cyanotriethylammonium (CTEA), and 1-cyano-4-dimethylaminopyridinium tetrafluoroborate (CDAP).

termination of each of the active and inactive species that are produced.[8] Three major products were found to be present: carbamates, imidocarbonates, and cyanate esters.[9] The last two are chemically active. Cyanate esters are stable below pH 4, whereas the imidocarbonates are most stable under basic conditions. By determining the amount of cyanate esters and imidocarbonates present on the resin, based on these differences in stability, it was possible to design and prepare carriers containing either functional group, thereby enabling prediction of the coupling capacity of activated resins toward ligands. Detailed analytical data of freshly activated Sepharose showed that 60–85% of the resin's total coupling capacity is due to the formation of the cyanate esters (Fig. 1) (cyanate ester 80%, imidocarbonates 20%).

Once this mechanism for activation by CNBr was established, it became possible to develop new and more efficient activation procedures.[10] Aside from increased efficiency, activation reagents based on CNBr have been developed that are nonvolatile solids and can be stored and safely handled without use of a fume hood, providing an attractive alternative to the highly hazardous CNBr[11a,b] (Fig. 2).

Activation Procedures

Most CNBr activation procedures are modifications of the originally described method,[5] in which inorganic bases, such as sodium hydroxide

[8] J. Kohn and M. Wilchek, *Anal. Biochem.* **115,** 375 (1981).

[9] J. Kohn and M. Wilchek, *Enzyme Microb. Technol.* **4,** 161 (1982).

[10] J. Kohn and M. Wilchek, *Biochem. Biophys. Res. Commun.* **107,** 878 (1982).

[11a] J. Kohn, R. Lenger, and M. Wilchek, *Appl. Biochem. Biotech.* **8,** 227 (1983).

[11b] J. Kohn and M. Wilchek, *FEBS Lett.* **154,** 209 (1983).

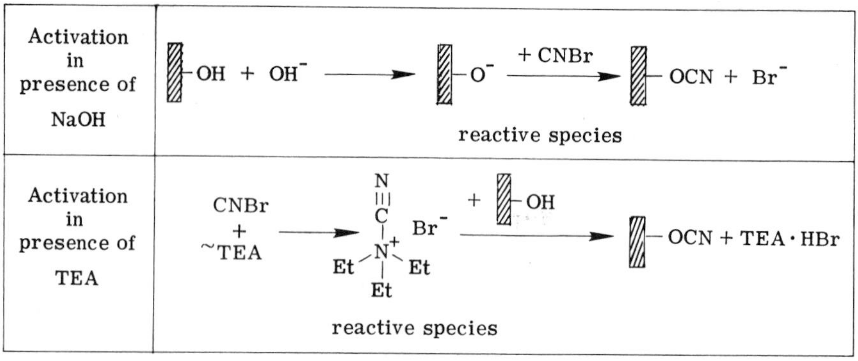

Activation in presence of NaOH	▨–OH + OH^- ⟶ ▨–O^- $\xrightarrow{+ CNBr}$ ▨–OCN + Br^- reactive species
Activation in presence of TEA	CNBr + ~TEA ⟶ N≡C–N^+(Et)(Et)(Et) Br^- $\xrightarrow{+ ▨–OH}$ ▨–OCN + TEA · HBr reactive species

FIG. 3. Comparison of the mode of action of NaOH and triethylamine (TEA). NaOH acts on the resin, causing an increase in the resin's nucleophilicity by formation of alkoxide ions. TEA is proposed as operating on CNBr by increasing the electrophilicity of the cyano moiety by complex formation.

or sodium carbonate, are used.[12] A new approach to the reaction of agarose with cyanogen bromide involves the use of triethylamine (TEA) as the "cyano-transfer" reagent at neutral pH[10] (Fig. 3). The procedure requires less than 10% of the usual amount of cyanogen bromide, and the resultant activated resins are free of imidocarbonates and carbamates, containing only active cyanate esters. Extremely high coupling capacities (75 μmol of ligand per gram of wet Sepharose 4B) are obtainable.

Method with Cyanogen Bromide and Triethylamine[10]

Wet Sepharose 4B (Pharmacia; 10 g) is washed and suspended in 10 ml of 60% acetone. The mixture is cooled to −15°. Depending on the desired degree of activation (Fig. 4), the required volume of CNBr solution (1 *M* in acetone) is added to the agarose suspension. With cooling and vigorous stirring, an equal volume of TEA solution (1.5 *M* in 60% acetone) is added dropwise during 1–3 min. Best results are achieved with a final molar ratio of CNBr to TEA of about 1 : 1.5. After 3 min, the entire reaction mixture is quickly poured into 100 ml of ice-cold washing medium (acetone : 0.1 *N* HCl, 1 : 1). The resin may be stored in washing medium at about 0° for about 1 hr without significant loss of activation. (During this period the coupling capacity of the activated resin can be determined.) The reaction can also be performed at 5° without significant loss in yield or purity.[10]

[12] S. C. March, I. Parikh, and P. Cuatrecasas, *Anal. Biochem.* **60,** 149 (1974).

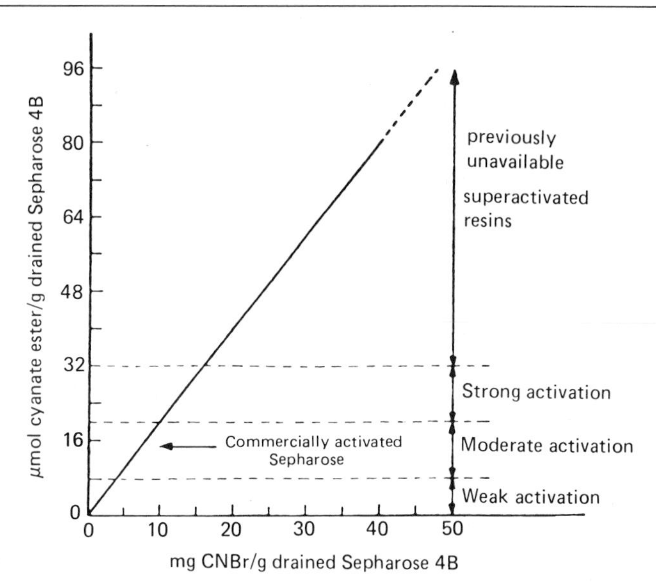

FIG. 4. Incorporation of active cyanate esters into Sepharose 4B as a function of the concentration of CNBr.

Method with p-Nitrophenyl Cyanate[11a,b]

Sepharose 4B (10 g), washed successively with acetone : H_2O (1 : 4, v/v), acetone : H_2O (1 : 1), and, finally, acetone : H_2O (7 : 3), is resuspended in 10 ml of acetone : H_2O (7 : 3) and cooled to 0°. To this solution, the desired volume of pNPC reagent [2.3 g of *p*-nitrophenyl cyanate (Sigma; pNPC) dissolved in 10 ml of acetone to yield 1.4 *M* pNPC] is added with vigorous stirring. This addition is followed by an equal volume of neat TEA. In this manner, a fivefold molar excess of base over activating agent is attained. After 10 min, the entire reaction mixture is poured into 200 ml of ice-cold acid-treatment medium (acetone : 0.5 *M* HCl, 1 : 1), and allowed to stand for 20 min at 0°. Thereafter, the resin is washed with ice-cold acetone : H_2O (1 : 1), followed by ice-cold water, and used for coupling as described for CNBr-activated resins.[1,3] For immediate coupling, the resin is washed with cold 60% acetone, followed by quick washings with cold 30% acetone, cold water, and the coupling medium.

Method with N-Cyanotriethylammonium (CTEA) and 1-Cyano-4-dimethylaminopyridinium Tetrafluoroborate (CDAP)

Drained Sepharose 4B (10 g) is washed sequentially with water, 30% acetone, and 60% acetone and resuspended in 10 ml of 60% acetone. The

TABLE I
AMOUNT OF ACTIVATING AGENT REQUIRED FOR ACTIVATION OF 10 g OF DRAINED SEPHAROSE 4B

Degree of activation	Approximate coupling capacity (μmol ligand/g resin)	CTEA activation[a]		CDAP activation[a]	
		CTEA[b] (mg)	TEA[b] (ml)	CDAP[c] (mg)	TEA[b] (ml)
Weak	5	60	0.6	25	0.2
Moderate	15	180	1.8	75	0.6
Strong	30	360	3.6	150	1.2

[a] The illustrative procedure was used as described.
[b] TEA, triethylammonium; TEA, triethylamine, 0.2 *M* aqueous solution.
[c] For convenient use, a stock solution of 1-cyano-4-dimethylaminopyridinium tetrafluoroborate (CDAP) in acetonitrile (0.1 g/ml) was prepared. This solution is stable at 4° for over a month.

suspension is cooled to 0°. Depending on the desired degree of activation, the required amount of activating agent is added, followed by a corresponding amount of TEA (see Table I). The TEA solution (0.2 *M* in water) is added dropwise with vigorous stirring. After 2 min, the entire reaction mixture is rapidly transferred into 100 ml of ice-cold washing medium (acetone : 0.1 *N* HCL, 1 : 1). The resin may be stored in this manner for over 1 hr without loss of activation. For coupling, the resin is washed with a large volume of cold water followed by one quick wash with the coupling medium. Coupling of ligand to CTEA- or CDAP-activated resins is performed as described for CNBr-activated resins.

For prolonged storage of CNBr-activated resins, the resin is prepared by extensive washing with storage medium (acetone : dioxane : H_2O, 60 : 35 : 5). When suspended in storage medium and stored airtight at −15 to −20°, the active cyanate ester content decreases at an approximate rate of 10% per month. After prolonged periods of storage, the resin may be reswollen for 5 min in cold washing medium and washed as described above for immediate coupling.

Determination and Characterization of Activated Resins

Determination of the expected coupling capacity of the resin is sometimes crucial for successful application of affinity chromatography columns. The determination of cyanate esters and imidocarbonates can be performed in about an hour. The results are accurate and reproducible and provide a quick method for resin characterization.[8]

Determination of Total Nitrogen Content

The following method does not require the laborious Kjeldahl procedure or Dumas microanalysis. A weighted amount of dry resin, about 15 mg, is placed in a 50-ml measuring bottle. H_2SO_4 (96%), 0.5 ml, is added, and the mixture is heated in an oven to between 100° and 120° for 15 min. Thereafter, 0.2 ml of 30% H_2O_2 is added slowly (vigorous reaction). The mixture clears, and heating is continued for an additional 90 min. The colorless solution is diluted to 50 ml with 0.2 *M* sodium acetate at pH 5.0. Aliquots of known volume are used for determination of ammonia by the ninhydrin reaction (see below); the amount measured is equal to the total nitrogen content of the resin.

Determination of Ammonia by Ninhydrin Reaction

Preparation of hydrindantin reagent: Hydrindantin (880 mg) and ninhydrin (4 g) are dissolved in 100 ml of Methyl Cellusolve. Unless stored airtight at −20°, the reagent will degrade during the course of a few days. To 1 ml of sample solution, containing between 0.05 and 0.8 μmol of NH_4^+, is added 1 ml of 2 *M* sodium acetate at pH 5.5 and 1.5 ml of hydrindantin reagent. After mixing, the solutions should be a light red. The stoppered test tubes are heated to 100° for 20 min. The contents of each tube are diluted without prior cooling to 10 ml with cold formaldehyde diluent (0.5% solution of formaldehyde in isopropanol : water, 1 : 1). The diluent reacts with excess hydrindantin so that a blue ninhydrin color appears within a few seconds. Absorbance is measured at 570 nm, and the amount of NH_4^+ in the sample is determined from a calibration curve.

Determination of Imidocarbonates[8]

Of all the nitrogen derivatives present on activated resins, only the imidocarbonates will release ammonia on mild acid hydrolysis. Thus, the amount of ammonia measured is equal to the amount of imidocarbonate on the resin.

A sample of activated resin containing about 1–20 μmol of imidocarbonates is placed in a 25-ml measuring bottle. Acid (5 ml of 0.1 *N* HCl) is added, and the mixture is heated at 40° for 30 min or allowed to remain at room temperature for 60 min. The volume is brought to 25 ml with 0.2 *M* sodium acetate at pH 5.5. After mixing, the suspension is allowed to settle and aliquots of the clear supernatant liquid are used for determination of NH_4^+ by the ninhydrin reaction.[13]

Determination of Cyanate Esters[8]

Cyanate esters are determined by the use of pyridine and either dimethylbarbituric acid or barbituric acid. With the possible exception of traces of triazines, cyanate esters are the only species present on activated resin giving rise to a characteristic purple color with barbiturates in pyridine. The reaction is highly sensitive (ε_{588} = 137,000 liter mol^{-1} cm^{-1}) so that as little as 5 nmol of cyanate ester can be determined. The method is also applicable to cyanuric chloride-activated polysaccharides. The color reaction can be used as a quick, qualitative test for the presence of activation of agarose; for quantitative determinations of cyanate esters directly on the activated resin; and as a quick test for the presence of excess CNBr in washings after activation.

Qualitative Test Reagent. Pyridine (14 ml) is mixed with 6 ml of water and 50 mg of barbituric acid or *N,N'*-dimethylbarbituric acid (Chemical Dynamics Corporation, South Plainfield, NJ), resulting in a clear, colorless solution that darkens slowly. For qualitative tests, coloration of the test reagent is not important. To 10–20 mg of dry polysaccharide, or to 0.1–1.0 ml of swollen polysaccharide, 1–2 ml of qualitative reagent is added. After shaking slightly, the presence of less than 5 nmol of cyanate ester can be detected by the formation of a red–purple color, which develops within 30 sec and reaches a maximum after 10 min.

Quantitative Test Reagent. N,N'-Dimethylbarbituric acid (500 mg; recrystallized from water) is suspended in 5 ml of water. Cold, freshly distilled pyridine (45 ml) is added and results in a clear, colorless solution. Since some discoloration is evident even after storage at −20°, it seems best to prepare the required amount of reagent mixture immediately prior to use.

Quantitative Determination of Cyanate Esters. To 5–20 mg of dry, activated resin, or an equivalent amount of wet resin, 5 ml of the quantitative test reagent is added. The mixture is warmed to 40° for 25 min in a closed test tube with vigorous stirring. After 25 min, the mixture is diluted to a convenient volume with water. For 10 mg of freshly activated resin, dilution to 250–500 ml is usual in order to reduce the absorbance of the purple solution to between 0.5 and 1.0. Filtered aliquots of the dilutions are used to measure the absorbance at about 588 nm (ε_{588} = 137,000).

[13] S. Moore and W. H. Stein, *J. Biol. Chem.* **211,** 907 (1954).

Coupling Capacity of an Activated Resin

Based on the assumption that coupling of ligand to cyanate esters and to imidocarbonates occurs simultaneously and independently, the total coupling capacity of a monovalent ligand to an activated resin is predictable. It is related to the amounts of cyanate esters and imidocarbonates present on the resin according to the following equation.

$$\text{Total coupling capacity} = A(\text{cyanate esters}) + B(\text{imidocarbonates})$$

Here, A and B represent the respective yields for the respective products of the coupling procedure. Based on values A and B of 0.8 and 0.15, respectively, the equation can be used to predict the coupling capacities of specific samples of activated agarose.

Newer Methods of Activation

Despite its popularity, the CNBr method of activation has several drawbacks. For example, coupling of amines to activated polysaccharides introduces N-substituted isourea bonds that are not completely stable, particularly in the presence of nucleophiles.[7] This small leakage of ligand can lead to erroneous findings and, over a period of time, results in the loss of capacity for adsorption of the desired protein. The isoureas are positively charged at physiological pH ($pK = 9.4$), allowing such columns to also exhibit ion-exchange properties that can interfere with the biospecificity of the adsorbent. In order to eliminate the difficulties inherent in the CNBr method, and provide other possibilities for binding ligands to carrier, several methods have been developed. One of these uses CNBr to activate polypyridine or other carriers containing the pyridine ring[14]; this is the reverse reaction of that used in the determination of the cyanic esters.[15]

Other new methods are based on well-established sugar chemistry in which activation is performed in anhydrous media. Sulfonyl chloride is used for activation in one such procedure, which is described in this volume [2].[16] For most of the other preparative procedures, the reactive species are activated carbonates or carbamates.[17,18] Upon reaction with

[14] F. Pittner, T. Miron, G. Pittner, and M. Wilchek, *J. Solid Phase Biochem.* **5,** 167 (1980).

[15] F. Pittner, T. Miron, G. Pittner, and M. Wilchek, *J. Am. Chem. Soc.* **102,** 2451 (1980).

[16] K. Nilsson and K. Mosbach, *Biochem. Biophys. Res. Commun.* **102,** 449 (1981).

[17] G. S. Bethell, J. S. Ayers, and W. S. Hancock, *J. Biol. Chem.* **254,** 2572 (1979).

[18] M. Wilchek and T. Miron, *Biochem. Int.* **4,** 629 (1982).

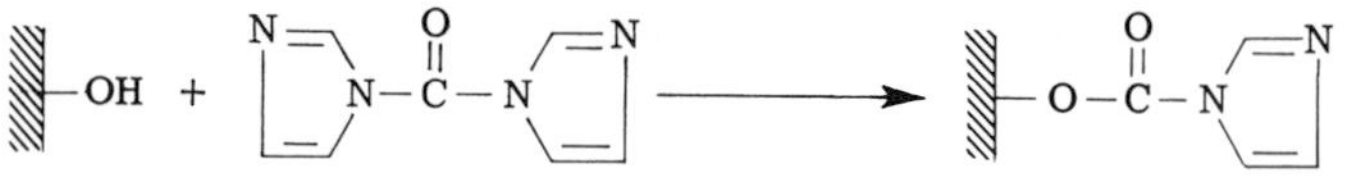

SCHEME 1.

amines, stable and uncharged carbamate (urethane) derivatives are obtained. The reagents include 1,1′-carbonyldiimidazole,[17] *p*-nitrophenylchloroformate,[18] *N*-hydroxysuccinimidochloroformate,[18] and tri- or pentachlorophenylchloroformate.[18] With the last three, the extent of activation and coupling can be measured spectrophotometrically.

Activation with 1,1′-Carbonyldiimidazole[17,19] *(Scheme 1)*

Sepharose CL-6B (3 g of moist cake) is washed sequentially with 20 ml each of water, dioxane–water (3 : 7), dioxane–water (7 : 3), and dioxane and is suspended in 5 ml of dioxane. 1,1′-Carbonyldiimidazole (120 mg) is added, and the suspension is shaken at room temperature for 15 min. It is washed with 100 ml dioxane and used immediately.

Cross-linked agarose was used, but normal agarose preparation is equally effective. The amount of reagent can be varied depending on the extent of activation required; up to 3 mmol of active groups can be introduced per gram of dry matrix. The isolated matrix is stable under anhydrous conditions and may be stored in dioxane. These activated carriers are also available commercially from Pierce Chemical Company.

Activation with Chloroformates[18] *(Scheme 2)*

Sepharose (4B or CL-4B) is washed with acetone–water mixtures of increasing concentration in acetone (1 : 3, 1 : 1, and 3 : 1) and, finally, with dry acetone. The gel (10 g) is suspended in 10 ml of dry pyridine at 4°; the chloroformate, dissolved in dry acetone, is added dropwise over a period of 2 min. The reaction mixture is stirred for an additional 15–30 min at 4°. The activated polymer is washed sequentially with acetone, methanol, and ether and then is dried in air. The activated carrier may be stored at 4° as a dry powder or in anhydrous solvents, such as isopropanol, acetone, or dioxane. Fresh preparations can be used without drying with ether but are washed with water before coupling.

Both Sepharose and cross-linked Sepharose can be used. However, cross-linked Sepharose appears to be more stable in many organic solvents. The extent of activation depends on the amount of chloroformate

[19] M. T. W. Hearn, E. L. Harris, G. S. Bethell, H. S. Hancock, and J. A. Ayers, *J. Chromatogr.* **218,** 509 (1981).

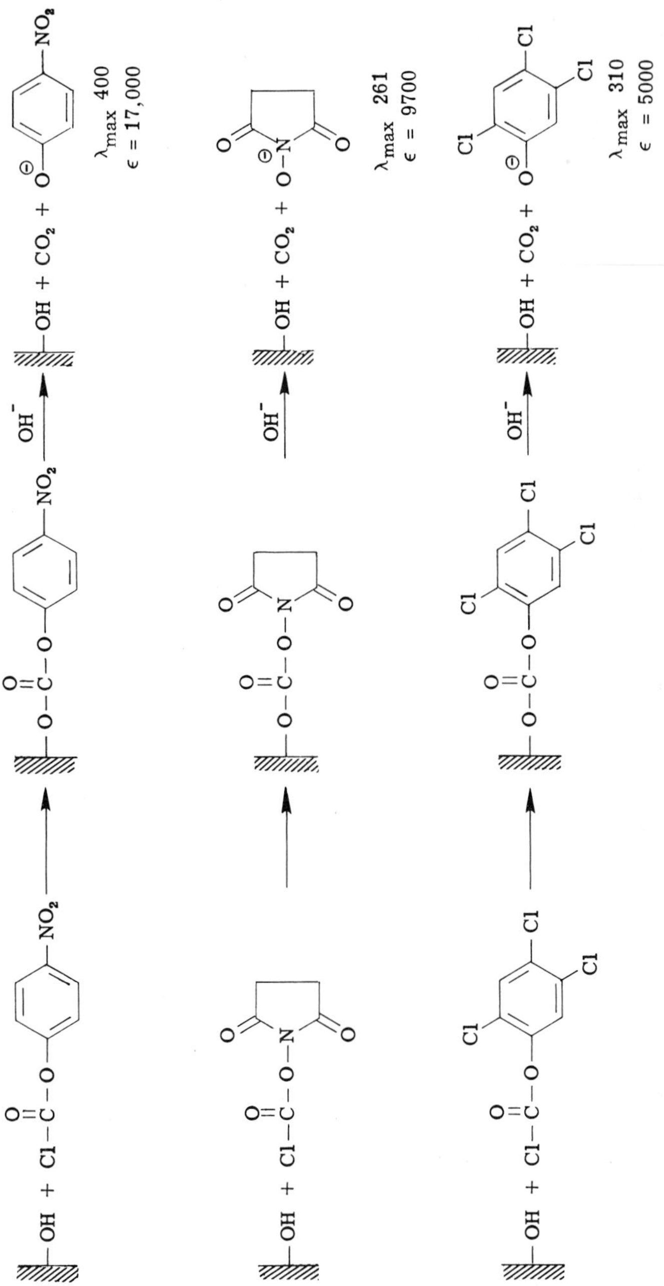

λ_{max} 400
$\epsilon = 17,000$

λ_{max} 261
$\epsilon = 9700$

λ_{max} 310
$\epsilon = 5000$

SCHEME 2.

added. Carrier containing up to 1.8 mmol of active groups per gram of polymer can be obtained. Yields of up to 30% can be assumed, based on the amount of chloroformate added. The swelling properties of the matrix in water change only nominally with low levels of substitution, although increasing substitution leads to decreased swelling of the gel. These activated carriers are commercially available from Sigma Chemical Company.

Activation of cellulose with the same reagents has also been described; activation was performed in dioxane at 60° or for prolonged periods at 25°.[20]

Determination of Active Groups on the Carriers

Active groups present on carriers prepared by reaction with *p*-nitrophenylchloroformate may be quantitated after hydrolysis of a weighed sample with 0.2 *N* NaOH for 15 min at room temperature. The free *p*-nitrophenylate is measured spectrophotometrically at 400 nm (ε = 17,000). The amount of hydroxysuccinimide on columns prepared with hydroxysuccinimide chloroformate is determined by nitrogen analysis or, after basic hydrolysis (0.2 *N* NH_4OH), the absorbance of the supernatant liquid may be measured spectrophotometrically at 261 nm (ε = 9700).[21,22] Active groups on preparations with trichlorophenylchloroformate are estimated either by chlorine analysis or spectrophotometrically at 310 nm (ε = 5000) as a function of trichlorophenol released after hydrolysis with 0.2 *N* NaOH.

No direct method is available for the determination of the imidazole-activated resins. Nitrogen analysis and titration have been used to confirm the presence of imidazole on carriers.[17]

Polyacrylhydrazidoagarose

Polyacrylhydrazidoagarose (PAHOS), prepared by the CNBr method, was described in this series.[23] Since then, this matrix has been applied to purification, particularly in the fields of immunology[24] and lectin receptors.[25] Although this material has greater stability when compared to nor-

[20] J. Drobnik, J. Labsky, H. Kudlvasrova, U. Sandek, and F. Svec, *Biotechnol. Bioeng.* **254,** 487 (1982).

[21] T. Miron and M. Wilchek, *Anal. Biochem.* **126,** 433 (1982).

[22] P. M. Abdalla, P. K. Smith, and G. Royer, *Biochem. Biophys. Res. Commun.* **87,** 736 (1979).

[23] M. Wilchek and T. Miron, this series, Vol. 34, p. 72.

[24] D. W. Sears, S. Young, P. H. Wilson, and J. E. Christiaansen, *J. Immunol.* **124,** 2641 (1980).

[25] R. Lotan and R. L. Nicolson, *Biochim. Biophys. Acta* **559,** 329 (1979).

mal CNBr-activated agarose, some leakage of ligands was observed. Owing to the general usefulness of hydrazide matrices, it appeared worthwhile to develop a better procedure for their preparation. Periodate was used to achieve an oxidized agarose from which hydrazides can be formed.[26] This system is presented separately, since it is not a CNBr preparation, nor are hydrazide columns, as such, a new development.

Preparation of PAHOS

Washed Sepharose 4B (10 g) is suspended in 30 ml of freshly prepared 0.25 *M* sodium periodate. The suspension is slowly stirred at room temperature for 3 hr in the dark; the oxidized Sepharose is then washed with cold water to remove excess periodate and its products. The gel is suspended in three volumes of polyacrylhydrazide solution in water (0.1–0.5%). Coupling is allowed to proceed overnight in the dark at room temperature with slow stirring. The conjugate is washed with 0.1 *M* sodium chloride until the washings are free of color after testing with trinitrobenzenesulfonic acid (TNBS).[26] The conjugate is reduced with 0.3 *M* sodium borohydride in 0.5 *M* Tris-chloride at about pH 8 for 3 hr at room temperature. The reduced gel is washed with water on a sintered-glass funnel. Although it is possible to prepare a carrier containing as much as 125 μmol of hydrazide per milliliter of Sepharose, it is suggested that derivatization be limited to 15–25 μmol of hydrazide per milliliter for normal usage. The PAHOS prepared by this method has many advantages; it is stable and colorless and is free of functional groups, such as carbamates and carbonates, that are produced by CNBr activation. These preparations may be stored at 4° for at least 2 years.

Derivatization of PAHOS

Matrix with Spacers. PAHOS can be used directly for coupling of compounds containing aldehyde groups such as glutaraldehyde or periodate-oxidized nucleic acids or can be modified further by reaction with a variety of functional groups that have been described.[23] Carriers bearing carboxyl groups, thiols, imidazoles, phenols, and amines can be prepared. These reactive functional groups can be used to couple ligand and protein via amide, thioester, and diazonium linkages among others.

Aldehyde-Containing PAHOS. PAHOS bearing a carbonyl group is prepared by suspending the resin in three volumes of 7% glutaraldehyde in water and slowly stirring for 2 hr at room temperature. Excess glutaraldehyde is removed by washing the gel with cold water on a sintered-

[26] T. Miron and M. Wilchek, *J. Chromatogr.* **215,** 55 (1981).

glass funnel until no further reaction of glutaraldehyde with nitrophenylhydrazide is detected.[26]

Carboxyl-Containing PAHOS. Free carboxyl-containing resins are prepared by acylation of the free hydrazide groups with an excess of succinic anhydride. The resulting succinylhydrazides provide carboxyl groups with long spacer arms. Succinic anhydride is usually added as 0.5 mmol per milliliter of packed hydrazido-Sepharose in an equal volume of water. The pH of the reaction mixture is maintained at 6.0 by addition of 1 *N* NaOH and allowed to proceed at room temperature for 3 hr with slow stirring. Completion of acylation is assayed with TNBS. The resin is washed with 0.1 *N* HCl and water.

Amino-Containing PAHOS. Compounds containing primary amino groups may be coupled to the aldehyde resin by incubation at pH 7.0 for 16 hr at room temperature or, alternatively, by coupling to carboxyl derivatives at pH 5.0 in the presence of a water-soluble carbodiimide. Coupling to carboxyl derivatives is also performed under anhydrous conditions using dioxane and dicyclohexylcarbodiimide. A large excess of diamine is used to decrease cross-linking in cases in which free amino-terminal groups are required. Such amino groups may be useful for further substitution with different ligands. Methods for the use of carbodiimide reagents have been detailed.[27]

Sulfhydryl-Containing PAHOS. Sulfhydryl groups are introduced by treating the resin with *N*-acetylhomocysteine thiolactone. The thiolactone (1 g) is added to a cold suspension of 10 ml of hydrazido-Sepharose in 20 ml of 1 *M* $NaHCO_3$. The reaction mixture is stirred at 4° for 16 hr, and the product is washed extensively with water and 0.1 *N* NaCl. The preparation can also be made by coupling of dithiodiglycolic acid in aqueous solution using a water-soluble carbodiimide or in dioxane using dicyclohexylcarbodiimide as described above for carboxyl coupling. After reaction, the disulfide resin is reduced with 0.2 *M* mercaptoethanol for 30 min.

Characterization of Hydrazides and Derivatives

Free hydrazide or free amino groups on conjugates may be assayed by reaction with an excess of TNBS. The quantity of unreacted TNBS is measured and subtracted from its initial concentration. Samples of washed gel (200 mg) are suspended in 1 ml of 1% TNBS, to which is added 1 ml of saturated sodium tetraborate. The suspension is stirred for 30 min at 24° and washed with 23 ml of 0.2 *M* NaCl. Aliquots of the washings, 100 μl, are incubated with 1.0 ml of 0.2 *M* adipodihydrazide and 0.9 ml of

[27] S. Bauminger and M. Wilchek, this series, Vol. 70, p. 151.

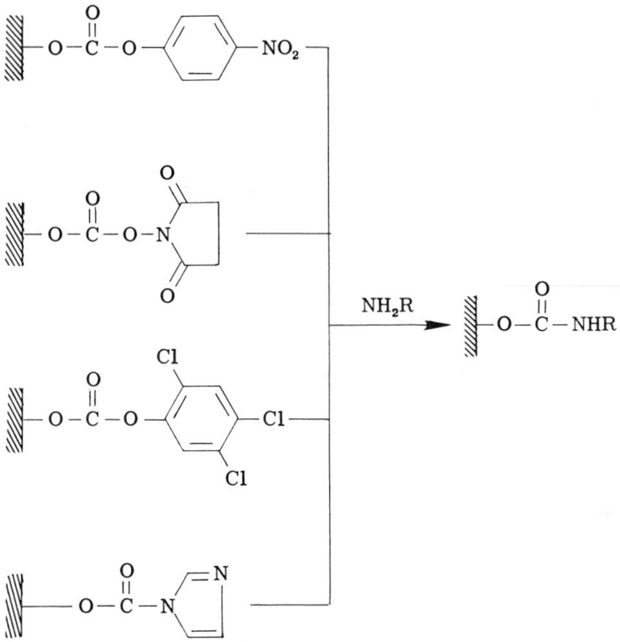

SCHEME 3. Structure of active carriers.

saturated borate for 15–20 min. Absorbance is recorded at 500 nm (ε = 16,500) and used for calculating the quantity of TNBS that had reacted with adipodihydrazide.

Free sulfhydryl groups on the resin are determined with an excess of 5,5′-dithiobis(2-nitrobenzoic acid). Samples of washed gel (200 mg) are suspended in 1 ml of 50 m*M* Ellman reagent in 0.1 *M* potassium phosphate, pH 7.5, for 30 min at room temperature and then filtered. The absorbance of the filtrate is recorded (ε_{412} = 13,600) and used for calculating the concentration of free sulfhydryl on the resin.

Coupling of Ligands to the Carriers

The major emphasis here has been on activation of the resin matrix. The second step in the process is that of coupling of ligands to the matrix. The structure formed on coupling of amino-containing ligands to CNBr-activated agarose is presented in Fig. 1; the products of coupling to carbonyldiimidazole and chloroformate-activated agarose are presented in Scheme 3.

The coupling of ligands is usually performed in 0.1–0.2 *M* sodium bicarbonate at pH 8–9 for 4 hr at room temperature or overnight at 4°. Precise directions for coupling can be found in Volume 34 of this series but are briefly summarized below.

Coupling Amino-Containing Ligands to CNBr or Carbonate-Activated Agarose[3]

Alkylamines, amino acids, peptides, and proteins may be coupled in 0.1 *M* sodium bicarbonate (pH 8.0–9.0) for 20 hr at 4° with gentle agitation. The residue is filtered and washed with 0.1 *M* $NaHCO_3$ and with distilled water. Excess active groups are removed by treatment for 5 min with 0.01 *N* NaOH or by reaction for 1 hr at room temperature with either 0.1 *N* NH_4OH or 0.1 *M* ethanolamine at pH 9.

Coupling of Nucleotides and RNA to PAHOS

Nucleosides, nucleotide phosphate coenzymes, and RNA possessing vicinal free hydroxyl groups are oxidized with periodate and coupled to PAHOS as described for other hydrazides.[25,26]

Coupling of Proteins and Lectins to PAHOS

The aldehyde-containing resin, obtained after glutaraldehyde treatment, is suspended in two volumes of 0.1 *M* potassium phosphate at pH 7.5, containing the appropriate protein (enzyme, lectins) at concentrations of 5–10 mg/ml, and stirred slowly for 16 hr at 4°. Conjugates are washed with cold 10 m*M* potassium phosphate–0.14 *M* NaCl at pH 7.4. Excess aldehyde groups are reduced by addition of 0.5 mg of $NaBH_4$ per milliliter of solution for 2 hr at 4° or by treatment with 0.05 *M* hydrazine hydrate at pH 8.0 for 2 hr at room temperature.

Determination of Coupled Ligand

The extent of ligand coupling can be estimated by several methods, including total hydrolysis, radioactivity, spectroscopy, and the like. The various different methods have been compiled in this series.[3]

Adsorption and Elution

Conditions for the purification of specific protein must be tailored to the properties of that protein; there is no all-purpose protocol that is applicable. It is clear, then, that those conditions that allow complex formation between the ligand and the protein must be manipulated with respect to ionic strength, temperature, pH, concentration, presence of metals, and flow rate.[3,28] For high-affinity systems, any of the carriers and even small amounts of ligand bound will be satisfactory. For low-affinity systems, large amounts of ligand must be bound to carrier.[29,30]

Elution

Elution of the adsorbed material from the column is the most important step in successful purification by affinity chromatography. Again, there is no protocol that serves as an example for elution, and individual conditions must be improvised. Several general strategies for elution, however, are commonly being used with success. These procedures include both specific and general elution conditions.

Specific elution refers to competition by hapten, ligands, or inhibitor in solution. These are the best conditions for elution and, when successful, strongly suggest that real affinity, bioaffinity, is involved. On the other hand, if apparently specific ligands fail to elute the desired protein, nonspecific means can be applied. Among the latter that have been used successfully are solvent or buffer changes, pH or ionic strength alterations, the application of chaotropic reagents, electrophoretic desorption, temperature changes, reversible denaturation, and destruction of the spacer arm.[31]

A good practice in attempting elution of the adsorbed material is the initial examination of several means of nonspecific elution. Nonspecific elution can be considered as an extra step in washing the column if the required compound is not eluted, i.e., it serves to remove contaminating proteins. Therefore, elution washing should be tried immediately after the sample has been applied and before a specific eluent is used. The washing should be started with a high salt concentration in the buffer so that species that bind because of their ion-exchange properties will tend to be eluted. On the other hand, it may occasionally be advisable to start with a very low salt concentration so as to elute species bound through hydrophobic interaction. If changes in ion strength are ineffective, different buffers could be tried—e.g., a shift from a Tris to a borate buffer. Purification is achieved by removing contaminating protein by these efforts or, conversely, by selective removal of the desired protein. The next step in such trials would be to change the pH, beginning with small changes

[28] P. O'Carra, S. Barry, and T. Griffin, this series, Vol. 34, p. 108.

[29] I. M. Chaiken, *Anal. Biochem.* **97,** 1 (1979).

[30] M. Wilchek and C. S. Hexter, *Methods Biochem. Anal.* **23,** 347 (1976).

[31] D. R. Absolom, *Sep. Purif. Methods* **10,** 239 (1981).

followed by using acid or base, e.g., with 0.1 *N* acetic acid or 0.5 *M* NH_4OH, respectively; such experiments are predicated upon stability requirements being met under these conditions. If elution of the desired species does not occur, treatment with ethylene glycol or dioxane or with an organic acid, such as propionic acid, may result in elution.

Specific elution is most desirable, so that competitive inhibitors, cofactors, or haptens must be tried. If neither specific inhibitors nor the nonspecific steps that have been outlined are effective, denaturing conditions using urea or guanidine are in order; these reagents sometimes result in reversible denaturation or in a time-dependent denaturation and, therefore, may be reversible if the eluent is collected in a dilution buffer immediately after leaving the chromatography column.

Binding to affinity columns may be so tight as to preclude recovery of an active enzyme or native protein. Under such circumstances ligands with lower affinity must be tried. Alternatively, columns containing a lower concentration of ligand, or ligand separated from the matrix by a spacer arm, may be successful. One approach is to bind the ligand by means of a hydrolyzable bond, either directly or to a destructible spacer arm. Esters[32,33] are readily hydrolyzed with mild bases; vicinal hydroxyl groups are cleared with periodate; and diazo bonds are reduced with dithionates.[34] Columns of this sort are of only limited utility, since they are destroyed by the "elution" procedure. All these strategies have been described in Volume 34 of this series.

Electrophoresis has been used as a direct method of elution.[35] The rationale behind this approach is that since proteins are charged at specific pH values, they will migrate toward the appropriate electrode; affinity forces should not be sufficiently strong to prevent this movement. This method was applied mostly to antigen–antibody systems even though success in other systems was also described. The procedure is mild and the yields are very good. Since all carriers cannot be used and electrophoretic equipment is required, the procedure may not have broad applicability.

Macromolecules Purified by Affinity Chromatography

Since the number of proteins that have been purified by affinity chromatography is very large, the methods that have been used previously may serve to suggest approaches to those attempting novel purifications.

[32] R. J. Brown, N. E. Swaisgood, and H. R. Horton, *Biochemistry* **18,** 4901 (1979).

[33] P. Singh, S. D. Lewis, and J. A. Shafer, *Arch. Biochem. Biophys.* **193,** 284 (1979).

[34] P. Singh, S. D. Lewis, and J. A. Shafer, *Arch. Biochem. Biophys.* **203,** 776 (1980).

[35] M. R. Morgan, P. J. Brown, M. J. Lieland, and P. D. Ocan, *FEBS Lett.* **87,** 239 (1978).

A large number of selected affinity methods are recorded and summarized in Tables II–IV with the intent of providing some perspective as to available methods.

Enzymes

Among the many proteins purified by affinity chromatography, the largest category consists of the enzymes. By reason of their discrete specificity, a huge variety of ligands are potentially available; these include competitive enzyme inhibitors, coenzymes, and, to a lesser extent, other substrates and cofactors. Purification of enzymes can be divided into two categories. One is directed at enzymes of narrow specificity, i.e., enzymes purified with the aid of specific inhibitors or substrates. The columns prepared with such ligands can be used only to purify the specific enzyme for which the column was tailored. At times, specific columns serve also for the purification of identical catalytic activities, but ones derived from different organisms. The second category is designed for the purification of a wide range of enzymes by using a general ligand, a ligand common to a larger group of proteins.[36] Since about 30% of the known enzymes require for their activity a coenzyme, i.e., a substrate common to an entire class of reactions, immobilization of coenzymes on a carrier will enable purification of a relatively large range of proteins on the same column. The coenzymes in more common use are NAD, NADP, ATP, and CoA, or derivatives of them which are themselves derivatives of adenine nucleotides. Some idea of the variation possible, as well as indications for specific enzymes, are presented in Table II, which is arranged alphabetically on the basis of the enzyme name. There is no attempt to be all inclusive, and many meritorious preparations have not been included.

Immunoaffinity Chromatography

Affinity chromatography has been in use for many years for the purification of antibodies by employing antigens bound to carriers. In recent years, there has been a large increase in the use of immobilized antibodies for purification of proteins, including enzymes. Under ideal conditions, antibody columns allow separation of specific proteins from crude mixtures by a one-step procedure. Both conventional and monoclonal antibodies have been used. A major difficulty in the use of immobilized antibodies is the high affinity of antibody for antigens, thereby making the recovery of active enzymes difficult owing to the drastic conditions re-

[36] K. Mosbach, *Adv. Enzymol. Relat. Areas Mol. Biol.* **46,** 205 (1978).

TABLE II
CONDITIONS FOR AFFINITY CHROMATOGRAPHY OF ENZYMES

Enzyme[a]	Ligand	Eluent
1. Acetylcholinesterase	Acridinium, pyridinum, Con A	Decamethonium, methyl-D-mannoside
2. Acetyl-CoA carboxylase	Biotin	NaCl
3. *N*-Acetylgalactosamine transferase	UDP, apomucin	NaCl
4. α-*N*-Acetylgalactosaminide-(2–6)-sialyltransferase	CDP	NaCl, CTP
5. *N*-Acetylglucosaminidase	Thio-*N*-acetylglucosamide	NaCl
6. *N*-Acetyl-β-hexosaminidase A	*N*-Acetyl-β-D-thioglucosamine	Borate
7. *N*-Acetyl-β-hexosaminidase	Con A, 2-acetamido-2-deoxy-β-glucosamine	α-Methyl glucoside, *N*-acetylglucosamine, pH, Triton X-100
8. Acid phosphatase	Con A	D-Mannose
9. Acid protease	Pepstatin	Urea gradient
10. Adenosine (phosphate) deaminase	Inosine	Adenosine, inosine
11. Adenosine 5′-phosphosulfate sulfohydrolase	AMP	AMP
12. Adenine phosphoribosyltransferase	AMP, GMP	Mg, AMP
13. Adenosine kinase	ATP, AMP	Adenosine, pH
14. Adenosinetriphosphatase	Calmodulin, ATP	EDTA, ATP
15. Adenylate cylase	Cholera toxin subunit GTP, ATP forskolin derivative	Tris forskolin
16. ADP-glucose pyrophosphorylase (glucose-1-phosphate adenylyltransferase)	6-Phospho-1-pyrophosphate	AMP
17. ADP-ribosyltransferase	Phenylagarose	Propylene glycol
18. Alanine aminotransferase	Cycloserine	P_i
19. D-Alanine carboxypeptidase	Penicillins	NaCl, hydroxylamine
20. Alanine dehydrogenase	NADP	NaCl
21. Alcohol dehydrogenase	AMP, NADP, ATP	NAD, NADH, pyrazole, KCl
22. Aldehyde dehydrogenase	AMP, NAD	NAD
23. Aldehyde reductase	NADP	KCl

TABLE II (*continued*)

Enzyme[a]	Ligand	Eluent
24. Aldose reductase	4-Carboxybenzaldehyde	P_i
25. Alkaline phosphatase	Phosphonic acid, Con A	P_i, α-methylmannoside
26. Amine oxidase	Con A	α-Methyl-D-glucoside
27. Aminolevulinate dehydratase (porphobilinogen synthase)	Aminophenyl, mercuric acetate	Dithiothreitol
28. Aminopeptidase	Leucylglycine derivatives	NaCl, Zn
29. α-Amylase	Glycogen	Glycogen
30. Anthranilate synthase	Anthranilate	pH
31. L-Arabinose kinase	ATP	NaCl
32. Arginase	Antiarginase, L-Arg	L-Arginine, NaCl
33. L-Arginine deaminase	L-Arg	L-Arg
34. Arylsulfatase A	Con A subunit	Sugar, Tris
35. Asparaginase	L-Asp	NaCl, Asp
36. Aspartase (aspartate ammonia-lyase)	L-Asp	NaCl, Asp
37. Aspartate-β-decarboxylase	L-Asp	NaCl
38. Aspartate transcarbamylase (aspartate carbamoyltransferase)	L-Asp	Asp, KCl
39. Carbamoyl-phosphate synthetase	Glu	KCl
40. Carbonic anhydrase (carbonate dehydratase)	Sulfonamides	Perchlorate, KI, KCN, NaCl
41. Carboxypeptidase A	Phenylpropionates, *p*-aminobenzyl succinate	KCl
42. Carboxypeptidase B	D-Arg, protease inhibitor	NaCl, pH
43. Carboxypeptidase N (arginine carboxypeptidase)	Aminobenzoylarginine	Guanidinoethyl-mercaptosuccinate
44. Catechol *O*-methyltransferase	Catechols	KCl
45. Cathepsin	Pepstatin	NaCl
46. Cellulase	Cellulose	Temperature
47. Choline acetyltransferase	Coenzyme A	Citrate-phosphate EDTA, dithioerythritol
48. Choline dehydrogenase	Choline	Dithiothreitol

(*continued*)

TABLE II (*continued*)

Enzyme[a]	Ligand	Eluent
49. Cholinephosphate cytidylyltransferase	Glycerolphosphocholine	NaCl
50. Chorismate synthase	P_i	P_i
51. Chorismate mutase-prephenate-dehydrogenase	Phenylalanine	Adamantane acetate
52. Chymotrypsin	D-Tryptophan methyl ester	Acetic acid
53. Citrate synthase	ATP	ATP
54. Collagenase	Collagen, Arg	NaCl
55. Collagen galactosyltransferase	Con A	α-Methyl-D-mannoside, ethylene glycol
56. Collagen glucosyltransferase	Collagen	Collagen peptides, ethylene glycol
57. Collagen glycosyltransferase	UDPglucuronate	Collagen peptides
58. Corrinoid enzyme	Pteroylglutamate	NaCl, pteroylglutamate
59. Creatine kinase	*p*-Mercuribenzoate	Mercaptoethanol, ATP, NaCl
60. Cyclic nucleotide phosphodiesterase	Phenylbutenolide troponin C	KCl
61. Cytochrome *c* oxidase	Cytochrome *c*	NaCl
62. Cytochrome reductase	NADP, ADP	NADP
63. Cytotoxic protease	Soybean inhibitor	pH 2
64. Deoxynucleotidyltransferase (exo)	Oligo(dT)	KCl
65. Deoxyribosylase	Adenines	Deoxyinosine, guanidine
66. Dihydrofolate reductase (tetrahydrofolate dehydrogenase)	Folate analogs	Folate, pH, dihydrofolate
67. Dihydropteridine reductase	1,2-Naphthoquinone, methotrexate	NADH, pH, P_i
68. Diol dehydrase	Adenosylcobalamin	Propane-1,2-diol
69. Dipeptidyl peptidase	4-Phenylbutylglycylproline	Ethylene glycol NaCl
70. DNA-dependent ATPase II	DNA	NaCl
71. DNA-dependent RNA polymerase	tRNA	$(NH_4)_2SO_4$
72. DNA ligase	DNA	KCl
73. DNA-nicking–closing enzyme	DNA	NaCl
74. DNA polymerase	DNA, pyrans, poly(N)	NaCl or KCl

TABLE II (*continued*)

Enzyme[a]	Ligand	Eluent
75. DNA polymerase β	DNA	NaCl
76. DNase	DNA, d-DNA	NaCl, KCl
77. DNA-unwinding enzyme II	DNA	NaCl
78. Dopamine β-hydroxylase (dopamine β-monooxygenase)	Con A	α-Methyl D-glucoside
79. Elastase	Elastin, $(Ala)_3$	Salt, Ala
80. Endonuclease	DNA	NaCl or KCl
81. Endopeptidase	Phenylbutylamine	NaCl
82. Enterokinase (enteropeptidase)	*p*-Aminobenzamidine	NaCl, ethylene glycol, benzamidine
83. Exonuclease I	DNA	NaCl
84. Fatty acid synthetase	Pantetheine NADP	pH, salt, NADP
85. Ferredoxin-$NADP^+$ reductase	Ferredoxin	NaCl
86. Ferredoxin-nitrate reductase	Ferredoxin	Dithionate
87. Ferrochelatase	Porphyrin	Mesoporphyrin
88. Flavokinase (riboflavin kinase)	Flavins	Riboflavin
89. FMN oxidoreductase	AMP, NADP	NAD, citrate
90. Formaldehyde dehydrogenase	AMP	NaCl
91. Formate dehydrogenase	AMP	NaCl
92. Formiminotransferase	Tetrahydrofolate	Formiminoglutamate
93. Formylmethionine aminopeptidase	*N*-Formylbestatin	NaCl
94. L-Fucose dehydrogenase	AMP	pH
95. α-L-Fucosidase	Fucosamine thiofucopyranoside	Fucose
96. Fumarase (fumarate hydratase)	Pyromelitate, 2-(5′-phenylpentyl) fumarate	Malate, citrate
97. α-Galactosidase	Galactopyranosylamine, Con A	D-Galactose, methyl-D-mannoside, NaCl
98. Acid D-galactosidase	Thiolactopyranoside	Galactonolactone
99. β-Galactosidase	Thiogalactoside	Thiogalactoside, ethylene glycol
100. β-Galactosidase A_2, A_3	Con A	α-Methylmannoside
101. Galactosaminidase (endo)	Antifreeze-glycoprotein	NaCl
102. β-Galactofuranosidase (exo)	Galactomannan	NaCl

(*continued*)

TABLE II (*continued*)

Enzyme[a]	Ligand	Eluent
103. β-Galactosidase (neutral)	Galactosylamine	Citrate
104. D-Galactose kinase	ATP	NaCl
105. Galactosylhydroxylysyl-glucosetransferase	Collagen, UDPglucose, Con A	Collagenpeptides, NaCl, α-methyl-D-glucoside, ethylene glycol
106. Galactosyltransferase	α-Lactalbumin UDP, mucin	UDPgalactose, NaCl, N-acetylglucosamine, EDTA, Triton X-100, sucrose
107. Glucokinase	2-Amino-2-deoxy-glucose	Glucose, KCl
108. Glucose isomerase	Xylitol	NaCl
109. Glucose-6-phosphate dehydrogenase	ADP, NADP	NADP
110. Glucose-6-phosphatase	Glucose 6-phosphate	KCl
111. α-D-Glucosidase	*p*-Aminophenyl-maltoside, phenylboronic acid	NaCl, sugars
112. Glucosyltransferase I, II	Decosanates	UDPglucose
113. Glutamate decarboxylase	Ethyl glutamate	NaCl
114. Glutamate dehydrogenase	NADP, 2-AMP	NaCl, NADP
115. Glutamate synthase	2′,5′-ADP	Glutamine, NADPH
116. γ-Glutamylcysteine synthase	ATP, cystamine	ATP, dithiothreitol
117. γ-Glutamyl hydrolase	Polyglutamate	NaCl
118. Glutathione reductase	ADP, glutathione	NADPH, NADP, KCl
119. Glutathione transferase	Bromosulfophthalein, glutathione	KSCN, bromosulfophthalein
120. Glyceraldehyde-3-phosphate dehydrogenase	NAD, AMP, ATP	NAD, NADH, glycerol 3-phosphate
121. Glycerol-3-phosphate dehydrogenase	ATP	NADH
122. Glycineamide ribonucleotide transformylase	Glycineamide, ribonucleotide	P_i
123. Glycogen phosphorylase	AMP	AMP
124. Glycosidases	Sugar derivative	Sugar derivative
125. Glyoxylase	Glutathione	pH
126. Glyoxylase I	Glutathione	Glutathione
127. Glyoxylase II	Glutathione	*S*-Octylglutathione
128. GM_1-ganglioside-β-galactosidase	Thiogalactoside, Con A	NaCl, α-methyl-D-mannoside
129. Gramicidin-S synthetase	Ornithine, D-phenylalanine	KCl

TABLE II (*continued*)

Enzyme[a]	Ligand	Eluent
130. GTP cyclohydrolase	GTP, dihydrofolate	GTP
131. Guanylate cyclase	GTP	EDTA, NaCl
132. Guanyloribonuclease (ribonuclease T_1)	Guanylyl-(2′,5′)-guanosine	pH
133. Guanylyltransferase (mRNA)	DNA	NaCl
134. Guanine aminohydrolase (guanine deaminase)	Guanine	NaCl
135. N^2-Guanine RNA-methyltransferase	*S*-Adenosylhomocysteine	NaCl
136. Hexokinase	Glucosamines	Glucose, MgATP
137. Hexosaminidase A,B	β-D-Glucosylamine	pH
138. Hexosaminidase A	Thiogalactoside 2-deoxy-1-thio-β-D-galactoside	Gluconolactone, *N*-acetylglucosaminolactone
139. Hexosaminidase P	*N*-Acetylglucosylamine	pH
140. Histaminase (amine oxidase)	Cadaverine	Heparin
141. L-Histidine 2-oxoglutarate aminotransferase (histidine aminotransferase)	Histidine	NaCl
142. Histidinol dehydrogenase	His, histamine, AMP	His, imidazole, AMP
143. Hyaluronidase	Con A, glycosaminoglycans	α-Methylglucoside, α-methylmannoside NaCl
144. 3-Hydroxyacyl-CoA dehydrogenase	NAD	P_i
145. ω-Hydroxy-fatty-acid NADP-oxidoreductase	NADP	KCl, NADP
146. 3-Hydroxy-3-methylglutaryl-CoA reductase	HHG-CoA	HMG-CoA, glycerol
147. 15-Hydroxyprostaglandin dehydrogenase	NAD	NADH
148. 3α-Hydroxysteroid dehydrogenase	NAD	P_i
149. 3β-Hydroxysteroid dehydrogenase	NAD	P_i
150. 20α-Hydroxysteroid dehydrogenase	11α-Hydroxyprogesterone	pH
151. 20β-Hydroxysteroid dehydrogenase	11α-Hydroxyprogesterone	Mercaptoethanol
152. 3β-Hydroxysteroid oxidase	Cholesterol	Triton X-100

(*continued*)

TABLE II (*continued*)

Enzyme[a]	Ligand	Eluent
153. Hypoxanthine-guanine phosphoribosyltransferase	GMP	KCl, GMP
154. Indolyl-3-alkane α-hydroxylase	Indole	Acetate
155. Inosinic acid dehydrogenase	8-Amino-GMP, AMP	IMP, XMP
156. Isocitrate dehydrogenase	AMP	NAD
157. α-Isopropylmalate isomerase	Leu	Glycerol
158. Kallikrein	Aprotinin, aminobenzamidine	Benzamidine
159. 2-Keto-3-deoxy-L-fuconate NAD oxidoreductase	NAD	NAD
160. 2-Kynurenine 3-hydroxylase (kynurenine 3-monooxygenase)	NADP	NADP
161. Lactate dehydrogenase	AMP, oxamate, ATP	NAD, NADH
162. Lipases (triacylglycerol lipase)	Palmitate, Con A, cholate	Detergent, α-methyl-D-glucoside, cholate
163. Lipoamide dehydrogenase (dihydrolipoamide reductase)	Lipoamide, NAD	Salt
164. Lipoprotein lipase	Heparin	NaCl
165. Lipoxygenase	Linolenate	NaCl
166. Lysine tRNA-synthetase	Lysine	Lysine
167. L-Lysine 6-aminotransferase	L-Lysylacetamidododecyl	NaCl
168. Lysozyme	Chitotetraose *N*-acetyl-β-glucosamide	NaCl
169. Lysyl oxidase	Collagen	NaCl, urea
170. Luciferase	FMN, benzyloxyanilin, benzylamine	FMN, ethylene glycol, P_i
171. Malate dehydrogenase	AMP, ADP, ATP, NAD	pH, NADH, NADP, NaCl
172. Malate thiokinase	ADP	CoA, ATP, malate
173. Malic enzyme	NADP, ADP	NADP
174. Malonyl-CoA decarboxylase	NADP	NaCl
175. Maltodextrin phosphorylase	Glycogen	NaCl

TABLE II (*continued*)

Enzyme[a]	Ligand	Eluent
176. α-D-Mannosidase	Mannosylamine, benzidine	Mannose, NaCl
177. Methionyl-tRNA synthetase	AMP, tRNA	L-Methionine
178. 3-Methyladenine DNA-glycosylase	DNA	NaCl
179. Methylmalonyl-CoA mutase	Vitamin B_{12}	Vitamin B_{12}
180. Methyltransferase mRNA (nucleoside-2′-)	ADP	NaCl
181. N^5-Methyltetrahydrofolate-homocysteine methyltransferase	Cobalamin	NaCl, photolysis
182. Monoamine oxidase (amine oxidase)	Tyramine	KCl
183. Myosin light-chain kinase	Light chain	P_i
184. Myosin kinase (I, II)	Calmodulin	EGTA
185. NAD kinase	NAD	NAD
186. NADPH-adrenodoxin reductase	Adrenodoxin	NaCl
187. NADPH-cytochrome *c* (*P*-450) reductase	ADP, NADP	AMP, 2-AMP, NADP
188. NADPH-ferredoxin reductase	Adrenodoxin	NaCl
189. NADPH-flavin oxidoreductase	Flavin (FMN)	FMN
190. NAD-protein ADP-ribosyltransferase	AMP	NH_4Cl
191. Neuraminidase	Tyrosyl *p*-nitrophenyloxamic acid, colominic acid	pH, acetate, NaCl
192. Neutral protease	*N*-Phenylphosphenyl-Phe-Phe	pH
193. Nitrite reductase	Ferredoxin	P_i
194. Nuclease (micrococcal)	DNA	KCl
195. Nuclease T	TDP	Guanidine
196. Nucleotidase	Con A, 5′-AMP	5′-AMP
197. Nucleotide phosphodiesterase	Polyhistidine	pH
198. Nucleotide phosphotransferase	AMP	H_2O
199. Nucleotide pyrophosphatase	AMP	NaCl

(*continued*)

TABLE II (*continued*)

Enzyme[a]	Ligand	Eluent
200. Ornithine decarboxylase	Pyridoxamine phosphate	Pyridoxal phosphate
201. Ornithine transcarbamylase	*N*-(Phosphonacetyl) L-Orn	Carbamoyl phosphate
202. Orotidine-5′-phosphate decarboxylase	Azauridine 5′-phosphate	NaCl
203. Orotate phosphoribosyltransferase	Orotidine 5-phosphate	Orotate-$MgCl_2$ or 5-phosphoribosyl-β-D-ribose diphosphate
204. Pepsin	Poly(L-lysine)	NaCl
205. Peroxidase	Con A	α-Methyl-D-mannoside
206. Phenylalanine ammonia-lyase	Phenylalanine	Phenylalanine
207. Phenylalanine hydroxylase	Pteridine derivative	Phenylalanine, pH
208. Phenylalanine-tRNA synthetase	RNA	KCl-glycerol
209. Phosphatidylglycerophosphate synthetase	CDP-diglyceride	CDP-diglyceride
210. Phosphodiesterase	Activator protein, regulatory protein	Ca^{2+}, chelating agent, electrophoresis
211. Phosphoenolypyruvate carboxykinase	GTP, GMP	KCl, IDP
212. Phosphofructokinase	AMP	KCl, ATP
213. 6-Phosphogluconate dehydrogenase	NADP, ADP	Citrate, NADP
214. Phosphoglucose isomerase	ATP	Glucose 6-phosphate
215. Phosphoglycerate kinase	ATP	ATP (Mg^{2+})
216. Phospholipase	Con A, heparin	α-Methyl-D-pyranoside, NaCl
217. Phospholipase A_2	Glycero-3-phosphocholine, phosphatidylcholine	EDTA
218. Phospholipase C	Lipoprotein	NaCl, urea
219. Phosphoprotein phosphatase	Protamine, histone	NaCl
220. Phosphorylase *a*	AMP	AMP
221. Phosphorylase *b*	AMP	AMP
222. Phosphorylase kinase	Calmodulin	Low Ca, EDTA
223. Poly(A) polymerase	ATP, DNA, poly(A), RNA	ATP, KCl

TABLE II (*continued*)

Enzyme[a]	Ligand	Eluent
224. Poly(adenosine diphosphate ribose) polymerase	DNA	NaCl
225. Polynucleotide phosphorylase (polyribonucleotide nucleotidyltransferase)	RNA	NaCl
226. Polynucleotide 5′-triphosphatase	ADP, poly(U)	NaCl
227. Post-proline cleaving enzyme	*Z*-Prolyl-D-alanine	*Z*-Prolylphenylalanine
228. Prolyl hydroxylase	Poly(L-proline), collagen	Poly(L-lysine), ethylene glycol
229. Propionyl-CoA carboxylase	Avidin	Biotin
230. Prostaglandin cyclooxygenase	Flurbiprofen	Flufenamic acid, ethylene glycol
231. Proteases	4-Phenylbutylamine, soybean trypsin inhibitor, D-tryptophanmethyl ester, etc.	NaCl, urea mercuric chloride Ca, etc.
232. Protein kinase (AMP- and GMP-dependent)	cAMP, histone, cGMP, catalytic subunit, phosvitin, EGF, casein	cAMP, NaCl, cGMP, guanidine NaCl, ethanolamine
233. Protein phosphatase (phosphoprotein phosphatase)	Histone	NaCl
234. Pteroyl-α-oligoglutamyl endopeptidase	Oligoglutamyl peptides	NaCl
235. Pullulanase	α-Dextrin	β-Dextrin
236. Purine nucleoside phosphorylase	Formycin B	Inosine
237. Pyridine dinucleotide transhydrogenase	NAD, 2′,5′-ADP	NADPH, 2′-AMP
238. Pyridoxal kinase	Pyridoxal	Pyridoxine
239. Pyridoxamine-5-phosphate oxidase	Phosphopyridoxyl	Pyridoxal phosphate
240. Pyruvate decarboxylase	Thiamin pyrophosphate	EDTA
241. Pyruvate dehydrogenase	Thiochrome	Salt

(*continued*)

TABLE II (*continued*)

Enzyme[a]	Ligand	Eluent
242. Pyruvate-UDP-*N*-acetylglucosaminetransferase	UDP	NaCl
243. Renin	Pepstatin, hemoglobin octapeptide	pH
244. Restriction endonuclease	DNA	P_i, KCl
245. Reverse transcriptase	Poly(U)	Poly(G)
246. Ribonuclease	CMP	NaCl
247. Ribonuclease F_1, F_2	5′-GMP	2′(3)-GMP
248. Ribonuclease H	Oligo(dT)	EDTA
249. Ribonuclease T_2	AMP	2′(3)-AMP, NaCl
250. Ribonuclease U_2	AMP	NaCl
251. Ribonucleotide reductase	dGTP	dGTP, urea
252. RNA ligase	2′,5′-ADP	Mg
253. tRNA nucleotidyltransferase	tRNA	NaCl-EDTA
254. RNA polymerase	DNA, rifamycin, heparin	Salts
255. RNA polymerase III	DNA	$(NH_4)_2SO_4$
256. RNA polymerase basic protein	RNA	$(NH_4)_2SO_4$
257. RNA polymerase (DNA dependent)	DNA	KCl
258. Salicylate hydroxylase (salicylate 1-monooxygenase)	*p*-Aminosalicylate	Salicylate
259. Sialidase (neuraminidase)	α-Acid glycoprotein	pH
260. Sialyl transferase (CMP-*N*-acetyl-neuraminate-galactosylglycoprotein sialyltransferase)	CDP	CDP
261. Sorbitol dehydrogenase (L-iditoldehydrogenase)	NAD	NAD
262. Spermine synthase	Spermine	Spermidine
263. Sphingomyelinase	Con A	α-Methyl-D-mannoside
264. Succinyl-CoA acetoacetyl-CoA transferase	Acetoacetyl-CoA	Acetoacetate
265. Succinyl-CoA synthetase	GDP	GDP
266. Succinate-semialdehyde dehydrogenase	AMP	AMP
267. Succinate thiokinase (succinyl-CoA synthetase)	3′,5′-ADP	CoA
268. Terminal deoxynucleotidyltransferase	Oligo(dT)	KCl

TABLE II (*continued*)

Enzyme[a]	Ligand	Eluent
269. Thioredoxin reductase	ADP	NADPH
270. Threonine deaminase (threonine dehydratase)	Val, Ile	KCl, K_3PO_4
271. Threonine dehydratase	AMP	AMP
272. Thrombin	Aminobenzamidine	Benzamidine
273. Thymidine kinase	Uracil, glycoprotein TMP	NaCl, thymidine TMP
274. Thymidylate synthase	Thymidylate analogs	P_i, Tris
275. Thyroid peroxidase	Tyrosine	pH
276. Transcarboxylase (methylmalonyl-CoA carboxyltransferase)	Avidin	pH
277. Transhydrogenase [NAD(P)$^+$ transhydrogenase]	NAD	NADH
278. Triglyceride lipase (triacylglycerol lipase)	Heparin	NaCl
279. Tryptophanase	Pyridoxal phosphate	Pyridoxal phosphate
280. Tryptophan 5-monooxygenase	Pteridine	NaCl
281. Tryptophan synthase	Indolepropionate, tryptophanol-P	Imidazole, indolepropanol-P
282. Tyrosinase	Con A	α-Methyl-D-mannoside
283. Tyrosine aminotransferase	Succinate	pH
284. Tyrosine phenol-lyase	Pyridoxal phosphate	Pyridoxal phosphate
285. UDPgalactose 4-epimerase	UDP	UMP, UDP
286. UDPgalactose glycoprotein galactosyltransferase (glycoprotein β-D-galactosyltransferase)	α-Lactalbumin	*N*-Acetylglucosamine
287. UDPglucose dehydrogenase	AMP	AMP
288. UDPglucuronyltransferase	UDP	UDPglucuronic acid
289. UDP-*N*-acetylenolpyruvylglucosamine reductase	NADP	NADPH
290. Urate oxidase	8-Aminoxanthine	Urate
291. Uricase (urate oxidase)	Urate, xanthine	Xanthine
292. Urokinase	β-Naphthamidine	pH
293. Valyl-tRNA synthetase	tRNA	KCl

References to TABLE II

[a] Key to references:

1. P. J. Morrod, A. G. Marshall, and D. G. Clark, *Biochem. Biophys. Res. Commun.* **63,** 335 (1975); P. Ott, B. Jenny, and V. Brodbeck, *Eur. J. Biochem.* **57,** 469 (1975); T. L. Rosenberry and J. M. Richardson, *Biochemistry* **16,** 3550 (1977); C. A. Reavil, M. S. Wooster, and D. T. Plummer, *Biochem. J.* **173,** 851 (1978); C. A. Reavil and D. T. Plummer, *J. Chromatogr.* **157,** 141 (1978); G. Webb and D. G. Clark, *Arch. Biochem. Biophys.* **191,** 278 (1978); A. S. Brooks, G. E. Tiller, and W. G. Struve, *Biochim. Biophys. Acta* **615,** 354 (1980); K. K. Parker, S. L. Chan, and A. J. Trevor, *Arch. Biochem. Biophys.* **187,** 322 (1978); T. H. Meedel, *Biochim. Biophys. Acta* **615,** 360 (1980).
2. J. S. Wolpert and M. L. Ernst-Fonberg, *Biochemistry* **14,** 1095 (1975).
3. M. Schwyzer and R. L. Hill, *J. Biol. Chem.* **252,** 2338 (1977); M. Sugiura, T. Kawasaki, and I. Yamashina, *J. Biol. Chem.* **257,** 9501 (1982).
4. J. E. Sadler, J. I. Rearick, and R. L. Hill, *J. Biol. Chem.* **254,** 5934 (1977).
5. L. R. Glasgow, J. C. Paulson, and R. L. Hill, *J. Biol. Chem.* **252,** 8615 (1977).
6. E. E. Grebner and I. Parikh, *Biochim. Biophys. Acta* **350,** 437 (1974).
7. D. K. Banerjee and D. Basu, *Biochem. J.* **145,** 113 (1975); R. G. Edwards, P. Thomas, and J. M. Westwood, *Biochem. J.* **151,** 145 (1975); M. Pokorny and C. P. J. Glaudermans, *FEBS Lett.* **50,** 66 (1975).
8. M. S. Saini and R. L. Van Etten, *Arch. Biochem. Biophys.* **191,** 613 (1978).
9. E. Hackenthal, R. Hackenthal, and U. Hilgengeldt, *Biochim. Biophys. Acta* **522,** 561 (1978); K. Yamamoto, N. Katsuda, and K. Kato, *Eur. J. Biochem.* **92,** 499 (1978).
10. H. Rosemeyer and F. Seela, *Anal. Biochem.* **115,** 339 (1981).
11. K. M. Rogers, G. F. White, and K. S. Dodgson, *Biochim. Biophys. Acta* **527,** 70 (1978).
12. A. S. Olsen and G. Milman, *Biochemistry* **16,** 2501 (1977); J. A. Holden, G. S. Meredith, and W. H. Kelley, *J. Biol. Chem.* **254,** 6951 (1979).
13. C. M. Andres and I. H. Fox, *J. Biol. Chem.* **254,** 11388 (1979); R. L. Miller, D. L. Adamczyk, and W. H. Miller, *J. Biol. Chem.* **254,** 2339 (1979); J. W. DeJong, E. Keijzer, M. P. Uitendaal, and E. Harmsen, *Anal. Biochem.* **101,** 407 (1980).
14. F. W. Hulla, M. Hodsel, S. Risi, and K. Dose, *Eur. J. Biochem.* **67,** 469 (1976); K. Gietzen, M. Tejcka, and H. U. Wolf, *Biochem. J.* **189,** 81 (1980); V. Niggli, J. T. Penniston, and E. Carafoli, *J. Biol. Chem.* **254,** 9955 (1979).
15. V. Bennett, E. O'Keefe, and P. Cuatrecasas, *Proc. Natl. Acad. Sci. U.S.A.* **72,** 33 (1975); N. I. Swislocki, T. Magnuson, and J. Tierney, *Arch. Biochem. Biophys.* **179,** 157 (1977); C. Homcy, S. Wrenn, and E. Haber, *Proc. Natl. Acad. Sci. U.S.A.* **75,** 59 (1978); T. Pfeuffer and H. Metzger, *FEBS Lett.* **146,** 369 (1982).
16. T. Haugen, A. Ishaque, A. K. Chatterjee, and J. Preiss, *FEBS Lett.* **42,** 205 (1974).
17. J. Moss, S. J. Stanley, and N. J. Oppenheimer, *J. Biol. Chem.* **254,** 8891 (1979).
18. T. K. Korpela, *J. Chromatogr.* **143,** 519 (1977); T. K. Korpela, A. E. Hinkkanen, and R. P. Raunio, *J. Solid Phase Biochem.* **1,** 215 (1977).
19. R. Yocum, P. M. Blumberg, and J. L. Strominger, *J. Biol. Chem.* **249,** 4863 (1974); M. Gorecki, A. Bar-eli, Y. Burstein, A. Patchornik, and E. B. Chain, *Biochem. J.* **147,** 131 (1975); H. H. Martin, W. Schilf, and C. Maskos, *Eur. J. Biochem.* **71,** 585 (1976); W. Schilf and H. H. Martin, *Eur. J. Biochem.* **105,** 361 (1980); J. Coyette, J. M. Ghuysen, and R. Fontana, *Eur. J. Biochem.* **88,** 297 (1978).

References to TABLE II (*continued*)

20. D. Keradjopoulos and A. W. Holldorf, *Biochim. Biophys. Acta* **570,** 1 (1979).
21. M. J. Comer, D. B. Craven, M. J. Harvey, A. Atkinson, and P. D. G. Dean, *Eur. J. Biochem.* **55,** 201 (1975); W. F. Bosron, T. K. Li, L. G. Lange, W. P. Dafeldecker, and B. L. Vallee, *Biochem. Biophys. Res. Commun.* **74,** 85 (1977); Bosron *et al., Proc. Natl. Acad. Sci. U.S.A.* **74,** 4378 (1977); C. N. Ryzewski and R. Pietruszko, *Arch. Biochem. Biophys.* **183,** 73 (1977); H. Von Bahr-Lindstrom, L. Andersson, K. Mosbach, and H. Jornvall, *FEBS Lett.* **89,** 293 (1978); A. J. L. Brown and C. Y. Lee, *Biochem. J.* **179,** 479 (1979); B. Tabakoff and J. P. von Wartburg, *Biochem. Biophys. Res. Commun.* **63,** 957 (1975); C. Y. Lee, D. Charles, and D. Bronson, *J. Biol. Chem.* **254,** 6375 (1979).
22. R. J. S. Duncan, *Biochem. J.* **161,** 123 (1977); N. J. Greenfield and R. Pietruszko, *Biochim. Biophys. Acta* **483,** 35 (1977); H. Nakayasu, K. Mihara, and R. Sato, *Biochem. Biophys. Res. Commun.* **83,** 697 (1978); K. Inoue, H. Nishimukai, and K. Yamasawa, *Biochim. Biophys. Acta* **569,** 117 (1979); K. A. Bostian and G. F. Betts, *Biochem. J.* **173,** 773 (1978); A. L. Bognar and E. A. Meighen, *J. Biol. Chem.* **253,** 446 (1978).
23. B. Tabakoff and J. P. von Wartburg, *Biochem. Biophys. Res. Commun.* **63,** 957 (1975).
24. P. F. Kador, D. Carper, and J. H. Kinoshita, *Anal. Biochem.* **114,** 53 (1981).
25. E. Mossner, M. Boll, and G. Pfleiderer, *Hoppe-Seyler's Z. Physiol. Chem.* **361,** 543 (1980); F. G. Lehmann, *Biochim. Biophys. Acta* **616,** 41 (1980); T. Komoda and Y. Sakagishi, *Biochim. Biophys. Acta* **445,** 645 (1976); M. Landt, S. C. Boltz, and L. G. Butler, *Biochemistry* **17,** 915 (1978); Y. Yokota, *J. Biochem.* (*Tokyo*) **83,** 1293 (1978); B. O'Keefe and J. E. Kinsella, *Int. J. Biochem.* **10,** 125 (1979).
26. J. J. Shiem, T. Tamaye, and K. J. Yasunobu, *Biochim. Biophys. Acta* **377,** 229 (1975); P. Turini, S. Sabatini, O. Befani, F. Chimenti, C. Casanova, P. L. Riccio, and B. Mondovi, *Anal. Biochem.* **125,** 249 (1982).
27. S. R. Chandrika and G. Padmanaban, *Biochem. J.* **191,** 29 (1980).
28. S. M. M. Basha, M. N. Horst, F. W. Bazer, and P. M. Roberts, *Arch. Biochem. Biophys.* **185,** 174 (1978); K. H. Rohm, *Hoppe-Seyler's Z. Physiol. Chem.* **363,** 641 (1982); C. Kettner, J. Rodriquez-Absi, G. I. Glover, and J. M. Prescott, *Arch. Biochem. Biophys.* **162,** 56 (1974).
29. R. Tkachuk, *FEBS Lett.* **52,** 66 (1975).
30. T. H. Grove and H. R. Levy, *Biochim. Biophys. Acta* **397,** 80 (1975).
31. P. H. Chan and W. Z. Hassid, *Anal. Biochem.* **64,** 372 (1975).
32. R. Tarrab, B. Perez, and C. Lopez, *Anal. Biochem.* **65,** 26 (1975); S. Traniello, R. Barsacchi, E. Magri, and E. Grazi, *Biochem. J.* **145,** 153 (1975).
33. T. Shibatani, T. Kakimoto, and I. Chibata, *J. Biol. Chem.* **250,** 4580 (1975).
34. K. A. Balasubramanian and B. K. Bachhawat, *Biochim. Biophys. Acta* **403,** 113 (1975); A. A. Farooqui and P. N. Srivastava, *Biochem. J.* **181,** 331 (1979); R. L. Van Etten and A. Waheed, *Arch. Biochem. Biophys.* **202,** 366 (1980).
35. T. Tosa, T. Sato, R. Sano, K. Yamamoto, Y. Matuo, and I. Chibata, *Biochim. Biophys Acta* **334,** 1 (1974).
36. Same as 35.
37. Same as 35.
38. G. S. J. Rao, M. S. Savithri, S. Seethalakshmi and N. A. Rao, *Anal. Biochem.* **95,** 401 (1979).

(*continued*)

References to TABLE II (*continued*)

39. A. T. H. Abdelal and J. L. Ingraham, *J. Biol. Chem.* **250,** 4410 (1975).
40. P. L. Whitney, *Anal. Biochem.* **57,** 467 (1974); P. J. Wistrand, S. Lindahl, and T. Wahlstrand, *Eur. J. Biochem.* **57,** 189 (1975); P. J. Wistrand and T. Wahlstrand, *Biochim. Biophys. Acta* **481,** 712 (1977); J. T. Johansen, *Carlsberg Res. Commun.* **41,** 73 (1976); R. S. Holmes, *Eur. J. Biochem.* **78,** 511 (1977).
41. G. Oshima and K. Nagasawa, *J. Biochem.* (*Tokyo*) **81,** 1285 (1977); L. B. Cueni, T. J. Bazzone, J. F. Riordan, and B. L. Vallee, *Anal. Biochem.* **107,** 341 (1980).
42. Y. Narahashi, K. Yoda, and S. Honda, *J. Biochem.* (*Tokyo*) **82,** 615 (1977); H. R. Trayer, M. A. Winstanley, and I. P. Trayer, *FEBS Lett.* **83,** 141 (1977).
43. T. H. Plummer and M. Y. Hurwitz, *J. Biol. Chem.* **253,** 3907 (1978).
44. R. T. Borchardt, C. F. Cheng, and D. R. Thakker, *Biochem. Biophys. Res. Commun.* **63,** 69 (1975); R. T. Borchardt and C. F. Cheng, *Biochim. Biophys. Acta* **522,** 49 (1978); P. A. Gulliver and K. F. Tipton, *Eur. J. Biochem.* **88,** 439 (1978).
45. K. Yamamoto, N. Katsuda, and K. Kato, *Eur. J. Biochem.* **92,** 499 (1978).
46. M. Nummi, M. L. Niku-Paavola, T. M. Enari, and V. Raunio, *Anal. Biochem.* **116,** 137 (1981).
47. R. L. Ryan and W. O. McClure, *Biochemistry* **18,** 5357 (1979).
48. H. Tsuge, Y. Nakano, H. Onishi, Y. Futamura, and K. Ohashi, *Biochim. Biophys. Acta* **614,** 274 (1980).
49. P. C. Choy and D. E. Vance, *Biochem. Biophys. Res. Commun.* **72,** 719 (1976).
50. T. W. Cole and F. H. Gaertner, *Biochem. Biophys. Res. Commun.* **67,** 170 (1975).
51. G. D. Smith, D. V. Roberts, and A. Daday, *Biochem. J.* **165,** 121 (1977).
52. F. Pochon and J. G. Bieth, *J. Biol. Chem.* **257,** 6683 (1982).
53. M. Lindberg and K. Mosbach, *Eur. J. Biochem.* **53,** 481 (1975).
54. C. C. Huang and M. Abramson, *Biochim. Biophys. Acta* **384,** 484 (1975); C. Gillet, Y. Eckhont, and G. Vaes, *FEBS Lett.* **74,** 126 (1977); I. Emod and B. Keil, *FEBS Lett.* **77,** 51 (1977).
55. L. Risteli, *Biochem. J.* **169,** 189 (1978).
56. H. Anttinen, R. Myllyla, and K. I. Kivinkko, *Eur. J. Biochem.* **78,** 11 (1977); R. Myllyla, H. Anttinen, L. Risteli, and K. I. Kivinkko, *Biochim. Biophys. Acta* **480,** 113 (1977).
57. R. Myllyla, H. Anttinen, L. Risteli, and K. I. Kivinkko, *Biochim. Biophys. Acta* **480,** 113 (1977).
58. F. K. Welty and H. G. Wood, *J. Biol. Chem.* **253,** 5832 (1978).
59. R. J. Boegman, *FEBS Lett.* **53,** 99 (1975); V. Madelian and W. A. Warren, *Anal. Biochem.* **64,** 517 (1975); N. Hall, P. Addis, and M. DeLuca, *Biochem. Biophys. Res. Commun.* **76,** 950 (1977); N. Hall, P. Addis, and M. DeLuca, *Biochemistry* **18,** 1745 (1979); C.-Y. Lee, L. H. Lazarus, D. S. Kabakoff, P. J. Russell, M. Laver, and N. O. Kaplan, *Arch. Biochem. Biophys.* **178,** 8 (1977).
60. D. M. Watterson and T. C. Vanaman, *Biochem. Biophys. Res. Commun.* **73,** 40 (1976); A. F. Prigent, G. Nemoz, and H. Pacheco, *Biochem. Biophys. Res. Commun.* **95,** 1080 (1980).
61. K. Bill, R. P. Casey, C. Brojer, and A. Azzi, *FEBS Lett.* **120,** 248 (1980); R. B. Gennis, R. P. Casey, A. Azzi, and B. Ludwig, *Eur. J. Biochem.* **125,** 189 (1982); T. Ozawa, M. Tanaka, and T. Wakabayshi, *Proc. Natl. Acad. Sci. U.S.A.* **79,** 7175 (1982).

References to TABLE II (*continued*)

62. J. D. Dignam and H. W. Strobel, *Biochemistry* **16,** 1116 (1977); D. A. Schafer and D. E. Hulguist, *Biochem. Biophys. Res. Commun.* **95,** 381 (1980).
63. V. B. Hatcher, M. S. Oberman, G. S. Lazarus, and A. I. Grayzel, *J. Immunol.* **120,** 665 (1978).
64. S. Okamura, F. Crane, H. A. Messner, and T. W. Mak, *J. Biol. Chem.* **253,** 3765 (1978).
65. J. Holguin and R. Cardinaud, *Eur. J. Biochem.* **54,** 505 (1975).
66. D. L. Peterson, J. M. Gleisner, and R. L. Blakley, *Biochemistry* **14,** 5261 (1975); D. Baccanari, A. Phillips, S. Smith, D. Sinski, and J. Burchall, *Biochemistry* **14,** 5267 (1975); S. V. Gupta, N. J. Greenfield, M. Poe, D. R. Makulu, M. N. Williams, B. A. Moroson, and J. R. Bertino, *Biochemistry* **16,** 3073 (1977); R. S. Gupta, W. F. Flintoff, and L. Siminovitch, *Can. J. Biochem.* **55,** 445 (1977); T. M. Lincoln, W. L. Dills, and J. D. Corbin, *J. Biol. Chem.* **252,** 4269 (1977); R. Ferone and S. Roland, *Proc. Natl. Acad. Sci. U.S.A.* **77,** 5802 (1980); R. L. Then, *Biochim. Biophys. Acta* **614,** 25 (1980).
67. C. G. H. Cotton and I. Jennings, *Eur. J. Biochem.* **83,** 319 (1978); S. Webber, T. L. Deits, W. R. Snyder, and J. M. Whiteley, *Anal. Biochem.* **84,** 491 (1978).
68. T. Toraya and S. Fukui, *J. Biol. Chem.* **255,** 3520 (1980).
69. R. M. Metrione, *Biochim. Biophys. Acta* **526,** 531 (1978).
70. E. Richet and M. Kohiyama, *J. Biol. Chem.* **253,** 7490 (1978).
71. B. Wittig and S. Wittig, *Biochim. Biophys. Acta* **520,** 598 (1978).
72. M. Katouzian and J. C. David, *Biochem. Biophys. Res. Commun.* **82,** 1168 (1978).
73. W. R. Bauer, E. C. Ressner, J. Kates, and J. V. Patzke, *Proc. Natl. Acad. Sci. U.S.A.* **74,** 1841 (1977).
74. D. J. Arndt-Jovin, T. M. Jovin, W. Bahr, A. H. Frischauf, and M. Marquardt, *Eur. J. Biochem.* **54,** 411 (1975); A. Matsukage, E. W. Bohn, and S. H. Wilson, *Biochemistry* **14,** 1006 (1975); L. A. Salzman and L. McKerlie, *J. Biol. Chem.* **250,** 5583 (1975); P. Chandra and L. K. Steel, *Biochem. J.* **167,** 513 (1977); A. Baer and W. Schiebel, *Eur. J. Biochem.* **86,** 77 (1978); R. A. Hitzeman and A. R. Price, *J. Biol. Chem.* **253,** 8518 (1978); H. Joenje and R. M. Benbow, *J. Biol. Chem.* **253,** 2640 (1978); W. Mastropaolo and C. A. Lang, *J. Biol. Chem.* **253,** 1978 (1978); C. A. Ross and W. J. Harris, *Biochem. J.* **171,** 231 (1978); J. G. Chirikjian, L. Rye, and T. S. Papas, *Proc. Natl. Acad. Sci. U.S.A.* **72,** 1142 (1975); M. J. Modak and S. L. Marcus, *J. Biol. Chem.* **252,** 11 (1977); G. Bauer, G. Jilek, and P. H. Hofschneider, *Eur. J. Biochem.* **79,** 345 (1977); A. Hizi and W. K. Joklik, *J. Biol. Chem.* **252,** 2281 (1977).
75. T. A. Kunkel, J. E. Tcheng, and R. R. Meyer, *Biochim. Biophys. Acta* **520,** 302 (1978).
76. G. Sabeur, P. J. Sicard, and G. Aubel-Sadron, *Biochemistry* **13,** 3203 (1974); H. Tanaka, I. Sasaki, K. Yamashita, K. Miyazaki, T. Matuo, J. Yamashita, and T. Hario, *J. Biochem. (Tokyo)* **88,** 797 (1980); H. Tanaka, I. Sasaki, K. Yamashita, Y. Matuo, J. Yamashita, and T. Horio, *J. Biochem. (Tokyo)* **91,** 1411 (1982).
77. M. Abdel-Monem, M. C. Chanal, and H. Hoffmann-Berling, *Eur. J. Biochem.* **79,** 33 (1977).
78. R. A. Rush, P. E. Thomas, S. H. Kindler, and S. Udenfriend, *Biochem. Biophys. Res. Commun.* **57,** 1301 (1974); R. P. Frigon and R. A. Stone, *J. Biol. Chem.* **253,** 6780 (1978).

(continued)

References to TABLE II (*continued*)

79. Y. Legrand, G. Pignaud, J. P. Caen, B. Robert, and L. Robert, *Biochem. Biophys. Res. Commun.* **63,** 224 (1975); K. Katagiri, T. Takeuchi, K. Taniguchi, and M. Sasaki, *Anal. Biochem.* **86,** 159 (1978); C. Gillet, Y. Eckkhont, and G. Vaes, *FEBS Lett.* **74,** 126 (1977).
80. M. Mechali and A. M. DeRecondo, *Eur. J. Biochem.* **58,** 461 (1975); S. Ljungquist, *J. Biol. Chem.* **252,** 2808 (1977); S. Riazuddin and L. Grossman, *J. Biol. Chem.* **252,** 6280 (1977); V. Bibor and W. G. Verly, *J. Biol. Chem.* **253,** 850 (1978).
81. K. D. Jany, H. Haug, G. Pfleiderer, and J. Ishay, *Biochemistry* **17,** 4675 (1978).
82. D. W. A. Grant and J. Hermon-Taylor, *Biochem. J.* **147,** 363 (1975); L. E. Anderson, K. A. Walsh, and H. Neurath, *Biochemistry* **16,** 3354 (1977); D. W. A. Grant, A. I. Magee, and J. Hermon-Taylor, *Eur. J. Biochem.* **88,** 183 (1978).
83. R. K. Ray, R. Rueben, I. Molineux, and M. Gefter, *J. Biol. Chem.* **249,** 5379 (1974).
84. F. A. Lornitzo, A. A. Qureshi, and J. W. Porter, *J. Biol. Chem.* **250,** 4520 (1975); A. A. Qureshi, M. Kim, F. A. Lornitzo, R. A. Jenik, and J. W. Porter, *Biochem. Biophys. Res. Commun.* **64,** 836 (1975); A. A. Qureshi, R. A. Jenik, M. Kim, F. A. Lornitzo, and J. W. Porter, *Biochem. Biophys. Res. Commun.* **66,** 344 (1975); W. I. Wood, D. O. Peterson, and K. Bloch, *J. Biol. Chem.* **253,** 2650 (1978).
85. M. Shin and R. Oshino, *J. Biochem.* (*Tokyo*) **83,** 357 (1978).
86. C. Manzano, P. Candau, and M. G. Guerrero, *Anal. Biochem.* **90,** 408 (1978).
87. K. Mailer, R. Poulson, D. Dolphin, and A. D. Hamilton, *Biochem. Biophys. Res. Commun.* **96,** 777 (1980).
88. A. H. Merrill and D. B. McCormick, *J. Biol. Chem.* **255,** 1335 (1980).
89. E. Jablonski and M. DeLuca, *Biochemistry* **16,** 2932 (1977).
90. L. Uotila and M. Koivusalo, *Arch. Biochem. Biophys.* **196,** 33 (1979).
91. Same as 90.
92. K. Slavik, V. Zizkovsky, V. Slavikova, and P. Fort, *Biochem. Biophys. Res. Commun.* **59,** 1173 (1974).
93. H. Suda, K. Yamamoto, T. Aoyagi, and M. Umezawa, *Biochim. Biophys. Acta* **616,** 60 (1980).
94. M. Endo and N. Hiyama, *J. Biochem.* (*Tokyo*) **86,** 1959 (1979).
95. D. Robinson and R. Thorpe, *FEBS Lett.* **45,** 191 (1974); J. A. Alhadeff, A. L. Miller, H. Wenaas, T. Vedvick, and J. S. O'Brien, *J. Biol. Chem.* **250,** 7106 (1975); D. J. Opheim and O. Touster, *J. Biol. Chem.* **252,** 739 (1977); J. A. Alhadeff and G. L. Andrews-Smith, *Biochem. J.* **187,** 45 (1980); S. F. Chien, and G. Dawson, *Biochim. Biophys. Acta* **614,** 475 (1980); R. S. Jain, A. Levy-Benshimol, C. A. Buck, and L. Warren, *J. Chromatogr.* **139,** 283 (1977).
96. S. Beeckmans and L. Kanarek, *Eur. J. Biochem.* **78,** 437 (1977); S. Chaudhuri and E. W. Thomas, *Biochem. J.* **177,** 115 (1979).
97. N. Harpaz, H. M. Flowers, and N. Sharon, *Eur. J. Biochem.* **77,** 419 (1977); J. W. Kusiak, J. M. Quirk, R. O. Brady, and G. E. Mook, *J. Biol. Chem.* **253,** 184 (1978).
98. A. L. Miller, R. G. Frost, and J. S. O'Brien, *Biochem. J.* **165,** 591 (1977).
99. H. Hamazaki and K. Hotta, *FEBS Lett.* **76,** 299 (1977); C. H. Kuo and W. W. Wells, *J. Biol. Chem.* **253,** 3550 (1978); P. V. Wagh, *Biochim. Biophys. Acta* **522,** 515 (1978).
100. R. G. Frost, E. W. Holmes, A. G. W. Norden, and J. S. O'Brien, *Biochem. J.* **175,** 181 (1978).

References to TABLE II (*continued*)

101. L. R. Glasgow, J. C. Paulson, and R. L. Hill, *J. Biol. Chem.* **252,** 8615 (1977).
102. M. Rietschel-Berst, N. H. Jentoft, P. D. Rich, C. Pletcher, F. Fang, and J. E. Gander, *J. Biol. Chem.* **252,** 3219 (1977).
103. Y. Ben-Yoseph, E. Shapira, D. Edelman, B. K. Burton, and H. L. Nadler, *Arch. Biochem. Biophys.* **184,** 373 (1977).
104. P. H. Chan and W. Z. Hassid, *Anal. Biochem.* **64,** 372 (1975).
105. H. Antinnen, R. Myllyla, and K. I. Kivirikko, *Biochem. J.* **175,** 737 (1978).
106. S. C. Magee, M. Ramswarup, and K. E. Ebner, *Biochemistry* **13,** 99 (1974); I. H. Fraser and S. Mookerjea, *Biochem. J.* **164,** 541 (1977); A. Bella, J. S. Whitehead, and Y. S. Kim, *Biochem. J.* **167,** 621 (1977); Y. Legrand, G. Pignaud, and J. P. Caen, *FEBS Lett.* **76,** 294 (1977); C. A. Smith and K. Brew, *J. Biol. Chem.* **252,** 7294 (1977); M. Pierce, R. D. Cummings, T. A. Cebula, and S. Roth, *Biochim. Biophys. Acta* **571,** 166 (1979); J. Mendicino, S. Sivakami, M. Davila, and V. Chandrasekaran, *J. Biol. Chem.* **257,** 3987 (1982).
107. M. J. Holroyde, J. M. E. Chesher, I. P. Trayer, and D. G. Walker, *Biochem. J.* **153,** 351 (1976).
108. Y. H. Lee, P. C. Wankat, and H. A. Emery, *Biotechnol. Bioeng.* **13,** 1639 (1976).
109. C.-Y. Lee, C. H. Langley, and J. Burkhart, *Anal. Biochem.* **86,** 697 (1978); A. Yoshida, *J. Chromatogr.* **114,** 321 (1975); E. Burgissor and J. L. Fauchere, *Helv. Chim. Acta* **59,** 760 (1976); A. Morelli, U. Benatti, G. F. Gaetani, and A. DeFlora, *Proc. Natl. Acad. Sci. U.S.A.* **75,** 1979 (1978); N. Kato, H. Sahm, H. Schutte, and F. Wagner, *Biochim. Biophys. Acta* **566,** 1 (1979); M. L. Dao, J. L. Watson, R. Delaney, and B. C. Johnson, *J. Biol. Chem.* **254,** 9441 (1979).
110. P. R. Reczek and C. A. Villee, *Biochem. Biophys. Res. Commun.* **107,** 1158 (1982).
111. J. Giudicelli, R. Emiliozzi, C. Vannier, G. DeBurlet, and P. Sudaka, *Biochim. Biophys. Acta* **612,** 85 (1980); T. A. Myohamen, V. Bouriotis, and P. D. G. Dean, *Biochem. J.* **197,** 683 (1981).
112. T. B. Breithaupt and R. J. Light, *J. Biol. Chem.* **257,** 9622 (1982).
113. T. Yamaguchi and Y. Matsumura, *Biochim. Biophys. Acta* **481,** 706 (1977).
114. W. Leicht, M. M. Werber, and H. Eisenberg, *Biochemistry* **17,** 4004 (1978); A. D. McCarthy, J. M. Walker, and K. F. Tipton, *Biochem. J.* **191,** 605 (1980); D. H. Watson and J. C. Wootton, *Biochem. J.* **167,** 95 (1977).
115. C. N. G. Schmidt and L. Jervis, *Anal. Biochem.* **104,** 127 (1980).
116. R. Sekura and A. Meister, *J. Biol. Chem.* **252,** 2599 (1977); G. Foure-Seelig and A. Meister, *J. Biol. Chem.* **257,** 5092 (1982).
117. M. Silink, R. Reddel, M. Bethel, and P. B. Rowe, *J. Biol. Chem.* **250,** 5982 (1975).
118. V. P. Pigiet and R. R. Conley, *J. Biol. Chem.* **252,** 6367 (1977); I. Carlberg and B. Mannervik, *Biochim. Biophys. Acta* **484,** 268 (1977); G. Krohne-Ehrich, R. H. Schirmer, and R. Untrucht-Crau, *Eur. J. Biochem.* **80,** 65 (1977); V. Boggaram, T. Brobjer, K. Larson, and B. Mannervik, *Anal. Biochem.* **98,** 335 (1979); I. Carlberg and B. Mannervik, *Anal. Biochem.* **116,** 531 (1981); J. Danner, H. M. Lenhoff, and W. Heagy, *Anal. Biochem.* **82,** 586 (1977).
119. A. Grahnen and I. Sjoholm, *Eur. J. Biochem.* **80,** 573 (1977); A. G. Clark and W. C. Dauterman, *Pept. Biochem. Physiol.* **17,** 307 (1982); P. C. Simons and D. L. V. Jagt, *Anal. Biochem.* **82,** 334 (1977); N. Pattinson, *Anal. Biochem.* **115,** 424 (1981).

(*continued*)

References to TABLE II (*continued*)

120. C. R. Lowe, *Eur. J. Biochem.* **73,** 265 (1977); J. D. Hillman, *Biochem. J.* **179,** 99 (1979); J. I. Harris, J. D. Hocking, M. J. Runswick, K. Suzuki, and J. E. Walker, *Eur. J. Biochem.* **108,** 535 (1980).
121. D. W. Niesel, G. C. Bewley, S. G. Miller, F. B. Armstrong, and C. Y. Lee, *J. Biol. Chem.* **255,** 4073 (1980).
122. C. A. Capereli, C. Chettur, L. Y. Lin, and S. J. Benkovic, *Biochem. Biophys. Res. Commun.* **82,** 403 (1978).
123. N. B. Sorenson and P. Wang, *Biochem. Biophys. Res. Commun.* **67,** 883 (1975).
124. L. R. Glasgow, Y. C. Paulson, and R. L. Hill, *J. Biol. Chem.* **252,** 8615 (1977); T. Mega and Y. Matsushima, *J. Biochem.* (*Tokyo*) **81,** 571 (1977).
125. M. V. Kester and S. J. Norton, *Biochim. Biophys. Acta* **391,** 212 (1975).
126. B. Oray and S. J. Norton, *Biochim. Biophys. Acta* **483,** 203 (1977); N. Elango, S. Janaki, and A. R. Rao, *Biochem. Biophys. Res. Commun.* **83,** 1388 (1978); A. C. Aronsson and B. Mannervik, *Biochem. J.* **165,** 503 (1977); A. C. Aronsson, G. Tibbelin, and B. Mannervik, *Anal. Biochem.* **92,** 390 (1979); B. Oray and S. J. Norton, *Biochim. Biophys. Acta* **611,** 168 (1980).
127. B. Oray and S. J. Norton, *Biochim. Biophys. Acta* **611,** 168 (1980).
128. J. K. Anderson, J. E. Mole, and H. J. Baker, *Biochemistry* **17,** 467 (1978).
129. K. Hori, T. Kurotsu, M. Kanda, S. Miura, A. Nozoe, and Y. Saita, *J. Biochem.* (*Toyko*) **84,** 425 (1978).
130. J. E. Cone, J. Plowman, and G. Guroff, *J. Biol. Chem.* **249,** 5551 (1974); R. L. Then, *Anal. Biochem.* **100,** 122 (1979).
131. D. L. Garbers, *J. Biol. Chem.* **251,** 4071 (1976); M. Nakane and T. Deguchi, *Biochim. Biophys. Acta* **525,** 275 (1978); D. L. Garbers, *J. Biol. Chem.* **254,** 240 (1979).
132. K. Ishiwata and H. Yoshida, *J. Biochem.* (*Tokyo*) **83,** 783 (1978).
133. G. Monroy, E. Spencer, and J. Hurwitz, *J. Biol. Chem.* **253,** 4481 (1978); S. Venkatesan, A. Gershowitz, and B. Moss, *J. Biol. Chem.* **255,** 2829 (1980).
134. J. D. Bergstrom and A. L. Bieber, *Prep. Biochem.* **8,** 275 (1978).
135. P. Izzo and R. Gantt, *Biochemistry* **16,** 3576 (1977).
136. D. Moser, L. Johnson, and C.-Y. Lee, *J. Biol. Chem.* **255,** 4673 (1980); C. L. Wright, A. S. Warsy, M. J. Holroyde, and I. P. Trayer, *Biochem. J.* **175,** 125 (1978); E. Kopetzki and K. D. Entian, *Anal. Biochem.* **121,** 181 (1982).
137. B. Geiger, Y. Ben-Yoseph, and R. Arnon, *FEBS Lett.* **45,** 276 (1974).
138. K. Sandhoff, E. Conzelmann, and H. Nehrkorn, *Hoppe Seyler's Z. Physiol. Chem.* **358,** 779 (1977); E. Conzelmann, K. Sandhoff, H. Nehrkorn, B. Geiger, and R. Arnon, *Eur. J. Biochem.* **84,** 27 (1978).
139. B. Geiger, E. Calet, and R. Arnon, *Biochemistry* **17,** 1713 (1978).
140. S. B. Baylin and S. Margolis, *Biochim. Biophys. Acta* **397,** 294 (1975).
141. A. J. Hacking and H. Hassall, *Biochem. J.* **147,** 327 (1975).
142. J. K. Keesey, R. Bigelis, and G. R. Fink, *J. Biol. Chem.* **254,** 7427 (1979).
143. C. H. Yang and P. N. Srivastava, *Biochim. Biophys. Acta* **391,** 382 (1975); P. N. Srivastava and A. A. Farooqui, *Biochem. J.* **183,** 531 (1979); M. Lyon and C. F. Phelps, *Biochem. J.* **199,** 419 (1981).
144. T. Shimakata, Y. Fujita, and T. Kusaka, *J. Biochem.* (*Tokyo*) **86,** 1191 (1979).
145. V. P. Agrawal and P. E. Kolattukudy, *Arch. Biochem. Biophys.* **191,** 452 (1978).
146. G. C. Ness, C. D. Spindler, and M. M. Moffler, *Arch. Biochem. Biophys.* **197,** 493 (1979); Z. H. Beg, J. A. Stonik, and H. B. Brewer, *Biochem. Biophys. Res. Commun.* **107,** 1013 (1982).

References to TABLE II (*continued*)

147. S. S. Braithwaite and J. Jarabak, *J. Biol. Chem.* **250,** 2315 (1975).
148. M. Shikita and P. Talalay, *Anal. Biochem.* **95,** 286 (1979).
149. Same as 148.
150. M. Mori and W. G. Wiest, *J. Steroid Biochem.* **11,** 1443 (1979).
151. F. Sweet and N. K. Adair, *Biochem. Biophys. Res. Commun.* **63,** 99 (1975).
152. T. Kamei, Y. Takiguchi, H. Suzuki, M. Matsuzaki, and S. Nakamura, *Chem. Pharm. Bull.* **26,** 2799 (1978).
153. S. H. Hughes, G. M. Wahl, and M. R. Capecchi, *J. Biol. Chem.* **250,** 120 (1975); J. A. Holden and W. N. Kelley, *J. Biol. Chem.* **253,** 4459 (1978); R. Schmidt, H. Wiegand, and U. Reichert, *Eur. J. Biochem.* **93,** 355 (1979).
154. G. Schmer and J. Roberts, *Biochim. Biophys. Acta* **527,** 264 (1978).
155. Y. D. Clonis and C. R. Lowe, *Eur. J. Biochem.* **110,** 279 (1980); P. E. Brodelius, R. A. Lannom, and N. O. Kaplan, *Arch. Biochem. Biophys.* **18,** 228 (1978); H. J. Gilbert, C. R. Lowe, and W. T. Drabble, *Biochem. J.* **183,** 481 (1979).
156. D. E. Nealon and R. A. Cook, *Biochemistry* **18,** 3616 (1979).
157. R. Bigelis and H. E. Umbarger, *J. Biol. Chem.* **250,** 4315 (1975).
158. R. Geiger, K. Mann, and T. Bettels, *J. Clin. Chem. Clin. Biochem.* **15,** 479 (1977); K. Uchida, M. Niimobe, M. Kato, and S. Fujii, *Biochim. Biophys. Acta* **614,** 501 (1980).
159. N. A. Nwokoro and H. Schachter, *J. Biol. Chem.* **250,** 6185 (1975).
160. Y. Nishimoto, F. Takeuchi, and Y. Shibata, *J. Chromatogr.* **169,** 357 (1979).
161. N. O. Kaplan, J. Everse, J. E. Dixon, F. E. Stolzenbach, C. Y. Lee, C. T. Lee, S. S. Taylor, and K. Mosbach, *Proc. Natl. Acad. Sci. U.S.A.* **71,** 3450 (1974); C. R. Lowe and K. Mosbach, *Eur. J. Biochem.* **52,** 99 (1975); W. R. Ellington and G. L. Long, *Arch. Biochem. Biophys.* **186,** 265 (1978); S. Huang, C. Y. Lee, and S. S. L. Li, *Int. J. Biochem.* **10,** 279 (1979); A. R. Place, D. A. Powers, and C. Y. Lee, *Anal. Biochem.* **83,** 636 (1977); L. Jervis, *Biochem. J.* **197,** 755 (1981).
162. Y. Horiuti and S. Imamura, *J. Biochem.* (*Tokyo*) **81,** 1639 (1977); C.-S. Wang, *Anal. Biochem.* **105,** 398 (1980).
163. J. Visser and M. Strating, *Biochim. Biophys. Acta* **384,** 69 (1975); J. Visser and M. Strating, *FEBS Lett.* **57,** 183 (1975).
164. C. Enholm, P. K. J. Kinnunen, J. K. Huttunen, E. A. Nikkila, and H. Ohta, *Biochem. J.* **149,** 649 (1975); D. Ganesan and H. B. Bass, *FEBS Lett.* **53,** 1 (1975); G. Bengtsson and T. Olivecrona, *Biochem. J.* **167,** 109 (1977).
165. J. C. Allen, C. Eriksson, and J. R. Galpin, *Eur. J. Biochem.* **73,** 171 (1977); T. Sekiya, T. Kajiwara, and A. Hatanaka, *Agric. Biol. Chem.* **42,** 677 (1978); I. Shahin, S. Grossman, and B. Sredni, *Biochim. Biophys. Acta* **529,** 300 (1978).
166. C. V. Dang and D. C. H. Yang, *Biochem. Biophys. Res. Commun.* **80,** 709 (1978).
167. T. Yagi, T. Yamamoto, and K. Soda, *Biochim. Biophys. Acta* **614,** 63 (1980).
168. J. Jolles and P. Jolles, *Eur. J. Biochem.* **54,** 19 (1975); E. Junowicz and S. E. Charm, *FEBS Lett.* **57,** 219 (1975).
169. A. S. Narayanan, R. C. Siegel, and G. R. Martin, *Arch. Biochem. Biophys.* **162,** 231 (1974).
170. C. A. Waters, J. R. Murphy, and J. W. Hastings, *Biochem. Biophys. Res. Commun.* **57,** 1152 (1974); J. C. Matthews, K. Hori, and M. J. Cormier, *Biochemistry* **16,** 85 (1977); B. R. Branchini, T. M. Marschner, and A. M. Montemurro, *Anal. Biochem.* **104,** 386 (1980); T. F. Holzman and T. O. Baldwin, *Biochemistry* **21,** 6194 (1982).

(*continued*)

References to TABLE II (*continued*)

171. D. Bout, H. Dupas, M. Capron, A. El-Gazawi, Y. Carlier, A. Delacourte, and A. Capron, *Immunochemistry* **15,** 633 (1978); I. P. Wright and T. K. Sundaram, *Biochem. J.* **177,** 441 (1979); C.-Y. Lee, C. H. Langley, and J. Burkhart, *Anal. Biochem.* **86,** 697 (1978); C.-Y. Lee, D. Charles, and D. Bronson, *J. Biol. Chem.* **254,** 6375 (1979); M. Mevarech, H. Eisenberg, and E. Neumann, *Biochemistry* **16,** 3781 (1977).
172. M. Elwell and L. B. Hersh, *J. Biol. Chem.* **254,** 2434 (1979).
173. T. Caldes, H. R. Fatania, and K. Dalziel, *Anal. Biochem.* **100,** 299 (1979); J. T. Chang and G. G. Chang, *Anal. Biochem.* **121,** 366 (1982).
174. Y. S. Kim and P. E. Kolattukudy, *Arch. Biochem. Biophys.* **190,** 234 (1978); Y. S. Kim, P. E. Kolattukudy, and A. Boss, *Arch. Biochem. Biophys.* **196,** 543 (1979).
175. K. H. Schachtele, E. Schlitz, and D. Palm, *Eur. J. Biochem.* **92,** 427 (1978).
176. N. C. Phillips, D. Robinson, and B. Winchester, *Biochem. J.* **153,** 579 (1976); P. V. Wagh, *J. Chromatogr.* **152,** 565 (1978); N. C. Phillips, B. G. Winchester, and R. D. Jolly, *Biochem. J.* **163,** 269 (1977).
177. G. Fayat, M. Fromant, D. Kahn, and S. Blanquet, *Eur. J. Biochem.* **78,** 333 (1977); O. Kellermann, A. Brevet, H. Tonetti, and J. P. Waller, *Eur. J. Biochem.* **88,** 205 (1978).
178. S. Riazuddin and T. Lindahl, *Biochemistry* **17,** 2110 (1978).
179. V. V. Murthy, E. Jones, T. W. Cole, and J. Johnson, *Biochim. Biophys. Acta* **483,** 487 (1977).
180. E. Barbosa and B. Moss, *J. Biol. Chem.* **253,** 7692 (1978).
181. K. Sato, E. Hiei, S. Shimizu, and R. H. Abeles, *FEBS Lett.* **85,** 73 (1978).
182. R. G. Dennick and R. J. Mayer, *Biochem. J.* **161,** 167 (1977); C. M. Buess, J. K. Price, B. D. Roberts, and W. R. Carper, *Experientia* **33,** 163 (1977).
183. E. M. V. Pires and S. V. Perry, *Biochem. J.* **167,** 137 (1977).
184. R. C. Bhalla, R. V. Sharma, and R. C. Gupta, *Biochem. J.* **203,** 583 (1982).
185. M. Tseng, B. G. Harris, and M. K. Jacobson, *Biochim. Biophys. Acta* **568,** 205 (1979).
186. T. Sugiyama and T. Yamano, *FEBS Lett.* **52,** 145 (1975).
187. J. L. Vermillion and M. J. Coon, *J. Biol. Chem.* **253,** 2694 (1978); Y. Yasukochi and B. S. Siler Masters, *J. Biol. Chem.* **251,** 5337 (1976); J. S. French and M. J. Coon, *Arch. Biochem. Biophys.* **195,** 565 (1979); K. M. Madyastha and C. J. Coscia, *J. Biol. Chem.* **254,** 2419 (1979); R. T. Mayer and J. T. Durrant, *J. Biol. Chem.* **254,** 756 (1979).
188. J. I. Pedersen and H. K. Godager, *Biochim. Biophys. Acta* **525,** 28 (1978).
189. G. A. Michaliszyn, S. S. Wing, and E. A. Meighen, *J. Biol. Chem.* **252,** 7495 (1977).
190. R. Skorko, W. Zillig, H. Rohrer, H. Fujiki, and R. Mailhammer, *Eur. J. Biochem.* **79,** 55 (1977).
191. B. Venerando, G. Tettamanti, B. Cestaro, and V. Zambotti, *Biochim. Biophys. Acta* **403,** 461 (1975); Y. Uchida, Y. Tsukada, and T. Sugimori, *J. Biochem.* (*Tokyo*) **82,** 1425 (1977).
192. B. Holmquist, *Biochemistry* **16,** 4591 (1977).
193. S. Ida, *J. Biochem.* (*Tokyo*) **82,** 915 (1977).
194. H. Tanaka, I. Sasaki, K. Yamashita, Y. Matuo, J. Yamashita, and T. Horio, *J. Biochem.* (*Tokyo*) **91,** 1411 (1982).

References to TABLE II (*continued*)

195. H. Taniuchi and J. L. Bohnert, *J. Biol. Chem.* **250,** 2388 (1975).
196. J. Dornand, J. C. Bonnafons, and J. C. Mani, *Eur. J. Biochem.* **87,** 459 (1978).
197. N. Miki, J. M. Baraban, J. J. Keirns, J. J. Boyce, and M. W. Bitensky, *J. Biol. Chem.* **250,** 6320 (1975).
198. E. F. Brunngraber and E. Chargaff, *Proc. Natl. Acad. Sci. U.S.A.* **74,** 3226 (1977).
199. E. Bischoff, T. A. Tran-Thi, and K. F. A. Decker, *Eur. J. Biochem.* **51,** 353 (1975).
200. R. J. Boucek and K. J. Lembach, *Arch. Biochem. Biophys.* **184,** 408 (1977).
201. N. J. Hoogenraad, T. M. Sutherland, and G. J. Howlett, *Anal. Biochem.* **101,** 97 (1980); M. Luisa de Martinis, P. McIntyre, and N. Hoogenraad, *Biochem. Int.* **3,** 371 (1981).
202. H. Rosemeyer and F. Seela, *J. Med. Chem.* **22,** 1545 (1979).
203. G. Dodin, *FEBS Lett.* **134,** 20 (1981).
204. P. F. Fox, J. R. Whitaker, and P. A. O'Leary, *Biochem. J.* **161,** 389 (1977).
205. A. Signoret and J. Crouzet, *Agric. Biol. Chem.* **46,** 459 (1982).
206. Y. Tanaka and I. Uritani, *J. Biochem.* (*Tokyo*) **81,** 963 (1977).
207. R. G. H. Cotten and P. J. Grattan, *Eur. J. Biochem.* **60,** 427 (1975); J. M. Al-Janabi, *Arch. Biochem. Biophys.* **200,** 603 (1980); S. Webber, G. Harzer, and J. M. Whiteley, *Anal. Biochem.* **106,** 63 (1980).
208. G. S. Swamy and D. T. N. Pillay, *Z. Pflanzenphysiol.* **93,** 403 (1979).
209. T. Hirabayashi, T. J. Larson, and W. Dowhan, *Biochemistry* **15,** 5205 (1976); T. J. Larson, T. Hirabayashi, and W. Dowhan, *Biochemistry* **15,** 974 (1976).
210. M. Miyake, J. W. Daly, and C. R. Creveling, *Arch. Biochem. Biophys.* **181,** 39 (1977); R. L. Kincaid and M. Vaughan, *Proc. Natl. Acad. Sci. U.S.A.* **76,** 4903 (1979).
211. G. Colombo, G. M. Carlson, and H. A. Lardy, *Biochemistry* **17,** 5321 (1978); B. S. T. Hung and R. Silverstein, *Prep. Biochem.* **8,** 421 (1978).
212. H. Hengartner and J. I. Harris, *FEBS Lett.* **55,** 282 (1975); J. Babul, *J. Biol. Chem.* **253,** 4350 (1978).
213. N. Kato, H. Sahm, H. Schutte, and F. Wagner, *Biochim. Biophys. Acta* **566,** 1 (1979).
214. C. Y. Lee, D. Charles, and D. Bronson, *J. Biol. Chem.* **254,** 6375 (1979).
215. G. W. K. Kuntz, S. Eber, W. Kessler, H. Krietsch, and W. K. Krietsch, *Eur. J. Biochem.* **85,** 493 (1978); B. Pegoraro and C. Y. Lee, *Biochim. Biophys. Acta* **522,** 423 (1978).
216. C. Ehnholm, W. Shaw, H. Greten, and W. V. Brown, *J. Biol. Chem.* **250,** 6756 (1975).
217. C. O. Rock and F. Snyder, *J. Biol. Chem.* **250,** 6564 (1975); R. M. Kramer, C. Wuthrich, C. Bollier, P. R. Allegrini, and P. Zahler, *Biochim. Biophys. Acta* **507,** 381 (1978).
218. T. Takahaski, T. Sugahara, and A. Ohsaka, *Biochim. Biophys. Acta* **351,** 155 (1974); C. Little, B. Aurebekk, and A. B. Otnaess, *FEBS Lett.* **52,** 175 (1975).
219. V. P. K. Titanji and S. Pahlman, *Biochim. Biophys. Acta* **525,** 380 (1978).
220. P. Gergely, A. G. Castle, and N. Crawford, *Int. J. Biochem.* **10,** 807 (1979).
221. S. Yonezawa and S. H. Hori, *Arch. Biochem. Biophys.* **181,** 447 (1977).
222. R. K. Sharma, S. W. Tam, D. M. Waisman, and J. H. Wang, *J. Biol. Chem.* **255,** 11102 (1980).

(*continued*)

References to TABLE II (*continued*)

223. M. Grez and J. Niessing, *FEBS Lett.* **77,** 57 (1977); J. R. Nevins and W. K. Joklik, *J. Biol. Chem.* **252,** 6939 (1977); D. Antoniades and O. Antonoglou, *Biochim. Biophys. Acta* **519,** 447 (1978).
224. K. Yoshsihara, T. Hashida, Y. Tanaka, H. Ohgushi, H. Yosihara, and T. Kamiya, *J. Biol. Chem.* **253,** 6459 (1978); T. Kristensen and J. Holtlund, *Eur. J. Biochem.* **88,** 495 (1978).
225. H. Soreg and U. Z. Littauer, *J. Biol. Chem.* **252,** 6885 (1977).
226. D. J. Tutas and E. Paoletti, *J. Biol. Chem.* **252,** 3092 (1977).
227. T. Yoshimoto, M. Fischl, R. C. Orlowski, and R. Walter, *J. Biol. Chem.* **253,** 3708 (1978).
228. L. Tuderman, E.-R. Kuutti, and K. I. Kivirikko, *Eur. J. Biochem.* **52,** 9 (1975); K. Tryggvason, K. Majamaa, J. Risteli, and K. I. Kivirikko, *Biochem. J.* **183,** 303 (1979).
229. R. A. Cravel, K. F. Lam, D. Mahuran, and A. Kronis, *Arch. Biochem. Biophys.* **201,** 669 (1980).
230. M. Hemler, W. E. M. Lands, and W. L. Smith, *J. Biol. Chem.* **251,** 5575 (1976).
231. G. Feinstein and A. Janoff, *Biochim. Biophys. Acta* **403,** 477 (1975); E. J. B. Fodor, H. Ako, and K. A. Walsh, *Biochemistry* **14,** 4923 (1975); S. Vavreinova and J. Turkova, *Biochim. Biophys. Acta* **403,** 506 (1975); D. Y. Twumasi and I. E. Liener, *J. Biol. Chem.* **252,** 1917 (1977).
232. W. L. Dills, J. A. Beavo, P. J. Bechtel, and E. G. Krebs, *Biochem. Biophys. Res. Commun.* **62,** 70 (1975); G. Shanker, H. Ahrens, and R. K. Sharma, *Proc. Natl. Acad. Sci. U.S.A.* **76,** 66 (1979); P. K. Tsung, T. Sakamoto, and G. Weissmann, *Biochem. J.* **145,** 437 (1975); G. Vauquelin, P. Geynet, J. Hanoune, and A. D. Strosberg, *Proc. Natl. Acad. Sci. U.S.A.* **74,** 3710 (1977); G. N. Gill, K. E. Holdy, G. M. Walton, and C. B. Kanstein, *Proc. Natl. Acad. Sci. U.S.A.* **73,** 3918 (1976); Y. Suzuki, E. Itugaki, H. Mori, and T. Hosoya, *J. Biochem.* (*Tokyo*) **81,** 1721 (1977); F. Farron-Furstenthal and J. R. Lightholder, *FEBS Lett.* **84,** 313 (1977); W. Thornburg, A. F. O'Malley, and T. J. Lindell, *J. Biol. Chem.* **253,** 4638 (1978); Y. Kitagawa and E. Racker, *J. Biol. Chem.* **257,** 4547 (1982).
233. P. K. Tsung, T. Sakamoto, and G. Weissmann, *Biochem. J.* **145,** 437 (1975).
234. P. K. Saini and I. H. Rosenberg, *J. Biol. Chem.* **249,** 5131 (1974).
235. B. S. Enevoldsen, L. Reimann, and N. L. Hansen, *FEBS Lett.* **79,** 121 (1977).
236. V. Zannis, D. Doyle, and D. W. Martin, *J. Biol. Chem.* **253,** 504 (1978); W. R. A. Osborne, *J. Biol. Chem.* **255,** 7089 (1980); D. A. Wiginton, M. S. Coleman, and J. J. Hutton, *J. Biol. Chem.* **255,** 6663 (1980).
237. W. M. Anderson and R. R. Fisher, *Arch. Biochem. Biophys.* **187,** 180 (1978); G. Voordouw, S. M. Vandervies, J. K. Eweg, C. Veeger, J. F. L. Van Breeman, and E. F. J. Van Bruggen, *Eur. J. Biochem.* **111,** 347 (1980).
238. C. D. Cash, M. Maitre, J. F. Rumigny, and P. Mandel, *Biochem. Biophys. Res. Commun.* **96,** 1755 (1980).
239. Same as 238.
240. B. A. Klyashchitsky, V. F. Pozdnev, V. K. Mitina, A. I. Voskoboyev, and I. P. Chernikevich, *Bioorg. Khim.* **6,** 1572 (1980).
241. J. Visser, M. Strating, and W. Van Dongen, *Biochim. Biophys. Acta* **524,** 37 (1978).
242. R. I. Zemmell and R. A. Anwar, *J. Biol. Chem.* **250,** 3185 (1975).

References to TABLE II (*continued*)

243. K. Murakami and T. Inagami, *Biochem. Biophys. Res. Commun.* **62,** 757 (1975); C. Devaux, J. Manard, P. Sicard, and P. Corvol, *Eur. J. Biochem.* **64,** 621 (1976); T. Inagami and K. Murakami, *J. Biol. Chem.* **252,** 2978 (1977); H. J. Chou, J. H. Sharper, and R. I. Gregerman, *Biochim. Biophys. Acta* **524,** 183 (1978); H. Yokosawa, T. Inagami, and E. Haas, *Biochem. Biophys. Res. Commun.* **83,** 306 (1978); T. Matoba, K. Murakami, and T. Inagami, *Biochim. Biophys. Acta* **526,** 560 (1978).
244. J. Reiser and R. Yuan, *J. Biol. Chem.* **252,** 451 (1977).
245. M. G. Sarngadharan, V. S. Kalyanaraman, R. Rahman, and R. C. Gallo, *J. Virol.* **35,** 555 (1980).
246. R. E. Scofield, R. P. Werner, and F. Wold, *Anal. Biochem.* **77,** 152 (1977).
247. H. Yoshida, I. Fukuda, and M. Hushiguchi, *J. Biochem.* (*Tokyo*) **88,** 1813 (1980).
248. J. G. Stavrianopoulos and E. Chargaff, *Proc. Natl. Acad. Sci. U.S.A.* **75,** 4140 (1978).
249. S. Kanaya and T. Uchida, *J. Biochem.* (*Tokyo*) **90,** 473 (1981).
250. T. Uchida and Y. Shibata, *J. Biochem.* (*Tokyo*) **90,** 463 (1981).
251. P. J. Hoffman and R. L. Blakely, *Biochemistry* **14,** 4804 (1975).
252. S. Sugiura, M. Suzuki, E. Ohtsuka, S. Nishikawa, H. Uemura, and M. Ikehara, *FEBS Lett.* **97,** 73 (1979).
253. P. Schofield and K. R. Williams, *J. Biol. Chem.* **252,** 5584 (1977).
254. R. R. Burgess and J. J. Jendrisak, *Biochemistry* **14,** 4634 (1975); K. Amemiya, C. W. Wu, and L. Shapiro, *J. Biol. Chem.* **252,** 4157 (1977); M. I. Goldberg, J. C. Perriard, and W. J. Rutter, *Biochemistry* **16,** 1655 (1977); K. H. Scheit and A. Stutz, *FEBS Lett.* **50,** 25 (1975); H. Sternbach, R. Engelhardt, and A. G. Lezius, *Eur. J. Biochem.* **60,** 51 (1975).
255. P. Hossenlopp, J. Sumegi, and P. Chambon, *Eur. J. Biochem.* **90,** 615 (1978); E. Gundelfinger, H. Saumweber, A. Dallendorfer, and H. Stein, *Eur. J. Biochem.* **111,** 395 (1980); T. M. Wandzilak and R. W. Benson, *Biochemistry* **17,** 426 (1978).
256. M. I. Goldberg, J. C. Perriard, and W. J. Rutter, *Biochemistry* **16,** 1648 (1977).
257. K. K. Cheung and A. Newton, *J. Biol. Chem.* **253,** 2254 (1978); P. A. Lowe, D. A. Hager, and R. R. Burgess, *Biochemistry* **18,** 1344 (1979); T. Nakayama, V. Williamson, K. Burtis, and R. H. Doi, *Eur. J. Biochem.* **88,** 155 (1978).
258. K. S. You and C. R. Roe, *Anal. Biochem.* **114,** 177 (1981).
259. M. J. Geison, *Biochem. J.* **151,** 181 (1975).
260. J. C. Paulson, W. E. Beranek, and R. L. Hill, *J. Biol. Chem.* **252,** 2356 (1977).
261. N. Leissing and E. T. McGuinness, *Biochim. Biophys. Acta* **524,** 254 (1978).
262. R. L. Pajula, A. Raina, and J. Kekoni, *FEBS Lett.* **90,** 153 (1978).
263. J. W. Callahan, P. Shankaran, M. Khalil, and J. Gerrie, *Can. J. Biochem.* **56,** 885 (1978).
264. A. Fenselau and K. Wallis, *Biochem. Biophys. Res. Commun.* **62,** 350 (1975).
265. D. J. Ball and J. J. Nishimura, *J. Biol. Chem.* **255,** 10805 (1980).
266. C. D. Cash, M. Maitre, L. Ossola, and P. Mandel, *Biochim. Biophys. Acta* **524,** 26 (1978).
267. S. Barry, P. Brodelius, and K. Mosbach, *FEBS Lett.* **70,** 261 (1976).
268. S. Okamura, F. Crane, H. A. Messner, and T. W. Mak, *J. Biol. Chem.* **253,** 3765 (1978).
269. V. P. Pigiet and R. R. Conley, *J. Biol. Chem.* **252,** 6367 (1977).

(*continued*)

References to TABLE II (*continued*)

270. K. Koerner, I. Rahimi-Laridjani, and H. Grimminger, *Biochim. Biophys. Acta* **397,** 220 (1975).
271. R. Bhadra and P. Datta, *Biochemistry* **17,** 1691 (1978).
272. W. H. Holleman and L. J. Weiss, *J. Biol. Chem.* **251,** 1663 (1976).
273. R. A. Madhav, M. L. Coetzee, and P. Ove, *Arch. Biochem. Biophys.* **200,** 99 (1980).
274. J. M. Whiteley, I. Jerkunica, and T. Deits, *Biochemistry* **13,** 2045 (1974); B. J. Dolnick and Y. C. Cheng, *J. Biol. Chem.* **252,** 7697 (1977); W. Rode, K. J. Scanlon, J. Hynes, and J. R. Bertino, *J. Biol. Chem.* **254,** 11538 (1979); C. K. Banerjee, L. L. Bennett, R. W. Brockman, B. P. Sani, and C. Temple, *Anal. Biochem.* **121,** 275 (1982).
275. K. Yamamoto and L. J. DeGroot, *J. Biochem.* (*Tokyo*) **91,** 775 (1982).
276. M. Berger and H. G. Wood, *J. Biol. Chem.* **250,** 927 (1975).
277. L. N. Y. Wu, R. M. Pennington, T. D. Everett, and R. R. Fisher, *J. Biol. Chem.* **257,** 4052 (1982).
278. C. Enholm, P. K. J. Kinnunen, J. K. Huttunen, E. A. Nikkila, and M. Ohta, *Biochem. J.* **149,** 649 (1975).
279. S. Ikeda, H. Hara, S. Sugimoto, and S. Fukui, *FEBS Lett.* **56,** 307 (1975).
280. H. Nakata and H. Fujisawa, *Eur. J. Biochem.* **122,** 41 (1982).
281. D. H. Wolf and M. Hoffmann, *Eur. J. Biochem.* **45,** 269 (1974); M. P. Gschwind, U. Gschwind, C. H. Paul, and K. Kirschner, *Eur. J. Biochem.* **96,** 403 (1979).
282. K. Nishioka, *Eur. J. Biochem.* **85,** 137 (1978).
283. P. Donner, H. Wagner, and H. Kroger, *Biochem. Biophys. Res. Commun.* **80,** 766 (1978).
284. Same as 281.
285. C. R. Geren and K. E. Ebner, *J. Biol. Chem.* **252,** 2082 (1977).
286. C. A. Smith and K. Brew, *J. Biol. Chem.* **252,** 7294 (1977).
287. C. R. Geren, C. M. Oloman, D. C. Primrose, and K. E. Ebner, *Prep. Biochem.* **7,** 19 (1977).
288. B. Burchell. *FEBS Lett.* **78,** 101 (1977); B. Burchell, *Biochem. J.* **173,** 749 (1978).
289. R. A. Anwar and M. Vlaovic, *Can. J. Biochem.* **57,** 188 (1979).
290. T. Watanabe and T. Suga, *Anal. Biochem.* **89,** 343 (1978); T. Watanabe and T. Suga, *Anal. Biochem.* **861,** 357 (1978).
291. F. Batista-Viera, R. Axén, and J. Carlsson, *Prep. Biochem.* **7,** 103 (1977).
292. B. Astedt, L. Holmberg, G. Wagner, P. Richter, and J. Ploug, *Thromb. Haemostasis* **42,** 924 (1979).
293. E. M. Clarke and J. R. Knowles, *Biochem. J.* **167,** 419 (1977).

TABLE III

PROTEINS PURIFIED BY IMMUNOAFFINITY CHROMATOGRAPHY ON ANTIBODY COLUMNS

Protein[a]	Eluent
1. Acetyl-β-glucosaminidase	NaCl
2. Adenosine deaminase	Urea
3. β-Adrenergic receptor	Propranolol
4. Alkaline phosphatase	High pH
5. α-Fetoprotein	Urea
6. α_2-Macroglobulin	Low pH
7. Aminopeptidase	Tris-HCl
8. Angiotensin I converting enzyme	$MgCl_2$
9. Astroglial protein	Acetic acid–urea
10. C1g, C1r, C1s subcomponents	EDTA, ethylene glycol
11. Complement (C5)	KBr
12. Carcinoembryonic antigen	Urea
13. Cardiotoxin	Acetic acid–NaCl
14. Catalase	Glycerol
15. Choriogonadotropin	$MgCl_2$
16. Cellulase	Low pH
17. Cyclic nucleotide phosphodiesterase	EGTA
18. Estrogen receptor	NaSCN
19. Factor VIII	NH_4SCN
20. Factor V	NaCl
21. Factor IX	NaCl, KSCN
22. Glucocorticoid receptor	Acetic acid
23. β-Glucuronidase	Urea
24. Hepatitis A virus	NaI
25. H-2K^k antigen	TNP-40, NaCl
26. HLA-B and B	Diethylamine pH 11.5
27. HCG–receptor complex	HCG
28. Insulin receptor	$MgCl_2$
29. Interferon	Ethylene glycol + citrate
30. Kallikreins	Guanidine
31. Legumin	KSCN
32. Lymphocyte	—
33. Monoamine oxidase	KSCN
34. Membrane glycoproteins	Diethylamine
35. Myoglobin	KCN-pH
36. Myosin	Guanidine
37. Neutrophil migration inhibition factor (NIF)	Urea
38. Phosphotyrosine proteins	Phenyl phosphate
39. Poliovirus	KSCN
40. Properdin	NaCl
41. Ribonuclease H	$MgCl_2$
42. Serine acetyltransferase	*O*-Acetylserine

(*continued*)

TABLE III (*continued*)

Protein	Eluent
43. Sm and RNP antigens	Urea, guanidine
44. Shiga toxin	MgCl
45. Somatostatin	Acetic acid
46. Terminal deoxynucleotidyltransferase	NH_4OH
47. Tetrodotoxin	pH
48. Thyrotropin	Guanidine
49. TSH	Glycine-HCl
50. Transplantation antigen	—
51. Trypsin inhibitor	Propionic acid
52. Urokinase	Glycine–HCl

[a] Key to references:

1. R. P. Erickson and R. Sandman, *Experientia* **33,** 14 (1977).
2. W. P. Schrader and A. R. Stacy, *J. Biol. Chem.* **252,** 6409 (1977).
3. C. M. Frazer and J. C. Venter, *Proc. Natl. Acad. Sci. U.S.A.* **77,** 7034 (1980).
4. J. J. Hughes and S. E. Charm, *Biotechnol. Bioeng.* **21,** 1439 (1979).
5. U. H. Stenman, M. L. Sutinen, R. K. Selander, K. Tontii, and J. Schroder, *J. Immunol. Methods* **46,** 337 (1981).
6. J. E. McEntire, *J. Immunol. Methods* **24,** 39 (1978).
7. H. Sjöström, O. Noren, L. Jeppesen, M. Stuan, B. Svensson, and L. Christiansen, *Eur. J. Biochem.* **88,** 503 (1978).
8. J. J. Lanzillo, R. P. Cynkin, and B. L. Fanburg, *Anal. Biochem.* **103,** 400 (1980).
9. D. C. Rueger, D. Dahl, and A. Bignani, *Anal. Biochem.* **89,** 360 (1978); D. Dahl, *Biochim. Biophys. Acta* **622,** 9 (1980).
10. Y. Mori, *J. Chromatogr.* **189,** 428 (1980).
11. R. A. Wetsel, M. A. Jones, and W. P. Kolb, *J. Immunol. Methods* **35,** 319 (1980).
12. E. Engvall, J. E. Shiveli, and M. Wronn, *Proc. Natl. Acad. Sci. U.S.A.* **75,** 1670 (1978).
13. E.-H. Wong, C.-L. Ho, and K.-T. Wang, *J. Chromatogr.* **154,** 25 (1978).
14. S. Webber and J. M. Whiteley, *Biochem. Biophys. Res. Commun.* **97,** 17 (1980).
15. C. Y. Lee, S. Wong, A. S. K. Lee, and L. Ma, *Hoppe Seyler's Z. Physiol. Chem.* **358,** 909 (1977).
16. U. Hakansson, L. G. Fagerstam, L. G. Pettersson, and L. Anderson, *Biochem. J.* **179,** 141 (1979).
17. R. Scott, R. S. Hansen, and J. A. Beavo, *Proc. Natl. Acad. Sci. U.S.A.* **79,** 2788 (1982).
18. B. Moncharmont, J.-L. Su, and I. Parikh, *Biochemistry* **21,** 6916 (1982).
19. J. L. Lane, H. Ekert, and A. Vafiadis, *Thromb. Haemostasis* **42,** 1306 (1979).
20. J. A. Katzmann, M. E. Neshiem, L. S. Hibbard, and K. G. Mann, *Proc. Natl. Acad. Sci. U.S.A.* **78,** 162 (1981).
21. A. H. Goodall, G. Kemble, D. P. O'Brien, E. Rawlings, F. Rotblat, G. C. Russell, G. Janossy, and G. D. Tuddenham, *Blood* **59,** 664 (1982).
22. H. J. Eisen, *Proc. Natl. Acad. Sci. U.S.A.* **77,** 3893 (1980).

References to TABLE III (*continued*)

23. F. E. Brot, C. E. Bell, and W. S. Sly, *Biochemistry* **17,** 385 (1978).
24. Y. Elkana, A. Thornton, and A. J. Zuckerman, *J. Immunol. Methods* **25,** 185 (1979).
25. J. E. Mole, F. Hunter, J. W. Paslay, A. S. Bhown, and J. C. Bennett, *Mol. Immunol.* **19,** 1 (1982).
26. P. Parham, *J. Biol. Chem.* **254,** 8709 (1979).
27. K. Metsikko and H. Rajaniemi, *FEBS Lett.* **106,** 193 (1979).
28. L. C. Harrison and A. Itin, *J. Biol. Chem.* **255,** 12066 (1980).
29. D. S. Secher and D. C. Burke, *Nature (London)* **285,** 446 (1980).
30. K. M. Gautvik, L. Johansen, K. Svindahl, K. Nustad, and T. B. Orstavik, *Biochem. J.* **189,** 153 (1980); R. Geiger and B. Clausnitzer, *Hoppe Seyler's Z. Physiol. Chem.* **362,** 1279 (1981).
31. R. Casey, *Biochem. J.* **177,** 509 (1979).
32. D. M. Raulet, P. D. Gottlieb, and M. J. Bevan, *J. Immunol.* **125,** 1136 (1980).
33. R. M. Denney, R. R. Fritz, N. T. Patel, and C. W. Abell, *Science* **215,** 1400 (1982).
34. E. N. Hughes and J. T. August, *J. Biol. Chem.* **257,** 3970 (1982).
35. H. K. B. Simmerman, C.-C. Wang, E. M. Horwitz, J. A. Berzofsky, and F. R. N. Good, *Proc. Natl. Acad. Sci. U.S.A.* **79,** 7739 (1982).
36. S. Sartore, L. D. Libera, and S. Schiaffino, *FEBS Lett.* **106,** 197 (1979).
37. R. H. Weisbart, A. J. Lusis, A. Kacena, L. Spolter, P. Eggena, and D. W. Golde, *Clin. Immunol. Immunopathol.* **22,** 408 (1982).
38. A. H. Ross, D. Baltimore, and H. N. Eisen, *Nature (London)* **294,** 654 (1981).
39. F. Brown, B. O. Underwood, and K. H. Fantes, *J. Med. Virol.* **4,** 315 (1979).
40. C. K. Ogle, J. D. Ogle, and J. W. Alexander, *Immunochemistry* **14,** 341 (1977).
41. W. Biusen, *J. Biol. Chem.* **257,** 7106 (1982).
42. P. A. Baecker and R. T. Wedding, *Anal. Biochem.* **102,** 16 (1980).
43. P. J. White, W. D. Gardner, and S. O. Hoch, *Proc. Natl. Acad. Sci. U.S.A.* **78,** 626 (1981); R. R. C. Buchanan, P. J. W. Venables, A. Morgan, N. A. Staines, P. R. Smith, and R. N. Maini, *Clin. Exp. Immunol.* **51,** 8 (1983).
44. A. D. O'Brien, G. D. LaVeck, D. E. Griffin, and M. R. Thompson, *Infect. Immun.* **30,** 170 (1980).
45. M. Lauber, M. Camier, and P. Cohen, *Proc. Natl. Acad. Sci. U.S.A.* **76,** 6004 (1979).
46. S. Srivastava, J. Y. H. Chan, and F. A. Siddiqui, *J. Biochem. Biophys. Methods* **2,** 1 (1980).
47. H. Nakayama, R. M. Withy, and M. A. Raftery, *Proc. Natl. Acad. Sci. U.S.A.* **79,** 7575 (1982).
48. F. Pekonen, D. M. Williams, and B. D. Weintraub, *Endocrinology (Baltimore)* **106,** 1327 (1980).
49. G. Ponsin and R. Mornex, *Acta Endocrinol.* **93,** 430 (1980).
50. G. C. DuBois, L. W. Law, and E. Apella, *Proc. Natl. Acad. Sci. U.S.A.* **79,** 7669 (1982).
51. R. S. Corfman and G. R. Reeck, *Biochim. Biophys. Acta* **715,** 170 (1982).
52. T. C. Wun, L. Ossowski, and E. Reich, *J. Biol. Chem.* **257,** 7262 (1982).

TABLE IV
RECEPTOR AND BINDING PROTEINS PURIFIED BY AFFINITY CHROMATOGRAPHY

Receptor[a]	Ligand	Eluent
1. Acetylcholine	Toxin	Decamethonium
2. α_2-Adrenergic	3-Benzazepine	Agonists, antagonists
3. β-Adrenergic	Alprenolol, acebutolol	Isoproterenol, alprenolol
4. Androgen	Heparin, DNA	Heparin, NaCl
5. Asialoglycoprotein	Asialo-orosomucoid	Galactose
6. Cyclic nucleotide	cAMP	cAMP, cGMP
7. 1,25-Dihydroxy-vitamin D	DNA, heparin, calciferol	KCl
8. Epidermal growth factor	Wheat germ agglutinin	*N*-Acetylglucosamine
9. Estradiol	Estradiol derivative oligo(dT)	Estradiol, KCl
10. Fc	IgG	Urea
11. Glucocorticoid	DNA, ATP	NaCl, pyridoxal phosphate
12. Glycine	Aminostrychnine	Glycine
13. Growth hormone	Growth hormone	Urea, MgCl
14. IgE	IgE, phenylarsonate IgE	Acetic acid, KSCN phenyl-azotyrosine
15. Insulin	Insulin	Guanidine, KSCN
16. Intrinsic factor	Intrinsic factor-B_{12}	EDTA
17. Macrophage (C3)	(C3)	Acetic acid
18. Opiate	Amidomorphine	Levorphanol
19. Progesterone	ATP	KCl
20. Prolactin	Prolactin	Urea, MgCl
21. Transcobalamin II	Transcobalamin II	EDTA
22. Transferrin	Transferrin	Transferrin, electrophoresis
23. Triiodothyronine (T_3)	T_3	T_3
24. TSH	TSH	KSCN
	Binding protein	
25. Actin	Deoxyribonuclease I, myosin	Guanidine ATP
26. Androgen	Dihydrotestosterone	Dihydrotestosterone
27. C-reactive	Phosphorylcholine	Ca^{2+}
28. C3b component	Factor H	NaCl
29. Calmodulin	Phenothiazine	EDTA
30. Corticosteroid-binding globulin	Androstene, cortisol	Cortisol
31. cAMP	cAMP	Urea, cAMP
32. Cytokinin	Benzyladenine isopentenyladenosine	NaOH, KCl
33. Dextran	Sephadex	Guanidine
34. DNA	DNA	NaCl
35. Elongation factors Tu, Ts	GDP	GDP

TABLE IV (*continued*)

Receptor[a]	Ligand	Eluent
36. Fibronectin (fragments)	Gelatin	Urea, arginine
37. Guanosine	GDP	GDP
38. Hemopexin	Heme, hematin	Gly-HCl, urea
39. Initiation factor elF-3	rRNA	KCl
40. Low- and high-density lipoproteins	Cholic acid	Triton X-100, hydroxylamine
41. Meromyosin	Actin	ADP, ATP
42. Messenger ribonucleoproteins	Oligo(dT)	Temperature change
43. Migration inhibitory factor	Glycolipids	KSCN
44. Myosin	ADP, actin	ADP, ATP
45. Neurophysin	Met-Tyr-Phe	Acetic acid
46. Nonhistone chromatin	DNA	NaCl
47. Penicillin	Ampicillin, 6-APA	Hydroxylamine
48. Phosphatidylcholine	Phosphocholine	Deoxycholate
49. Retinal	Prealbumin	NaCl
50. Riboflavin	Flavin	NaCl
51. r-Ribosomal proteins	rRNA	KCl-EDTA
52. t-Ribosomal proteins	tRNA	Urea-LiCl
53. mRNA cap	m^7GDP	KCl
54. Serotonin	Serotonin	Ca^{2+}
55. Sex steroid	Dihydrotestosterone	Dihydrotestosterone
56. Tubulin	Lactoperoxidase	NaCl
57. Vimentin	DNA	NaCl
58. Z-DNA	Poly(dG-dC)	NaCl

[a] Key to references:

1. J. P. Merlie, J.-P. Changeux, and F. Gros, *J. Biol. Chem.* **253,** 2882 (1978); D. Bartfeld and S. Fucs, *Biochem. Biophys. Res. Commun.* **89,** 512 (1979); N. Kalderon and I. Silman, *Biochim. Biophys. Acta* **465,** 331 (1977).
2. J. W. Regan, N. Barden, R. J. Lefkowitz, M. G. Carson, R. M. DeMarinis, A. J. Krog, K. G. Holden, W. O. Matthews, and J. P. Hieble, *Proc. Natl. Acad. Sci. U.S.A.* **79,** 7223 (1982).
3. M. G. Caron, Y. Srinivasan, J. Pitha, K. Kociolek, and R. J. Lefkowitz, *J. Biol. Chem.* **254,** 2923 (1979); G. Vauguelin, P. Geynet, J. Hanoune, and A. D. Strosberg, *Eur. J. Biochem.* **98,** 543 (1979); C. J. Homcy, S. G. Rockson, J. Countaway, and D. A. Egan, *Biochemistry* **22,** 660 (1983).
4. E. Mulder, J. A. Foekens, M. J. Peters, and H. J. Van Der Molen, *FEBS Lett.* **97,** 260 (1979); S. T. Hiremath, R. M. Loor, and T. Y. Wang, *Biochem. Biophys. Res. Commun.* **97,** 981 (1980).
5. W. E. Pricer and G. Ashwell, *J. Biol. Chem.* **251,** 7539 (1976); A. L. Schwartz, A. Marshak-Rothstein, D. Rup, and H. F. Lodish, *Proc. Natl. Acad. Sci. U.S.A.* **78,** 3348 (1981).

(*continued*)

References to TABLE IV (*continued*)

6. J. Ramseyer, C. B. Kanstein, G. M. Walton, and G. N. Gill, *Biochim. Biophys. Acta* **446,** 358 (1976).
7. J. W. Pike and M. R. Haussler, *Proc. Natl. Acad. Sci. U.S.A.* **76,** 5485 (1979); S. Sharpe, C. J. Hillyard, M. Szelke, and I. MacIntyre, *FEBS Lett.* **75,** 265 (1977).
8. R. A. Hock, E. Nexo, and M. D. Hollenberg, *Nature (London)* **277,** 403 (1979); S. Cohen, G. Carpenter, and L. King, *J. Biol. Chem.* **255,** 4834 (1980).
9. G. Redeuilh, R. Richard-Foy, C. Secco, V. Torelli, R. Bucourt, E. E. Baulieu, and H. Richard-Foy, *Eur. J. Biochem.* **106,** 481 (1980); S. Thrower, C. Hall, L. Lim, and A. N. Davison, *Biochem. J.* **160,** 271 (1976); P. Hubert, J. Mester, B. Dellacherie, J. Neel, and E. E. Baulieu, *Proc. Natl. Acad. Sci. U.S.A.* **75,** 3143 (1978).
10. T. Suzuki, R. Sadasivan, G. Wood, and W. Bayer, *Mol. Immunol.* **17,** 491 (1980).
11. H. J. Eisen and W. H. Alinsmann, *Biochem. J.* **171,** 177 (1978); O. Wrange, J. C. Duke, and J. A. Gustafsson, *J. Biol. Chem.* **254,** 9284 (1979); V. K. Moudgil and J. K. John, *Biochem. J.* **190,** 809 (1980).
12. F. Pfeiffer, D. Graham, and H. Betz, *J. Biol. Chem.* **257,** 9389 (1982).
13. M. J. Waters and H. G. Friesen, *J. Biol. Chem.* **254,** 6815 (1979).
14. A. Kulczycki and C. W. Parker, *J. Biol. Chem.* **254,** 3187 (1979); D. H. Conrad and A. Froese, *J. Immunol.* **120,** 429 (1978); J. Kanellopoulos, G. Rossi, and H. Metzger, *J. Biol. Chem.* **254,** 7691 (1979).
15. S. Jakobs, E. Hazum, Y. Schechter, and P. Cuatrecasas, *Proc. Natl. Acad. Sci. U.S.A.* **76,** 4918 (1979); S. T. Hiremath, P. F. Pilch, and P. Czech, *J. Biol. Chem.* **255,** 1732 (1980).
16. S. Yamada, H. Itaya, O. Nakazawa, and M. Fukuda, *Biochim. Biophys. Acta* **496,** 571 (1977); G. Marcoullis and R. Grasbeck, *Biochim. Biophys. Acta* **499,** 309 (1977); I. Kouvonen and R. Grasbeck, *Biochem. Biophys. Res. Commun.* **86,** 358 (1979).
17. R. J. Schneider, A. Kulczycki, S. K. Law, and J. P. Atkinson, *Nature (London)* **290,** 789 (1981).
18. J. M. Bidlack, L. G. Abood, P. O. Gyimah, and S. Archer, *Proc. Natl. Acad. Sci. U.S.A.* **78,** 636 (1981).
19. V. K. Moudgil and D. O. Toft, *Proc. Natl. Acad. Sci. U.S.A.* **73,** 3443 (1976); J. B. Miller and D. O. Toft, *Biochemistry* **17,** 173 (1978).
20. D. S. Liscia and B. K. Vanderhaar, *Proc. Natl. Acad. Sci. U.S.A.* **79,** 5930 (1982).
21. P. A. Seligman and R. H. Allen, *J. Biol. Chem.* **253,** 1766 (1978).
22. J. A. Fernandez-Pol and D. J. Klos, *Biochemistry* **19,** 3904 (1980); B. S. Stein and H. H. Sussman, *J. Biol. Chem.* **258,** 2668 (1983).
23. J. W. Apriletti, N. L. Eberhard, K. R. Latham, and J. D. Baxter, *J. Biol. Chem.* **256,** 12094 (1981).
24. G. F. Fenzi, E. Macchia, L. Bartalena, F. Manzani, and A. Pincheria, *FEBS Lett.* **88,** 292 (1978).
25. C. D. Strader, E. Lazarides, and M. A. Raftery, *Biochem. Biophys. Res. Commun.* **92,** 365 (1980); C. L. Leonardi and R. W. Rubin, *Anal. Biochem.* **118,** 58 (1981).
26. N. A. Musto, G. L. Gunsalus, and C. W. Bardin, *Biochemistry* **19,** 2853 (1980).
27. J. E. Volanakis, W. L. Clements, and R. E. Schrohenloher, *J. Immunol. Methods* **23,** 285 (1978).
28. J. D. Scott and J. E. Fothergill, *Biochem. J.* **205,** 575 (1982).
29. G. A. Jamieson and T. C. Vanaman, *Biochem. Biophys. Res. Commun.* **90,** 1048 (1979); C. R. Caldwell and A. Haug, *Anal. Biochem.* **116,** 325 (1981).

References to TABLE IV (*continued*)

30. K. E. Mickelson and U. Westphal, *Biochemistry* **18,** 2685 (1979); D. K. Mahajan, R. B. Billiar, and A. B. Little, *J. Steroid Biochem.* **13,** 67 (1980).
31. J. M. Trevillyan and M. L. Pall, *J. Biol. Chem.* **257,** 3978 (1982).
32. K. Yoshida and T. Takegami, *J. Biochem.* (*Tokyo*) **81,** 791 (1977); C. Chen, O. K. Melitz, B. Petschow, and R. L. Eckert, *Eur. J. Biochem.* **108,** 379 (1980).
33. M. M. McCabe, R. H. Hamelik, and E. E. Smith, *Biochem. Biophys. Res. Commun.* **78,** 273 (1977).
34. S. O. Hoch and E. McVey, *J. Biol. Chem.* **252,** 1881 (1977).
35. G. R. Jacobson and J. P. Rosenbusch, *FEBS Lett.* **79,** 8 (1977).
36. L.-H. E. Hahn and K. M. Yamada, *Proc. Natl. Acad. Sci. U.S.A.* **76,** 1160 (1979); M. Vuento and A. Vaheri, *Biochem. J.* **183,** 331 (1979).
37. D. Ricquier, C. Gervais, J. C. Kader, and P. Hemon, *FEBS Lett.* **101,** 35 (1979).
38. J. Suttner, F. Hrkal, and F. Vodrazka, *J. Chromatogr.* **131,** 453 (1977); K. W. Olsen, *Anal. Biochem.* **109,** 250 (1980).
39. O. Nygard and P. Westermann, *Biochim. Biophys. Acta* **697,** 263 (1982).
40. A. Wichmann, *Biochem. J.* **181,** 691 (1979).
41. H. R. Trayer, M. A. Winstanley, and I. P. Trayer, *FEBS Lett.* **83,** 141 (1977).
42. S. K. Jain, M. G. Pluskal, and S. Sarkar, *FEBS Lett.* **97,** 84 (1979).
43. D. Y. Lin, J. R. David, and H. G. Remold, *Nature* (*London*) **296,** 78 (1982).
44. M. A. Winstanley, D. A. P. Small, and I. P. Trayer, *Eur. J. Biochem.* **98,** 441 (1979).
45. I. M. Chaiken, *Anal. Biochem.* **97,** 302 (1979).
46. T. L. Thomas and G. L. Patel, *Proc. Natl. Acad. Sci. U.S.A.* **73,** 4369 (1976).
47. T. Tamura, H. Suzuki, Y. Nishimura, J. Mizoguchi, and Y. Hirota, *Proc. Natl. Acad. Sci. U.S.A.* **77,** 4499 (1980); H. Amanuma and J. L. Strominger, *J. Biol. Chem.* **255,** 11173 (1980).
48. L. I. Barsukov, C. W. Dam, L. D. Bergelson, G. I. Muzja, and K. W. A. Wirtz, *Biochim. Biophys. Acta* **513,** 198 (1978).
49. G. Fex, P. A. Albertsson, and B. Hansson, *Eur. J. Biochem.* **99,** 353 (1979).
50. J. A. Froehlich, A. H. Merril, C. O. Clogett, and D. B. McCormick, *Comp. Biochem. Physiol.* **66,** 397 (1980).
51. N. Ulbrich, A. Lin, and I. G. Wool, *J. Biol. Chem.* **254,** 8641 (1979); H. R. Burrell and J. Horowitz, *Eur. J. Biochem.* **75,** 533 (1977).
52. M. Yukioka and K. Omori, *FEBS Lett.* **75,** 217 (1977).
53. N. Sonenberg, K. M. Rupprecht, S. M. Hecht, and A. J. Shatkin, *Proc. Natl. Acad. Sci. U.S.A.* **66,** 4345 (1979).
54. A. Rotman, *Brain Res.* **146,** 141 (1978).
55. P. H. Petra and J. Lewis, *Anal. Biochem.* **105,** 165 (1980).
56. B. Rousset and J. Wolff, *J. Biol. Chem.* **266,** 11677 (1980).
57. W. J. Nelson, C. E. Vorgias, and P. Traub, *Biochem. Biophys. Res. Commun.* **106,** 1141 (1982).
58. A. Nordheim, P. Tesser, F. Azorin, Y. H. Kwon, A. Moller, and A. Rich, *Proc. Natl. Acad. Sci. U.S.A.* **79,** 7729 (1982).

quired for elution of such avid pairs. It is not unusual to find eluents that consist of chaotrophs (5 *M* KSCN), low pH (2.2) or high pH (11.5) buffers, urea (3.5–8 *M*) or guanidine (6 *M*). Despite such harsh conditions, it is of interest that antigenicity is retained in most cases. Because of the harsh conditions, methods have been developed in which antibodies to the contaminating proteins are prepared and the protein is purified by "reverse immunoadsorption"; i.e., contaminating proteins are adsorbed while the required protein is unretarded and comes through in the eluent.[37] Another approach utilizes immobilized protein A[38,39] or antihapten antibody[40,41] to bind antibody–antigen or hapten antibody–antigen complexes; it is a purified complex that is eluted, not the antigen.

The introduction of monoclonal antibodies, i.e., homogeneous immunoglobulins directed against a specific antigenic determinant,[42,43] decreased the possibility of contamination with antibodies directed against proteins other than that desired (see this volume [24]). Since the number of monoclonal antibodies produced against a specific protein are many, selection of low-affinity antibodies for immobilization is desirable. The lower affinity allows milder conditions for elution and less possibility of irreversible denaturation. Monoclonal antibodies are being adopted rapidly for purification in all fields of biochemistry. Indeed, if the trend continues and mild conditions for elution become the norm, a shift from affinity chromatography to immunoaffinity chromotography will become apparent. Some of the proteins purified by antibodies, more recently with monoclonal antibodies, are summarized in Table III.

Lectins

Lectins are sugar-binding, cell-agglutinating proteins of nonimmune origin and of wide distribution. They are being utilized extensively as macromolecular carbohydrate-specific reagents. The ability to combine with carbohydrates specifically and reversibly enables their purification on immobilized sugars[44]; conversely, immobilization of the lectin itself

[37] J. A. Weare, J. T. Gafford, N. S. Ur, and E. G. Erdos, *Anal. Biochem.* **123,** 310 (1982).
[38] D. M. Gorsten and J. J. Marchalans, *J. Immunol. Methods* **24,** 305 (1978).
[39] C. Schneider, R. A. Newman, D. R. Sutherland, U. Asser, and M. F. Greaves, *J. Biol. Chem.* **257,** 10766 (1982).
[40] M. Wilchek and M. Gorecki, *FEBS Lett.* **31,** 149 (1973).
[41] J. Kanellopoulos, G. Rossi, and N. Metzger, *J. Biol. Chem.* **254,** 7691 (1979).
[42] G. Kohler and C. Milstein, *Nature (London)* **256,** 495 (1975).
[43] J. W. Goding, *J. Immunol. Methods* **39,** 285 (1980).
[44] H. Lis and N. Sharon, *J. Chromatogr.* **215,** 361 (1981).

allows isolation of a variety of carbohydrate contaminating compounds.[45] The use of lectins for affinity chromatography of glycoconjugates is accomplished readily, since glycoconjugates do not interact very strongly with lectins ($K = 10^2$ to 10^4) and, therefore, can be displaced readily from affinity columns by specific sugars at neutral pH. Of the many lectins available, two are used frequently in affinity purification of carbohydrates: concanavalin A and *Ricinus communis* agglutinin. Since lectins are generally used at the early stages of purification of glycoproteins, and require additional procedures for complete purification, details of specific methods are not included. The general method for their use and application has been reviewed.[45]

Receptor-Binding Proteins

A receptor is a molecule that recognizes a specific chemical entity, binds to it, and initiates a series of biochemical events resulting in a characteristic physiological response.[46] The interaction between hormone, neurotransmitter, or drug and the respective receptor is selective and specific. Due to the selectivity, interaction between the ligand and its receptor is one of high affinity, among the highest known among biological interactions. Therefore, harsh elution conditions are being used to elute receptors from columns containing a selective affinity ligand.

In overcoming some of these difficulties, the use of monoclonal antibodies is being introduced. The rationale behind this approach depends on the ability to prepare monoclonal antibodies against cell surface antigens. The monoclonal antibody will inhibit binding of the ligand, e.g., a hormone, to the receptor; will interact specifically with the receptor; and can be used, therefore, to isolate the receptor. In Table IV examples are presented of the use of specific ligands and monoclonal antibodies for the isolation of receptors and other binding proteins.

[45] R. Lotan and G. L. Nicolson, *Biochim. Biophys. Acta* **559,** 329 (1979).

[46] M. D. Hollenberg and P. Cuatrecasas, *Methods Cancer Res.* **12,** 317 (1976).

[2] Immobilization of Ligands with Organic Sulfonyl Chlorides

By KURT NILSSON and KLAUS MOSBACH

A number of methods for binding biomolecules to solid phases are in use, many of which are summarized in Volumes 34 and 44 of this series, as well as in this volume [1]. Few, however, can satisfy all of the following requirements: high yields of coupled product at neutral pH, mildness (i.e., avoidance of denaturation or changes in the properties of the biomolecule), and stability of linkages between support and ligand. Organic sulfonyl chlorides, such as *p*-toluenesulfonyl chloride (tosyl chloride) and 2,2,2-trifluoroethanesulfonyl chloride (tresyl chloride), can be used to convert hydroxyl groups into good leaving groups (sulfonates) that, on reaction with nucleophiles, allow stable linkages to be formed between the nucleophile and the initial hydroxyl group-carrying carbon. These reagents appear to be suitable for the immobilization of enzymes and affinity ligands to supports bearing a hydroxyl group, such as agarose, cellulose, glycophase glass, and others.[1-3] The activation and coupling of the ligand to the support are thought to involve the following steps.

Activation:

$$\text{Support-CH}_2\text{OH} + \text{R-SO}_2\text{Cl} \rightarrow \text{Support-CH}_2\text{OSO}_2\text{-R}$$

Coupling:

$$\text{Support-CH}_2\text{OSO}_2\text{-R} + \text{H}_2\text{N-ligand} \rightarrow \text{Support-CH}_2\text{-NH-ligand} + \text{HOSO}_2\text{-R}$$
$$\text{Support-CH}_2\text{OSO}_2\text{-R} + \text{HS-ligand} \rightarrow \text{Support-CH}_2\text{-S-ligand} + \text{HOSO}_2\text{-R}$$
$$(\text{R} = \text{CH}_2\text{CF}_3 \text{ or } \text{C}_6\text{H}_4\text{CH}_3)$$

The influence of the R group on the reactivity of the sulfonate has been studied with soluble organic molecules containing hydroxyl groups.[4] It was found that the relative reactivities for $CH_3C_6H_4$—, CF_3CH_2—, and CF_3 sulfonate esters are 1 : 100 : 4000. Tosylates and tresylates are most convenient reagents for enzyme and ligand immobilization.

Unlike tresylation, the introduction of the tosyl ester can be conveniently followed with UV spectroscopy ($\varepsilon_{261} = 480\ M^{-1}\ \text{cm}^{-1}$). In addition, tosyl chloride is inexpensive, a distinct advantage over tresyl chloride.

[1] K. Nilsson and K. Mosbach, *Eur. J. Biochem.* **112,** 397 (1980).
[2] K. Nilsson, O. Norrlöw, and K. Mosbach, *Acta Chem. Scand. Ser. B* **B35,** 19 (1981).
[3] K. Nilsson and K. Mosbach, *Biochem. Biophys. Res. Commun.* **102,** 449 (1981).
[4] R. K. Crossland, W. E. Wells, and V. J. Shiner, Jr., *J. Am. Chem. Soc.* **93,** 4217 (1971).

ISBN 0-12-182004-1

The drawback to the method lies in its lower reactivity, being only slightly more reactive than that of epoxy groups. Tosyl chloride constitutes an alternative to tresyl chloride when the protein or affinity ligand to be coupled is stable at pH 9–10.5, i.e., the optimal range for coupling of primary amino groups to tosyl-Sepharose.[2]

Tresyl chloride forms sulfonates much more readily than tosyl chloride. It is possible to obtain about 0.6 mmol of sulfonates per gram of dry Sepharose 6B in 10 min using 2.5 mmol of tresyl chloride, whereas 12 mmol of tosyl chloride is needed for the same degree of substitution. The reactivity of the resulting Sepharose tresyl groups is very high, allowing a 75–100% yield in protein coupling after 1 hr at pH 7.5 in the cold. This efficiency has been found to be equal to or even greater than that of CNBr-activated Sepharose.[3]

The relationship between the degree of support activation (sulfonation) obtained and the amount of sulfonyl chloride used during the activation process seems to be almost linear over a broad substitution range (100–1000 μmol of ester per gram of dry Sepharose), thereby allowing ready prediction and reproducibility of the degree of activation in contrast to CNBr. Highly substituted gels do not diminish in coupling capacity over a period of several weeks when stored under acidic conditions (1 m*M* HCl) in the cold. Only a few percent of the sulfonates are lost owing to hydrolysis when a tresylated gel is stored at pH 7.5 in the cold overnight, in sharp contrast to CNBr-activated Sepharose, which must be used directly after activation. Freeze-dried tresylated preparations (such as silica) have been stored for a year in a desiccator without losing coupling capacity.

Thiols and primary amino groups are the most reactive nucleophiles with sulfonate esters on gels, thiols showing the highest reactivity. Amino acid analyses of bound proteins indicate that imidazole and tyrosine hydroxyl groups can also displace the esters. This broad range of reactive nucleophiles can be a great advantage when a very tight binding between protein and support is desired, i.e., for stabilization or "freezing" of the protein conformation. Furthermore, one is not restricted to amino group-carrying spacers when constructing affinity ligands. With the use of thiol spacers, uncharged S—C bonds are obtained. The sulfonyl chloride method mentioned above can also be utilized for activation of water-soluble polymers, which are very useful in biochemical separations. Polyethylene glycol (PEG) is frequently used in phase partitioning systems for the separation of biomolecules and cells[5,6]; tresyl chloride can be used for

[5] P.-Å. Albertsson, "Partition of Cell Particles and Macromolecules." Wiley, New York, 1960.

[6] F. Bückmann, G. Johansson, and M. Morr, *Macromol. Chem.* **182,** 1379 (1981).

activation of PEG and for subsequent coupling of PEG to proteins, affinity ligands, and cells.[7]

Experimental Procedures

Preparation of Tresylated Supports (Activation)

Activation is performed in a water-free solvent in order to avoid hydrolysis of sulfonyl chloride. Pyridine is added to neutralize liberated protons. Higher yields are obtained when using dried solvents rather than commercially available solvents. It is especially important to use dried solvents when low levels of activation are desired. The activation procedure should be performed in a well-ventilated hood.

In a typical procedure, 10 g (wet) Sepharose 4B is washed successively with 100 ml of each of the following: 30 : 70 and 70 : 30 of acetone : water (v/v), twice with acetone, and three times with dry acetone (dried with a molecular sieve overnight, 25 g 4A per liter of acetone). The gel is then transferred to a dried beaker containing 3 ml of dry acetone and 150 μl of dry pyridine (dried with a molecular sieve). During magnetic stirring, 100 μl of tresyl chloride (Fluka AG, Buchs, Switzerland; also available from Fluorochem Ltd., UK) is added dropwise. After 10 min at room temperature the gel is washed twice with 100 ml of each of the following: acetone, 30 : 70 and 50 : 50; 70 : 30 of 5 m*M* HCl : acetone (v/v); and 1 m*M* HCl. The product is stored at 4° until used. The amount of introduced tresyl groups is determined on freeze-dried gel by elemental analysis for sulfur. The procedure described above will yield about 450 μmol of tresyl groups per gram of dry gel.

Cellulose is treated in a similar manner. Glass and silica, however, must first undergo a "coating reaction" before they can be activated. In this procedure a hydrophilic layer consisting of glycerylpropyl groups is introduced to the surface of these particles.[8] Glass treated in this manner is commercially available from Pierce, U.S.A. (Glycophase glass) and silica from Merck, West Germany (LiChrosorb Diol). In a typical activation, 2 g of the dry support is washed three times with 50 ml each of dried acetone. The moist gel, about 5 g, is added to a dry beaker containing 2.5 ml of dry acetone and 130 μl of pyridine. With magnetic stirring, tresyl chloride (90 μl) is added to the suspension. After 15 min at 0° the gel is washed as described above for Sepharose. This procedure will introduce

[7] K. Nilsson and K. Mosbach, to be published.

[8] F. E. Regnier and R. Noel, *J. Chromatogr. Sci.* **14,** 316 (1976).

about 130 μmol of tresyl groups per gram of dry silica (300-Å pores). If the activated preparations are not used directly for coupling, they should be washed with water, 50 : 50 (v/v) water : acetone, and acetone and dried for storage.

Preparation of Tosylated Supports

Sepharose is transferred to dry acetone as has been described above. Tosyl chloride (0.6 g) is dissolved in 3 ml of dry acetone, and 8 g of moist Sepharose CL-6B (in dry acetone) is added. After addition of 1 ml of pyridine, the reaction is continued for 1 hr at room temperature with continuous magnetic stirring. The activated gel is transferred to 1 m*M* HCl after the washing steps described for preparation of tresylated supports. The gel is now ready for use. This procedure will introduce about 0.6 mmol of tosyl groups per gram of dry Sepharose. The degree of substitution by tosyl groups is determined by adding 40 mg of tosyl-Sepharose to one cuvette and 40 mg of unreacted Sepharose CL-6B to a reference cuvette, both in a solution (1 ml) of glycerol–water, 87 : 13, (v/v). The difference spectrum is recorded between 300 and 250 nm. The extinction coefficient for ethyl tosylate at 261 nm (480 M^{-1} cm^{-1}) is used to calculate the tosyl group content. Alternatively, the degree of activation may be determined by elemental analysis for sulfur. Tosylated gels can be stored for several weeks in 1 m*M* HCl at 4° without losing coupling capacity.

Preparation of Tresylated Polyethylene Glycol (PEG)

PEG-4000 (4 g) is dissolved in 10 ml of dichloromethane.[7] Pyridine (250 μl) and tresyl chloride (220 μl) are added at 0°. The reaction is allowed to continue at 20° for 1.5 hr before the solvent is evaporated to dryness under reduced pressure. The product is dissolved in 60 ml of ethanol containing 250 μl of concentrated HCl and is precipitated overnight at −18°. The supernatant liquid is discarded after centrifugation, and another 60 ml of ethanol containing 50 μl of concentrated HCl is added. The precipitation procedure is repeated six times, or until no pyridine is detected in the UV (255 nm). After precipitation in ethanol without HCl, and drying, about 2.5 g of white crystalline product is generally obtained. Elemental analysis for sulfur indicates that about 1.6 mmol of tresyl groups is formed per millimole of PEG; i.e., 80% of the hydroxyl groups in PEG had been transformed into tresyl esters. The same principle can be used for tresylation of PEG-6000; with PEG-6000, methanol is preferable to ethanol.

Coupling of Affinity Ligands and Proteins

The procedures for coupling proteins or ligands to sulfonated supports are the same as for CNBr-activated supports. Any buffer that does not contain strong nucleophiles can be used in the coupling step, i.e., phosphate, HEPES, carbonate, and so forth. After coupling, the gel is preferentially treated with Tris–HCl buffer to remove unreacted tresyl groups that might interfere in the subsequent use of the product. The time necessary for this treatment varies with the number of tresyl groups that had been added. Treatment with 0.2 *M* Tris–HCl, pH 7.5–8.5, for 5 hr at 4° is usually sufficient. For higher activated gels, longer incubation periods are recommended. Tosyl-activated supports have to be treated at higher pH and temperature for longer time periods (pH 9–10, 20–40°, 15 hr) in order to remove unreacted sulfonates. Mercaptoethanol is found to be more efficient than Tris buffer for removing tosyl groups.

Similar procedures can be applied to glass and silica. If, however, the activated glass or silica has been freeze dried, it is first washed with cold coupling buffer, transferred to the coupling vessel, and deaerated under reduced pressure before addition of the protein or ligand solution.

Examples

Coupling of N^6-(6-Aminohexyl)-5′-AMP to Tresyl-Sepharose. The AMP analog (Sigma, 16 mg) was dissolved in 1 ml of 0.2 *M* sodium phosphate containing 0.5 *M* sodium chloride, pH 8.2 (coupling buffer) and added to 1 g of wet tresyl-Sepharose that had been briefly washed with cold coupling buffer.[3] Coupling proceeded with gentle agitation by tumbling for 16 hr at 4°. The gel was treated with 0.2 *M* Tris, pH 8.5, for 5 hr (20°) and washed with 0.2 *M* sodium acetate–0.5 *M* NaCl, pH 3.5, 0.5 *M* NaCl, distilled water, and 0.2 *M* phosphate, pH 7.5. The amount of bound AMP analog was determined from the UV spectrum obtained with 40 mg of AMP–gel and 40 mg of unreacted Sepharose added to 1 ml of glycerol–water (87 : 13) in the sample and reference cuvettes, respectively (ε_{268} = 17,600). The content of the bound analog could also be determined from elemental analysis of nitrogen on a sample of gel that had not been treated with Tris–HCl.

Coupling of Soybean Trypsin Inhibitor (STI) to Tresyl-Sepharose. Soybean trypsin inhibitor (type 1-S, Sigma, 10 mg) was dissolved in 0.2 *M* sodium phosphate at pH 7.5.[3] Tresyl-Sepharose was briefly washed with the cold coupling buffer and added to the protein solution. Coupling proceeded with gentle agitation overnight at 4°. The gel was treated with

Tris–HCl and washed as described above. The amount of bound STI could be determined either by UV measurements, elemental analysis of nitrogen, or amino acid analysis (see Table II).

Coupling of Hexokinase and Trypsin to Tresyl-Sepharose. Hexokinase (yeast, type III, Sigma, 10 mg) was immobilized to 1 g of tresyl-Sepharose, by the method used for trypsin inhibitor, in 0.2 *M* HEPES at pH 7.0, containing 15 m*M* $MgCl_2$.[3] After coupling, the gel was washed as above for trypsin inhibitor, omitting the Tris and acetate buffers. The amount of bound enzyme was determined by amino acid analysis. Assay of enzyme activity was done in a cuvette with a magnetic stirring device (20 mg of gel in 3 ml of assay medium) as outlined in the Worthington manual.[9] Trypsin (bovine pancreas, type III, Sigma, 10 mg), was coupled and washed, using the same procedure as for trypsin inhibitor.[3]

Coupling of N^6-(6-Aminohexyl)-5′-AMP to Tosyl-Sepharose. Tosyl-Sepharose (0.7 g) was washed briefly with 0.25 *M* $NaHCO_3$ at pH 10.5, and added to 0.3 ml of coupling buffer containing 20 mg of the AMP–analog.[1] Coupling was allowed to proceed with gentle agitation overnight at 40°. The gel was then treated with 0.8 *M* mercaptoethanol, pH 10, for 15 hr at 40° and washed as recommended for the tresyl-coupled analog. The AMP content (5 μmol/g wet gel) was determined from the UV spectrum as described for tresyl-Sepharose. To minimize interference from possible remaining tosyl groups, the absorption at 290 nm was used (ε_{290} = 3300 M^{-1} cm^{-1}).

Coupling of Soybean Trypsin Inhibitor (STI) to Tosyl Sepharose. The protein (115 mg), was dissolved in 3 ml of 0.25 *M* $NaHCO_3$ at pH 9.5, and 10 g of the activated gel, briefly washed with coupling buffer, was added.[1] Coupling was allowed to proceed for 20 hr with gentle agitation at 20°. Remaining interfering tosyl groups were removed by adding 20 ml of 0.8 *M* Tris–HCl at pH 9.5. After 30 hr at 20°, the gel was washed as described above for coupling of STI to tresyl-Sepharose. In this way about 3.5 mg of the inhibitor was bound per gram of wet gel.

Coupling of Horse Liver Alcohol Dehydrogenase to Tosyl-Sepharose. Alcohol dehydrogenase from horse liver (Sigma) was immobilized, by the same method as was trypsin inhibitor, in 0.2 *M* sodium phosphate at pH 7.5.[1] After coupling, the gel was washed three times with 10 gel volumes of 0.1 *M* $NaHCO_3$ at pH 8.5, 0.5 *M* NaCl, and 0.1 *M* sodium phosphate at pH 7.5. The amount of enzyme bound was determined by amino acid analysis.

[9] "Worthington Enzyme Manual" (L. A. Decker, ed.), p. 66. Worthington, Freehold, New Jersey, 1977.

TABLE I
ACTIVATION OF DIFFERENT SUPPORTS WITH TRESYL CHLORIDE[a]

Support	Tresyl chloride used (mmol/g dry support)	Tresyl groups introduced[b] (mmol/g dry support)	Yield (%)
Sepharose 4B	0.70	0.11	16
Sepharose 4B	2.70	0.45	17
Sepharose 4B	4.50	0.80	18
Sepharose 4B	9.00	1.30	15
Sepharose CL-6B	2.00	0.45	23
Sepharose CL-6B	4.00	0.98	25
Sepharose CL-6B	6.00	1.10	18
Diol-silica 300[c]	0.16	0.05	31
Diol-silica 300	0.41	0.13	31
Diol-silica 300	2.45	0.40	16
Separon HEMA[d]	0.44	0.14	32
Separon HEMA	2.50	0.78	31

[a] Data are reproduced, with permission, from Nilsson and Mosbach[3] and Nilsson and Larsson.[11]

[b] The activation time was 10 min for Sepharose and 15 min for silica and Separon HEMA.

[c] Glycerlypropyl-silica with 300-Å pores; 10-μm particles; about 400 μmol of diol groups per gram of dry weight.[11]

[d] A special type of hydroxyethyl methacrylate for HPLC (Chemapol, Prague) with 1000-Å pores; 7–10 μm particles; 2.2 mmol of hydroxyethyl groups per gram of polymer.

Coupling and Activity Yield

Table I shows the degrees of activation obtained when Sepharose, glycerylpropyl-silica, and hydroxyethyl methacrylate were activated with varying amounts of tresyl chloride.[3,10,11] As can be seen, the activation level can be adjusted over a very broad range simply by changing the amount of sulfonyl chloride.

Because of the high reactivity of tresylated supports, a low activation level [i.e., 150–300 μmol of tresyl groups per gram of dry Sepharose or 50–100 μmol per gram of dry silica (100–300 Å)] is effective in obtaining adequate yields of coupled products (Table II). As indicated in Table II, 90% coupling of protein A was achieved with 150 μmol of tresyl groups per gram of dry Sepharose[12] and almost 100% yield was obtained with

[10] K. Nilsson and K. Mosbach, unpublished results, 1981.

[11] K. Nilsson and P.-O. Larsson, *Anal. Biochem.*, in press (1983).

[12] M. Ramstorp, K. Nilsson, R. Mosbach, and K. Mosbach, in preparation.

TABLE II
COUPLING OF PROTEINS TO DIFFERENT SUPPORTS ACTIVATED WITH TRESYL CHLORIDE[a]

Support	Tresyl groups on activated support (mmol/g dry support)	Ligand	Time for coupling[b] (hr)	pH	Ligand bound (mg/g dry support)	Yield (%)
Sepharose 4B	0.15	Protein A	15	8.5	70	90
Sepharose 4B	1.30	Concanavalin A	15	7.5	360	97
Sepharose 4B	1.30	STI[c]	15	7.5	195	68
Sepharose 4B	1.30	STI	15	8.5	211	74
Sepharose 4B	1.30	STI	1.5	8.5	200	70
Sepharose CL-6B	1.30	Nucleotide[d]	15	8.2	87	32
Diol-silica 100[e]	0.45	STI	20	8.0	250	83
Diol-silica 300	0.13	STI	20	8.0	115	33
Diol-silica 300	0.40	STI	20	8.0	280	80
Diol-silica 300	0.40	STI	1	8.0	220	63
Diol-silica 1000	0.06	STI	20	8.0	46	92
Diol-silica 100	0.45	Nucleotide[d]	20	7.5	13	80
Separon HEMA	0.14	Protein A	12	8.0	10	100

[a] Data are reproduced, with permission, from Nilsson and Mosbach,[3] Nilsson and Larsson,[11] and Ramstorp *et al.*[12]

[b] All couplings were performed at 4° except for protein A and for the AMP analog to diol-silica, which were at room temperature. The estimated amount of diol in the silanized silicas was 450, 400, and 50 μmol per gram of support for 100-, 300-, and 1000-Å pore silicas, respectively. The maximum theoretical loadings of STI[c] on diol-silicas 100, 300, and 1000 were estimated to be 250, 270, and 50 mg, respectively.

[c] STI, soybean trypsin inhibitor.

[d] N^6-(6-Aminohexyl)-5′-AMP.

[e] Glyceryl-propyl-silica (10 μm).

hydroxyethyl methacrylate. The method is rapid, even at neutral pH. If, however, very high protein loadings are desired, the activation level has to be increased. Thus, almost all of the available surface of glyceryl-propyl-silica 100 and 1000 could be covered with soybean trypsin inhibitor by using the tresyl chloride activation method (250 and 50 mg of protein per gram of dry gel, respectively). The amount of tresyl groups on these highly activated silicas corresponded to the original amount of diol groups (Table II).[11]

Trypsin was bound with a 70% yield at pH 8.2 to tresyl-Sepharose, whereas hexokinase was bound with a 53% yield at pH 7.0.[3] The specific activities of bound trypsin and hexokinase relative to the free enzyme were 33 and 26%, respectively. These values are in agreement with results

TABLE III
COUPLING OF HORSE LIVER ALCOHOL DEHYDROGENASE TO DIOL-SILICA 1000[a]

Tresyl chloride for activation (μmol/g dry silica)	Enzyme added (mg/g dry silica)	Enzyme coupled[b] (mg/g dry silica)	Relative activity bound to free enzyme[c] (%)
100	21	11	100
300	21	21	95

[a] Data are reproduced, with permission, from Nilsson and Larsson.[11]
[b] Immobilization was in 0.25 *M* phosphate at pH 8.0 for 20 hr at room temperature together with 2 m*M* NADH and 100 m*M* isobutyramide.
[c] Activity was measured with 9 m*M* NAD and 9 m*M* ethanol in 0.25 *M* sodium phosphate at pH 7.8.

obtained when the enzymes were immobilized to CNBr-activated Sepharose.[10] Apart from the enzymes listed above under Examples, a number of other enzymes have been immobilized in high yields to tresyl-Sepharose, such as T4 DNA ligase,[13] horse liver alcohol dehydrogenase,[14] and chymotrypsin.[10]

Direct quantitation of remaining intrinsic activity after immobilization to Sepharose is difficult because of diffusional hindrances caused by the relatively large Sepharose particles (40–120 μm). A better assessment would be possible by immobilization to smaller particles. As shown in Table III, virtually complete retention of added horse liver alcohol dehydrogenase activity was obtained after covalent attachment to tresyl chloride-activated glycerylpropyl-silica (1000-Å pores, 10 μm), despite quite a high loading of enzyme.[11] Furthermore, the dissociation constants of NADH for free and immobilized alcohol dehydrogenase were very similar (the dissociation constants were determined from Scatchard plots of bound radioactive nucleotide to the enzyme gels[11]). This appears to be a unique demonstration of complete retention of catalytic activity and binding characteristics of an enzyme after immobilization.

[13] L. Bülow and K. Mosbach, *Biochem. Biophys. Res. Commun.* **107,** 458 (1982).
[14] M.-O. Månsson, N. Siegbahn, and K. Mosbach, *Proc. Natl. Acad. Sci. U.S.A.* **80,** 1487 (1983).

Tosyl-Sepharose was found to bind horse liver alcohol dehydrogenase in high yield (60%) at pH 7.5, with the same retention of activity as tresyl-Sepharose.[1] The high efficiency with tosyl-Sepharose was, in this case, attributed to the many free thiol groups on the surface of this enzyme. Thiol compounds have been shown to couple about twice as efficiently as primary amino group-containing compounds to both tosyl- and tresyl-Sepharose.[2,10] Very large amounts of ligand can be bound when coupling takes place in DMF; about 40 μmol of hexylamine per gram of wet Sepharose was introduced when the ligand was allowed to react at 60° in DMF with tosyl-Sepharose.[2] Details of activation and coupling yields by the tosyl chloride method have been presented.[1,2]

Affinity Chromatography

To facilitate the removal of interfering sulfonates after coupling and to minimize the risk of altering the properties of the immobilized protein, it is recommended that the level of activation be kept quite low, between 150 and 400 μmol of sulfonate groups per gram of dry Sepharose and 50–150 μmol of sulfonate groups per gram of dry glycerylpropyl-silica (100- or 300-Å pores). Some examples of affinity chromatographic behavior after immobilization of ligands to sulfonyl chloride-activated supports are given below.

N[6]-(6-Aminohexyl) analogs of 5′-AMP and 2′,5′-ADP, bound to tosyl- or tresyl-Sepharose,[1,10] showed similar affinity characteristics as when they were immobilized to CNBr-activated Sepharose.[15] For example, 5′-AMP and 2′,5′-ADP analogs coupled to tosyl-Sepharose, specifically bound NAD^+-dependent lactate dehydrogenase, and NADH-dependent 6-phosphogluconate dehydrogenase, respectively, from a mixture of the two enzymes.[1] Furthermore, the immobilized AMP analog was applied for the single-step purification of lactate dehydrogenase from crude beef heart extract.[1] The AMP analog binding capacity for lactate dehydrogenase, when bound to tresyl-Sepharose (3 μmol of ligand per gram of wet support), was 6.5 mg per gram of wet support.[10]

Soybean trypsin inhibitor, immobilized to tosyl- or tresyl-Sepharose, showed intact affinity properties, similar to CNBr-coupled gels.[1,3] The capacity of a preparation containing 11 mg of inhibitor bound to tresyl-Sepharose (1 ml) was 9 mg of trypsin.[3]

The lectin concanavalin A (Con A) efficiently bound (97% yield) to tresyl-Sepharose (Table II) with retained affinity properties.[3] On application of commercially available horseradish peroxidase (grade II, Boehringer) to a column containing Con A–Sepharose, impurities were

[15] P. Brodelius, P.-O. Larsson, and K. Mosbach, *Eur. J. Biochem.* **47,** 81 (1974).

unretarded and bound peroxidase was completely eluted with buffer containing 0.11 *M* mannose. The $A_{403} : A_{280}$ ratio was 3.15, which corresponds well with literature data for the pure homogeneous peroxidase.[16]

Protein A was also found to bind very efficiently to tresyl-Sepharose with retained binding characteristics for immunoglobulin G (IgG).[12] Care was taken not to immobilize protein A with too many linkages, since difficulties have been encountered in eluting all of the bound IgG under the latter circumstances. Protein A was, therefore, coupled to Sepharose CL-4B with 150 μmol of tresyl groups per gram of dry support at pH 8.5 overnight at room temperature. The yield for coupling was 90% (Table II). The gel, 1 ml containing 2.5 mg of protein A, was packed in a column, and 12.5 mg of human serum IgG was applied at pH 7. Immunoglobulin lacking specificity for protein A passed through unretarded (0.5 mg). The bound IgG was completely eluted on lowering the pH to 2.2 with 0.2 *M* glycine–HCl. The capacity of the column was very high; i.e., 36 mg of IgG. Under these circumstances leakage was observed of 0.05% of protein A immobilized to tresyl-Sepharose when maintained at pH 7.0 and 37° for 8 days. In general, affinity ligands bound by single-point attachment to tresyl-Sepharose show slightly more leakage under these conditions. It may well be that this slight leakage is due to instability of the support chains per se. These questions are presently under investigation.[3]

High-Performance Liquid Affinity Chromatography

N^6-(6-Aminohexyl)-5′-AMP (16 mg per gram of dry silica) was coupled to tresyl-activated glycerylpropyl-silica (LiChrosorb Diol, 5 μm, 100-Å pores, 450 μmol of tresyl groups per gram of dry support). As can be seen from Table II, very high yields of bound ligand (80%) were obtained when coupling proceeded overnight at room temperature and pH 7.5.[3] The AMP–silica was used in an HPLC system allowing highly rapid separations to be achieved (see Fig. 1).

Concanavalin A also has been bound to tresyl chloride-activated silica and used in an HPLC system for the successful separation of glucosides.[17]

The possibility of utilizing an enzyme, horse liver alcohol dehydrogenase, immobilized to tresylated silica for affinity separations on an analytical scale, together with its use in the screening of K_d for retained material, was demonstrated.[11] Details of the utilization of these materials for affinity HPLC are presented in this volume [10].

[16] H. Theorell and A. C. Maehly, *Acta Chem. Scand.* **4,** 422 (1950).
[17] A. Borchert, P.-O. Larsson, and K. Mosbach, *J. Chromatogr.* **244,** 49 (1982).

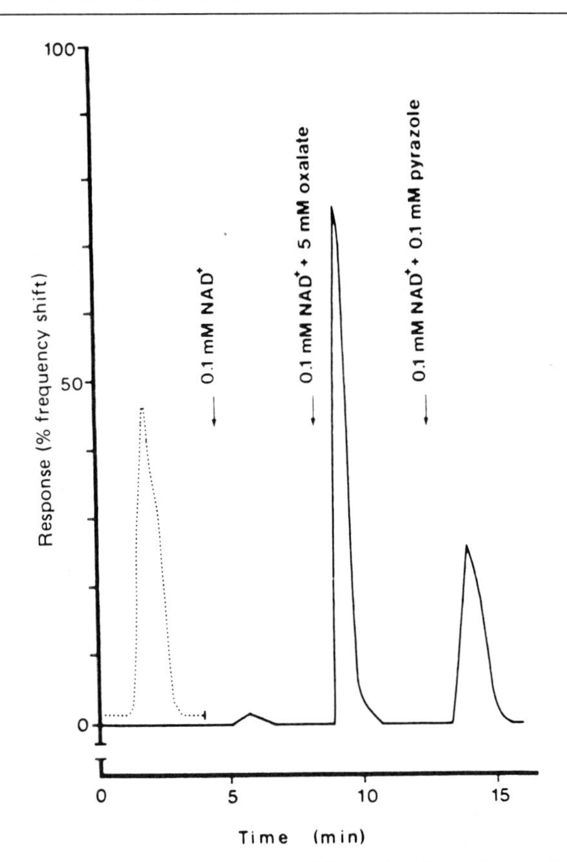

FIG. 1. Separation of bovine serum albumin (25 μg), beef heart lactate dehydrogenase (5 μg), and horse liver alcohol dehydrogenase (25 μg) on N^6-(6-aminohexyl)-AMP-silica by the high-performance liquid affinity chromatography technique. The sample (450 μl) was applied at room temperature to a column (0.5 × 10 cm) containing 29 μmol of AMP analog per gram of dry support. The flow rate was 1 ml/min, which gave a pressure drop over the column of 7 MPa. The column effluent (0.05 M sodium phosphate, pH 7.5) was monitored at 280 nm and then mixed, on-line, with assay reagents for dehydrogenases and measured at 340 nm. Reproduced, with permission, from Nilsson and Mosbach.[3]

Stabilization of Enzyme Configuration

Horse liver alcohol dehydrogenase (Boehringer), 5 mg per gram of dry gel, was immobilized in its ternary complex conformation with NADH and isobutyramide to tresyl chloride-activated glycerylpropyl-silica with 100-Å pores (LiChrosorb Diol, Merck) at pH 8.0, 20° overnight.[18] After

[18] K. Nilsson and K. Mosbach, in preparation.

rigorous washing, an enzyme preparation was obtained having lower dissociation constants (50%) for NADH and NAD when compared with a preparation immobilized together with ADP–ribose, which is known not to induce a conformational change. Immobilization of the enzyme to silica with a low amount of tresyl groups does not yield such an effect. The binding of AMP, however, was the same for all preparations; this is expected, since the part of the dehydrogenase responsible for AMP binding does not undergo gross conformational change.[19] The activity of the "frozen" preparation is also lower than that for the other preparations, possibly owing to a decrease in the rate of dissociation of the coenzyme.

The stability of enzymes in organic solvents can be increased by their immobilization to tresyl-Sepharose. The higher the activation of the support by treatment with increasing tresyl chloride concentration, the better the stability. This was shown for the synthesis of *N*-acetyl-L-tyrosine ethyl ester from *N*-acetyl-L-tyrosine and ethanol[20] and for peptide synthesis.[18] Thus, after 3 days in 50% DMF–0.2 *M* $NaHCO_3$ at pH 9, the remaining activity for synthesis of *N*-acetyl-L-phenylalanine glycinamide from *N*-acetyl-L-phenylalanine methyl ester and glycinamide was 40% for chymotrypsin immobilized to highly activated tresyl-Sepharose, compared to 10% when immobilized to Sepharose with a low activation level.[18]

A stabilizing effect on the activity of T4 DNA ligase after immobilization to tresyl-Sepharose was also reported.[13] The stabilization was greater with tresyl- than with CNBr-bound enzyme. The sulfonate method has been used as a facile and effective means for immobilization of proteins (IgG, Con A, lactate dehydrogenase) to glycerylpropyl-coated silicon plates.[21] The interactions of the immobilized proteins with added biomolecules have been studied by ellipsometry.[21]

Comments

Tresyl chloride can be used with advantage for coupling of pH-sensitive ligands or proteins with high yields, even at neutral pH. In the pH range 9 to 10.5, tosylated supports, similar to epoxy gels but easier to prepare, are effective for coupling ligands that are stable in this range. We consider the sulfonyl chloride method a facile one for immobilization, with advantages over the widely used CNBr method. The degree of activation is easy to predict and to reproduce, and the activated gel does not lose coupling capacity when stored for several weeks in aqueous media.

[19] H. E. Eklund and C.-I. Brändén, *J. Biol. Chem.* **254,** 3458 (1979).

[20] T. Mori, K. Nilsson, P.-O. Larsson, and K. Mosbach, in preparation.

[21] C. F. Mandenius, S. Welin, B. Danielsson, I. Lundström, and K. Mosbach, *Anal. Biochem.*, in press (1983).

The tresyl chloride method is ideally suited for the activation of sorbents applicable to high-performance liquid affinity chromatography and for activation of polyethylene gycol. In addition to amino groups, thiol groups react efficiently with sulfonate esters, enabling coupling of such a group carrying affinity ligands and proteins.

[3] Hydrophobic Chromatography

By SHMUEL SHALTIEL[1]

Until the late 1960s, commonly used methods for the separation of proteins from each other were based on differences in their size, their net charge, their solubility, and their stability to high temperature or to extreme pH values. An important step forward was made with the development of affinity chromatography as a general method for protein purification.[2-5] This method makes use of the biorecognition of enzymes, antibodies, and receptors for their biospecific ligands, covalently linked to an inert matrix backbone. In some instances, the purification was significantly improved upon interposing a hydrocarbon chain (an "arm") between the ligand and the matrix.[2] It was assumed that these arms relieve the steric restrictions imposed by the matrix backbone and allow an increased flexibility and availability of the ligand.[6] It was also assumed that such hydrocarbon chain arms do not damage the inert nature of the matrix, a condition that must be fulfilled to preserve a biospecific adsorption. With water-soluble proteins, the latter assumption seemed reasonable at the time, as it had been found in the case of lysozyme that "most of the markedly nonpolar and hydrophobic side chains . . . are shielded from the surrounding liquid by more polar parts of the molecule" and that, as predicted by Sir Eric Rideal and Irving Langmuir, "lysozyme is quite well described as an oil drop with a polar coat."[7]

[1] Incumbent of the Hella and Derrick Kleemkan Chair in Biochemistry at the Weizmann Institute of Science, Rehovot, Israel. This chapter was written while serving as a Scholar-in-Residence at the Fogarty International Center of the National Institutes of Health, Bethesda, Maryland.

[2] P. Cuatrecasas, M. Wilchek, and C. B. Anfinsen, *Proc. Natl. Acad. Sci. U.S.A.* **61,** 636 (1968).

[3] P. Cuatrecasas and C. B. Anfinsen, *Annu. Rev. Biochem.* **40,** 259 (1971).

[4] W. B. Jakoby and M. Wilchek, eds., this series, Vol. 34.

[5] M. Wilchek, T. Miron, and J. Kohn, this volume [1].

[6] P. Cuatrecasas, *J. Biol. Chem.* **245,** 3059 (1970).

[7] D. C. Phillips, *Sci. Am.* November, p. 78 (1966).

METHODS IN ENZYMOLOGY, VOL. 104

ISBN 0-12-182004-1

In the course of studies on the enzymes involved in glycogen metabolism, an attempt was made to coat beaded agarose with glycogen for use in the biospecific purification of these enzymes and for study of their assembly onto the glycogen organelle. The procedure selected for the preparation of the column materials consisted of activating agarose with CNBr[8] and allowing it to react with an α,ω-diaminoalkane to obtain an ω-aminoalkyl agarose, then binding of CNBr-activated glycogen to the ω-aminoalkyl agarose.

Unexpectedly, two glycogen-coated agarose preparations that differed only in the length of the hydrocarbon chains bridging the ligand and the agarose bead (Seph-C_8-glycogen and Seph-C_4-glycogen[9]) exhibited a different behavior toward glycogen phosphorylase *b*. Whereas the column material with eight-carbon-atom bridges adsorbed the enzyme, that with four-carbon-atom bridges did not even retard it.[10,11] Control experiments revealed that both the adsorption of the enzyme by Seph-C_8-glycogen and its exclusion by Seph-C_4-glycogen could be reproduced with columns that contained no glycogen whatsoever (Seph-C_8-NH_2 and Seph-C_4-NH_2[12]), strongly indicating that free ω-aminooctyl chains (which had not reacted with glycogen) were mainly responsible for the retention of phosphorylase on Seph-C_8-glycogen and suggesting that the very length of the hydrocarbon chain determines the retention power of these columns.

Homologous Series of Alkylagaroses (Seph-C_n)[13]

The above-mentioned difference in the adsorption properties of Seph-C_8-NH_2 and Seph-C_4-NH_2 could arise from steric causes. For example, it could be argued that the column with the longer hydrocarbon chain was effective in the retention of phosphorylase *b* because the amino groups at the tip of the hydrocarbon chains could reach out farther and, being positively charged at neutral pH, could interact electrostatically with less accessible, negatively charged groups in the protein molecule.

In an attempt to approach the problem in a systematic manner, a homologous series of alkylagaroses (Seph-C_n) were synthesized that had no amino groups at the tip of their hydrocarbon chains; these were studied

[8] R. Axén, J. Porath, and S. Ernbäck, *Nature (London)* **214,** 1302 (1967).

[9] Seph-C_n-glycogen represents Sepharose 4B activated with CNBr, allowed to react with an α,ω-diaminoalkane *n* carbon atoms long, then coupled with CNBr-activated glycogen.

[10] Z. Er-el, Y. Zaidenzaig, and S. Shaltiel, *Biochem. Biophys. Res. Commun.* **49,** 383 (1972).

[11] S. Shaltiel, this series, Vol. 34, p. 126.

[12] Seph-C_n-NH_2 represents Sepharose 4B activated by CNBr and allowed to react with an α,ω-diaminoalkane *n* carbon atoms long.

[13] Seph-C_n represents Sepharose 4B activated by CNBr and allowed to react with an α-alkylamine *n* carbon atoms long.

for their ability to retard or retain a mixture of glycogen phosphorylase *b* and glyceraldehyde-3-phosphate dehydrogenase.[10,11] As used here, the term "homologous series of columns" refers to a set of columns that are identical in all structural respects, including ligand density, net charge, and ultrastructure. Such homologous series can be considered as series of consecutive controls in which each column differs from the preceding one in the series only in having hydrocarbon chains one carbon atom longer.

When mixtures of D-glyceraldehyde-3-phosphate dehydrogenase and glycogen phosphorylase *b* were passed through such Seph-C_n columns (n = 1–6), it was found[10] that, under the conditions used, the dehydrogenase was not adsorbed or retarded by any of the columns. However, whereas Seph-C_1 and Seph-C_2 excluded phosphorylase *b*, Seph-C_3 slightly retarded this enzyme, and higher alkylagaroses retained it. Elution of phosphorylase *b* was possible with a "deforming buffer" (0.4 *M* imidazolium citrate, pH 7.0) that had previously been shown[14,15] to have a pronounced, fully reversible effect on the structure and conformation of this enzyme, e.g., dissociating it into monomers and loosening its grip on its cofactor, pyridoxal 5′-phosphate. The binding of phosphorylase *b* to a higher member in the series (Seph-C_6) was so tight that the enzyme could be recovered only under very drastic eluting conditions, e.g., 0.2 *M* CH_3COOH, which detached the protein from the column but at the same time denatured it (Fig. 2 of Er-el *et al.*[10]).

Interestingly, under the same experimental conditions, different proteins displayed different adsorption patterns on the Seph-C_n columns.[10,11] For example, under the conditions used to retain phosphorylase *b* ($n \geq 4$), D-glyceraldehyde-3-phosphate dehydrogenase was not even retarded on any of the columns tested (n = 1–6). Thus, while Seph-C_1 and Seph-C_2 would not distinguish between the two enzymes and would exclude both of them, Seph-C_3 would begin to resolve them, and Seph-C_4 would separate them efficiently,[10] indicating that the ability to discriminate between two proteins, and thereby resolve and separate them, is a function of the number of carbon atoms in the hydrocarbon chains of the columns.

The original publication on the use of homologous series of alkyl agaroses[10] for the purification of glycogen phosphorylase *b* thus indicated that:

1. The retention of proteins on Seph-C_n depends on the number of carbon atoms in the hydrocarbon chains.
2. The resolution power of Seph-C_n columns depends on n, suggesting that it is mainly derived from differences in the size and lipophilicity

[14] S. Shaltiel, J. L. Hedrick, and E. H. Fischer, *Biochemistry* **5,** 2108 (1966).
[15] J. L. Hedrick, S. Shaltiel, and E. H. Fischer, *Biochemistry* **8,** 2422 (1969).

of available hydrophobic patches or pockets in the various proteins.

3. Using a series of Seph-C_n columns, it is possible to delicately adjust the tightness of adsorption of a given protein and avoid an overly strong retention, which then requires drastic conditions for elution and, subsequently, might denature the eluted protein.
4. It is possible to resolve and purify proteins on the basis of a new criterion, i.e., the hydrophobic nature of their surface. This was illustrated in the preparative isolation of glycogen phosphorylase *b* from crude muscle extract on a Seph-C_n column, with a purification factor of 50- to 100-fold in one step.
5. It is possible to use specific "deforming agents" as efficient and delicate eluents in detaching the adsorbed proteins from the column materials.

These findings set the stage for the development of *hydrophobic chromatography*[10,11,16]—a general, systematic approach to the purification of water-soluble, as well as lipophilic, proteins. This approach makes use of homologous series of alkylagaroses and their derivatives (Seph-C_n-X where X = H, NH_2, COOH, OH, C_6H_5, etc.; Fig. 1) to achieve resolution and purification of proteins and cells. Each member in these series offers flexible hydrophobic arms or "yardsticks" that interact with accessible hydrophobic "patches" or "pockets" in the various proteins, retaining only some proteins out of a mixture. Further resolution is then achieved by gradually changing the nature of the eluent. Work in the author's laboratory and elsewhere has demonstrated the wide applicability of this type of chromatography in protein purification[10,11,16–28] and in separating cells and probing their surface.[29–31]

[16] S. Shaltiel and Z. Er-el, *Proc. Natl. Acad. Sci. U.S.A.* **70,** 778 (1973).
[17] R. J. Yon, *Biochem. J.* **126,** 765 (1972).
[18] B. H. J. Hofstee, *Biochem. Biophys. Res. Commun.* **50,** 751 (1973).
[19] Z. Er-el and S. Shaltiel, *Isr. J. Med. Sci.* **9,** 528 (1973).
[20] S. Shaltiel, *Metab. Interconvers. Enzymes, Int. Symp., 3rd, 1973, Abstr.,* p. 62 (1973).
[21] S. Shaltiel, G. F. Ames, and K. D. Noel, *Arch. Biochem. Biophys.* **159,** 174 (1973).
[22] S. Shaltiel, U.S. Patent 3,917,527 (Appl. #360303, 1973).
[23] S. Shaltiel, *Metab. Interconvers. Enzymes, Int. Symp., 3rd, 1973,* p. 379 (1974).
[24] B. H. Hofstee, *Anal. Biochem.* **52,** 430 (1973).
[25] S. Hjertén, *J. Chromatogr.* **87,** 325 (1973).
[26] J. Porath, L. Gundberg, N. Fornstedt, and J. Olsen, *Nature (London)* **245,** 465 (1973).
[27] R. A. Rimerman and G. W. Hatfield, *Science* **182,** 1268 (1973).
[28] S. Shaltiel, *in* "Chromatography of Synthetic and Biological Polymers" (R. Epton, ed.), Vol. 2, p. 13. Horwood, Chichester, England, 1978.
[29] G. Halperin and S. Shaltiel, *Biochem. Biophys. Res. Commun.* **72,** 1497 (1976).
[30] G. Halperin and S. Shaltiel, *in* "Affinity Chromatography" (O. Hoffmann-Ostenhof *et al.*, eds.), p. 307. Pergamon, Oxford, 1978.
[31] S. Shaltiel and G. Halperin, *Hoppe-Seyler's Z. Physiol. Chem.* **359,** 324 (1978).

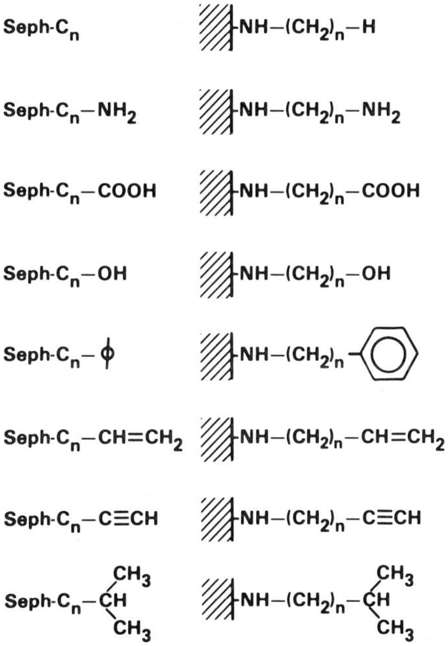

Seph-C_n — $NH-(CH_2)_n-H$

Seph-C_n-NH_2 — $NH-(CH_2)_n-NH_2$

Seph-C_n—COOH — $NH-(CH_2)_n-COOH$

Seph-C_n—OH — $NH-(CH_2)_n-OH$

Seph-$C_n-\phi$ — $NH-(CH_2)_n-C_6H_5$

Seph-$C_n-CH=CH_2$ — $NH-(CH_2)_n-CH=CH_2$

Seph-$C_n-C\equiv CH$ — $NH-(CH_2)_n-C\equiv CH$

Seph-$C_n-CH(CH_3)_2$ — $NH-(CH_2)_n-CH(CH_3)_2$

FIG. 1. A variety of homologous series of alkylagarose derivatives (Seph-C_n-X) which can be used for chromatography. Additional series can be formed by using other functional groups, by placing the functional group at different fixed positions along the hydrocarbon chain, or by having more than one functional group on the chain. Such a series of column materials will draw its adsorption and discrimination power not only from hydrophobic interactions, but also from other interactions involving group X (e.g., Seph-C_n-NH_2, which will be positively charged at neutral pH). However, it will be possible gradually to increase the contribution of the hydrophobic interactions by increasing the number of carbon atoms in the hydrocarbon chains.

Preparation of Column Materials

Synthesis of Seph-C_n (n = 1–12) by the CNBr Procedure (Type I, Fig. 2).[10,11] Sepharose 4B (Pharmacia) is activated by reaction with CNBr[32] at pH 10.5–11 and 22°. One gram of CNBr dissolved in 1 ml of dioxane is added to each 10 g (wet weight[33]) of agarose suspended in 20 ml of water. Activation is initiated by raising the pH of the mixture to 11 with 5 *N* NaOH. The reaction is allowed to proceed for 5 min with gentle swirling, while maintaining the pH between 10.5 and 11 with 5 *N* NaOH. Activation is terminated by addition of crushed ice, filtration, and wash-

[32] R. Axén, J. Porath, and S. Ernbäck, *Nature* (*London*) **214,** 1302 (1967).

[33] Before weighing, the agarose is placed on a Büchner funnel to drain off excess water until the very first signs of dryness (cracks on the gel surface) appear.

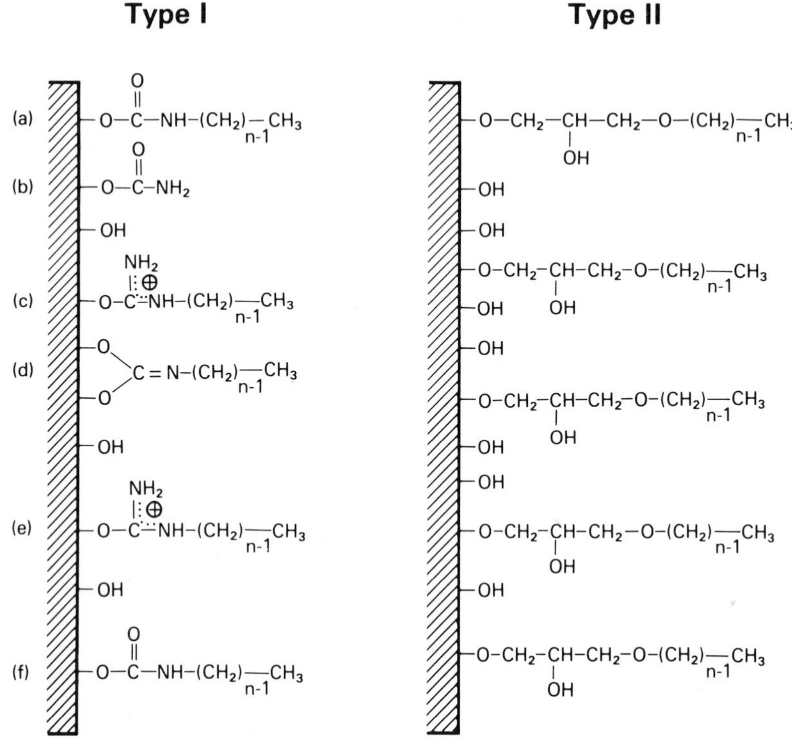

FIG. 2. Structure of alkylagaroses prepared by the CNBr procedure[10,11] (type I) and by the alkylglycidyl ether procedure (type II).[36] Groups (a) and (f) denote a substituted carbamate; (b), a nonsubstituted carbamate formed in the course of the activation step; (c) and (e), substituted isourea linkages that are protonated at neutral pH and, therefore, positively charged; (d) a substituted carbonate; n denotes the number of carbon atoms in each hydrocarbon chain. The characterization of these column materials is described by Halperin *et al.*[34]

ing of the gel with ice-cold water (100 ml). The appropriate normal α-alkylamine (4 mol per mole of CNBr used for the activation of the agarose) is dissolved in 20 mol of a solvent composed of equal volumes of N,N'-dimethylformamide and 0.1 M $NaHCO_3$ (pH 9.5), and the pH is adjusted, if necessary, to 9.5. With the higher members in the series ($n \geq 8$) a gel or a precipitate may form; the mixture should be warmed to about 60° to obtain a clear solution, cooled rapidly, and mixed with the activated agarose before precipitation recurs. The activated agarose is suspended in 20 ml of 0.1 M $NaHCO_3$ (pH 9.5) and mixed with the alkylamine solution. The reaction is allowed to proceed at room temperature (20–25°) for 10 hr with gentle swirling. The alkylagarose is filtered and washed with five times its volume of each of the following: water, 0.2 *M* CH_3COOH, water, 20 m*M* NaOH, water, dioxane–water (1 : 1), 50 m*M* CH_3COOH, and with about 20 volumes of water. Before use, each column is washed and equilibrated with the buffer chosen for chromatography.

An alternative procedure used lately in our laboratory[34] is similar except that the coupling step is carried out in dioxane–water (95 : 5, v/v), allowing the reaction to proceed for 20 hr at pH 9 and 22°. The column materials are washed successively with 2–3 volumes of each of the following dioxane–water mixtures (v/v)—95 : 5; 80 : 20; 60 : 40; 40 : 60; and 20 : 80. The resulting gel is transferred into 0.1 *M* $NaHCO_3$, pH 9.5, shaken for 1 hr at room temperature, then washed with about 20 volumes of water and suspended in the chosen buffer.

It should be emphasized that the same synthetic procedure should be used throughout the series in order to make the columns identical in all respects except for the length of their alkyl side chains. In systematic studies aimed at establishing a standard purification procedure, it is essential that the ligand and charge density of the column materials be identical, as determined by the procedure described below.

Synthesis of Seph-C_n-NH_2 (n = 2–12) by the CNBr Procedure.[11,16] This homologous series is prepared by the procedure described above for the Seph-C_n (type I) series.

Synthesis of Seph-C_n (Type II) by the Alkylglycidyl Ether Procedure. The synthesis of glycidyl ethers from the appropriate alcohols is carried out according to Ulbrich *et al.*,[35] and their coupling to agarose (Sepharose 4B) as described by Hjertén *et al.*[36]

Properties and Characterization of the Column Materials

Determination of Dry Weight. The dry weight of the column materials to be characterized with respect to their ligand and charge density is determined by washing the gel with 20 volumes of deionized water, weighing three 1-g samples (wet weight[33]), and drying them under reduced pressure at 120° to constant weight. The average of the triplicate is taken as the dry weight of that particular column material.

Determination of the Ligand Density of Columns with Radioactively Labeled Ligands. Samples of column materials (0.3 g, wet weight[33]) are dissolved each in 1 ml of 6 *N* HCl, and 10 ml of Triton–toluene scintillator is added. The mixture is shaken until a clear solution is formed, then

[34] G. Halperin, M. Breitenbach, M. Tauber-Finkelstein, and S. Shaltiel, *J. Chromatogr.* **215,** 211 (1981).

[35] V. Ulbrich, J. Makeš, and M. Jurečeh, *Collect. Czech. Chem. Commun.* **29,** 1466 (1964).

[36] S. Hjertén, J. Rosengren, and S. Påhlman, *J. Chromatogr.* **101,** 281 (1974).

counted in a scintillation counter (scintillation liquid composed of 0.1 g of 1,4-bis[2-(5-phenyloxazolyl)]benzene, 8 g of 2,5-diphenyloxazole, 666 ml of toluene, and 333 ml of Triton X-100). The efficiency of radioactivity counting is determined by counting six identical samples of the radioactively labeled free ligand (e.g., *n*-[1-^{14}C]dodecyl alcohol), three samples alone, and three in the presence of 0.3 g (wet weight) of the appropriate nonradioactively labeled column material dissolved each in 1 ml of 6 *N* HCl as described above. The experimentally determined counting efficiency is then used to correct the values obtained in the ligand density determination.

Determination of Ligand Densities by Nuclear Magnetic Resonance (NMR). The method used in our laboratory for the NMR determination of ligand densities in gels obtained by the epoxy coupling (type II columns) was that described by Rosengren *et al.*[37] The spectra are taken in [$^2H_6$]dimethyl sulfoxide using an HFX-10 Bruker 90 MHz NMR spectrophotometer connected to a 1080-Nicolet computer (FT mode). Tetramethylsilane is used as an internal reference. This method was also adapted to gels obtained by the CNBr coupling procedure after appropriate modification[34] of the equation for the calculation of the molar ratios (MR) of alkyl chains to galactose units.

For the type II columns,[37]

$$\mathrm{MR}_{\text{Type II}} = \frac{\text{moles of alkyl chains}}{\text{moles of galactose units}} = \frac{(B + C)/[2(n-2)+3]}{\{A - [7(B+C)]/[2(n-2)+3]\}/7}$$

For the definition of A, B, C, and n used in this equation, see Rosengren *et al.*[37] and Fig. 3.

Based on the structural differences between the type I and type II column materials,

$$\mathrm{MR}_{\text{Type I}} = \frac{\text{moles of alkyl chains}}{\text{moles of galactose units}} = \frac{(B + C)/[2(n-2)+3]}{\{A - [2(B+C)]/[2(n-2)+3]\}/7}$$

For both type I and type II column materials, the ligand densities are calculated from the following equation:

$$\text{Ligand density} = \frac{\text{moles of alkyl chains}}{\text{grams of wet gel}} = \left[\frac{\text{dry weight of 1 g of gel}}{(\mathrm{MR} \times \mathrm{MW}) + 153}\right] \mathrm{MR}$$

[37] J. Rosengren, S. Påhlman, M. Glad, and S. Hjertén, *Biochim. Biophys. Acta* **412,** 51 (1975).

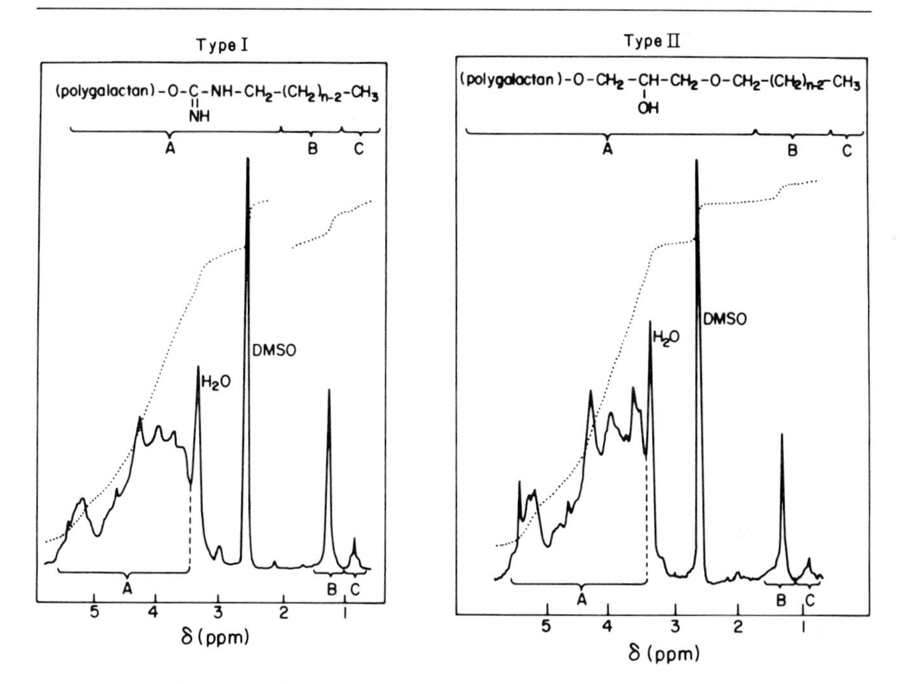

FIG. 3. Nuclear magnetic resonance (NMR) spectra of Seph-C_{10} prepared by the CNBr procedure (type I, left panel) and by the glycidyl ether procedure (type II, right panel). The ligand densities of the column materials used were 13 and 8.5 μmol per gram of wet gel, respectively. Data are from Halperin *et al.*[34]

where MW stands for the molecular weight of the ligand residue and 153 represents the average molecular weight of the carbohydrate residues in the agarose.

The ligand density values determined for the same gel by ^{14}C incorporation and by NMR are in good agreement (see Table I of Halperin *et al.*[34]).

Determination of the Charge Density.[38–41] The charge density of the columns is determined by potentiometric titrations. Column materials to be titrated are thoroughly washed with 1 *M* KCl and suspended in the same KCl solution so that a 3-ml sample of the suspension will contain 0.5 ml (settled volume) of column material. An automatic titrimeter is convenient [Radiometer instrument (TTTIC) with microtitration kit

[38] Z. Er-el, Ph.D. Thesis, The Weizmann Institute of Science, 1975.

[39] G. Halperin, Ph.D. Thesis, The Weizmann Institute of Science, 1978.

[40] S. Shaltiel, *Proc. FEBS Meet.* **40,** 117 (1975).

[41] S. Shaltiel, G. Halperin, Z. Er-el, M. Tauber-Finkelstein, and A. Amsterdam, *in* "Affinity Chromatography" (O. Hoffmann-Ostenhof *et al.*, eds.), p. 141. Pergamon, Oxford, 1978.

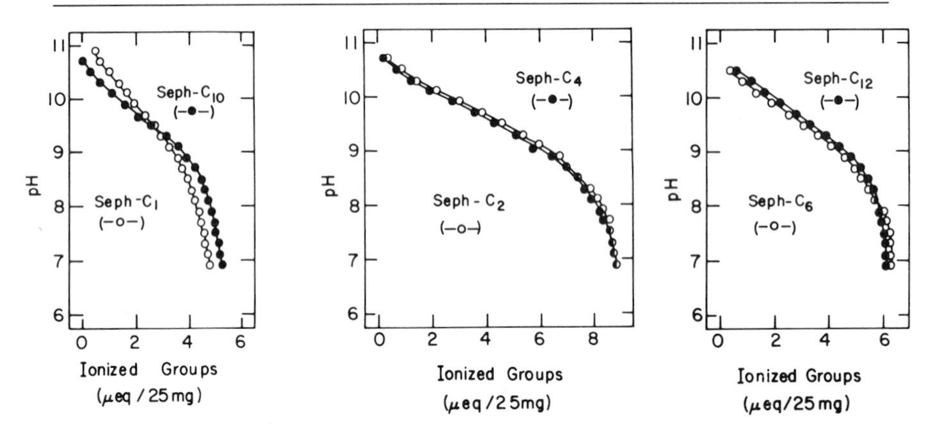

FIG. 4. Potentiometric titrations of arbitrarily chosen members of the Seph-C_n series, carried out on samples (25 mg) of dried column materials. Data are from Halperin *et al.*[39] and Shaltiel *et al.*[41]

(TTA 31), a temperature-controlled compartment (pHA 924), and a recorder (SBR 2C)]. The titration chamber at 25° is maintained under nitrogen during the procedure. The pH is adjusted to 6.5 with 0.2 *N* HCl, and the automatic titration is initiated with 50 m*M* KOH up to pH 11.5. Blanks of unmodified Sepharose 4B from the same batch are similarly titrated. Figure 4 represents several examples of such titrations with different members of the Seph-C_n series. The results shown in each titration curve represent the difference between the titration curves obtained with the tested column materials and those obtained with the blanks of Sepharose 4B. From these curves, one can readily obtain the charge density of the column materials at the pH used for the chromatographic separation (positive charges arising from the N-substituted isourea linkages, Fig. 2). Furthermore, by plotting $\Delta eq/\Delta pH$ vs pH one obtains an apparent pK_a value for the various members in the Seph-C_n series; the pK_a value determined for all the members in the series is 9.7.[29]

Repeated Use, Stability, Capacity, and Flow Rates of Type I Columns. Alkylagaroses and their derivatives can be used repeatedly; a Seph-C_4 column has been used more than 20 times; and a Seph-C_5-NH_2 column, more than 30 times. When suspended in a neutral or acidic aqueous solution at 4°, the columns may be stored for months, losing less than 5% of the ligand within 25 days[42] (Fig. 5). They become considerably more labile at alkaline pH values. Covering the column suspensions with tolu-

[42] S. Shaltiel, *in* "Enzyme Engineering" (E. K. Pye and H. H. Weetall, eds.), Vol. 3, p. 321. Plenum, New York, 1978.

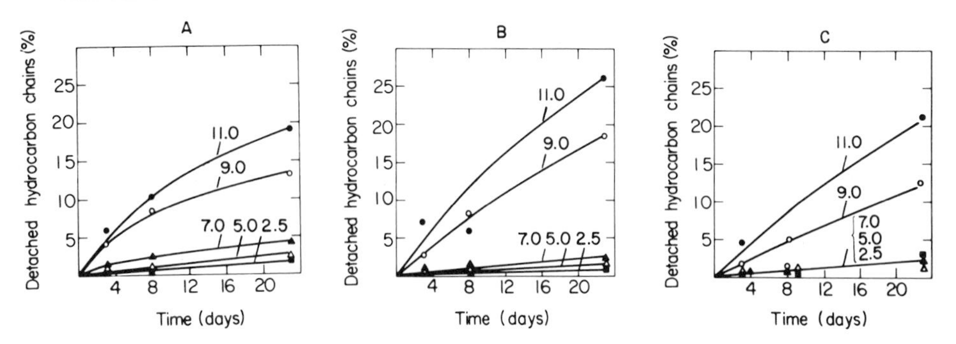

FIG. 5. Stability of Seph-C_n-X column materials at various pH values. Three column materials were synthesized with radioactively labeled hydrocarbon chains: Seph-C_4, Seph-C_4-NH_2, and Seph-C_1-COOH. Samples (300 mg, wet weight) of column materials were suspended in 1 ml of the appropriate buffer and kept at 4°. At the indicated time, aliquots of the supernatant fluids as well as of the centrifuged column materials were removed and counted, and the percentage of detached hydrocarbon chains was calculated. The 50 m*M* buffers used included sodium acetate (pH 2.5 and 5.0), potassium phosphate (pH 7.0), sodium carbonate (pH 9.0), and glycine–NaOH (pH 11.0). Data are from Shaltiel.[42]

ene prolongs shelf life. Between runs it is recommended that the columns be washed sequentially with water, 1 *M* NaCl, and water. Additional washing with the series of eluents described at the end of the preparative procedure of the columns (see above) or with 1 *M* urea (pH 7) can be attempted (preferably on a Büchner funnel) if clogging causes a considerable decrease in the capacity or the flow rate of the column. The columns have a good capacity. For example, Seph-C_4 binds up to 10 mg of glycogen phosphorylase *b* per milliliter of settled volume and the column can be used at flow rates of 0.5–2 ml/min.

Ultrastructural Identity. In order to establish that the differences observed in the resolution power of the columns arise mainly from differences in the hydrophobicity of their ligands, it is necessary to demonstrate that they constitute a homologous series, i.e., that the column materials are identical in all structural respects except for the length of their hydrocarbon chains. Since the reaction between CNBr-activated agarose and the various alkylamines is carried out at quite a high concentration of the ligand,[10,11] one cannot overlook the possibility that the alkylamine itself, occupying an increasing proportion of the reaction mixture volume as the molecular weight (M_r) of the ligand increases, might affect differently the properties of the medium in which the reaction takes place and thereby cause structural differences in the column material.

Morphological studies under a phase-contrast light microscope show that activation of agarose with CNBr and subsequent reaction with alkylamines do not significantly affect the size and shape of the beads.[43] The hydrocarbon chains in these column materials are evenly distributed both on the surface and within the beads, as shown, for example, by autoradiography of Seph-C_4-NH_2 with tritiated hydrocarbon chains.[41] Examination of several bead slices (1-μm sections) reveals that the autoradiographic grain density, and thus the density of the hydrocarbon chains, does not fluctuate by more than $\pm 15\%$ throughout the bead.

Furthermore, untreated agarose (Seph-C_0) and arbitrarily chosen derivatized agaroses (Seph-C_4, Seph-C_4-NH_2, and Seph-C_8-NH_2) are indistinguishable in their ultrastructure.[41] Thin sections of the abovementioned column materials embedded in Epon reveal a network skeleton with variable pore sizes that are distributed at random throughout the bead.[44]

Selecting a Column for a Given Purification

The Exploratory Kit. The selection of a suitable substituted agarose within a given homologous series, e.g., Seph-C_n, is achieved by means of an exploratory group of column materials designed for this purpose.[11,22,23] This "kit" is composed of a set of Pasteur pipettes (0.5×5 cm), each containing a different member of the homologous series, e.g., Seph-C_n, $n = 1$–10. In addition, two control columns are included: one of unmodified agarose (Seph-C_0) and the other (Seph-C_{0N}) of agarose that has been activated with CNBr and then treated with ammonia in an attempt to simulate, in part, the functional groups introduced by the activation with CNBr and the subsequent coupling with an amine. A similar kit of column materials can be prepared for any other series of the Seph-C_n-X type.

When aliquots of a protein mixture, an extract of muscle, are applied to each of this group of columns, some of the proteins become adsorbed and some are excluded. If the mixture of unadsorbed protein from each of the columns is then subjected to polyacrylamide gradient gel electrophoresis in the presence of sodium dodecyl sulfate, it can be seen (Fig. 6) that increasing the hydrocarbon chain length results in the exclusion of fewer proteins, indicating a greater number of proteins being adsorbed onto the columns. Type I and type II columns behave in a very similar manner in this respect, provided they have comparable ligand densities.[34]

Adsorption–Elution Profiles. Densitometric scans of the gels and measurement of the area of each peak (denoted A to O and corresponding to a protein band) allow presentation of an adsorption profile for each of the

[43] A. Amsterdam, Z. Er-el, and S. Shaltiel, *Isr. J. Med. Sci.* **10,** 1580 (1974).
[44] A. Amsterdam, Z. Er-el, and S. Shaltiel, *Arch. Biochem. Biophys.* **171,** 673 (1975).

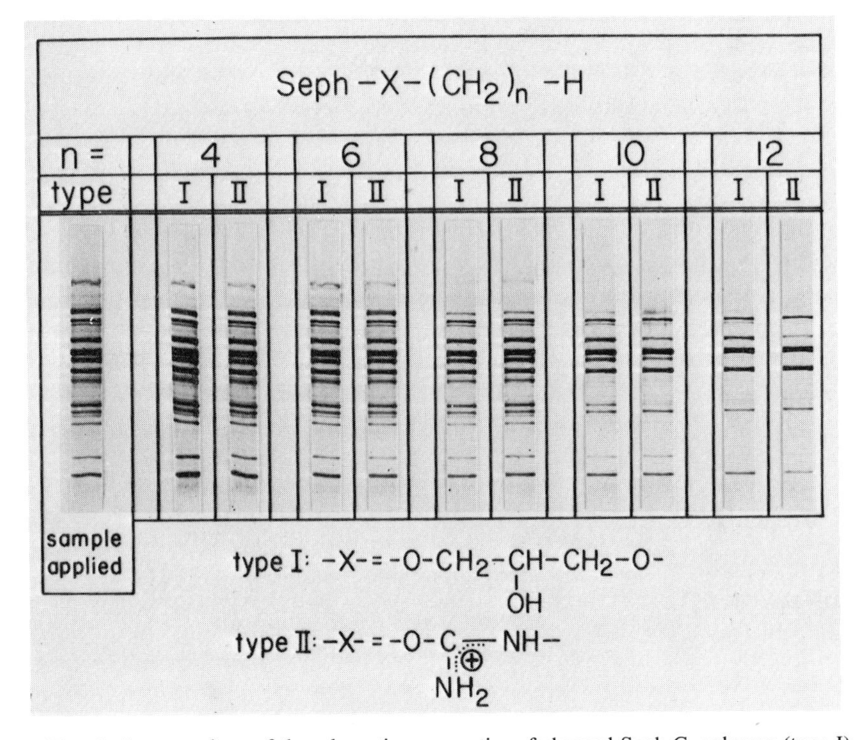

FIG. 6. A comparison of the adsorption properties of charged Seph-C_n columns (type I) with those of electrically neutral Seph-C_n columns (type II). The columns compared here had identical hydrocarbon chains (equal number of carbon atoms per hydrocarbon chain) and very similar ligand densities (10 ± 1 μmol per gram of gel). Samples (0.4 ml containing 4–5 mg of protein) of crude muscle extract (dissolved in a buffer composed of 10 m*M* sodium phosphate and 90 m*M* NaCl, pH 7), were applied each on a column (ca. 0.6 × 4 cm, gel volume 1 ml), equilibrated and run (22°) with the same buffer. The excluded protein (first 2 ml) was collected, and an aliquot (10 μl) of this fraction was subjected to electrophoresis on a polyacrylamide gradient gel, 7 to 20% in the presence of sodium dodecyl sulfate. Note the gradual disappearance of some of the proteins (i.e., their gradual adsorption onto the columns) with increasing hydrocarbon chain length (n), and the remarkable similarity in the adsorption properties of the charged (type I) and the electrically neutral (type II) column materials having an equal value of n. Data are from Halperin *et al.*[34]

protein bands (percentage of protein excluded as a function of the number of carbon atoms per hydrocarbon chain; see Fig. 7). It can be seen, for example, that protein band A is adsorbed on Seph-C_6, band K on Seph-C_8, band I on Seph-C_{10}, and band G on Seph-C_{12}. It should be emphasized that the adsorption profiles are characteristic of these protein bands under a given set of conditions. These include pH, temperature, ionic strength, ionic composition of the buffer, concentration of apolar solvents

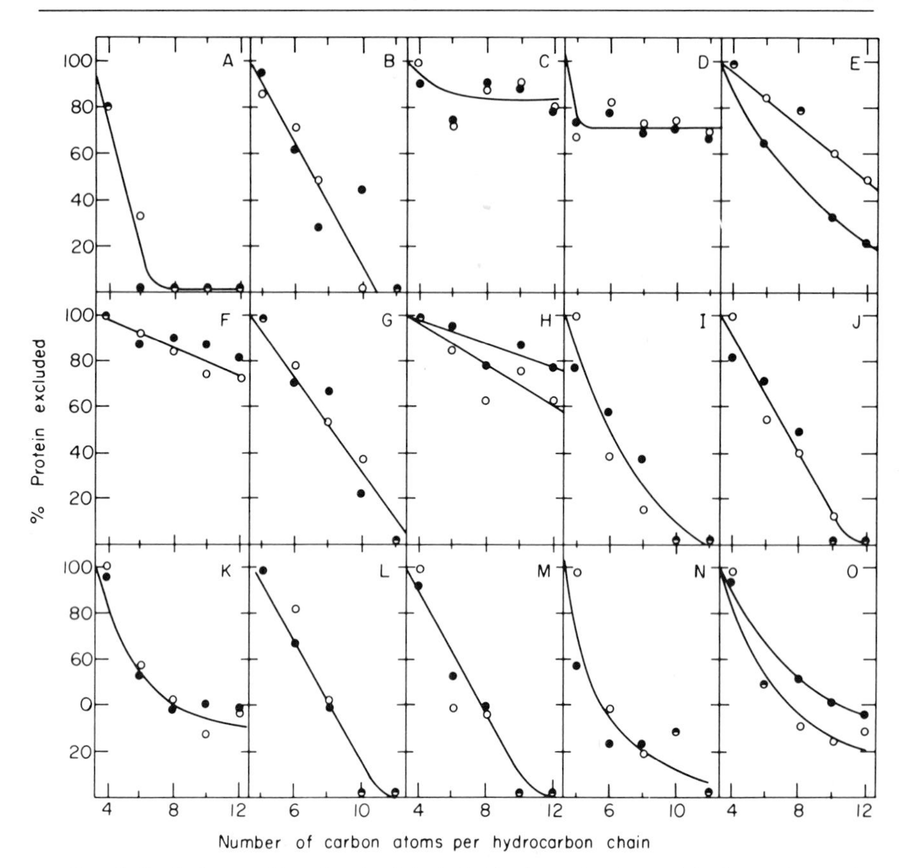

FIG. 7. Exclusion profiles of each of the individual proteins (A to O) in the crude muscle extract depicted in Fig. 6. These profiles were obtained from densitometric scans of each of the gels shown in Fig. 6 by measuring the relative area of the peaks (cutting out the peaks, weighing them, and dividing these weights by that of 1-cm^2 areas of the same paper). ●, Type I columns; ○, type II columns. The 100% value for each protein band is taken from the densitometric scan of the sample applied on the columns (Fig. 6, first panel on the left). Data are from Halperin *et al.*[34]

or of deforming agents.[10,11,16] The presence during chromatography of biospecific ligands[20] (coenzymes, metal ions, substrates, inhibitors, and allosteric effectors), which often affect the conformation of their target proteins, may also affect these adsorption profiles.

If, under a certain set of conditions, a specific protein is adsorbed onto hydrocarbon chains n carbon atoms long, a change in one of the parameters listed above may either allow adsorption on a column with shorter hydrocarbon chains, or promote exclusion from Seph-C_n and require a

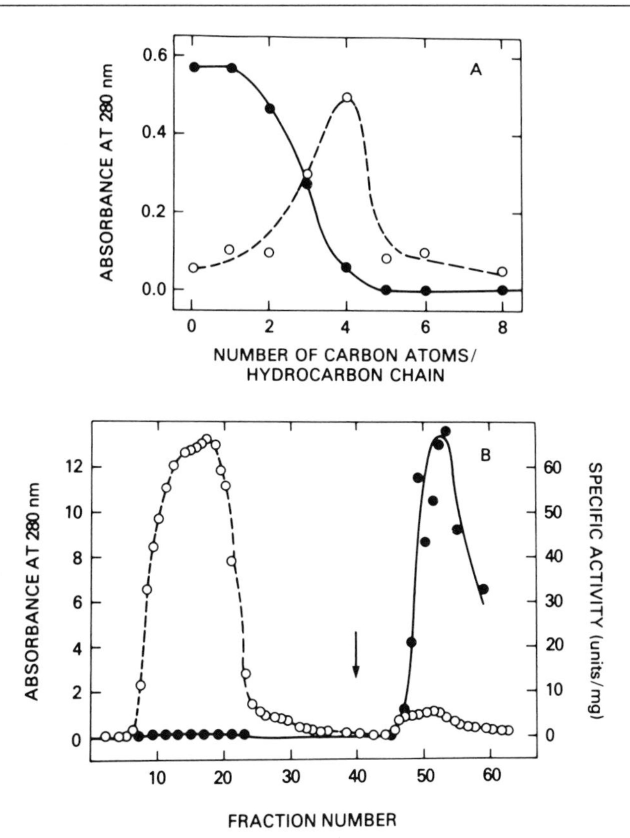

FIG. 8. Panel A: Adsorption profile (●) of glycogen phosphorylase *b* on a Seph-C_n kit equilibrated at 22° with 50 m*M* sodium β-glycerophosphate and 1 m*M* EDTA at pH 7.0. Elution (○) was attempted with a "deforming buffer," 0.4 *M* imidazole, adjusted to pH 7 with citric acid. Aliquots of the AMP-free enzyme (1 mg in 0.1 ml) were applied to each column. The first 2 ml was collected, the deforming buffer was applied, and an additional 2 ml was collected afterwards. Panel B: Preparative-scale purification of glycogen phosphorylase *b* from muscle extract on Seph-C_4. Thirty milliliters of the extract with phosphorylase activity of 0.7 unit/mg were applied to a Seph-C_4 column (16 × 1.1 cm) equilibrated at 22° with the glycerophosphate buffer. Fractions of 2.6 ml were collected. Absorbance (○) and specific activity (●) were monitored. Nonadsorbed protein was removed by washing and (arrow) elution with the "deforming buffer" was initiated. The phosphorylase-specific activity was measured after dialysis of the fractions against 100 volumes of glycerophosphate buffer. Data are from Er-el *et al.*[10] and Shaltiel.[28]

higher member of the series for adsorption. An example of an adsorption–elution profile of a homogeneous protein, glycogen phosphorylase *b*, on the Seph-C_n series, is given in Fig. 8A. As the length of the hydrocarbon chains increases, less protein is excluded from the column. Under the specific conditions of the experiment, Seph-C_0 and Seph-C_1 exclude all the protein, Seph-C_2 and Seph-C_3 retard it (only part of the protein is found in the first 2-ml fraction collected), whereas Seph-C_4 and higher members of the series retain the protein.[10,11] By passage of a 0.4 *M* imidazolium–citrate buffer, a "deformer" for glycogen phosphorylase *b*,[14,15] it is possible to elute the enzyme quantitatively from the Seph-C_4 column, but not from higher members in the series. In fact, recovery of the enzyme from Seph-C_6 was achieved only with a drastic eluent, 0.2 *M* CH_3COOH,[10] which displaces the enzyme but also denatures it.

Preparative Purification. On the basis of such results it seems obvious that the column material of choice within this series for the extraction of glycogen phosphorylase *b* would be Seph-C_4, where adsorption is complete and desorption is possible under mild conditions. The preparative purification of glycogen phosphorylase *b* from crude rabbit muscle extract (Fig. 8B) demonstrates clearly that the results obtained with the exploratory kit do indeed provide valid guidance for preparative-scale purification: when the muscle extract is applied to Seph-C_4, more than 95% of the total protein is excluded from the column, with essentially no phosphorylase activity. Upon elution with 0.4 *M* imidazolium citrate buffer, a small amount of protein is eluted with very high phosphorylase *b* activity (Fig. 8B). However, the specific activity of phosphorylase *b* is not constant throughout all of the eluted fractions, indicating that some protein impurities coelute with the enzyme. Nevertheless, a 60- to 100-fold purification is obtained in one step with greater than 95% recovery of enzyme activity.

The column of choice need not be the one with the shortest hydrocarbon chain that retains the desired enzyme. For example, in the case of maltodextrin phosphorylase and the Seph-C_n-NH_2 series, it was found[45] that the enzyme was retained on Seph-C_4-NH_2, as was about 70% of the total protein applied (Fig. 1 in Thanner *et al.*[45]). However, maltodextrin phosphorylase could be eluted from higher members in the series by adding NaCl (final concentration 1 *M*) to the elution buffer. In this instance, Seph-C_{10}-NH_2 was preferred for preparative purification, since it releases this enzyme before releasing other proteins in the mixture.[45] In fact, an even better resolution can be achieved if, instead of the abrupt increase in ionic strength, a shallow salt gradient is applied. On a preparative scale, this method resulted in an eightfold purification of maltodextrin phosphorylase with an 80% yield.[45]

Optimalizing the Resolution of Two Enzymes. For the resolution of two enzymes, it is necessary to obtain the characteristic adsorption profile of each by monitoring their concentration in the excluded fractions. If

[45] F. Thanner, D. Palm, and S. Shaltiel, *FEBS Lett.* **55,** 178 (1975).

the two differ in their adsorption profiles, it is usually possible to select a column that retains one enzyme while excluding the other. For example, when an extract of rabbit muscle is subjected to chromatography on Seph-C_n-NH_2, the columns with short chains ($n = 2$ and $n = 3$) exclude glycogen synthase I, but retain the enzyme on Seph-C_4-NH_2, from which it can be eluted with a linear NaCl gradient.[16] A column of Seph-C_4-NH_2 retains synthase I but excludes glycogen phosphorylase *b*, which in this column series and under the same conditions requires hydrocarbon chains 5 or 6 carbon atoms long for retention (Fig. 2 in Shaltiel and Er-el[16]). Therefore, it is possible[16] to isolate glycogen synthase by passage of muscle extract through Seph-C_4-NH_2, and then to extract phosphorylase *b* by subjecting the excluded proteins to chromatography on Seph-C_6-NH_2, setting the stage for the principle of consecutive fractionation[46] described below.

In a preparative experiment,[16] 70 ml of rabbit muscle extract containing 1750 mg of protein and a synthase activity of 0.2 unit per milligram of protein was applied on a Seph-C_4-NH_2 column (2.4 × 10 cm) equilibrated with a buffer composed of 50 m*M* sodium β-glycerophosphate, 50 m*M* 2-mercaptoethanol, and 1 m*M* EDTA at pH 7.0. After the unadsorbed protein was removed by washing, a linear NaCl gradient in the same buffer was applied. The I form of the synthase was purified 20- to 50-fold in one step. Since Seph-C_4-NH_2 discriminates very efficiently between this enzyme and hemoglobin, the major protein in erythrocytes, glycogen synthase could be isolated from human erythrocytes with a purification factor of 600-fold in one step[28,38,47] (Fig. 9).

Selecting the Series. The exploratory kits of columns are very useful for deciding not only which column in the series is likely to provide the optimal purification, but also upon the most suitable series of column materials. The resolution of clostripain and collagenase[46] is a case in point. Both enzymes have very similar adsorption profiles on the Seph-C_n series (n = 1–7, Fig. 10A), both of them being retained on Seph-C_6. However, they exhibit different adsorption profiles with the Seph-C_n-NH_2 series: whereas clostripain is essentially retained on Seph-C_4-NH_2, collagenase is excluded by this column (Fig. 10B). Under the conditions of the experiment,[46] a higher member in the series (Seph-C_7-NH_2) is required to retain the enzyme completely. Moreover, the binding of the enzymes to Seph-C_4-NH_2 and to Seph-C_7-NH_2 is quantitatively reversed in both cases when 1 *M* NaCl is included in the eluent (Fig. 10B).

On the basis of this exploratory experiment, it is clear that it is advis-

[46] M. R. Kula, D. Hatef-Haghi, M. Tauber-Finkelstein, and S. Shaltiel, *Biochem. Biophys. Res. Commun.* **69,** 389 (1976).

[47] N. Bashan, Ph.D. Thesis, The Hebrew University, Jerusalem, 1974.

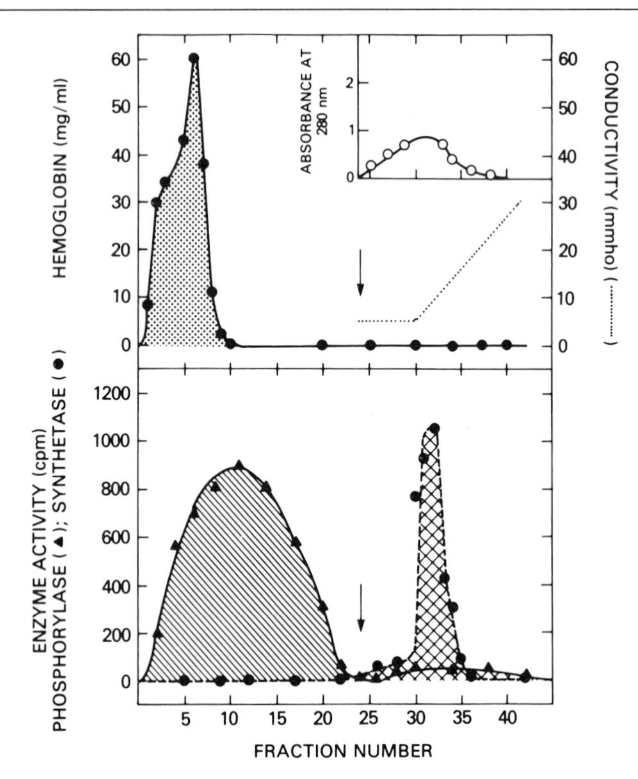

FIG. 9. Purification of glycogen synthase from erythrocytes on Seph-C_4-NH_2. Hemolysate, 2 g of protein in 20 ml, was applied to a column (2 × 10 cm) equilibrated at 23° with the glycerophosphate buffer described in the legend to Fig. 8B. Unadsorbed protein was removed by washing, and (arrow) an NaCl gradient (in the same buffer) was applied. Hemoglobin (●——●), adsorption (○), glycogen phosphorylase activity (▲), glycogen synthase activity (●---●), and conductivity (· · · ·) were monitored for fractions of 6 ml. Data are from Er-el[38] and Shaltiel.[28]

able to resolve the two enzymes by passage of the crude preparation through Seph-C_4-NH_2 in order to extract clostripain, and then to extract collagenase by applying the excluded proteins to Seph-C_7-NH_2. The results depicted in Fig. 11 show that this consecutive use of two columns from the Seph-C_n-NH_2 series brings about resolution and purification of both enzymes.[46]

Means of Elution. Systematic studies aimed at the optimalization of elution from alkylagaroses have shown that proteins can be desorbed from such columns by a variety of means. These include polarity-reducing agents, specific deformers, mild detergents, low concentration of denaturants, alteration in pH or temperature, and changes in ionic strength and

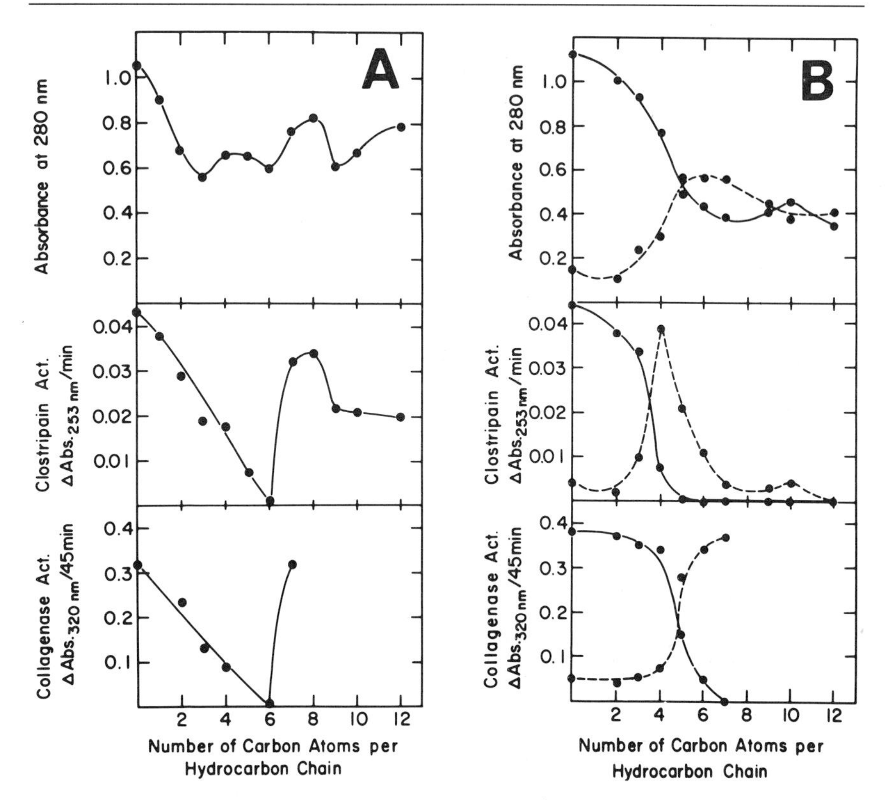

FIG. 10. Adsorption–elution profiles of collagenase and clostripain on homologous series of (A) alkylagarose columns (Seph-C_n) and (B) ω-aminoalkyl agarose columns (Seph-C_n-NH_2). Exploratory kits of column materials (n = 0–12) were equilibrated at 20° with 50 m*M* Tris chloride containing 2 m*M* dithioerythritol at pH 7.5. Aliquots (0.1 ml containing 1 mg of protein) of the crude lyophilized collagenase preparation dissolved in the same buffer were applied to each column (0.4 × 5 cm). Columns were washed with 2 ml of the buffer. In the Seph-C_n-NH_2 series (B), which discriminated between the two enzymes, each of the columns was subsequently washed with 2 ml of the same buffer containing 1 *M* NaCl (---). Data are from Kula *et al.*[46]

ionic composition. Since the availability of hydrophobic crevices or patches on the surface of a protein appear to depend on its conformation, and since the retention of proteins by alkylagaroses depends largely on the lipophilicity, size, shape, and number of these crevices and patches, the above-mentioned means of elution may function either by directly disrupting the hydrophobic interactions between the column material and the protein or by changing the conformation of the protein.[11,28] In fact, both mechanisms may often operate simultaneously. High selectivity in elution may also be achieved by using biospecific ligands such as

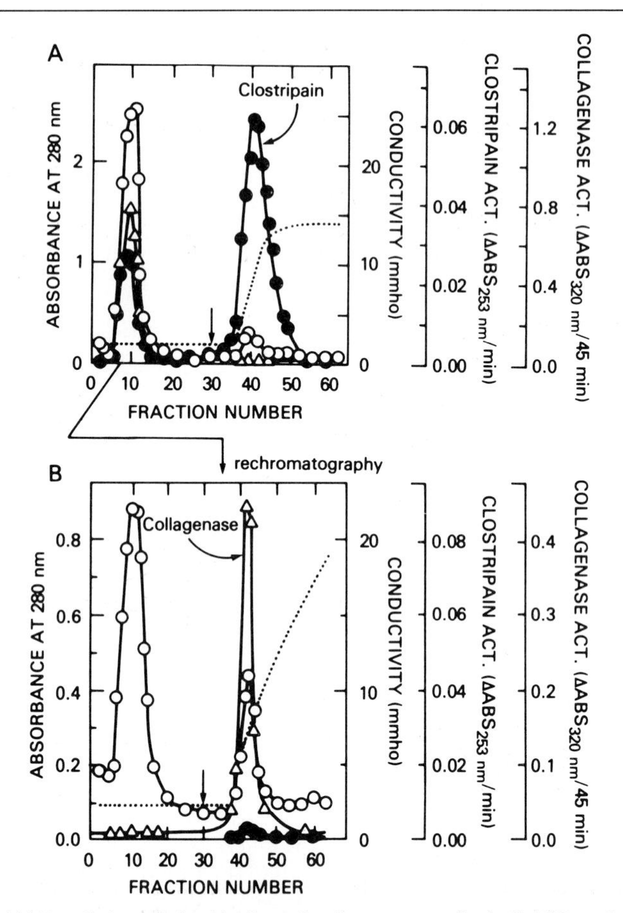

FIG. 11. (A) Resolution of clostripain and collagenase on Seph-C_4-NH_2 and purification of clostripain. A crude, lyophilized collagenase preparation (10.7 mg) in 50 m*M* Tris chloride and 2 m*M* dithioerythreitol at pH 7.5, was applied to a column (1.5 × 6 cm) equilibrated at 20° with the same buffer. Unadsorbed protein was removed by washing, and (arrow) a gradient up to 1 *M* NaCl (in the same buffer) was applied at a flow rate of 85 ml/hr. Fractions of 1.3 ml were collected. Absorbance (○), conductivity (······), collagenase activity (△), and clostripain activity (●) were monitored. (B) Purification of collagenase on Seph-C_7-NH_2. Fractions 7 through 10 from the Seph-C_4-NH_2 column (see panel A) were combined, of which 4.1 ml was applied to a Seph-C_7-NH_2 column (1.5 × 6 cm), which was washed and eluted as described above.

coenzymes, substrates, specific metal ions, and allosteric effectors, which often bring about ligand-induced conformational changes in proteins.[20,41]

Glycogen phosphorylase *b* can be eluted from Seph-C_4 by polarity-reducing compounds such as ethylene glycol. Upon increasing the con-

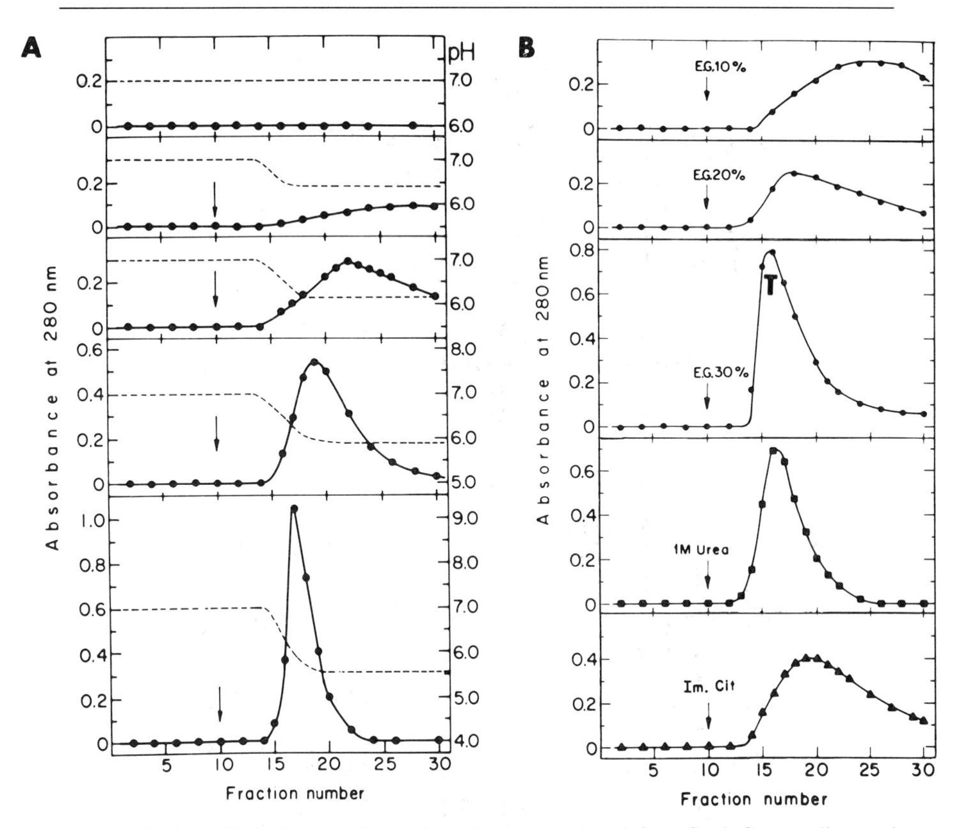

FIG. 12. (A) Effect of pH on the elution of phosphorylase *b* from Seph-C_4. An aliquot of the enzyme (4.7 mg) was applied to each of five columns (0.9 × 8 cm) equilibrated at 22° with a glycerophosphate buffer at pH 7.0. Elution was attempted with glycerophosphate buffers acidified to different pH values with glacial acetic acid. Absorbance (●——●) and pH (---) were monitored. (B) Effect of ethylene glycol (E.G.) (10, 20, and 30%), as well as 1 *M* urea and 0.4 *M* imidazolium citrate (Im. Cit.) on the desorption of phosphorylase *b* from Seph-C_4. The experiment was conducted as described in the legend to panel (A) except for the indicated additions to the eluting buffer. Data are from Er-el[38] and Shaltiel.[40]

centration of this glycol in the irrigating solvent to 30%, the enzyme emerges from the column as a rather narrow peak[40] (Fig. 12B). Alternatively, the enzyme can be desorbed from the same column by simply lowering the pH of the eluting buffer down to 5.6 (Fig. 12A), which is known to bring about distinct structural changes in this enzyme.[48] Addition of urea (1 *M*) to the irrigating buffer also promotes desorption of the enzyme (Fig. 12B). However, from the point of view of preserving the

[48] M. Cortijo, I. Z. Steinberg, and S. Shaltiel, *J. Biol. Chem.* **246,** 933 (1971).

native conformation of this enzyme (highest catalytic activity and unaltered response to regulatory metabolites), the optimal eluent was found to be its specific "deformer," 0.4 *M* imidazolium citrate, pH 7.0.

The term "deformer" had been introduced[14] to describe a compound that brings about a localized, rather limited, conformational change in a protein, a change that can be fully and readily reversed by removal of the compound. Different proteins will have different specific deformers,[14,15] and numerous examples of enzymes that are exceptionally sensitive to specific cations, anions, or uncharged compounds are recorded in the literature. Their effect on the structure of the enzyme is often reflected in its reversible inactivation (affecting the K_m or the V_{max} or both), its aggregation state, or its binding constant for certain ligands. Very often, such compounds are mentioned in reporting the optimalization of the assay procedure for a given enzyme and are referred to as noncompetitive inhibitors whose presence in the assay medium is to be avoided. Some of these compounds may act as specific deformers and should therefore be screened with the exploratory kit to determine effectiveness as eluents.[49,50]

Useful Features of Hydrophobic Chromatography by the Homologous Series Approach

Hydrophobic chromatography was not introduced to replace other methods for the purification of proteins, but rather to provide an additional criterion for their resolution. Indeed, protein mixtures that have already been resolved by procedures based on differences in stability, solubility, charge, size, or biospecific ligand recognition can be further resolved by alkylagaroses.

Among the most important features of this approach is the capability for gradually adjusting the adsorption forces and thereby avoiding an overly strong retention, the latter often dictating rather drastic conditions at the elution step. This delicate adjustment is particularly important in the purification of labile regulatory enzymes and receptors, whose native conformation, catalytic activity, or ability to recognize and respond to specific biological signals may be lost in the course of purification.

Homologous series of alkylagaroses can be used as probes for the detection of structural changes in proteins, if these are reflected in the size or distribution of hydrophobic crevices or patches. Regulatory proteins

[49] S. Shaltiel, G. B. Henderson, and E. E. Snell, *Biochemistry* **13,** 4330 (1974).
[50] G. B. Henderson, S. Shaltiel, and E. E. Snell, *Biochemistry* **13,** 4335 (1974).

that are designed for assuming more than one stable conformation and thus regulating the rate of metabolic processes, may be good candidates for resolution and probing by hydrophobic chromatography.[23] This is illustrated in the separation of the *a* from the *b* form of glycogen phosphorylase.[51] The two forms are identical in their amino acid sequence, except for the state of phosphorylation of a unique serine residue in each of the enzyme protomers ($M_r \sim 97,000$). Phosphorylase *a* is retained by Seph-C_1, whereas hydrocarbon chains four carbon atoms long are necessary to retain the *b* form of the enzyme under the same conditions.

A potential use of these column materials may be in the purification and study of lipophilic membrane-bound proteins that have accessible hydrophobic regions used for their localization within the membrane (see also this volume [18]).

An important feature of the homologous series approach is that it provides the basis for the development of automatic consecutive fractionators for the purification of proteins,[46] for example, by sequential passage of a crude extract through Seph-C_1, Seph-C_2, Seph-C_3, . . . , Seph-C_n. Each column would be expected to retain a few proteins from the mixture, which would be resolved by consecutive detachment as shown in the resolution of enzymes involved in the regulation of glutamine metabolism.[52] Some of the proteins may be purified by being excluded from all the members used in a given series of columns.[21] Consecutive extraction of proteins may thus lead to maximal utilization of expensive, scarcely available tissues, or to the simultaneous identification and determination of several components from one biological fluid or tissue.

The use of homologous series of alkylagaroses is not restricted to proteins. They may also provide a useful tool in cytology. Erythrocytes, for example, can be efficiently adsorbed onto and desorbed from Seph-C_n columns under physiological, isotonic conditions.[29,30] This reversible adsorption can be carried out with essentially no physical entrapment or lysis of the cells, as indicated by the fact that over 95% of the cells applied can be recovered. The erythrocytes eluted from such columns and those yet to be applied were found to be morphologically indistinguishable and to exhibit an identical osmotic fragility profile, suggesting that they are not significantly damaged by the adsorption–desorption procedure. Furthermore, erythrocytes from different sources exhibit different adsorption profiles on such column series, illustrating potential use in discriminating between closely related cells and in probing cell surfaces.[29,30]

[51] Z. Er-el and S. Shaltiel, *FEBS Lett.* **40,** 142 (1974).

[52] S. Shaltiel, S. P. Adler, D. Purich, C. Caban, P. Senior, and E. R. Stadtman, *Proc. Natl. Acad. Sci. U.S.A.* **72,** 3397 (1975).

On the Mechanism of Action of Alkylagaroses

In their native, biologically functional conformation, water-soluble proteins are folded so as to bury as many as possible of their hydrophobic side chains in the interior of the molecule and to expose as many as possible of their polar, charged side chains to interaction with water. This seems to be a major driving force in the folding of such proteins into their three-dimensional conformation.[53] Nevertheless, complete burying of all hydrophobic groups is generally not achieved, leaving some hydrophobic groups exposed at the surface of the protein.[53] Together with hydrophobic components of charged amino acids, such as the $(\text{—}CH_2\text{—})_n$ stretches of lysyl and glutamyl residues or the phenyl ring of tyrosines, these may form hydrophobic "patches" or "pockets" at the surface of the molecule. A sufficiently large hydrophobic patch may constitute a binding site for the hydrocarbon chains implanted on the hydrophilic agarose matrix. Owing to the flexibility of the normal hydrocarbon chains, they can accommodate themselves within such sites (Fig. 13) and form "hydrophobic bonds," freeing "ordered" water molecules and allowing them to interact with each other.[54–56]

It seems reasonable to assume that the available hydrophobic patches and pockets of different proteins will vary in number, size, shape, and lipophilicity and that these variations will be reflected in the relative affinities of different proteins for a specific alkylagarose. It is the properties of such patches, and perhaps their distribution on the surface of different proteins, that most likely play a major role in the resolution of proteins on alkylagaroses. This is corroborated by the following experimental observations:

1. The retention of proteins by alkylagaroses is greatly affected by the length of the hydrocarbon chains attached to the matrix. Different members of a given homologous series of alkylagaroses (which are indistinguishable in the size of the beads, their shape, and their ultrastructure and having a similar charge density and hydrocarbon chain density) differ dramatically in their capacity to bind a given protein under a given set of conditions. For example, both the Seph-C_2 and Seph-C_4 columns used in the experiment depicted in Fig. 8A had a charge density of ~8.8 μeq/ml of settled column material at pH 7, and both had a hydrocarbon chain density of ~32 μmol/ml (settled volume), yet, under the same conditions, Seph-C_2 excluded glycogen phosphorylase *b* while Seph-C_4 retained it.

[53] C. Tanford, "The Hydrophobic Effect," Wiley (Interscience), New York, 1980.
[54] G. S. Hartley, "Aqueous Solutions of Paraffin-Chain Salts." Hermann, Paris, 1936.
[55] H. S. Frank and M. W. Evans, *J. Chem. Phys.* **13,** 507 (1945).
[56] W. Kauzmann, *Adv. Protein Chem.* **14,** 1 (1959).

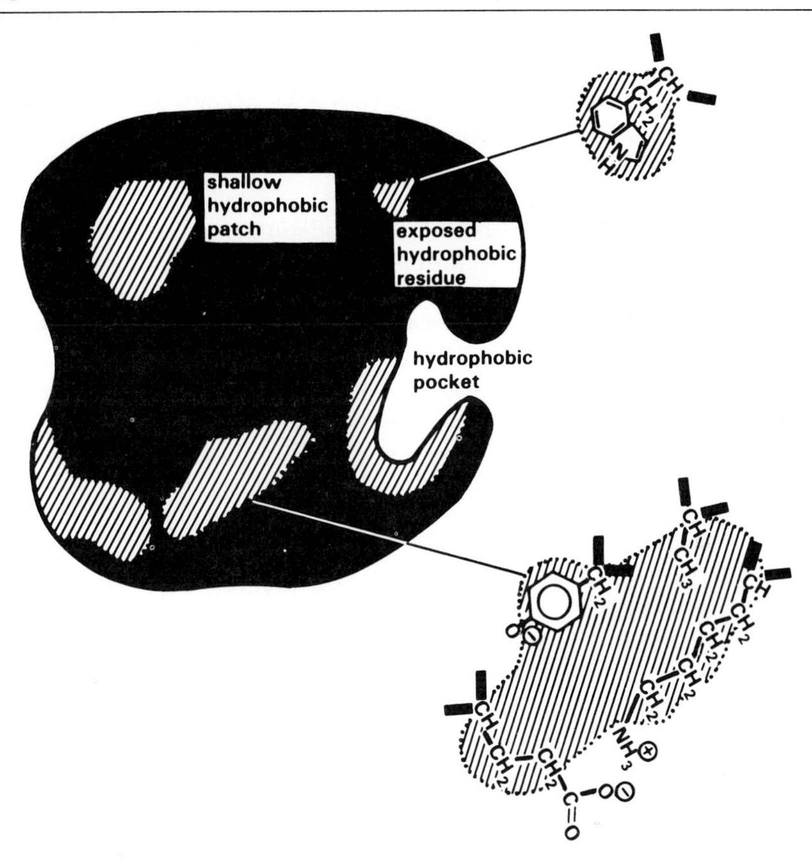

FIG. 13. Schematic representation of a protein molecule with an exposed hydrophobic amino acid residue (tryptophan) and hydrophobic "pockets" or "patches" (hatched). The scheme also illustrates how hydrophobic constituents of several nonhydrophobic (water soluble) amino acids bearing hydrophobic functional groups can together form a site capable of accommodating a hydrophobic hydrocarbon chain.

The Seph-C_2 column can be regarded, in this case, as a very close control for the Seph-C_4 column, and the dramatic difference in retention power has to be associated with the fact that each hydrocarbon chain in Seph-C_4 is two carbon atoms longer than each hydrocarbon chain in Seph-C_2.

A further illustration of the importance of the length of the hydrocarbon chain for retention is the observation that the tightness of binding of a given protein usually, but not always, increases upon increasing the number of carbon atoms in the hydrocarbon chain, as judged from the requirement for more drastic conditions in order to detach the protein. Phosphorylase *b*, for example, can be detached with deforming agents from

$Seph-C_4$, but not from $Seph-C_6$ or $Seph-C_8$, under similar experimental conditions (Fig. 8A).

2. There is no direct correlation between the M_r of a protein and the member in a homologous series of alkylagaroses required to retain it.[11,52] Thus, glutamate synthase (800,000), the P_{IID} protein involved in the regulation of glutamine metabolism (50,000), the uridyl-removing enzyme (130,000), glutamine synthase (600,000), and glutamine synthase adenyl transferase (130,000) are all retained by $Seph-C_5-NH_2$ under a given set of conditions. Moreover, even the order of detachment of the proteins from the column bears no relation to their size.[52]

3. An increased capacity of a specific column material need not necessarily coincide with a parallel increase in charge density. In a series of $Seph-C_4$ columns varying in charge density and in hydrocarbon chain density, the increase in capacity for glycogen phosphorylase *b* was roughly proportional to the increase in chain density, rather than to the increase in charge density.[38] For example, upon raising the charge density about 2.5-fold and the chain density 11.1-fold, the capacity of the column for the enzyme increased about 10-fold. Similarly, whereas a $Seph-C_4$ column retained glycogen phosphorylase *b* (~10 mg/ml of settled column material), a DEAE-Sephadex column with a 5-fold higher charge density retained, under the same conditions, a negligible amount of the enzyme.

4. Both negatively and positively charged proteins may, but need not, be retained on alkylagaroses, which are known to be positively charged at neutral pH. For example, lysozyme (p*I* 10.5–11.0) is increasingly retarded on the $Seph-C_n$ series ($n > 5$) as the pH of the eluent is lowered (Fig. 13 in Shaltiel *et al.*[41]). Similarly, the calf thymus histone H3 (p$I > 9.0$) is retained on $Seph-C_9$ at pH 4 and low salt concentration.[57] On the other hand, the histidine-binding protein *J*, which is negatively charged at neutral pH (p*I* 5.5) is excluded at pH 7.2 by all $Seph-C_n-NH_2$ columns tested ($n = 4–12$), even at low ionic strength.[21] It seems, therefore, that in order to achieve retention of a protein on alkylagaroses, conditions should be found under which hydrophobic regions in the protein become available for interaction with the hydrocarbon side chains of these columns.

5. Increasing the number of charged groups per hydrocarbon chain often decreases the retention of an oppositely charged protein. For example, side chains four carbon atoms long suffice to retain glycogen phosphorylase *b* in the alkylagarose series, whereas chains six carbon atoms long are needed for retention of this enzyme (under the same conditions) in the ω-aminoalkyl series, which, having an additional $—NH_3^+$ group at the tip of the chains, should have been more effective in binding a protein

[57] H. A. Arfmann and S. Shaltiel, *Eur. J. Biochem.* **70,** 269 (1976).

with a negative net charge at neutral pH. These results are not surprising, however, if we keep in mind that by implanting a charged group at the tip of the hydrocarbon chains we not only allow the column to participate more efficiently in ionic interactions with the negatively charged protein, but also reduce the hydrophobic character around this ionized group.

6. Upon comparing the adsorption profiles of 15 protein bands on 2 types of Seph-C_n columns, 1 (type I, Fig. 2) having a net positive charge (due to protonation of the substituted isourea linkages) and the other (type II, Fig. 2) devoid of electrical charge, it was found[34] that companion columns of the 2 types retained the same set of proteins and to the same extent (Figs. 6 and 7). This would suggest that the contribution of the charges in the type I columns is practically negligible and that hydrophobic interactions contribute dominantly to the retention and discrimination power.[10,11,16]

One of the questions, which has been the subject of some controversy, has to do with the relative contribution of hydrophobic and ionic interactions during chromatography on Seph-C_n prepared by the CNBr procedure (type I, Fig. 2). The major argument brought in favor of a considerable involvement of ionic interactions in this process[18,58] was based on the fact that it is sometimes possible to obtain an elution of proteins by increasing the ionic strength or lowering the pH of the irrigating buffer.

Generally speaking, the conditions required for elution of a molecule can provide an indication as to the type of interaction involved in its retention. This is simple and clear cut when dealing with a rigid molecule, as to which it is justifiable to assume that the structure of the molecule is not affected by the change in conditions introduced for the purpose of elution. However, when dealing with proteins that do not have a rigid structure and which are known to readily undergo conformational changes as a result of a change in their environment, such assumptions cannot be made, and the mechanism of adsorption cannot be deduced from the conditions required for elution.

It is now established that the native conformation of proteins is maintained by, and is therefore dependent on, a large number of intramolecular interactions (ionic, hydrophobic, hydrogen bond, etc.). Therefore, on changing the environment of a protein it is impossible to avoid perturbation resulting from such interactions and a subsequent change in its conformation. Numerous reports are available describing pronounced conformational changes caused by moderate modulations of the above-mentioned parameters. The free catalytic subunit of rabbit muscle cAMP-dependent protein kinase, to cite one example, undergoes a distinct

[58] L. Hammar, S. Påhlman, and S. Hjertén, *Biochim. Biophys. Acta* **403,** 554 (1975).

conformational change (detected by a 4.5-fold *increase* in the reactivity of one sulfhydryl and a concomitant 3.8-fold *decrease* in the reactivity of another sulfhydryl in the enzyme)[59,60] when the ionic strength is increased from 0.03 M to only 0.22 M. Thus, the effect observed on changing the eluent will be the net result of the influence of this change on the conformation of the protein (possibly also of the column material), and on the interactions between the hydrocarbon chains of the column and the hydrophobic sites on the protein surface. Because of such phenomena, an increase in ionic strength may sometimes promote retention, sometimes promote elution, or have no effect at all.

The experiment depicted in Figs. 6 and 7 clearly shows that structurally comparable Seph-C_n columns, charged and noncharged, display a remarkable qualitative and quantitative similarity in their adsorption and resolution properties. Furthermore, when an increase in ionic strength enhances the binding of a protein to a Seph-C_n column, the enhancement occurs whether the column is charged (type I) or neutral (type II), and it occurs to the same extent (cf. Fig. 8 in Halperin *et al.*[34]). Similarly, altering the pH of the irrigating buffer has identical consequences for the retention properties of a Seph-C_n column, whether or not it is charged.[34,61] These results lead to the suggestion[10,11,16] that the mechanism of action of Seph-C_n columns involves mainly hydrophobic interactions between the hydrocarbon chains offered by the columns and accessible hydrophobic crevices or patches on the surface of proteins.

Acknowledgments

Part of the work described in this chapter was supported by grants from the US–Israel Binational Science Foundation (No. 3131) and from the Israeli Academy of Sciences (No. A-9).

[59] A. Kupfer, J. S. Jiménez, and S. Shaltiel, *Biochem. Biophys. Res. Commun.* **96,** 77 (1980).
[60] J. S. Jiménez, A. Kupfer, V. Gani, and S. Shaltiel, *Biochemistry* **21,** 1623 (1982).
[61] S. Shaltiel and G. Halperin, *Proc. FEBS Meet.* **52,** 441 (1979).

[4] Affinity Chromatography on Immobilized Dyes

By CHRISTOPHER R. LOWE and JAMES C. PEARSON

The development of group-specific or general-ligand affinity adsorbents in which the immobilized ligand interacts specifically and reversibly with a wide range of complementary proteins has proved to be a valuable addition to the techniques available for enzyme purification.[1-3] In recent years, some of the more widely used ligands for application in this type of affinity chromatography have been a variety of reactive triazine-based textile dyes immobilized to agarose and other matrices.[3-8] In particular, immobilized Cibacron Blue F3G-A appears to be an especially effective adsorbent for the purification of pyridine nucleotide-dependent oxidoreductases, phosphokinases, coenzyme A-dependent enzymes, hydrolases, acetyl-, phosphoribosyl-, and aminotransferases, RNA and DNA nucleases and polymerases, restriction endonucleases, synthetases, hydroxylases, glycolytic enzymes, phosphodiesterases, decarboxylases, sulfohydrolases, a number of blood proteins including serum albumin, clotting factors, serum lipoproteins, transferrin, and complement factors, and a host of other seemingly unrelated proteins including interferon and phytochrome.[3,6,9] Other triazine dyes such as Procion Red H-E3B, Procion Red H-8BN, Procion Yellow MX-8G, Procion Scarlet MX-G, Procion Green H-4G, and Procion Brown MX-5BR have also been found to be suitable ligands for the selective purification of individual pyridine nucleotide-dependent dehydrogenases, phosphokinases, plasminogen, carboxypeptidase G2, alkaline phosphatase, and L-aminoacyl-tRNA synthetases.[8,10-18]

[1] C. R. Lowe and P. D. G. Dean, *FEBS Lett.* **14,** 313 (1971).

[2] K. Mosbach, H. Guilford, R. Ohlsson, and M. Scott, *Biochem. J.* **127,** 625 (1972).

[3] C. R. Lowe, *in* "An Introduction to Affinity Chromatography: Laboratory Techniques in Biochemistry and Molecular Biology" (T. S. Work and E. Work, eds.). North-Holland Publ., Amsterdam, 1979. See also Vol. 34 of this series.

[4] R. L. Easterday and I. M. Easterday, *Adv. Exp. Med. Biol.* **42,** 123 (1974).

[5] S. T. Thompson, K. H. Cass, and E. Stellwagen, *Proc. Natl. Acad. Sci. U.S.A.* **72,** 669 (1975).

[6] P. D. G. Dean and D. H. Watson, *J. Chromatogr.* **165,** 301 (1979).

[7] C. R. Lowe, D. A. P. Small, and A. Atkinson, *Int. J. Biochem.* **13,** 33 (1981).

[8] J. Baird, R. F. Sherwood, R. J. G. Carr, and A. Atkinson, *FEBS Lett.* **70,** 61 (1976).

[9] E. Gianazza and P. Arnaud, *Biochem. J.* **201,** 129 (1982).

[10] Y. D. Clonis and C. R. Lowe, *Biochim. Biophys. Acta* **659,** 86 (1981).

[11] A. J. Turner and J. Hryszko, *Biochim. Biophys. Acta* **613,** 256 (1980).

[12] D. H. Watson, M. J. Harvey, and P. D. G. Dean, *Biochem. J.* **173,** 591 (1978).

METHODS IN ENZYMOLOGY, VOL. 104

ISBN 0-12-182004-1

The use of synthetic dyes such as Cibacron Blue F3G-A and dyes of the Procion series as ligands for affinity chromatography offers several advantages over the more conventional immobilized coenzyme and other biological group-specific media. For example, the protein-binding capacities of immobilized dye adsorbents exceed those of natural biospecific media by factors of 10 to 100 and the low capital cost, general availability, and ease of coupling to matrix materials represent major advantages of dyes for large-scale affinity chromatography.[7,19] Furthermore, synthetic dyes are largely resistant to chemical and enzymic degradation and the triazine bond, an essential feature of many reactive dyes, is less prone to solvolysis than the isouronium linkage introduced during CNBr activation of polysaccharides.[3,20] In addition, the characteristic spectral properties of the dyes permits ready monitoring of ligand concentrations and identification of column materials.[7,21] Finally, the general applicability and reusability of the adsorbents together with the ease of elution of bound proteins, often with good recoveries, ensures excellent prospects for the further exploitation of affinity chromatography on immobilized dyes. This chapter describes the chemical nature of reactive dyes, procedures for their immobilization to matrix materials and the subsequent use of the dyed adsorbents in affinity chromatography and related techniques. In addition, advice is offered on the selection of the appropriate dye and matrix, and recommendations are made as to operational parameters for the adsorption and selective desorption of proteins from immobilized dyestuffs in order to achieve optimal purification or resolution of the desired proteins.

Structure and Chemistry of Reactive Dyes

The reactive triazine dyes exploited in the purification of a host of proteins by affinity chromatography were originally developed at Imperial Chemical Industries (ICI) in the 1950s and were destined primarily for the textile industry. The Procion range of reactive dyes is composed of

[13] C. R. Lowe, M. Hans, N. Spibey, and W. T. Drabble, *Anal. Biochem.* **104,** 23 (1980).
[14] V. Bouriotis and P. D. G. Dean, *J. Chromatogr.* **206,** 521 (1981).
[15] Y. D. Clonis, M. J. Goldfinch, and C. R. Lowe, *Biochem. J.* **197,** 203 (1981).
[16] C. J. Bruton and A. Atkinson, *Nucleic Acids Res.* **7,** 1579 (1979).
[17] P. Hughes, C. R. Lowe, and R. F. Sherwood, *Biochim. Biophys. Acta* **700,** 90 (1982).
[18] E. E. Farmer and J. S. Easterby, *Anal. Biochem.* **123,** 373 (1982).
[19] M. D. Scawen, J. Darbyshire, M. J. Harvey, and A. Atkinson, *Biochem. J.* **203,** 699 (1982).
[20] C. R. Lowe, *Int. J. Biochem.* **8,** 177 (1977). See also Wilchek *et al.*, this volume [1].
[21] Y. Hey and P. D. G. Dean, *Chem. Ind.* (*London*) **20,** 726 (1981).

various polysulfonated chromophores in a comprehensive range of shades linked either to reactive dichlorotriazinyl functional groups by an aminoether bridge (Procion MX dyes) or to the less reactive monochlorotriazinyl group (Procion H, HE, or P dyes). Dichlorotriazinyl dyes are readily prepared from chromophores containing a primary or secondary amino group by reaction with an equimolar proportion of cyanuric chloride (1,3,5-*sym*-trichlorotriazine) for 1–2 hr at pH 6–7 and 0–5°. Monochlorotriazinyl dyes are conveniently prepared by replacing the second chlorine of the triazinyl ring of dichlorotriazinyl dyes by a nonlabile substituent such as an alkoxy, aryloxy, amino, arylamino, or *o*,*m*,*p*-sulfoanilino group at pH 6–7 and 30–40°. An alternative, and often preferred, procedure is to introduce the nonlabile substituent into the triazine ring first, whence the resulting substituted cyanuric chloride is condensed with an appropriate chromophoric base.[22]

Chlorotriazine dyes encompass a complete range of shades in both the dichlorotriazine (Procion MX) and monochlorotriazine (Procion H, HE, or P) series. The reactive dyes available commercially from ICI[23] are derived from three principal types of chromophore: azo, anthraquinone, and phthalocyanine. The anthraquinone chromophore is responsible for bright blue shades with absorption maxima in the range of 600–630 nm, the phthalocyanines producing bright turquoise shades. Green dyes are derived from mixed chromophores of the anthraquinone-stilbene, anthraquinone-azo, or phthalocyanine-azo classes, whereas the majority of the yellow, orange, and red with absorption maxima in the range 380–450 nm are derived from the azo class. Rubine, violet, blue, navy, deep brown, or black dyes are invariably composed of 1 : 1 or 1 : 2 metal complexes of *o*,*o*′-dihydroxyazo or *o*-carboxy-*o*′-hydroxyazo chromophores. Figure 1 illustrates the structures of representative dyes. Structures of the majority of the commercial dyes are undisclosed, and therefore this figure includes only examples of dyes to be found in the public domain and patent literature.

[22] The chemistry and detailed preparation procedures for a number of reactive dyes are described in "The Chemistry of Synthetic Dyes" (K. Venkataraman, ed.), Vol. 6. Academic Press, New York, 1972.

[23] The chlorotriazine dyes mentioned in this chapter are available from Imperial Chemical Industries p.l.c. (Organics Division, P.O. Box 42, Hexagon House, Blackley, Manchester, M9 3DA, U.K.) under the trade name Procion. Other reactive dyes based on 2,4,5-trichloropyrimidinyl groups are available from Ciba-Geigy and Sandoz, respectively, under the commercial names Reactone and Drimarene; Bayer markets dyes based on the 2,3-dichloroquinoxaline reactive group (Levafix E dyes) and on alkylsulfonyl-substituted pyrimidines (Levafix P range); Hoechst produces a range of Remazol dyes containing vinyl sulfone, sulfatoethyl sulfone, or β-chloroethylsulfone groups. The relative reactivities of different commercial ranges of reactive dyestuffs are given in J. A. Fowler and W. J. Marshall, *J. Soc. Dyers Colourists* **80,** 858 (1964).

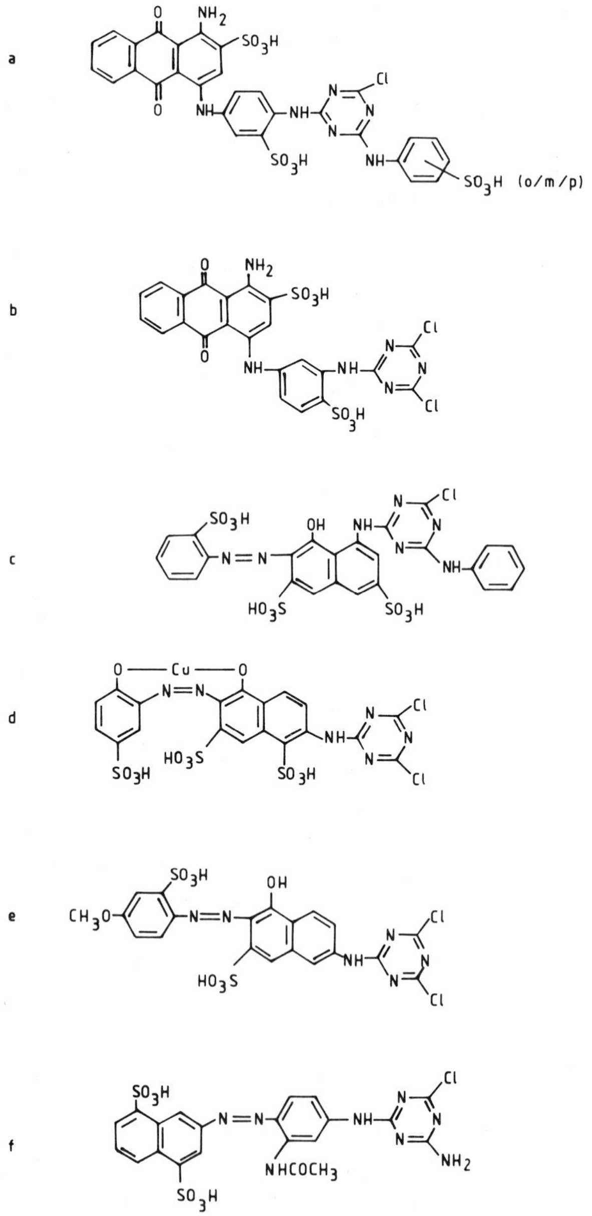

FIG. 1. Structures of several representative triazine dyes: (a) Cibacron Blue F3G-A; *o*-isomer: C.I. 61211; Procion Blue H-B; *m*, *p*-isomers: C.I. 61211. (b) Procion Blue MX-R: C.I. 61205. (c) Procion Red H-3B: C.I. 18159. (d) Procion Rubine MX-B: C.I. 17965. (e) Procion Scarlet MX-G: C.I. 17908. (f) Procion Yellow H-A: C.I. 13245.

Reactive dye nomenclature is generally based on the commercial name and Colour Index (C.I.) generic name and number,[24] although some confusion may arise when a dye is synthesized by both of the major manufacturers, ICI and Ciba. For example, the yellow azoic dye, reactive yellow 3 (C.I. 13245), may also be referred to as Procion Yellow H-A (ICI) or Cibacron Yellow RA (Ciba). The use of different isomeric substituents in dye manufacture can create further confusion: although the anthraquinone dye Cibacron Blue F3G-A (Ciba) contains an *o*-aminobenzenesulfonate substituted on the triazine ring (Fig. 1) and the equivalent product from ICI, Procion Blue H-B, a meta and para mixture of isomers, both products share the same Colour Index generic name (reactive blue 2) and number (C.I. 61211). Some caution should be exercised, therefore, in the interpretation of data obtained with dyes from different commercial sources.

Purification and Properties of Reactive Dyes

Commercially supplied dye powders usually contain a number of additives to ensure adequate operational storage and handling properties. Thus, dye powders incorporate a buffer such as phosphate, since the stability of reactive chlorotriazine dyes is maximal at pH 6.4. Diluents, such as sodium chloride, and traces of surface-active agents are also generally added to standardize the shade in different dye batches and to produce a nondusty product, respectively. The crude dye powder may be washed on a sintered funnel with diethyl ether prior to dissolution in distilled water and precipitation of the dye with ethanol, acetone, or, as the K^+ salt, with potassium acetate to remove some of these contaminants.[7,25] Virtually all of the commercially available dyes are chromophorically heterogeneous when examined by thin-layer chromatography[17,26,27] or high-performance liquid chromatography.[7,28] The principal

[24] "Colour Index," 3rd Ed. The Society of Dyes and Colourists, Bradford, U.K., 1971.

[25] A. Atkinson, J. E. McArdell, M. D. Scawen, R. F. Sherwood, D. A. P. Small, C. R. Lowe, and C. J. Bruton, *in* "Affinity Chromatography and Related Techniques" (T. C. J. Gribnau, J. Visser, and R. J. F. Nivard, eds.), p. 399. Elsevier, Amsterdam, 1982.

[26] P. Hughes, R. F. Sherwood, and C. R. Lowe, *Biochem. J.* **205,** 453 (1982).

[27] J. E. C. McArdell, A. Atkinson, and C. J. Bruton, *Eur. J. Biochem.* **125,** 361 (1982).

[28] A typical high-performance liquid chromatography system would comprise chromatography on a C_{18} Bondapak (5-μm) column at 1000 psi (70 MPa) and 1 ml min^{-1} in an isocratic system, such as water : acetonitrile : acetic acid : triethylamine (65 : 35 : 1.0 : 0.5, by volume) (unpublished observations).

chromophores may also be purified by conventional column chromatography on silica gel,[29] cellulose,[30] alumina,[27] or Sephadex LH-20.[7,28,31]

A typical procedure currently used in our laboratory for the routine purification of the triazine dyes Procion Red H-8BN, Procion Yellow H-A, Procion Green H-4G, and Cibacron Blue F3G-A by preparative thin-layer chromatography[17,28] is performed on kieselgel 60 plates (20 × 20 cm; 2-mm thickness; E. Merck, Darmstadt, FRG). The plates are dried at 80°, and crude dye (20 mg) in deionized water (0.5 ml) is loaded at 20–22° as a narrow band across the base of the plate. Ascending chromatography in 2-butanol–1-propanol–ethyl acetate–water (20 : 40 : 10 : 30 by volume) for Cibacron Blue F3G-A and Procion Yellow H-A or 2-butanol–1-propanol–ethyl acetate–water (20 : 35 : 10 : 35 by volume) for Procion Green H-4G and Procion Red H-8BN is performed overnight. The principal dye band is carefully scraped from the plate and suspended in deionized water; silica gel is removed by filtration through Whatman No. 1 filter paper, and the filtrate is lyophilized. Purified dyes should be stored dry after lyophilization in sealed vials protected from light.

Applications in affinity chromatography do not normally require highly purified dyes, as most of the minor chromophoric impurities do not covalently attach to support matrices. For analytical applications, however, or where precise quantitative information on dye–protein interactions is required, the exclusive use of purified preparations is recommended. One unfortunate consequence of the impurity of commercial dye powders and batch-to-batch variation is that either definitive molar absorption coefficients are not available or the literature contains widely different values for the same supposed dye. Table I lists the spectral properties of some of the more commonly exploited triazine dyes.

Immobilization of Reactive Dyes to Matrices

Chlorotriazine dyes have been covalently attached to a variety of support matrices including agarose,[3,8–18] Sephadex,[4,32,33] beaded cellulose,[34] metal oxides,[35] polyacrylamide,[36] Sephacryl S-200,[6] Spheron,[6]

[29] B. H. Weber, K. Willeford, J. G. Moe, and W. Piszkiewicz, *Biochem. Biophys. Res. Commun.* **86,** 252 (1979).

[30] R. A. Edwards and R. W. Woody, *Biochem. Biophys. Res. Commun.* **79,** 470 (1977).

[31] L. A. Haff and R. L. Easterday, *in* "Theory and Practice in Affinity Techniques" (P. V. Sundaram and F. Eckstein, eds.), p. 23. Academic Press, New York, 1978.

[32] G. Sand, H. Brocas, and C. Erneux, *Acta Endocrinol.* **82,** S204 (1976).

[33] F. Qadri and P. D. G. Dean, *Biochem. J.* **191,** 53 (1980).

[34] D. Mislovicova, P. Gemeiner, L. Kuniak, and J. Zemek, *J. Chromatogr.* **194,** 95 (1980).

[35] C. R. Lowe, *FEBS Lett.* **106,** 405 (1979).

[36] H.-J. Böhme, G. Kopperschläger, J. Schulz, and E. Hofmann, *J. Chromatogr.* **69,** 209 (1972).

TABLE I
PROPERTIES OF SOME TRIAZINE DYES

Reactive dye	M_r (Na salt)	λ_{max} (nm)	Molar absorption coefficient (M^{-1} cm^{-1})
Procion Yellow H-A	578.5	385	8,900
Procion Red H-E3B	—	522	30,000[a]
Procion Brown MX-5BR	588.2	530	15,000
Procion Red H-8BN	801.2	546	21,300
Cibacron Blue F3G-A	773.5	610	13,600
Procion Blue MX-R	635.9	600	4,100
Procion Green H-E4BD	—	630	20,800
Procion Green H-4G	1760.1	675	57,400

[a] A molar absorption coefficient of 11,300 M^{-1} cm^{-1} is quoted in the product literature of the Amicon Corporation, Lexington, MA.

glass,[37] microparticulate silica,[38–40] and agarose–polyacrylamide (Ultrogel) copolymers.[6] Of these, agarose would appear to be the best general-purpose matrix at present, although the cross-linked dextran Sephadex G-200 has been reported to give better capacities for proteins of relatively low molecular weight.[8,33] Triazine dyes may be covalently attached to support matrices either directly by reaction with hydroxyl groups on the matrix backbone or indirectly with spacer molecules or dextran–dye conjugates. However, since comparisons of the relative properties of directly and indirectly coupled triazine dye affinity adsorbents have given contradictory results, the ease and simplicity with which triazine dyes may be coupled directly to polysaccharides makes this approach the universally accepted procedure. The following protocol has been found to work consistently well in our hands.[13,15,17]

Sepharose 4B (5 g) is thoroughly washed on a sintered-glass funnel with 1–5 liters of distilled water under reduced pressure. Surplus water is removed by gentle suction, and the moist gel cake is transferred to a stoppered glass vial containing dye (50 mg) in distilled water (5 ml). The gel suspension is tumbled for 30 min, NaCl solution (22% w/v, 1 ml) is added to give a final concentration of 2% (w/v) NaCl,[41] and the gel suspen-

[37] P. A. Anderson and L. Jervis, *Biochem. Soc. Trans.* **6,** 263 (1978).

[38] C. R. Lowe, M. Glad, P. O. Larsson, S. Ohlson, D. A. P. Small, A. Atkinson, and K. Mosbach, *J. Chromatogr.* **215,** 303 (1981). See also this volume [2].

[39] D. A. P. Small, C. R. Lowe, and A. Atkinson, *J. Chromatogr.* **216,** 175 (1981).

[40] S. Rajgopal and M. Vijayalakshmi, *J. Chromatogr.* **243,** 164 (1982).

[41] This procedure "salts out" the dye onto the matrix backbone and thereby facilitates preferential reaction between the chlorotriazine ring of the dye and matrix-borne nu-

sion is gently tumbled for an additional 30 min prior to adding solid Na_2CO_3 to a final concentration of 1% (w/v). The gel is tumbled for a period of time that depends on the type of dye and the incubation temperature. Typically, dichlorotriazinyl dyes require 2–4 hr at room temperature at 20°, whereas satisfactory coupling of monochlorotriazinyl dyes may be achieved by incubation for 3–5 days at 20°, for 2 days at 37°, or for 2 hr at 60°.

Other alkalies, such as NaOH, may be substituted for Na_2CO_3 in the above protocol. A final concentration of 0.01 *M* NaOH in the coupling step is optimal for dichlorotriazinyl dyes and 0.05–0.2 *M* for monochlorotriazinyl dyes.[25,42] Under the above conditions the immobilization of dichlorotriazinyl dyes, such as Procion Red MX-2B, is rapid and virtually complete in 2 hr at 25–35°. The reaction with agarose displays a distinct optimum for alkali concentration at 10 m*M*, and at higher concentrations rapid hydrolysis of the second chlorine on the triazine ring ensues and gives a sharp decline in the yield. Residual unreacted chlorines in the coupled dye may be converted to hydroxyl groups by incubating the matrix at pH 8.5 and room temperature for 2–3 days, or to amino groups by reaction with 2 *M* NH_4Cl at pH 8.5 for 4 hr at 20°.[42] Immobilization of monochlorotriazinyl dyes, such as Procion Red H-3B, is slow at room temperature and plateaus after about 3 days. At 60° the reaction rate is 3–4 times that at 20°. The amount of dye immobilized to agarose is relatively insensitive to alkali concentrations in the range of 0.05–0.2 *M* although there is a slight fall at higher concentrations (0.4 *M*). The amount of dye incorporated into the matrix is a linear function of dye concentration up to 2 mg of dye per milliliter; above this value the relationship becomes progressively nonlinear, with concentrations above 4 mg of dye per milliliter resulting in barely detectable increases in incorporation.

The reactivity of the relatively inert monochlorotriazinyl dyes may be enhanced by a factor of four- to eightfold in the presence of certain tertiary amines, hydrazines, tetrazenes, or hydrazones when used in the dyeing solution in quantities of 0.1–10% (w/w) of the dye employed.[43] Pyridine, trimethylamine, *N*,*N*′-dimethylhydrazine, tetramethyltetrazene, and 1,4-diazabicyclo[2.2.2]octane (DABCO) are especially effective auxiliary compounds in this respect.

---

cleophiles rather than solvolysis. Although significant coupling can be achieved in the absence of NaCl, optimal dye coupling requires a salt concentration greater than 0.25 *M*.

[42] A. Atkinson, P. M. Hammond, R. D. Hartwell, P. Hughes, M. D. Scawen, R. F. Sherwood, D. A. P. Small, C. J. Bruton, M. J. Harvey, and C. R. Lowe, *Biochem. Soc. Trans.* **9,** 290 (1981).

[43] D. Hildebrand, *in* "The Chemistry of Synthetic Dyes" (K. Venkataraman, ed.), Vol. 6, p. 361. Academic Press, New York, 1972.

After covalent attachment of the reactive dye, the gel is copiously washed on a sintered-glass funnel with distilled water, 1 *M* KCl, and distilled water again until the washings are free of color. Gels are stored as a moist cake or as a suspension in water at or below 4° in the presence of a small amount of microbial growth inhibitor, such as 0.02% NaN_3.

Determination of Immobilized Dye Concentration

The immobilized dye concentration is best determined by acid hydrolysis of the agarose gel.[13,15,17] Gel sucked moist on a sintered funnel (30 mg) is incubated at 37° with 0.6 ml of 5 *M* HCl for 5 min, whence 2.4 ml of 2.5 *M* sodium phosphate at pH 7.5 is added, and the absorbance of the resulting solution is determined at the λ_{max} of the dye.[44] The dye concentration on the gel is calculated from the known molar absorbance coefficients (Table I) and should typically lie in the range of 1–10 μmol of dye per milliliter (or gram wet weight) of gel. Where the structures of the dyes are undisclosed, it is common practice to quote immobilized dye concentrations as milligrams of dye coupled per milliliter (or gram wet weight) of gel.

Considerations Relating to the Selection of Dyes as Ligands for Affinity Chromatography

There are few guidelines on which to base a rational choice of a dye for a putative protein purification. A number of reactive dyes, including Cibacron Blue F3G-A, Procion Red H-E3B, Procion Blue MX-4GD, Procion Green H-E4BD, and Procion Red H-8BN, have been shown to possess the general ability to bind to a wide range of NAD- and NADP-dependent dehydrogenases[3,6,7,13,17,18,25,42] and aminoacyl-tRNA synthetases.[45] Apart from these examples, however, the task of selecting a suitable immobilized dye remains very much an empirical process involving the testing of a number of dyes for the ability to bind the protein in question. Often it is adequate simply to test one or two of the more commonly used dyes, since these are found to give satisfactory results in a surprisingly large number of cases.

The suitability of a range of dyes may be conveniently tested by establishing a series of small columns, each containing 500 mg moist weight, of

[44] These conditions refer to conventional agarose adsorbents. Cross-linked matrices (Sepharose 4B-CL, 6B-CL) require higher temperatures or longer hydrolysis times to effect dissolution of the gel. In all cases the final solution is buffered at pH 7.5 in order to avoid problems of color change in some dyes at extremes of pH.

[45] C. J. Bruton and A. Atkinson, *Nucleic Acids Res.* **7**, 1579 (1979).

agarose-bound dye equilibrated with an appropriate buffer, in Pasteur pipettes plugged with glass wool.[46] A small sample containing the protein of interest is applied to each column, and the flow is interrupted for 10 min to allow equilibration of the sample with the dye matrix. Nonadsorbed protein is purged with 5–10 ml of equilibration buffer. Elution of specifically bound protein is then effected with 1 M KCl in the same buffer and collected in a single fraction. Both the void and the eluate fractions are assayed for protein and activity-enabling recoveries, purification factors, and capacities to be speedily assessed.[7] The results from such a screening program generally suggest several dyes as being worthy of further study. For example, the dye screen may indicate that a tandem system involving both "negative" and "positive" chromatography steps might be worth considering. In this case, the sample is applied first to a "negative" column that does not bind the protein of interest but adsorbs other proteins. Subsequently, the void volume eluate is applied directly to a second "positive" column. The latter column binds the desired protein and on elution yields purified protein. The preliminary screening operation may also suggest when sequential immobilized dye adsorbents may best be employed[10] to enhance purification.

A number of examples of such dye-screening procedures have been published[10,13,45] although in many cases their value is diminished by insufficient characterization of the dye matrix in terms of the concentration of immobilized dye.[47,48] The concentration of dye immobilized on agarose matrices is quite variable[7,10,13] and must be taken into account when comparing different immobilized dye adsorbents. An alternative and preferred procedure for screening dyes involves the characterization of dye–protein interactions in terms of kinetic inhibition constants or dissociation constants for a range of dyes in free solution.[49,50] As a rule of thumb, there is a general trend of protein–dye dissociation constants (K_d) across the color spectrum. Thus, blue and green dyes tend to bind more tightly to pig heart lactate dehydrogenase ($K_d < 8\ \mu M$) than yellow dyes ($K_d > 36\ \mu M$), and orange and red dyes display intermediate affinities ($10\ \mu M < K_d < 33\ \mu M$).[50] These observations account for the popularity of dyes such as Cibacron Blue F3G-A, Procion Blue MX-4GD, Procion Green H-E4BD,

[46] Disposable polypropylene columns (Econo-columns) comprising a 0.8 × 4 cm column holding up to 2 ml of chromatographic medium and including an integral 10-ml reservoir, porous polyethylene bed support disk, and Luer column tip are available from Bio-Rad Laboratories, Richmond, CA. These products offer an effective replacement for Pasteur pipettes and other such improvised columns.

[47] V. Bouriotis and P. D. G. Dean, *J. Chromatogr.* **206,** 521 (1981).

[48] K. G. McFarthing, S. Angal, and P. D. G. Dean, *Anal. Biochem.* **122,** 186 (1982).

[49] A. Ashton and G. Polya, *Biochem. J.* **175,** 501 (1978).

[50] Y. D. Clonis and C. R. Lowe, *Biochem. J.* **191,** 247 (1980).

TABLE II
REPORTED OPERATIONAL PARAMETERS FOR ADSORPTION OF PROTEINS TO IMMOBILIZED DYES

Condition	Data
Equilibration buffers	
Molarity	5–100 m*M*
pH	5.5–9.0
Composition	Tris, phosphate, acetate, triethanolamine, tricine, HEPES, MOPS, bicarbonate
Temperature	4–50°
Additives	EDTA, KCl, NaCl, NH_4Cl, $(NH_4)_2SO_4$, 2-mercaptoethanol, dithiothreitol, cysteine, thioglycerol, phenylmethylsulfonyl fluoride, urea, sucrose, glycerol, ethylene glycol, detergents, chloramphenicol, amphotericin B
Metal ions	Na, K, Mg, Ca, Sr, Ba, Zn, Cu, Co, Ni, Mn, Fe, Al, Fe, Cr
Sample applied per bed volume	0.03–950 mg/ml

Procion Red H-E3B, and Procion Red H-8BN for dye–ligand chromatography.

Operational Parameters in Immobilized Dye Chromatography

The successful operation of a chromatographic separation on immobilized dyes requires that careful consideration be given to a number of variables, such as sample size, flow rate, buffer composition, pH, ionic strength, and temperature. In this respect, chromatography on dye adsorbents closely parallels the approach used in conventional biospecific affinity chromatography.[3] However, the mixed electrostatic–hydrophobic characteristics of most textile dyes suggest that manipulation of the pH, ionic strength, and temperature of the column irrigants is a worthwhile exercise in improving adsorption and subsequent elution of complementary proteins. Table II lists a selection of conditions that have been used to adsorb proteins to immobilized-dye adsorbents. Generally speaking, irrigant buffers would have molarities in the range of 5–100 m*M*, pH values of 5.5–9.0,[51] and a variety of additives depending on the system in question. The effect of temperature on the adsorption to immobilized-dye adsorbents is equally unpredictable; sometimes raising the temperature may increase affinity in one instance and lower it in another. Neverthe-

[51] The affinity of proteins for immobilized dyes is generally reduced on raising the pH of the equilibrating buffer,[17] although there are exceptions.[21]

TABLE III
REPORTED ELUENTS OF PROTEINS FROM IMMOBILIZED TRIAZINE DYE AFFINITY ADSORBENTS

Method	Data
Nonspecific	
Molarity	0.025–6 *M*
Salts	NaCl, KCl, $CaCl_2$, NH_4Cl, $(NH_4)_2SO_4$
Polyols	Glycerol, ethylene glycol
Chaotropes	Urea, NaSCN, KSCN
Chelating agents	EDTA, 2,2′-bipyridyl, pyridine dicarboxylic acid
Detergents	0.1% (w/v) SDS, Triton X-100
Specific elutants	
Molarity	0.001–25 m*M*
Composition	Coenzymes, nucleotides, polynucleotides, ternary complexes, substrates, dyes
Electrophoretic desorption	—

less, operating temperatures in the range of 0–50° have been exploited; room temperature often gives more satisfactory results than 4° provided that the stability of the protein is not compromised.

Table III lists a number of methods that may be used to desorb bound proteins from immobilized dye adsorbents. Two particularly favored techniques that have been universally exploited in dye chromatography are pH and salt elution. Biomolecules adsorbed to immobilized dye matrices may be eluted by increasing the ionic strength either in steps, pulses, or gradients[9,13,17,52] or by raising the concentration of buffer ions.[45] In addition, many substrates, coenzymes, and other biospecific ligands have been used to elute bound proteins, although the precise conditions selected remain empirical. In some instances, a suitable combination of substrates and coenzymes may elute the desired protein in good yield, whereas the individual components alone may be singularly ineffective. For example, the elution of *Escherichia coli* adenylosuccinate synthetase from immobilized Procion Blue H-B is efficiently executed by quaternary complex formation with IMP, GTP, and L-aspartate, but with much lower yields with individual substrates or with combinations of pairs of substrates.[10]

The excellent durability of both dyes and agarose supports toward chemical and biological degradation permits the immobilized dye columns to be reused many times. However, repeated use of the columns with crude protein extracts can lead to fouling with denatured proteins, lipids, lipoproteins, and other lipophilic materials. Regular rejuvenation of the

[52] P. G. H. Byfield, S. Copping, and W. A. Bartlett, *Biochem. Soc. Trans.* **10,** 104 (1982).

columns with 8 *M* urea in 0.5 *M* NaOH, detergents (1%, w/v, sodium dodecylsulfate or Triton X-100), ethylene glycol (<50%, v/v), or potassium thiocyanate (1–3 *M* KSCN) is recommended. In extreme cases of lipid fouling, washing the gel with a chloroform–methanol solution will generally restore the adsorbent to its original utility.

Examples of the Application of Immobilized Dyes in Affinity Chromatography

The literature abounds with examples in which immobilized dyes have been exploited to purify individual proteins.[3,6,7,53] The following examples have been selected to illustrate specific facets of affinity chromatography on immobilized dyes.

Purification of Inosine 5'-Monophosphate Dehydrogenase from E. coli on Immobilized Procion Yellow MX-8G[13]

A glass microcolumn (5 × 50 mm) packed with 0.5 g (moist weight) of Sepharose 4B-bound Procion Yellow MX-8G (2.1 mg of dye per gram of wet gel; 2.8 μmol of dye per gram of wet gel) is equilibrated at 4° with 20 m*M* potassium phosphate at pH 7.4 containing 15% (v/v) ethylene glycol.[54] A cell-free extract of *E. coli* (0.4 ml) that had been previously dialyzed against 100 volumes of 20 m*M* potassium phosphate at pH 7.0 is applied to the top of the column. The flow is interrupted for 10 min to allow adsorption of the protein; nonadsorbed protein is removed by washing with the equilibration buffer (7.5 ml). Elution is effected with a linear gradient of KCl (0 to 1 *M*; 20-ml total volume) in 20 m*M* potassium phosphate, pH 7.4, containing 15% (v/v) ethylene glycol. Fractions (1.5 ml) are collected at a flow rate of 7.5 ml/hr. Under these conditions, IMP dehydrogenase is quantitatively eluted with a 14-fold overall purification factor to yield a product of approximately 90% purity.

The Large-Scale Purification of E. coli Adenylosuccinate Synthetase Using Consecutive Dye–Agarose Affinity Adsorbents[10]

A cell-free extract from a 40-liter culture of *E. coli* (M_r 1068) is partially purified by adsorption and elution from DEAE-cellulose using a NaCl gradient,[10] and the pooled fractions, containing adenylosuccinate synthetase activity, are immediately precipitated with $(NH_4)_2SO_4$. Dialy-

[53] I. G. I. Pesylakas, O. F. Sudzhyuvene, and A. A. Glemzha, *Appl. Biochem. Microbiol.* **17,** 337 (1981).

[54] Ethylene glycol (15% v/v) is included throughout the column irrigants to reduce nonspecific hydrophobic interactions between the protein and the dyed agarose.

sis against 1000 volumes of 50 mM Tris–HCl buffer, pH 7.5, containing 1.5 mM $MgCl_2$ and 2.0 mM dithiothreitol, yields a preparation showing a 7.6-fold increase in specific activity over the cell-free extract and an overall 90% yield.

The dialyzed preparation (7.1 ml, 4800 units of adenylosuccinate synthetase activity; 284 mg of protein) is applied to a column (1.1 × 16.5 cm; 16 g, moist weight, of gel) of agarose-bound Procion Red H-3B (1.4 μmol of dye per gram, moist weight, of gel), equilibrated with the same Tris buffer, and is used for dialysis. The sample is applied to the column, flow is stopped for 5 min to allow equilibration, and nonadsorbed protein is removed by washing with the Tris buffer (90 ml). Adenylosuccinate synthetase activity is eluted with a linear gradient of KCl (0 to 1 M; 120 ml total volume) and collected at a flow rate of 30 ml/hr. Pooled, active fractions are concentrated to a 2-ml volume through an ULVAC ultrafiltration membrane and dialyzed overnight against 1 liter of the Tris buffer. The dialysate (2.8 ml; 2128 units of enzyme activity; 16.8 mg of protein) is applied to a column (1.1 × 13.5 cm; 12.8 g of gel, moist weight) of agarose-bound Procion Blue H-B (0.8 μmol of dye per gram of gel, moist weight) equilibrated with 50 mM Tris-HCl at pH 7.5 containing 1.5 mM $MgCl_2$ and 2.0 mM dithiothreitol. Flow is interrupted for 5 min, and nonadsorbed protein is removed with 60 ml of the equilibration buffer. Adenylosuccinate activity is eluted with 50 ml of equilibration buffer containing IMP (4 mM), GTP (2 mM), and L-aspartate (10 mM) at a flow rate of 40 ml/hr. Adenylosuccinate synthetase is pooled and concentrated (1.8 ml) to give the final preparation composed of 3.6 mg of protein and 1829 units of activity.

Purification of Yeast Hexokinase by Mg^{2+}-Promoted Binding to Procion Green H-4G[15]

Preliminary studies on the inactivation of yeast hexokinase by a range of triazine dyes showed that, of those tested (Procion Green H-4G, Blue H-B, Turquoise H-A, Turquoise H-7G, Brown H-2G, Green H-E4BD, Red H-E3B, Yellow H-5G, and Yellow H-A), the enzyme was inactivated most rapidly by Procion Green H-4G. Almost complete protection from inactivation was afforded by 20 mM ATP, suggesting that the inactivation by Procion Green H-4G was active-site directed. A most interesting property of this dye was that in the presence of 10 mM $MgCl_2$ the rate of inactivation by Procion Green H-4G dropped to 20% of that recorded in the absence of Mg^{2+}. The effect was unique for this dye and associated with a fivefold decrease in the apparent dissociation constant of the dye–enzyme complex.[15] These unique properties of Procion Green H-4G were exploited in the purification of yeast hexokinase by affinity chromatography on Procion Green H-4G–Sepharose 4B.

A concentrated crude yeast extract (50 mg in 1 ml of 30 mM Tris–HCl at pH 7.5 containing 10 mM $MgCl_2$) is dialyzed overnight against 1 liter of the same buffer at 4°. A sample (400 μl; 24.3 units of hexokinase activity; 10.5 mg of protein) of the extract is applied to a column (10 × 30 mm; 2.5 g, of gel moist weight) of Sepharose 4B-immobilized Procion Green H-4G (2.6 μmol of dye per gram of gel, moist weight) equilibrated with the Tris buffer. The column is washed with about 12 ml of the same buffer. Elimination of $MgCl_2$ from the equilibration buffer elutes approximately 7.5% of the applied hexokinase activity, whereas a pulse (5 ml) of 20 mM ATP–10 mM $MgCl_2$ is required to elute the remaining enzyme activity. The peak fraction eluted after the ATP pulse displayed a specific activity sevenfold greater than that of the crude yeast extract.

Other examples of metal ion-promoted protein adsorption to immobilized triazine dyes have been examined for potential exploitation in affinity chromatography[17,26] and high-performance liquid affinity chromatography.[38,39] In particular, low concentrations of first row transition metal ions such as Zn^{2+}, Co^{2+}, Mn^{2+}, Ni^{2+}, Cu^{2+} and, to a lesser extent, the group IIA ions, Ca^{2+} and Mg^{2+}, promote the binding of proteins to a number of immobilized triazine dye affinity adsorbents.[17] For example, Zn^{2+} promotes binding of carboxypeptidase G2, alkaline phosphatase, and yeast hexokinase to immobilized Procion Red H-8BN, Procion Yellow H-A, and Cibacron Blue F3G-A, respectively. The binding of ovalbumin to immobilized Cibacron Blue F3G-A and Procion Orange MX-G is selectively enhanced in the presence of Al^{3+}. In the case of ovalbumin and alkaline phosphatase, the effect is almost totally specific for the protein, metal ion, and dye, whereas with carboxypeptidase G2 and hexokinase, metal ions such as Co^{2+}, Ni^{2+}, Mn^{2+}, Cu^{2+}, Ca^{2+}, and Mg^{2+} also promote binding to varying degrees; almost all other monovalent and trivalent metal ions appear to be ineffectual. Proteins bound to immobilized triazine dyes by metal ion-promoted interactions can subsequently be eluted with appropriate chelating agents of the amine, aminocarboxylate, or substituted pyridine classes.[17]

The Purification of 3-Hydroxybutyrate Dehydrogenase and Malate Dehydrogenase from Rhodopseudomonas spheroides[7,19]

A cell-free extract from *R. spheroides* is applied to a 1.5-liter column of agarose-bound Procion Red H-3B; 3-hydroxybutyrate dehydrogenase is eluted with 1 M KCl, and malate dehydrogenase is subsequently recovered by including 2 mM NADH in the elution buffer. Pooled and dialyzed fractions containing malate dehydrogenase activity are applied to a column (1 liter) of immobilized Procion Blue MX-4GD, and the enzyme is eluted with a linear gradient of KCl (0 to 700 m*M*; 8 liters) to yield 1 g of homogeneous enzyme in 70% overall yield. The fractions containing 3-hydroxybutyrate dehydrogenase activity eluted from Procion Red H-3B are applied to a column (1 liter) of immobilized Procion Blue MX-4GD and subsequently eluted with 2 m*M* NADH to yield 300 mg of homogeneous enzyme in 80% overall yield.[55]

Other Applications of Triazine Dyes in Biotechnology

Dye columns have been exploited for the removal of contaminants from almost homogeneous protein preparations,[56] especially the removal of serum albumin from samples of IgG,[57] the production of apoenzymes,[58] the resolution of functionally active from inactive proteins[59] and isoenzymes,[38] and the fractionate multienzyme complexes.[60] Reactive dyes bound to microparticulate silica have been extensively used to resolve enzymes and other proteins by analytical[38,39] and preparative[61] high-performance liquid affinity chromatography. Triazine dyes have been used as chromogenic substrates for enzymes[62] and lectins,[63] as ligands for novel electrochemical sensors,[35,64] as histochemical stains and colored protein markers,[7] and as active-site probes and irreversible affinity labels.[15,50,65]

Comments

The three principal characteristics of reactive dyes, namely their ability to bind a plethora of proteins, their spectral properties, and their reactivity with nucleophilic groups, ensure a wide range of applications

[55] This protocol for the purification of 3-hydroxybutyrate dehydrogenase compares favorably with the conventional scheme involving nine separate steps with an overall yield of 9%.

[56] M. Dao, J. Watson, R. Delaney, and B. Johnson, *J. Biol. Chem.* **254,** 9441 (1979).

[57] J. Travis, J. Bowen, D. Tewksbury, D. Johnson, and R. Pannell, *Biochem. J.* **157,** 301 (1976).

[58] S. T. Thompson, R. Cass, and E. Stellwagen, *Anal. Biochem.* **72,** 293 (1976).

[59] P. Menter and W. Burke, *Fed. Proc. Fed. Am. Soc. Exp. Biol.* **38,** 673 (1979).

[60] R. DeAbreu, A. Dekok, and C. Veeger, *FEBS Lett.* **82,** 89 (1977).

[61] D. A. P. Small, A. Atkinson, and C. R. Lowe, *J. Chromatogr.* **266,** 151 (1983).

[62] B. Klein and J. A. Foreman, *Clin. Chem.* **26,** 250 (1980).

[63] B. S. Rathaur, G. S. Khatri, K. C. Gupta, G. K. Narang, and N. K. Mathur, *Anal. Biochem.* **112,** 55 (1981).

[64] C. R. Lowe and M. J. Goldfinch, *Biochem. Soc. Trans.* **11,** 448 (1983).

[65] D. A. P. Small, C. R. Lowe, A. Atkinson, and C. J. Bruton, *Eur. J. Biochem.* **128,** 119 (1982).

for these dyes. The purification of proteins by biospecific affinity chromatography is possibly the broadest and best documented application to date, although their potential in other areas of biotechnology should not be underestimated.[7] This growing interest in the application of triazine dyes in preparative and analytical biotechnology has prompted investigations into the basis of the protein–dye interactions. Thus, it has been suggested that the polyanionic aromatic dye chromophores mimic the overall shape, size, and charge distribution of the naturally occurring biological heterocycles, such as the nucleotides and coenzymes. Indeed, studies by induced circular dichroism,[30,66] X-ray crystallography,[67] and irreversible enzyme inactivation[15,50,65] all confirm this suggestion. However, the available evidence indicates that Cibacron Blue F3G-A and other triazine dyes are not highly specific analogs of nucleotide coenzymes and that there is some degree of latitude in binding reactive dyes to active-site regions of proteins. Nevertheless, reactive dyes clearly have a bright future in preparative and analytical biochemistry.

[66] R. A. Edwards and R. W. Woody, *Biochemistry* **18,** 5197 (1979).

[67] J. F. Beillmann, J.-P. Samama, C.-I. Brändén, and H. Eklund, *Eur. J. Biochem.* **102,** 107 (1979).

[5] Displacement Chromatography of Proteins

By Elbert A. Peterson and Anthony R. Torres

Whereas the separation of substances in elution chromatography, the original and most widely used form of chromatography, depends upon differences among the sample components with respect to their ability to compete with the eluent for sites on the adsorbent, displacement chromatography is based on direct competition among the adsorbed components, themselves, including substances of intermediate affinity that may be introduced as spacers. This competition gives rise to a train of contiguous bands having successively higher affinities for the adsorbent, all being driven through the column at the same velocity by the continuous addition of a solution containing a substance that has the highest affinity in the system. In the classical displacement train, first described by Tiselius[1] for sugars chromatographed on a column of carbon with phenol as the "de-

[1] A. Tiselius, *Ark. Kemi Mineral. Geol.* **16A,** n18 (1943).

ISBN 0-12-182004-1

veloper,'' and observed by others in the displacement chromatography of hydrocarbons,[2] fatty acids,[3] steroids,[4] and amino acids and peptides,[5,6] the separated components emerge as contiguous plateaus, each higher in concentration than the preceding one when plotted on a mass basis. The concentration of each is a reflection of the affinity of that component for the adsorbent under the prevailing conditions and serves to identify the substance. The area under the plateau is proportional to the amount, reducing the determination of quantity to a measure of height and length (concentration and volume). So long as the concentration of the developer remains constant, increasing the amount of any component in the sample increases the length of its plateau but does not affect the concentration at which it emerges.

Displacement chromatography has been applied to separations of such widely different classes of substances as amino acids,[7,8] rare earths,[9] and oligonucleotides[10] on ion exchangers, and it has been used on reversed-phase columns designed for high-performance liquid chromatography to fractionate a variety of organic substances on a preparative scale.[11]

From the beginning, it was recognized that displacement chromatography offered several advantages over elution chromatography: (1) the elimination of tailing caused by convex isotherms, (2) a high capacity resulting from high concentrations attainable in the separated bands and maintained by the band-sharpening that prevented widening of bands once the train had been fully developed, (3) the potential for spacing the components, and (4) the ease of measuring the quantity of each component in analytical applications. The mixing that occurred at the boundaries of the contiguous plateaus was later overcome by including other substances with intermediate affinities to serve as spacers that could readily be removed from the sample components after chromatographic separation.[5,8] Application of these principles has been limited by the problem of finding suitable spacers and developers, although the search is aided in the case of small molecules by the fact that the positions of the components in the train can be determined from their isotherms,[1] as measured on the adsorb-

[2] S. Claesson, *Ann. N.Y. Acad. Sci.* **49,** 183 (1948).
[3] R. T. Holman and L. Hagdahl, *J. Biol. Chem.* **182,** 421 (1950).
[4] J. G. Hamilton, Jr., and R. T. Holman, *Arch. Biochem. Biophys.* **36,** 451 (1952).
[5] A. Tiselius and L. Hagdahl, *Acta Chem. Scand.* **4,** 394 (1950).
[6] J. Porath, *Acta Chem. Scand.* **6,** 1237 (1952).
[7] S. M. Partridge and R. C. Brimley, *Biochem. J.* **51,** 628 (1952).
[8] D. L. Buchanan, *J. Biol. Chem.* **229,** 211 (1957).
[9] F. H. Spedding and J. E. Powell, *in* "Ion Exchange Technology" (F. C. Nachod and J. Schubert, eds.), Chap. 15. Academic Press, New York, 1962.
[10] G. G. Brownlee, F. Sanger, and B. G. Barrell, *J. Mol. Biol.* **34,** 379 (1968).
[11] C. Horvath, A. Nahum, and J. H. Frenz, *J. Chromatogr.* **218,** 365 (1981).

ent to be employed in chromatography. In ion exchange, the dissociation constants of the organic acids and bases serve as a guide.[8]

The situation is quite different in the ion-exchange chromatography of macromolecular polyelectrolytes such as proteins. Under pH and salt conditions that permit adsorption of most of the proteins in a biological mixture, only a few, if any, will be in finite adsorption equilibrium with the adsorbent; the others will be either not adsorbed or so tightly bound by multiple bonds that they will not migrate appreciably in the initial buffer in the absence of displacers. In the latter case adsorption isotherms will not be helpful. Also, the number of different molecular species in a natural sample is likely to be very large, with few, if any, present in quantities sufficient to form plateaus. Nevertheless, all the other advantages of displacement over elution cited above for small molecules apply to the displacement chromatography of proteins, provided suitable systems can be developed to achieve it. Two approaches have been taken to accomplish this: one is the use of the complex mixtures of ampholytes that have been developed for isoelectric focusing, and the other involves the use of polyanions prepared specifically for the displacement chromatography of proteins. In the discussion of these systems and their applications, care will be taken to distinguish elution from displacement.[1] Although the word "elute," according to its dictionary meaning, "to wash away," is a term broad enough to cover any chromatographic process, its use will be confined to the process in which the sample components compete with eluent molecules, which have lower affinity for the adsorbent than their own, a process that has been historically associated with the word "elution." When, however, substances are caused to migrate along the column because of competition with substances having affinities for the adsorbent that are greater than their own, the word "displace" will be used. At the same time, it must be kept in mind that both mechanisms can operate in different parts of the chromatogram. If the concentration of the developer is so low that the developer front moves more slowly than the rates at which low-affinity components are eluted by the background salt, these will move ahead as eluted peaks.

Ampholyte Displacement Chromatography

Leaback and Robinson[12] introduced a procedure in which the ampholytes developed for isoelectric focusing are used as displacers of proteins on ion exchangers. By this means they were able to separate two forms of the B variant of pig epididymal β-*N*-acetyl-D-hexosaminidase that could

[12] D. H. Leaback and H. K. Robinson, *Biochem. Biophys. Res. Commun.* **67**, 248 (1975).

not be resolved by isoelectric focusing. An excellent separation was obtained when 1 ml of the partially purified enzyme preparation was applied to a 45 × 11 mm column of CM-cellulose (Whatman CM-52) that had been equilibrated at 5° with 0.02 *M* Tris-HCl at pH 7.6, and 4% (w/v) of pH 8–10 Ampholine (LKB) was passed through. Efforts to resolve the two forms by elution from a similar column with a shallow gradient of NaCl were unsuccessful. A gradient of increasing pH would have been more appropriate for this comparison in view of the fact that the first isoenzyme emerged in the steep pH gradient imposed by the Ampholine solution. These authors concluded that the two proteins were displaced selectively by amphoteric ions having closely similar isoelectric points, and this explanation has largely been accepted by others who have used similar procedures in a variety of applications.[12–19]

Figure 1 shows one such application reported by Young and Webb,[13] the separation of human serum albumin from α-fetoprotein on DEAE-cellulose (Whatman DE-52) with 4% (w/v) pH 4–6 Ampholine in 10 m*M* sodium phosphate at pH 7.8. The emergence of the α-fetoprotein in a number of peaks and shoulders led the authors to suggest that this protein occurs in multiple forms, as do many serum proteins, frequently based on differences in the content of sialic acid.[20] Although this is likely to be the case, the possibility that this multiplicity might be the result of the formation of complexes between certain ampholytes and the protein should be considered. Since the peaks shown in the figure are based on immunological assays, no determination of the presence or the absence of other protein in the background can be made. Page and Belles-Isles[14] employed a similar procedure as a final step in the purification of mouse α-fetoprotein from hepatoma tissue and obtained a product that gave a single band on electrophoresis after treatment with sodium dodecyl sulfate.

Although the ampholyte solution absorbs too strongly at 280 nm to permit monitoring at that wavelength,[18] absorbance at 415 nm was used by Chapuis-Cellier *et al.*[15] to delineate the profiles of hemoglobin variants (stabilized with KCN) that were separated with 1.2% pH 7–9 Ampholine on DE-52 columns. Relatively smooth pH gradients were generated by passing the Ampholine, adjusted to pH 7.5, 7.8, or 8.0, through columns

[13] J. L. Young and B. A. Webb, *Anal. Biochem.* **88,** 619 (1978).
[14] M. Page and M. Belles-Isles, *Can. J. Biochem.* **56,** 853 (1978).
[15] C. Chapuis-Cellier, A. Francina, and P. Arnaud, *Protides Biol. Fluids* **27,** 743 (1980).
[16] A. Francina, E. Dorleac, and H. Cloppet, *J. Chromatogr.* **222,** 116 (1981).
[17] J. L. Young and B. A. Webb, *Protides Biol. Fluids* **27,** 739 (1980).
[18] J. L. Young and B. A. Webb, *Sci. Tools* **25,** 54 (1978).
[19] J. L. Young, B. A. Webb, D. G. Coutie, and B. Reid, *Biochem. Soc. Trans.* **6,** 1051 (1978).
[20] L. Anderson and N. G. Anderson, *Proc. Natl. Acad. Sci. U.S.A.* **74,** 5421 (1977).

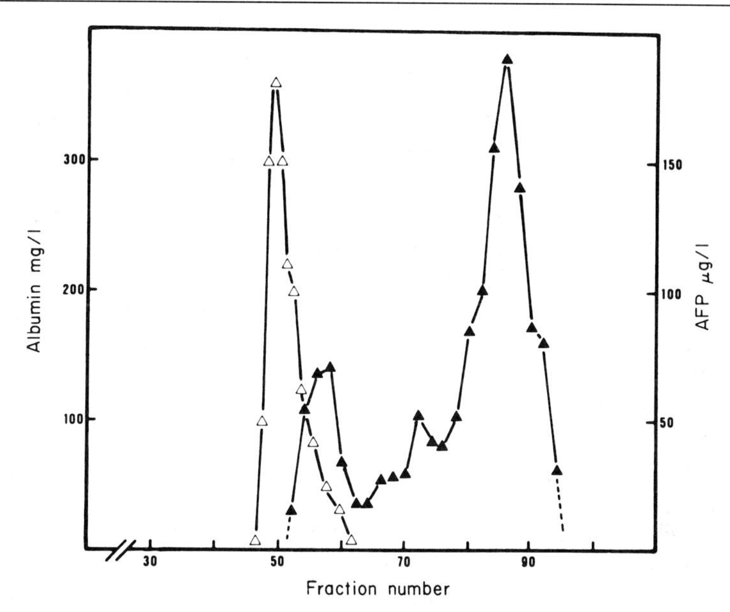

FIG. 1. Ampholyte displacement chromatography of amniotic fluid on DEAE-cellulose. A 1 : 4 dilution of amniotic fluid was dialyzed against 10 m*M* sodium phosphate, pH 7.8, and 3 ml was applied to a 10 × 0.9 cm column of DE-52 equilibrated with the same buffer. Then a 4% (w/v) solution of pH 4–6 Ampholine in the pH 7.8 buffer was pumped into the column at 13.5 ml/hr. Fractions (0.625 ml) were assayed for α-fetoprotein (AFP) (▲) and albumin (△) by immunological methods. From Young and Webb.[13]

equilibrated with 0.05 *M* Tris-HCl at pH 8.9, and the separated hemoglobins migrated within these gradients. Under such conditions the separation process appeared to be essentially chromatofocusing. Francina *et al.*[16] found the relative mobilities of hemoglobin variants that were separated on an anion-exchange paper (Whatman DE-81) to be linearly related to the isoelectric points when either 1.2 or 4% concentrations of pH 6–8 Ampholine (not adjusted in pH) were used, a result compatible with either displacement or elution (including chromatofocusing in its most general sense). However, Young and Webb[17] have shown that when 4 or 6% solutions of Ampholine are used, a region of relatively constant pH is formed in which protein peaks can migrate, with narrower peaks of other proteins appearing where the pH changes sharply at either end of this region. Moreover, they found that the pH did not have to fall to a value at or near the isoelectric point of a protein in order for that protein to be released from the adsorbent. Also, in other work[18] in which the same authors determined the positions of a number of proteins in a chromatogram of maternal serum by immunological means, they observed that not all of these emerged in the order of their isoelectric points. Again, the result is compatible with both displacement and elution mechanisms, although not generally expected in chromatofocusing in its special sense, where salt is absent or very low in concentration. However, all these mechanisms are affected by the presence of background salt, which will cause the release of the protein from an anion exchanger above its isoelectric point, where selection is made on the basis of the number of remaining effective charges. The isoelectric point is merely a condition where positive and negative charges are balanced; it does not reflect differences in density, distribution, or total numbers of charges, all of which affect the number of effective bonds that can be made with the adsorbent at pH values away from the isoelectric point. Since adsorption involves interactions with regions of the protein rather than the entire surface of the molecule, taken as a whole, as is the case in the electrophoretic methods that are used to determine the isoelectric point, one can expect some proteins that have unusual concentrations of like charge in certain regions of their molecules to remain adsorbed at their isoelectric points (in low salt) or even beyond it.

Decreasing the concentration of a displacement developer, increasing the background salt concentration, or altering the pH can convert a displacement process into an elution process, at least in part. Therefore, to identify a displacement mechanism it is necessary to determine, with relatively pure proteins so that plateaus can be formed under suitable circumstances, the shapes of the bands and how they are affected by changes in the sample load and in the concentration of the displacers, as has been done in the case of the carboxymethyldextrans that are described in the next section.

Carboxymethyldextran Displacers

Principle

Any soluble polyanion having a sufficient number of negative charges per molecule to match those offered to a cluster of charges on an anion exchanger by a given protein can be expected to compete with that protein in a displacement train, though differences in the spacing of the charges may give one or the other an advantage. If the polyanion can offer an additional negative charge, it is likely that it will move the protein ahead of it on a column that has sufficient resolving power to discriminate between them. Any series of soluble polyanions that do not interact by nonionic mechanisms with the adsorbent or with each other and have sufficient charge to be adsorbed should form a displacement train based on differences in their ion-exchange affinities for the adsorbent, i.e., on differences in the number of effective negative charges per molecule. It is also reasonable to assume that proteins added to the column will find positions in that train according to their own abilities to compete, as all the components of the train migrate through the column in finite adsorption equilibrium with the adsorbent. Such a system of polyanions can be made up of molecules of uniform molecular weight with varying densities of charge, or of varying molecular weight with uniform density of charge, or molecules that are heterogeneous in both respects. To facilitate the characterization of the polyanions and the control of their affinities, the first of these alternatives was chosen, in principle. Specifically, carboxymethyldextrans (CM-Ds)[21] were selected to constitute the polyanion series because dextran is hydrophilic and is commercially available in relatively narrow ranges of molecular weight, and carboxymethyl groups can be readily attached to it by reaction with monochloroacetate under alkaline conditions. The number of carboxymethyl groups incorporated per average dextran chain (or per unit of mass) can be controlled by limiting the amount of monochloroacetate used, the time of reaction, or the temperature of reaction. The amount of water and alkali in the reaction mixture also affects the incorporation.

Actually, a CM-D that has been prepared under fixed conditions of reaction is, nevertheless, heterogeneous with respect to the affinity of its constituent molecules, and this heterogeneity is very much greater in the low-affinity range than at higher levels of substitution. In addition to the variation in affinity due to heterogeneity in molecular weight, statistical variation in the degree of substitution is to be expected in an incompletely substituted product, and differences in the reactivity of highly branched and sparsely branched dextran chains affect the distribution of the chargeable groups among the chains. Although the chosen starting material is nominally 10,000 in molecular weight, the lowest in the series offered by Pharmacia Fine Chemicals, it contains substantial quantities of dextran smaller than 5000 or larger than 15,000, according to the manufacturer's analysis. The unfractionated product might therefore appear to be in the category of the third alternative: heterogeneous with respect to both molecular weight and density of charge. However, the very high-molecular-weight components that might lead to excessive bonding, high viscosity,

[21] E. A. Peterson, *Anal. Biochem.* **90,** 767 (1978).

and difficulty of migration in gel electrophoresis are excluded. Unfractionated preparations may include so wide a range of affinities that only a small fraction accurately represents the value that is used to characterize the preparation, whatever index of affinity may be employed. Consequently, they are inefficient in spacing proteins that differ only slightly in affinity. However, these widely overlapping products can be used in combination to provide a partial or full-range continuum of affinity, permitting a scan of a mixture of proteins or a preliminary determination of the affinity index of the CM-D that follows or precedes a protein of interest in a displacement train.

Since the ability to space the proteins in the chromatogram with appropriate displacers is a major advantage of displacement chromatography, it is desirable to have at hand a series of preparations each of which represents a narrow segment of the whole spectrum of potentially required affinities, so that proteins with closely similar affinities for the adsorbent can be caused to emerge from the column well separated. Such preparations also make possible the focusing of the resolving power of the system on the narrow range of affinities represented by a single protein of interest and its neighboring impurities.

Preparation of CM-Ds

In early work,[21] a number of preparations of CM-D were produced by mixing monochloroacetic acid, water, dextran, and lithium hydroxide in various proportions and allowing the resulting solution to stand, sealed at room temperature, for periods ranging up to 72 hr. A purification procedure involving precipitation of the CM-D by ethanol provided the solid product, initially in the free-acid form, separated from lithium chloride, glycolic acid, and residual unreacted monochloroacetic acid. Very highly substituted products had to be precipitated with acetone instead of ethanol because of their solubility in the latter. The CM-Ds were finally converted to their potassium salts (which gave better precipitates than the sodium salts) for characterization and storage because long-term storage of the solid in its free-acid form at 4° was found to produce changes in its solubility, apparently as the result of cross-linking by ester bonds.

This method of synthesis, which has the advantage of requiring only the simplest of laboratory equipment and is relatively efficient in the use of reactants, is still employed to prepare the relatively high-affinity CM-Ds that serve as final displacers. At their high level of substitution, the original product is usually sufficiently homogeneous for this purpose. However, CM-Ds that are to be used as spacers are now made by a reaction procedure that first produces a series of preparations that represent overlapping segments of a continuum of affinities, populated with molecules at every level of substitution below a chosen limit. These are later fractionated by displacement chromatography in water on DEAE-cellulose to obtain a series of narrow-range CM-Ds that are suitable even for spacing proteins that have closely similar affinities.

The initial synthesis is carried out under conditions that produce an approximately linear time vs reaction curve,[21] the reaction mixture being withdrawn continuously[22] or at short intervals[23] to receivers that contain acid to stop the reaction. Thus, several separate preparations are produced that together compose a distribution of molecules that have been exposed to progressively longer reaction periods. After purification by alcohol precipitation, these are fractionated by displacement chromatography on a DEAE-cellulose column. In order to avoid the disturbing effects of the rather high concentrations of displaced counterion that are developed when high concentrations of CM-D are used,[21] the adsorbent in its free-base form is titrated in water with the free-acid form of the CM-D preparation that has the lowest level of substitution. This results in the binding of an amount of CM-D that is comparable to the dry weight of the adsorbent itself. This complex of adsorbent and low-affinity CM-D is packed into a column, and solutions of the potassium salts of CM-Ds having progressively higher affinity are subsequently pumped in. The CM-D is the only anion, provided the preparations employed are free of chloride and organic acids.

Figure 2 illustrates the shape of the refractive index profile of the CM-D after such a fractionation on DEAE-cellulose (Whatman DE-23). It is essentially the mass profile of a CM-D displacement train in water, demonstrating that, in the absence of competing salt, the capacity of DEAE-cellulose for CM-D increases as the affinity of the CM-D decreases, i.e., as the number of carboxyl groups on the dextran molecule decreases. This is attributable to the fact that a larger mass of dextran must accompany the carboxyl groups that are necessary to saturate the adsorbent with low-affinity CM-D.[21] There is a high degree of conformational accommodation of both the charged adsorbent chain and the CM-D to the spacing of charges on the other, leading to a very high efficiency in the matching of charges on the CM-D with those on the adsorbent. Since some of the components of the train were applied after earlier, low-affinity ones had emerged, each period during which a different preparation was being pumped in represented a segment of the train in which a different CM-D with a different affinity was serving as a "final displacer."

[22] E. A. Peterson and A. R. Torres, *Anal. Biochem.* **130,** 271 (1983).

[23] A. R. Torres and E. A. Peterson, *J. Biochem. Biophys. Methods* **1,** 349 (1979).

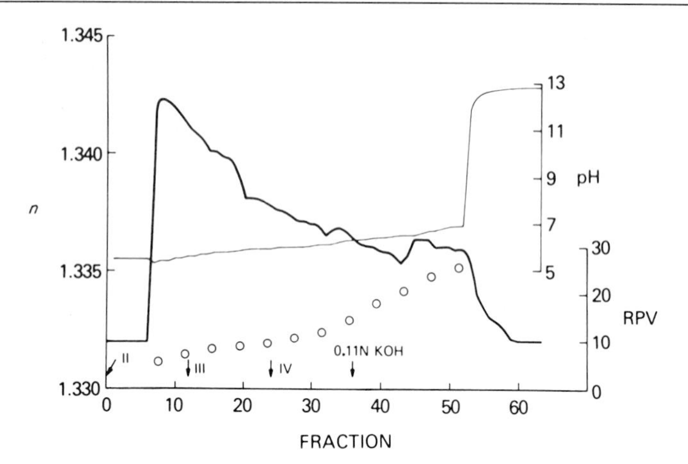

FIG. 2. Fractionation of carboxymethyldextran (CM-D) preparations into narrow-range CM-D by displacement chromatography. A water slurry of DE-23 in the free-base form was adjusted to pH 5.6 with a salt-free 6% solution of CM-D (RPV = 9) in the free-acid form, and this suspension was packed into a column to form a 115-ml bed. Water solutions of the potassium salts of CM-D preparations were then pumped successively into the column at 50 ml/hr: about 150 ml each of 6% II (RPV = 14), 3.6% III (23), and 3% IV (27). CM-D remaining on the column was displaced by 0.11 N KOH. The volume of each fraction was 12.5 ml, but pairs were pooled before determination of the reciprocal of the pellet volume (RPV). The heavy line represents the refractive index (n); the thin line, the pH; and the circles, RPV. From Peterson and Torres.[22]

Therefore, the successive preparations of CM-D were applied at concentrations that would drive the various segments at approximately the same rate in order to minimize changes in the shape of the train. The raised plateau at the end of the train resulted from the use of a concentration of KOH that was about 20% too high in displacing the final CM-D from the column.

Recovery of the fractionated CM-D, all of which emerged in the form of its potassium salt, was quantitative: 34 g of CM-D was applied and 34 g was recovered, including 2.6 g of unadsorbed material and 1.2 g of high-affinity CM-D that was recovered in the alkaline end fractions after removal of alkali by treatment with the acid form of Bio-Rex 70 (Bio-Rad). The solutions were stored in the frozen state, without preservatives.

Characterization of the CM-D

The open circles (RPV) in Fig. 2 represent values obtained in an assay that provides an indirect measure of the carboxyl group content of the CM-D, and they therefore serve as an affinity index. In early work[21,23] the

CM-D preparations were characterized by a specific absorption index, $A_{220}/\Delta n$ (where Δn is the difference between the refractive index of the CM-D solution and that of water), based on the strong absorbance of the carboxyl group at that wavelength. Values of $A_{220}/\Delta n$ were found to be essentially proportional to the NaCl concentration at which preparations of CM-D were eluted from DEAE-cellulose, and they served to characterize pure CM-D with respect to affinity. However, this index could not be used to evaluate the affinity of CM-D emerging in effluent fractions containing proteins separated by displacement because of the overwhelming absorption, at this wavelength, by proteins and other substances introduced with the sample. Fortunately, the observation that the extent of shrinkage of DEAE-Sephadex A-50 when saturated with CM-D was inversely related to the $A_{220}/\Delta n$ of the CM-D provided the basis for an assay that could be applied to both pure CM-D and that in protein-containing fractions.[22]

To 2.5-ml portions of 1% solutions of CM-D in Bauer–Schenck tubes (3-ml glass centrifuge tubes with long, narrow tips graduated from 0 to 0.4 ml in 0.004-ml divisions) are added 0.50-ml portions of a 1% suspension of DEAE-Sephadex A-50 containing sufficient acetate buffer to bring the mixture to pH 5. After equilibration on a rotator for 1 hr, the suspension is centrifuged for 20 min at 2000 rpm in a Model PR-2 International centrifuge, and the volumes of the pellets are read to the nearest microliter. The reciprocal relationship of these volumes to the $A_{220}/\Delta n$ values suggests that the differences arise from a limitation of the extent of shrinkage by the effective volume of the adsorbed CM-D molecules as the electrostatic forces that caused the initial extensive swelling of the DEAE-Sephadex A-50 beads are strongly diminished or eliminated by saturation of the adsorbent with CM-D. Since, at saturation, approximately the same number of charges on the adsorbent are utilized by CM-D of different levels of substitution,[21] the extent of this limitation is a direct function of the equivalent weight of the CM-D employed. The reciprocal of the pellet volume (RPV) is therefore a direct function of the carboxyl content of the CM-D and, accordingly, serves as an index of affinity.

The assay is carried out at pH 5, where the affinity of CM-D for DEAE-Sephadex A-50 is very strong, since the adsorbent is maximally charged and the affinity of most proteins is very weak or nil because of an adverse balance of positive and negative charges. This evaluation of the CM-D can therefore be carried out on protein-containing fractions from a displacement chromatogram as well as on pure CM-D. It is essential that an excess of CM-D be provided, since the DEAE-Sephadex beads must be saturated. A larger amount of low-affinity CM-D is required than high-affinity for this purpose, but the concentrations of CM-Ds in the effluent fraction of a fully developed displacement train on DEAE-cellulose at moderate pH and at low salt are automatically adjusted by the displacement process to provide the necessary levels so long as the concentration of the final displacer is, itself, high enough (e.g., 0.5%) to saturate the beads in the assay.

Protein Separations

Principle. When human serum or other biological mixtures containing colored anionic proteins are applied to a column of fibrous DEAE-cellulose under conditions permitting adsorption, the proteins arrange themselves in a series of colored bands. As additional sample enters the column, each band widens and shifts position to accommodate the expanding widths of the bands behind it, providing evidence that protein–protein displacement is occurring. This phenomenon has even been used to separate the various components of bovine plasma albumin, the fractions being recovered by sectioning the column and eluting the protein from each section.[24] The CM-Ds provide a means of continuing the displacement process that starts with the application of the protein sample, causing the otherwise immobilized protein bands to move through the column in a displacement train of CM-D that also spaces them to achieve useful separations. The electrophoretic evaluation of every fraction in the chromatogram is facilitated by the fact that the conductivity of the CM-D is low enough not to interfere with electrophoresis, though its mobility is so high that it migrates with the discontinuity in disc gel electrophoresis, as revealed by the highly refractile band observed at that position. Therefore, all of the chromatographic fractions can be applied directly to the gel without dialysis. Since displacement chromatograms provide fractions at suitable concentrations, concentrating procedures have not been necessary.

The serum proteins composed too complex a mixture to allow confirmation that the operating mechanism was displacement, apart from the observation that the array of colored bands moved down the column at constant velocity and constant width.[21] Attention was therefore turned to the study of "pure" proteins such as purified ovalbumin and β-lactoglobulin with the purpose of determining whether they would behave chromatographically in the manner observed by Tiselius[1] in the displacement development of mixtures of small, reversibly adsorbed molecules. Experiments with 50 mg of ovalbumin on a 1.4-ml column of DE-52 showed that the protein emerged as a plateau when followed by 0.1% CM-D (an unfractionated preparation having an RPV of 18) and as a peak when a 0.2%

[24] R. W. Hartley, Jr., E. A. Peterson, and H. A. Sober, *Biochemistry* **1**, 60 (1962).

solution of the same CM-D was used. Twice as much of the 0.1% CM-D was required as 0.2%, in agreement with the displacement requirement for saturating the column. When the load was increased to 100 mg, and followed by the 0.1% CM-D, the emerging plateau was twice as wide as when 50 mg was used and its height was the same. The position of the rear margin of both plateaus was the same; the extra load was added to the front of the plateau. These observations[23] clearly establish displacement as the mechanism operating in these experiments. However, attempts to demonstrate protein–protein displacement with a mixture of 25 mg each of ovalbumin and β-lactoglobulin A, displaced by 0.1% CM-D having an RPV of 20, failed. The proteins emerged with little resolution. A good separation was achieved when the experiment was repeated with an unfractionated CM-D having an intermediate affinity (RPV = 15), introduced before the final displacer to serve as a spacer between the proteins. The two proteins emerged as separated peaks, not plateaus. The individual loads (25 mg) were too small to maintain a plateau against the margin-mixing effects of passage through a bed of particles. However, the CM-D used as a spacer was undoubtedly a more efficient displacing agent than the larger and more rigid protein molecules.

Spacing. The potential for spacing chromatographed proteins is one of the great attractions of displacement. This offers the possibility of inserting sufficient nonprotein substance between two difficultly resolvable proteins without excessively widening the peaks, as would be the case if a longer column or a flatter gradient were used in elution chromatography. This requires that the column be capable of distinguishing each protein from the spacer between them, a requirement that is greater in terms of resolving power than simply distinguishing one protein from the other. However, any deficiency in such resolution will leave one or more of the proteins contaminated with some of the spacer CM-D instead of with the other protein. So long as enough spacer is used, the volume of effluent separating the peaks will prevent their being mixed by the dispersive mechanisms associated with passage through the adsorbent bed, and the column can be made as long as is necessary to achieve the separation without widening the bands.

Figure 3 illustrates the use of a narrow-range CM-D as a spacer in separating an artificial mixture of the genetic variants of β-lactoglobulin on DE-52.[22] These variants, A and B, differ in their number of carboxyl groups: A has 56 and B has 54 per molecule of 36,000 daltons,[25] and their isoelectric points are about 0.1 pH unit apart.[26] In panel A of Fig. 3, 120 mg

[25] K. A. Piez, E. W. Davis, J. E. Folk, and J. A. Gladner, *J. Biol. Chem.* **236,** 2912 (1961).
[26] J. Bours, *Sci. Tools* **20,** 29 (1973).

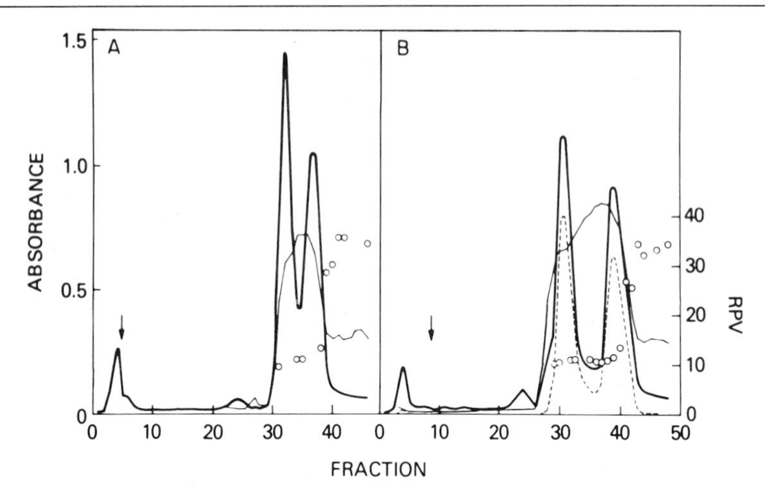

FIG. 3. Spacing of β-lactoglobulins A and B with different amounts of carboxymethyldextran (CM-D) on DEAE-cellulose. In each experiment a 7-ml column of DE-52, equilibrated with 20 m*M* sodium phosphate, pH 6, was used to chromatograph 20 mg of an equal mixture of the A and B variants applied in 2 ml of the starting buffer. In panel A, 10 ml of a 1.2% solution (120 mg) of narrow-range CM-D having an RPV of 11 was used as spacer; in panel B, 20 ml of the same solution (240 mg) was used. In both cases, the final displacer was a 0.5% solution of unfractionated CM-D with an RPV of 29. The flow rate was 10 ml/hr, and the fraction volume was 2.5 ml. All solutions contained the pH 6 phosphate buffer. The heavy line delineates absorbance at 280 nm, the thin line absorbance at 625 nm after reaction of 10-μl samples with anthrone as a measure of carbohydrate, and the dashed line absorbance at 750 nm after reaction of 50-μl samples in the Lowry assay. The circles represent RPV and the arrows indicate the points at which the final displacer was started. From Peterson and Torres.[22]

of narrow-range CM-D having an RPV of 11 was used to space A and B (10 mg of each) in displacement chromatography on a 7-ml column of DE-52 initially equilibrated with 20 m*M* sodium phosphate, pH 6.0. The spacer CM-D emerged between and within the protein peaks, as shown by the thin solid line representing the anthrone reaction[27] and the open circles indicating RPV values of CM-D in the fractions. In panel B of Fig. 3 twice as much spacer was employed and the peak centers are almost twice as far apart. Lowry determinations[28] of the protein content (dashed line) indicated that the recovery was 98%. Gel electrophoresis identified the first peak as the B form and the second as A. In both experiments, the RPV value of the CM-D that emerged between the proteins was 11.

[27] T. A. Scott, Jr. and E. H. Melvin, *Anal. Chem.* **25,** 1656 (1953).

[28] O. H. Lowry, N. J. Rosebrough, A. L. Farr, and R. J. Randall, *J. Biol. Chem.* **193,** 265 (1951).

The amount of spacer CM-D used in these experiments was substantial, and a means to reduce the requirement would be desirable. One approach involves increasing the molarity of the background buffer concentration. In one set of experiments,[22] raising the molarity of the buffer from 20 to 40 m*M* sodium phosphate resulted in a 60% increase in the separation of the peaks of A and B because the capacity of the adsorbent for the more lightly adsorbed CM-D and protein was much more strongly reduced than its capacity for the more tightly bound CM-D and protein, which was hardly affected at all. The increase in buffer concentration is advantageous in that it affords better control of the pH and provides a flux of small ions that can be expected to speed equilibration, but since loss of capacity diminishes the resolving power of the column, an optimum buffer concentration must exist for each separation, depending on the affinity of the protein to be purified. Careful adjustment of the buffer concentration is likely to be a useful option.

An increase in the pH of the buffer affects the protein and the CM-Ds quite differently. Although the charge on the CM-Ds is not affected by such a change, the positive charge on most anion exchangers used for protein chromatography is diminished and their capacity for adsorbing any given polyanion of fixed charge is reduced. On the other hand, the increase in net negative charge on the proteins is generally more effective than the decrease in the positive charge on the adsorbent, so their affinities usually increase. This has two effects: a rise in pH can significantly lower the capacity of a column for binding CM-D without diminishing its capacity for binding protein, and a specific CM-D will match a specific protein in affinity only at a certain pH. As the pH rises, the affinity of that protein will be matched by a CM-D with a somewhat higher RPV value. The lowering of the capacity for CM-D also lowers the resolving power of the column for CM-D, but if the spacers used are sufficiently homogeneous in affinity, this should not be a handicap.

Experiments with DEAE-Trisacryl M (LKB) provide a special example of this in separations of the β-lactoglobulin A and B in 20 m*M* sodium phosphate at pH 6.0. It was found that only 25 mg of spacer CM-D (RPV = 11) was required for this column to provide one-third more spacing of A and B than was obtained on DE-52 with 240 mg under the same conditions, despite the fact that the former contained twice as many milliequivalents of ion-exchange capacity as the latter. The explanation appears to depend on two effects: there was an unusually high rise in pH (0.7 unit) as a result of the collapse of electrostatic shielding of the chargeable groups on the adsorbent as the CM-D entered the beads, and the effect of this pH rise on the capacity for CM-D was unusually great because approximately half of the chargeable basic groups in this adsorbent are very weak (pK' =

6.2). Thus, the diminished capacity for CM-D rendered the spacer much more effective in moving the first peak ahead of the second. There was too little spacer in the fractions to permit determination of RPV values by our usual procedure.

On DE-52, CM-D is apparently held on many sites that are not accessible to proteins or not effective for binding them. On the other hand, DEAE-BioGel A (DEAE-agarose made by Bio-Rad), with an ion-exchange capacity about one-tenth that of DE-52 for a given volume of packed bed, has about 40% of the capacity of DE-52 for adsorbing β-lactoglobulin, but its capacity for adsorbing a given CM-D, relative to that of DE-52, is much smaller. The ratio of the number of extra CM-D-binding sites to the number of sites that can bind proteins as well as CM-D must be much smaller in this adsorbent. It therefore offers a means of reducing the amount of CM-D required for a given spacing of the separated proteins. Experimental trial has confirmed that only about one-fourth as much spacer is needed to separate A and B on this adsorbent as is needed on DE-52, and the peaks are very sharp.[22] The same spacer CM-D (RPV = 11) serves to separate these proteins at pH 6 on such widely different adsorbents as DE-52, DEAE-Trisacryl M, DEAE-BioGel A, and DEAE-Sephacel, although each of these adsorbents requires a different salt concentration in gradient elution. However, only the β-lactoglobulin variants have been studied on all of these adsorbents, so a similar conclusion regarding proteins in general cannot be drawn as yet. Because of its low capacity for adsorbing CM-D, one cannot expect DEAE-BioGel A to contribute adequately to the further resolution of the spacer CM-Ds, so the spacers employed with this adsorbent must be sufficiently homogeneous to do the job. For more difficult separations, a small precolumn packed with a more effective adsorbent (about one-fourth the size normally used) could presumably serve to fractionate further the reduced amount of spacer that is needed. The DEAE-BioGel A column would then function as a spacing enhancer as well as contribute to the separation of the proteins from the CM-D.

Resolution. Experience has shown that columns packed with DE-52, DEAE-Trisacryl M, or DEAE-Sephacel are capable of resolving CM-Ds that differ very slightly in RPV, thus establishing positions representing affinity values that will be recognized by proteins in the sample. The latter two adsorbents are particularly effective. The resolution of the CM-Ds by displacement on a given column may well be better than the column's resolution of proteins by elution, for the CM-Ds are relatively small, flexible chains that have access to or at least can utilize more adsorption sites than the proteins. Moreover, in contrast to the situation in elution, where migration depends on a progressive decrease in the capacity of the column for the adsorbed substances as the eluent increasingly dominates, the capacity of the column in displacement chromatography remains the same throughout the experiment. For any specific component, the same number of sites are available per unit of volume to be competed for with members of the train that are rapidly selected by the process to resemble that component more and more with respect to affinity for those sites. The CM-Ds should be able to compete with their equals and inferiors among the proteins no matter what sites on the ion exchanger may be involved, so long as the binding forces are ionic. The tailing that arises from convex isotherms, and is only partially compensated for by the relatively flat gradients required for high resolution in elution chromatography, is not a problem in displacement.

In the isolation of minor protein components such as enzymes and protein markers in disease, the column is sometimes deliberately overloaded with respect to proteins having lower affinities than the desired protein in order to be able to apply a larger amount of the latter. Even though the separation may be completed by elution chromatography, it is displacement that is relied upon to achieve the fractionation that permits the overloading to occur without loss of the desired protein. However, CM-Ds are more efficient displacers, better results being obtained when they are used. In unpublished work, the authors undertook to isolate a minor protein from the serum of a person afflicted with psoriasis. This protein was known only as a small spot appearing in two-dimensional electrophoresis gels with an incidence correlated with the disease. The first step in its purification was the displacement chromatography of 6 ml of the dialyzed serum on a 7-ml column of DEAE-Sephacel at pH 7, using narrow-range spacer CM-Ds with RPV values of 5.5 to 7.0. Fractions were examined for the presence of the psoriasis-associated protein by two-dimensional electrophoresis, since its position in such a gel was the only criterion by which it could be recognized. The CM-D did not interfere with electrophoresis in either dimension.[29] The protein emerged with CM-D having an RPV of 6.5 at the very end of a highly compressed group of peaks and shoulders representing most of the protein applied. The fractions that contained the desired product were pooled (less than 12 ml), and chromatographed on a similar column at pH 8, with spacer CM-Ds ranging in RPV from 6.7 to 9. The psoriasis-associated protein emerged early in a series of crowded peaks, the CM-D having an RPV of 7.0. Almost pure, it was among those components least affected by the rise in pH; most of the others required CM-D with an RPV of 8 or more. The CM-Ds in the displacement train provided a continuous array of positions

[29] A. R. Torres, G. G. Krueger, and E. A. Peterson, *Clin. Chem.* **28,** 998 (1982).

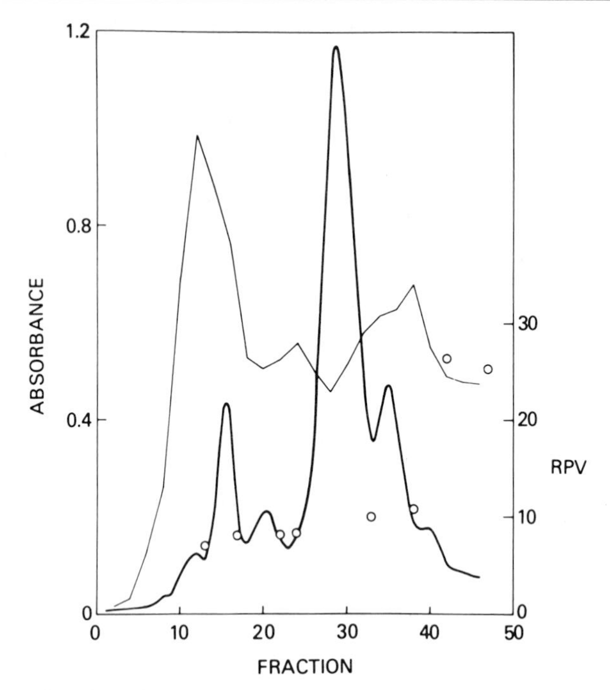

FIG. 4. Fractionation of ovalbumin by displacement chromatography on DEAE-cellulose. A solution of 50 mg of ovalbumin in 2 ml of 10 mM sodium phosphate, pH 7.5, was applied to a 7-ml column of DE-52 equilibrated with the same buffer. Four narrow-range carboxymethyldextrans (CM-Ds) were used as spacers: 4 ml of 1.9% CM-D (RPV = 7.1), 9 ml of 1.8% (7.7), 8 ml of 1.6% (8.3), and 5 ml of 1.4% (9.2), totaling 434 mg of CM-D. The final displacer was a 0.5% solution of unfractionated CM-D with an RPV of 27. The heavy line represents absorbance at 280 nm; the thin line, absorbance at 625 nm after reaction of 10-μl samples with anthrone; and the circles, RPV.[27] From Peterson and Torres.[22]

along the abscissa of the chromatogram that represented affinities that could be matched by the proteins in the sample. The larger the amount of CM-D in a given affinity range, the wider was the spacing of these positions and the greater the amplification of the differences in affinity induced in the proteins by the change in pH.

Another model protein mixture has provided an opportunity to demonstrate the resolving power of the CM-D displacement system. Figure 4 illustrates the separation obtained when 50 mg of a highly purified commercial ovalbumin preparation, claimed by the supplier (Sigma) to be 99% ovalbumin on the basis of electrophoretic analysis, was chromatographed on a 7-ml column of DE-52 in 10 mM sodium phosphate at pH 7.5, using four narrow-range CM-Ds (RPV = 7–9, total 434 mg).[22] Three forms of ovalbumin were expected: A_1, A_2, and A_3, containing two, one, and no phosphate groups, respectively,[30] with isoelectric points of neighboring forms differing by about 0.1 pH unit.[31] However, six definite peaks appear in the A_{280} profile, and each is represented by protein that migrates within the range attributable to ovalbumin in the patterns obtained by electrophoresis in polyacrylamide gels (not shown). The appearance of dimers at the end of the chromatogram, in fraction 40, is not unexpected, but the presence of so many components in purified ovalbumin was not anticipated. However, Iwase *et al.*[32] have reported that differences in carbohydrate composition as well as phosphate content provide a basis for heterogeneity in gel electrophoresis of ovalbumin, though not in isoelectric focusing. They concluded from the latter observation that the electrophoretic heterogeneity contributed by the carbohydrate was not the result of differences in charge. This suggests that a partial masking of the charges on the protein by carbohydrate chains, leading to intermediate affinities, may be the basis of some of the chromatographic heterogeneity seen in Fig. 4. Vanecek and Regnier[33] have obtained gradient elution profiles of ovalbumin of similar complexity on an analytical scale, employing a commercial HPLC ion-exchanger (see this volume [8]). However, relative to the size of the column, the protein load was one-fiftieth of that used in Fig. 4.

Removal of CM-Ds from Isolated Proteins. Separation of the protein from the CM-D that accompanies it in the fractions of a displacement chromatogram is likely to occur whenever those fractions are subjected to further purification by other procedures, e.g., electrophoresis, gel filtration (depending on the molecular weight of the protein), or chromatography on other types of adsorbents. If the pH can be safely lowered to render the protein cationic, passage of the fraction through a small column of CM-cellulose will result in the adsorption of the protein while the CM-D passes through unadsorbed. Sharp elution with a suitable salt solution will then yield the protein in highly concentrated form. On the other hand, one can use a column of DEAE-cellulose or similar anion exchanger, at a pH that makes the protein cationic, to bind the CM-D and allow the protein to pass through unadsorbed. This procedure promises to be the basis of a convenient recycling of the spacer CM-D, a matter that is likely to be of interest in large-scale separations.

The high-affinity CM-D that saturates the column after all the proteins

[30] G. E. Perlmann, *J. Gen. Physiol.* **35,** 711 (1952).

[31] M. B. Rhodes, P. R. Azari, and R. E. Feeney, *J. Biol. Chem.* **230,** 399 (1958).

[32] H. Iwase, Y. Kato, and K. Hotta, *J. Biol. Chem.* **256,** 5638 (1981).

[33] G. Vanecek and F. E. Regnier, *Anal. Biochem.* **109,** 345 (1980).

have been driven from it is readily recovered by displacing it with a dilute solution of alkali.

Scale of Operation. Displacement chromatography of proteins can be carried out at any scale. Its classical advantages of high capacity and concentrated bands are particularly attractive for preparative work, and its inherent band-sharpening may be of special value in the multiple columns used in very large-scale separations by utilizing entry into each successive column to correct any skewing that might have occurred in passage through the preceding one. Large-scale separations of proteins by the procedures described in this chapter have not yet been attempted, but there is reason to believe that fewer problems will be encountered in scaling up displacement separations than in scaling up those based on elution chromatography, since effective systems will be smaller. The usefulness of displacement on a repetitive, large-scale basis will, however, depend on the refinement of procedures for recycling the spacers as well as the final displacers. Proper recycling of the spacers can be expected to improve their performance inasmuch as they should become more homogeneous with respect to affinity than the initial supply.

At the other end of the scale, in analytical applications such as high-performance liquid chromatography (HPLC), displacement chromatography offers the advantage of requiring very much smaller columns for a given load and relatively simple apparatus. The potential for spacing the peaks for maximum efficiency is also attractive. Although experience in this area is meager, preliminary trials in collaboration with Dr. Vernon Alvarez have been promising. Thus, a very good separation of the A and B forms of β-lactoglobulin was achieved on a 0.1-ml silica-based ion-exchange column (Aquapore AX-300, sold as a microbore guard column by Brownlee) in 35 min, applying solutions containing 0.4 mg of the protein, 3 mg of the CM-D spacer (RPV = 12), and 8 mg of the final displacer (RPV = 30) at 5.5 ml/hr with a Pharmacia P-3 peristaltic pump. The entire chromatogram encompassed only 3.2 ml, and the concentrations of the two bands were high. After the separation the column was rapidly stripped of CM-D with a salt solution before reequilibration with the buffer for reuse.

Given the objectives of analytical HPLC, one is unlikely to use protein samples or conditions that will produce plateaus instead of peaks. However, it must be kept in mind that in displacement the position of a peak is not fixed, as in elution, by standardizing the conditions under which the samples are run. Any substance in the sample with an affinity higher than that of a given protein will add to the displacement of that protein, and a very marked change in the distribution of low- and high-affinity components of a sample might produce a significant shift in its position. However, by developing a standard CM-D train that contains many times as much CM-D as the quantity of protein in the samples, one should be able to make such shifts insignificant.

In the work of Francina *et al.*[16] mentioned above, microgram quantities of hemoglobin were applied to DE-81 paper strips for separation by ampholytes. Presumably, thin-layer chromatography of other proteins on other thin-layer media is feasible, since spacers can be easily applied. However, procedures for detecting enzyme activity must take into account the fact that much of the enzyme in a displaced band is not immobilized on the adsorbent and can readily be washed away by the reagent solution. Also, the dyes that are used to stain proteins are bound by positively charged adsorbents, and produced intensely colored backgrounds.

Where the objective is the separation of isoenzymes to provide information of clinical interest, the use of microcolumns from which individual forms of the enzyme can be displaced by carefully selected narrow-range spacers is more promising. The suppression of tailing and the control of band width characteristic of displacement are of importance in such applications.

[6] High-Performance Liquid Chromatography: Care of Columns

By C. TIMOTHY WEHR

The high-performance liquid chromatography (HPLC) columns now used in protein isolation and characterization represent a considerable investment by the user compared to the low-pressure supports they are replacing. Careful use should allow these columns to give satisfactory performance from several months to over a year. This review is meant to provide information on the maintenance and troubleshooting of commercial HPLC columns to enable the user to achieve maximum performance and column lifetime. In addition to the specific recommendations on routine operation and column repair described below, a troubleshooting guide for defective column performance is outlined in Table I. Since the column is an integral part of a larger system, symptoms caused by pump, injector, or detector conditions may be misdiagnosed as column failure and are included in the table. Only the chromatographic modes commonly used in protein or peptide separations (size exclusion, ion exchange, re-

METHODS IN ENZYMOLOGY, VOL. 104

ISBN 0-12-182004-1

versed phase) are considered; information on the use of adsorption or normal-phase HPLC columns has been presented.[1-3]

Column Installation

High-performance liquid chromatography columns consist of microparticulate (3- to 50-μm particle diameter) materials based on silica or organic polymers (resins) most commonly packed at high pressure into stainless steel tubing. Although most HPLC packings have the high mechanical strength needed to withstand the pressure drops of 150–5000 psi encountered at typical flow rates, column beds will not tolerate undue mechanical shock. For this reason, columns should be handled carefully during installation and removal and should be stored so as to avoid shocks that might cause disruption of the packed bed. The majority of commercially available HPLC columns are "universal," that is, they may be installed on any HPLC pump or injector either directly or with a simple adaptor. However, when fitting a new column to an existing system, care must be taken to ensure that the connections between injector and column do not introduce dead volume. Inlet nut and ferrule must be mated with the column terminator, unions and connectors should be zero- or low-dead volume type, and connecting tubing should be of low internal diameter (~0.3 mm). Connections should be sufficiently tight to prevent leaks (approximately finger-tight plus a quarter-turn). If overtightening is required to prevent leakage, it suggests a mismatch between terminator and inlet nut or ferrule that can be remedied only by replacement. Fitting mismatch can result (in addition to leakage) in band broadening or carryover due to excessive dead volume or unswept voids. Since there is to date no standardization of terminators and fittings among manufacturers, it is recommended that, in cases where columns from different sources are used on the same HPLC system, separate transfer lines be fabricated for each column type using fittings supplied by the column manufacturer.

Most commercial HPLC columns are shipped containing an organic solvent designated in the column installation instructions. If this solvent is not compatible with the mobile phase solvent, the column must be washed with an intermediate solvent prior to use. For example, reversed-phase columns shipped in methanol or acetonitrile should be washed with water before introduction of buffers or salts. Columns shipped in hexane (e.g.,

[1] L. R. Snyder and J. J. Kirkland, "Introduction to Modern Liquid Chromatography," 2nd ed., pp. 782–823. Wiley, New York, 1979.

[2] F. M. Rabel, *J. Chromatogr. Sci.* **18,** 394 (1980).

[3] D. J. Runser, "Maintaining and Troubleshooting HPLC Systems," pp. 69–90. Wiley, New York, 1981.

nitrile or amino columns) must be washed with propanol or tetrahydrofuran before introduction of methanol, acetonitrile, or aqueous solvents. When a new column is installed, it should not be connected to the detector until several milliliters of wash solvent or mobile-phase solvent have passed through the column to prevent residual packing solvent or silica fines from entering the detector.

The trend in HPLC continues to be toward the use of high efficiency columns packed with supports of 5-μm particle diameter or less. These columns may be of standard 25- or 30-cm lengths for difficult separations requiring very high efficiency. Short columns 4–15 cm in length and packed with 5-μm or sub-5-μm materials may be operated at normal flow rates (~1.0 ml/min) to achieve moderate efficiency and low solvent consumption or at high flow rates (2–5 ml/min) to obtain very short analysis times. The user should be aware that as column efficiency increases and column length decreases, the extra-column contributions to band broadening will become more significant. Use of these columns demands that extra-column dead volume be kept to a minimum by reducing the length of transfer lines, using small inner diameter (i.d.) tubing, and installing appropriate guard columns. With microbore HPLC columns of 2-mm i.d. or less, extra-column effects are sufficiently severe that a conventional liquid chromatograph would require extensive modifications, including use of low-volume injectors, microbore transfer lines, and detectors with micro flow cells (<5 μl) and rapid time constants (<100 msec).

Column Testing

A newly purchased HPLC column should always be tested upon receipt using the manufacturer's suggested test compounds and separation conditions. Manufacturers generally test columns to minimize batch-to-batch variations in stationary-phase characteristics and to verify that individual columns meet performance specifications; test compounds are selected that effectively probe such stationary-phase characteristics as bed efficiency, selectivity, free silanol content, and contamination by trace metals. Test compounds are usually small molecules selected for their stability, low toxicity, and availability in the user's laboratory. In some cases, column packings designed expressly for chromatography of proteins may be batch-tested by the manufacturer with a protein test mixture to check selectivity or recovery. The user should test a column upon receipt to verify that the column meets the published specifications and has not been damaged in shipment; in some cases, failure to test a column within a given time period may void the manufacturer's warranty. Variances of up to 10% from the manufacturer's test results should not be considered indicative of a defective column since they may reflect extra-column effects, different flow accuracy, or small differences in solvent composition. In addition to confirming column integrity, testing a new column provides a baseline for monitoring column performance during use. In some cases, it may be advisable also to test a new column with compounds more chemically related to the user's samples. Since interactions of polypeptides with HPLC stationary phases are in some respects different from those of small molecules,[4] it is unlikely that a manufacturer's test results will adequately probe such column characteristics as protein selectivity and recovery. A test sample made up of standard proteins or peptides may be more relevant in establishing a performance baseline and monitoring column aging.

Column Operation

Sample and Mobile-Phase Preparation

The most common cause of early column failure is lack of care in solvent selection and preparation and inadequate cleanup of biological samples. Solvents may contain dissolved impurities that can be retained on the stationary phase and gradually alter the phase characteristics or elute as spurious or "ghost" peaks. Undissolved particulates in the mobile phase can plug system hydraulic components or the column head, resulting in excessive operating pressure. The first step in avoiding these problems is to use solvents of high purity (HPLC grade or spectroscopic grade) and filter them prior to use through an appropriate micropore (0.22 or 0.45 μm) filter. Where pumps with ball-and-seat inlet check valves are used, solvents must be thoroughly degassed prior to use to prevent loss of pump prime by cavitation. Degassing is also advisable when using stationary phases susceptible to oxidation, e.g., amino phases, or when using low UV detection ($\leq$200 nm), where absorbance by dissolved oxygen can give rise to noisy or drifting baselines. Degassing mobile-phase solvents will minimize bubble formation due to outgassing in the detector cell, although this can be prevented by creating 20–100 psi resistance on the outlet side of the detector with a restrictor or with about 3 meters of 0.3 mm i.d. tubing.

Since peptides and protein fragments may not always contain chromophores absorbing at the commonly used detection wavelengths (254 and 280 nm), low UV detection in the 205- to 220-nm range is often used. Many of the mobile-phase modifiers (alcohols, organic acids) employed in pep-

[4] G. Vanecek and F. E. Regnier, *Anal. Biochem.* **121,** 156 (1982).

tide chromatography have UV cutoffs in this spectral range, and some baseline offset in gradient elution must be expected. Solvent impurities, however, become a severe problem at these wavelengths, giving rise to noisy baselines or spurious peaks that may be incorrectly diagnosed as column or pump disorders. Water is the most notorious offender, often containing trace organic impurities that become concentrated on reversed-phase columns and appear as spurious peaks during gradient elution. This can even be a problem with commercial HPLC-grade water whose optical purity is specified at longer wavelengths, e.g., 254 nm. Trace impurities are best removed by passing the water or aqueous mobile phase through an activated charcoal or reversed-phase column prior to use. The same procedure can be used to remove UV-absorbing impurities from mobile-phase additives such as buffer salts and ion pair agents.

Solvent impurities are also a problem in ion-exchange chromatography employing phosphate gradient elution. Phosphate salts contain UV-absorbing impurities that cause severe baseline offset at high detector sensitivity. This problem is compounded by the tendency of these impurities to collect on the column and elute at high ionic strength; thus, the baseline offset increases with column use. Solvent purification techniques have been devised using recrystallization and ion-exchange cleanup[5] or passage through a chelating resin,[6] but none of them completely eliminates the difficulty. The best remedy is periodic stripping of adsorbed impurities from the column (see Column Repair, below).

Where impurities in the weak (A) solvent in gradient elution can be removed by passage through an appropriate adsorbent bed, a stripper column packed with the adsorbent material can be placed in-line on the outlet side of the "A" pump in a multipump gradient HPLC system. However, the breakthrough volume of the stripper column must be accurately determined to avoid elution of impurities during an analysis.

Where mixtures of an aqueous buffer and an organic solvent are used, as in reversed-phase chromatography, the user must be careful to avoid precipitation of buffer salts in the system. Premixed solvents should be filtered through a micropore filter. If aqueous–organic mixtures are proportioned from separate reservoirs by the HPLC pumping system, precipitation may occur in hydraulic components leading to reduced seal life or component failure. If precipitation occurs in the column bed, column failure is a certainty. The surest means for avoiding buffer–organic solvent incompatibility is to use as the strong eluent a premixed combination of buffer and organic modifier filtered prior to use. Neither the column nor

[5] H. W. Shmukler, *J. Chromatogr. Sci.* **8,** 581 (1970).
[6] G. Karkas and G. Germerhausen, *J. Chromatogr.* **214,** 267 (1981).

the HPLC system should be stored while containing buffers, and when a system is converted from one application to another, e.g., from ion-exchange to reversed-phase chromatography, the entire system should be flushed with water before solvent and column are changed.

Column failure may arise from the gradual accumulation of particulate material originating from wear of the system seals or microbial growth in the mobile-phase solvents. This can be avoided by installation of an in-line filter either before the injector or between injector and column; the latter arrangement will protect against particulates originating from the sample or from wear of the injector seal. In-line filters, available from several manufacturers, have easily replaceable frits in the 0.5- to 2-μm pore size range and add only a minimal amount of dead volume to the system. Microparticulate precolumns and guard columns also serve as effective solvent filters.

A number of mobile-phase additives may cause column failure by irreversibly changing the stationary-phase characteristics or by accelerating column degradation. Ion-pairing agents or detergents that have bulky hydrophobic groups (e.g., camphorsulfonic acid, sodium dodecyl sulfate) may irreversibly partition into the stationary phase of steric exclusion and reversed-phase columns, changing the phase chemistry and reducing apparent pore volume. In applications requiring the uses of these components, it is recommended that the column be dedicated for this use. Anionic detergents cannot be used as mobile-phase components with anion-exchange columns. Cationic alkylamines used as ion-pairing agents or competing bases for silanol complexation (e.g., triethylamine, tetramethylammonium salts) tend to accelerate dissolution of the silica support in silica-based columns when used at neutral pH. It is recommended that these agents not be used at a mobile-phase pH of 6 or greater and that they be flushed from the column immediately afterward. The use of halide salts with stainless steel HPLC pumping systems and columns has long been a point of concern in protein purification, particularly since many ion-exchange procedures developed on carbohydrate-based gels employ chloride eluents as high as 1–2 *M*. Actually, the 316 stainless steel used in most HPLC components is reasonably resistant to chloride at neutral pH, and component lifetime should not be significantly reduced if such solutions are flushed from the system after use. Chloride at acidic pH should *not* be used in HPLC systems. If the investigator is uncertain about the solvent compatibility of a particular instrument or component, the manufacturer should be consulted.

It is generally recommended that samples injected onto HPLC columns be as free as possible of contaminating material to minimize interferences in detection and to prevent unnecessary adsorption of sample components on the column. This often is not possible when HPLC is used for purification of polypeptides from biological fluids or extracts. However, samples containing significant amounts of lipids should be extracted with a nonpolar solvent (ether, hexane) prior to injection. Size-exclusion HPLC provides a rapid means of removing high-molecular-weight material before injection onto ion-exchange or reversed-phase columns. Samples should always be filtered to remove particulates, and guard columns should always be used to prevent strongly retained components from reaching the analytical column.

Chromatographic resolution will be affected by sample volume and sample concentration. In steric exclusion chromatography, sample volume should be less than 1% of the column permeation volume (e.g., a sample volume of about 100 μl for a standard 8-mm × 25-cm analytical column), and sample loads of between 1 and 5 mg can typically be applied. In ion-exchange and reversed-phase chromatography using gradient elution, the sample must be applied in a weak solvent (typically the initial or A solvent); here sample volume is not critical and several milliliters or more can be applied to an analytical column if necessary. Because of the high capacity of ion-exchange and reversed-phase HPLC supports, relatively large sample loads, 5–15 mg total protein or greater, can be injected, and often analytical columns can be used for semipreparative applications if gradient elution is employed. When such columns are operated isocratically, resolution will be more sensitive to sample volume, sample concentration, and sample matrix effects, particularly for low k' compounds.

Guard Columns

Guard columns are used to protect the analytical column from particulate material or strongly retained contaminants in the sample matrix; they are by far the most effective means of preserving and extending the life of the analytical column. Guard columns are placed between the injector and analytical column and are discarded or repacked often to ensure that contaminants are not eluted. The frequency of replacement will vary depending on column capacity, solvent strength, sample type, and sample load; typically, the life of a guard column is between 10 and 50 injections. Selection of the appropriate guard column packing and column configuration must take into account two considerations. First, the packing should be similar in chemical structure to the analytical stationary phase so that the selectivity of the system is not altered. This is not always possible, since not all manufacturers supply guard column analogs of analytical packings and, when they are available, phase characteristics of the guard material may not be identical to the analytical packing. If a low-capacity guard material is used, selectivity mismatch is not as critical as with a high-capacity material. In steric exclusion chromatography, mismatches in guard column pore diameter and pore volume can affect protein elution profiles. The second consideration in selecting a guard column is loss of efficiency; ideally, this should not exceed 10% in order to preserve overall resolution. This is an important point when short, high-efficiency analytical columns are used, since extra-column contributions to band broadening are more significant. Here, particular care should be taken if a high-capacity microparticulate guard column is used. A high-capacity guard column should have a height equivalent of a theoretical plate (HETP) value similar to that of the analytical column. For example, coupling of a 3-cm guard column packed with 10-μm material to a 15-cm, 5-μm analytical column can result in a 30 to 40% loss in efficiency.

Two types of guard column packings are currently available: pellicular materials and fully porous microparticulate packings. Pellicular packings consist of solid-core beads (usually 37–40 μm in diameter) with the stationary phase bonded to or polymerized on the surface as a "pellicle." Because of their large particle size, pellicular packings are easily packed in dry form by the user without any special equipment; replacement cost for material in a 4 $\times$ 50 mm guard column is about $4. Since the surface area of a pellicular packing is small compared to a porous microparticle, capacity is quite low. Thus, such columns must be repacked frequently, but their effect on selectivity is not large. Two manufacturers (Scientific Systems, Inc., State College, PA, and Upchurch Scientific, Bremerton, WA) have introduced low-volume guard columns that couple directly to the analytical column; these can be packed with pellicular materials and used with 5-μm and sub-5-μm analytical columns to achieve low efficiency losses of 5% or less. The need for frequent replacement should be offset by the ease of repacking.

Microparticulate guard columns are packed with the same material used in analytical columns and are designed for situations in which high efficiency is required. Because they must be slurry-packed, they are usually obtained from the manufacturer as prepacked columns or cartridges. Brownlee Laboratories (Santa Clara, CA) offers prepacked disposable guard cartridges that fit into a reusable holder, and they are available with a variety of 5- and 10-μm reversed-phase, ion-exchange, and steric-exclusion packings. Since microparticulate guard columns are of high capacity, a close selectivity match with the analytical column is important, and dead volume in connectors and transfer lines must be minimized to reduce extra-column band broadening. When an exact selectivity match is not possible, a guard column support that is less retentive than the analytical packing should be used. In some cases, saturated microparticulate guard columns can be regenerated off-line by washing with a series of strong solvents.

Guard columns can also be used for the concentration or cleanup of samples. In a typical application, the guard column replaces the sample loop in a six-port rotary injection valve. The sample is introduced through the injection port and concentrated on the guard column; poorly retained contaminants may be eluted with a weak solvent, followed by valve rotation and introduction of the sample onto the analytical column with a stronger solvent.

Saturator Columns

A major limitation of silica-based HPLC supports is the instability of silica under alkaline conditions. Ion-exchange and size-exclusion chromatography of native proteins usually requires the use of mobile phases in a pH range of 7 to 8, conditions that reduce the lifetime of silica columns. Strong anion-exchange (SAX) columns with bonded quaternary ammonium phases are particularly vulnerable to high pH. Column life can be extended by saturating the mobile phase with dissolved silica to suppress degradation of the analytical support. This can be done by adding silica to the mobile phase during preparation or, more commonly, by installing an in-line saturator or solvent-conditioning precolumn before the injector, which is packed with porous silica. Large-particle (30–50 μm) silica is generally used for ease of packing and to reduce system pressure. Although it has been reported that use of a saturator may extend column life more than 10-fold,[7] lifetime extensions of 2- to 5-fold are more typical using a 4 $\times$ 300 mm precolumn. The precolumn should be maintained at the same temperature as the analytical column and should be topped off periodically to replace dissolved packing. Use of a saturator precolumn carries several disadvantages. First, a silica-saturated mobile phase may precipitate if allowed to stand in the system; the saturator column should be removed and the HPLC flushed prior to shutdown. Second, dissolution of the silica packing can generate fines that, if passed through the exit frit, can plug the injector, transfer lines, or analytical column. It is recommended that a small-pore (0.5-μm) frit be installed in the precolumn outlet terminator. Third, voids created by dissolution of the precolumn packing will increase the dead volume of the system, leading to poor reproducibility in gradient elution. The precolumn should be inspected and repacked often to prevent this. Fourth, the precolumn can act as a trap, concentrating solvent impurities that later elute when a stronger solvent is intro-

[7] J. G. Atwood, G. J. Schmidt, and W. Slavin, *J. Chromatogr.* **171,** 109 (1979).

duced. The analytical column should be removed initially during solvent changeover to prevent its contamination. Finally, the presence of silicates in isolated sample components may alter biological activity or interfere with subsequent chemical characterization.

Gradient Elution

Gradient elution is employed in the vast majority of HPLC applications in protein isolation and characterization. Most commercial gradient HPLC systems have sufficiently high solvent proportioning precision to provide retention time reproducibility of a few percent or better in gradient elution. However, to achieve this level of performance, the user must make sure that the column is reequilibrated at initial conditions at the beginning of each analysis and that the gradient elution protocol including column regeneration and equilibration is exactly repeated for each trial. When solvents of widely different strengths are used, the column should be regenerated with a reverse gradient to initial conditions, followed by equilibration with 5–10 column volumes of the initial solvent. Regeneration and equilibration can be done at elevated flow rates to reduce turnaround time. In ion-exchange chromatography, column equilibration can require up to 50 column volumes or more if initial and final solvents differ in pH. It is not uncommon that ion-exchange columns, even when stored in the initial solvent, will exhibit retention time variations in the first analysis of the day and stabilize over successive runs. Operation of reversed-phase columns with ion-pairing agents or buffers will also require extended equilibration periods. For maximum reproducibility, the concentration of the ion-pairing agent should be kept constant across the solvent gradient.

Operation Limits

Column supports and stationary phases differ in their limitations to pressure, flow rate, temperature, and pH. This information should be supplied in installation instructions delivered with the column and should be reviewed by the user to ensure that the column warranty is not voided. Some column packings are incompatible with certain mobile phases—for example, amine phases react with carbonyl groups; therefore, aldehydes or ketones cannot be used as mobile-phase modifiers. In cases where the user is unsure of incompatibility with an unusual solvent, the manufacturer should be contacted.

Column Storage

For short periods, e.g., overnight, it is best to store silica-based columns with an organic solvent or mixtures of an organic solvent and water. Columns should not be stored with buffer or salt solutions; if it is necessary to leave buffer in the column for equilibration overnight, a low flow of about 0.1 ml per minute should be maintained to prevent precipitation within the column or within hydraulic components of the system. For long-term storage, silica-based columns should be flushed with water if they have been used with buffers or salts and then filled with an organic solvent (methanol or acetonitrile). Since stationary phases can act as substrates for microbial growth, aqueous solvents are not recommended for storage. Column terminators should be tightly capped as drying may cause changes in bed geometry, and the storage solvent should be indicated on a label attached to the column. Columns based on polystyrene resins or hydrophilic organic polymers should be stored with solvents recommended by the manufacturers. Columns should be kept in a place in which they will not be exposed to potential shock or extremes of temperature.

It is good practice to maintain a log in which column use and performance are detailed. This should include the type of column and its serial number, along with initial test data; the history of its analytical use, including sample type, mobile-phase conditions, and operator; test data from periodic performance checks; details on column repair.

Troubleshooting the Column

The most common problems encountered in the operation of HPLC columns are outlined in Table I, along with suggested diagnoses and treatments. Several of these problems occur frequently in chromatography of proteins and peptides, particularly in the reversed-phase mode.

Spurious or "ghost" peaks can arise from mobile-phase impurities, from sample carryover in the injector or transfer lines, or from elution of adsorbed sample components in subsequent runs. Mobile-phase impurities are a common source of ghost peaks in gradient elution when detection in the low UV is used, and they typically originate from the water. This can be diagnosed by pumping water through the column for varying time periods, followed by blank gradients. An increase in ghost peak height as a function of the volume of water used is indicative of impure water. Mobile-phase additives (ion-pair agents, buffers, salts) can be checked individually in the same manner. Spurious peaks arising from

TABLE I
TROUBLESHOOTING HPLC COLUMNS

Circumstances	Possible cause	Remedy
	A. Abnormal operating pressure	
Rapid increase in pressure	Plug in detector, column, injector, or chromatograph	Working back from detector to pump, break each fitting to localize source of resistance. If localized to column, flush or replace frit. Otherwise, backflush or replace plugged component
Gradual increase in pressure	1. Plugged inlet frit or column head	1. Replace frit; remove top of column bed and repack
	2. Plugged outlet frit	2. Replace frit
Reduced pressure	Leak or defective check valve	Tighten fittings; with volatile solvents, slow leaks can be detected by cold fitting; check pump seals and check valves
Fluctuating pressure	Defective hydraulic system	Check inlet and outlet check valves, pulse damper, etc.
	B. Decreasing retention time	
Pressure increasing; all peaks including solvent front affected	Flow rate too high	Check flow volumetric rate; repair or reset flow
Efficiency also decreasing, solvent front unaffected	Losing stationary phase	1. Guard column is consumed 2. Replace inlet frit and top off 3. Replace column
Column pressure decreasing, solvent front unaffected	Column temperature increasing	Control column temperature
Retention time varies with injection volume	Injection solvent decreasing column activity	Use starting mobile phase as injection solvent
Large sample	Column overload	Use smaller sample
Gradient elution	Column recondition incomplete	Use longer reconditioning
	Regeneration not reproducible	Use more care in reconditioning program
Gradient elution or proportioned solvents	Increase in percentage organic modifier	Control solvent composition, prepare new mixture, check performance of solvent proportioning system

TABLE I (*continued*)

Circumstances	Possible cause	Remedy
Prolonged use	Column contaminated	1. Flush with strong solvent 2. Replace precolumn 3. Replace analytical column
Spiking sample with authentic knowns gives different T_r than standards alone.	Components in sample or injection solvent are altering chromatography (matrix effect)	1. Prepare standard in same controlled matrix 2. Remove matrix interference by sample cleanup 3. Use internal standard technique
Ion pair or ion exchange chromatography	Concentration of ion pair reagent too low; buffer pH or ionic strength incorrect	Increase concentration of ion pair agent; check buffer composition
Retention time of some (not all) peaks changes with sample amount	Sample changing with concentration	Dilute or concentrate sample to minimize effects; change sample solvent
	C. Increasing retention time	
Pressure lower than normal, solvent front also affected	Flow rate too low	Check volumetric flow rate; reset flow or repair
Backpressure increasing, flow rate OK, poor peak shape, solvent front unaffected	Column contaminated	1. Replace precolumn 2. Replace inlet filter
Liquid found or cold fitting; flow rate lower than expected	Leak or defective check valve	Tighten fittings; slow leaks of volatile solvents can be detected by cold fitting. Inspect check valve
Reversed-phase chromatography	Decrease in concentration of nonpolar solvent by evaporation or incorrect preparation	Control solvent composition
Gradient elution or isocratic with proportioned solvents	One solvent not flowing owing to: 1. Empty reservoir 2. Plugged inlet frit 3. Air bubble in line 4. Inlet valve plugged or damaged 5. Programming error	 1. Fill reservoir 2. Clean frit 3. Purge line 4. Consult manual 5. Check program

(*continued*)

TABLE I (*continued*)

Circumstances	Possible cause	Remedy
Widely varying environmental temperature	Column temperature change	Control column temperature
Gradient elution	Variations in column regeneration program	Use more care in reproducing reconditioning program
Spiking sample with authentic knowns gives different retention time than standards alone	Components in sample or injection solvent are altering the chromatography (matrix effects)	1. Prepare standards in same matrix 2. Remove matrix interference by sample cleanup 3. Use internal standard technique
Column pressure rising	Column temperature decreasing	Control column temperature
	Tubing starting to plug	Working from detector to pump, break each fitting and determine if there is excessive resistance to flow anywhere but in the column; if found, investigate and repair
Ion-pair or ion-exchange chromatography	Concentration of ion pair reagent too high; buffer pH or ionic strength incorrect	Reduce concentration of ion pair agent; change buffer composition
	D. Decreasing resolution	
Trend observed over several runs	Slow buildup of contaminants on column	See Part B above
Retention times unchanged	Column efficiency decreasing	Check column head for void; refill if necessary; replace column or precolumn
Gradient elution or proportioned solvents	Efficiency unchanged, retention or selectivity changed	Pump malfunction; confirm that each pump is delivering correct amount of solvent
Retention time shorter	Flow too high	Check volumetric flow rate; reset flow or repair
Gradient elution	Regeneration program inadequate	Lengthen regeneration
Changed or rechanged solvent reservoir recently	Change in solvent composition	Use more care in preparing eluent
Large sample load	Column overload	Dilute sample

TABLE I (*continued*)

Circumstances	Possible cause	Remedy
	E. Peak tailing	
All peaks affected	Void at top of column	1. Fill in void with suitable packing or filler 2. Avoid high flow rates, excessive pressures, high mobile phase pH 3. Replace column
All peaks affected	Poorly packed column	Confirm by comparison with good column. Repair column or replace column
All peaks affected	Frits contaminated or partially plugged	Replace fittings or frits
Most or all peaks affected	Adsorbed contaminants on column giving mixed mechanism separations	1. Flush with strong solvent 2. Use precolumn; replace as required
Reversed-phase chromatography; only a few peaks affected	Acidic or basic samples interacting with silanol groups	Add competing base, e.g., 0.1% $(CH_3)_4NCl$
One peak affected	Unresolved minor component	Improve resolution: longer column, different mobile phase
Large sample load	Column overloaded	May be acceptable; decrease sample size
One peak affected	Sample decomposing on column	1. Decrease column temperature 2. Decompose sample to stable product 3. Change mobile phase 4. Change column packing
Reversed-phase ion-pair chromatography	Slow reaction kinetics	1. Use column with monolayer coverage 2. Increase concentration of ion pairing agent
Ion-exchange chromatography	Secondary adsorption effects	1. Increase column temperature 2. Add low concentration of organic modifier such as isopropyl alcohol to solvent
Chromatography of surfactants	Micelle formation	Use more polar mobile phase
All peaks affected, particularly early eluting peaks	Dead volume introduced in instrument	Remove dead volume, inspect fittings, use 0.010-in. tubing, keep lines short

(*continued*)

TABLE I (*continued*)

Circumstances	Possible cause	Remedy
	F. Fronting peaks	
All peaks affected	Poorly packed column	Replace or repack column
	G. Double or split peaks	
All peaks affected	Void or channel in column bed	1. Check for void at top of column, fill if found 2. Replace column
All peaks affected	Inlet frit plugged or contaminated	Replace or clean frit
	H. Ghost peaks	
Reversed-phase gradient elution	Impurities in water	1. Confirm by pumping varying volumes of aqueous solvent prior to gradient 2. Use HPLC grade water 3. Clean up water by passage over reversed-phase support
Gradient or isocratic elution	Sample carryover	Flush injector with strong solvent
Gradient or isocratic elution	Elution of sample components on column	1. Repeat elution program with blank injections 2. Flush column with strong solvent
	I. Truncated peak	
Concentrated or large sample	Detector overloaded	1. Decrease sample size 2. Use shorter path length cell 3. Use different wavelength
	J. Negative peaks	
Near V_p (size exclusion chromatography) or solvent front (other modes)	1. Refractive index of injection solvent different than in mobile phase 2. Temperature fluctuation associated with injection	1. Use same solvent for sample and mobile phase 2. Water jacket detector flow cell and column
	K. Drifting baseline	
	1. Late eluting components 2. Temperature fluctuation	1. Flush column with strong solvent 2. Control ambient temperature or detector flow cell temperature

TABLE I (*continued*)

Circumstances	Possible cause	Remedy
	3. Dissolved oxygen in mobile phase (low UV detection)	3. Degas mobile phase
	4. Washout of previous solvent in pump or column	4. Purge entire system with fresh solvent
	5. Residual immiscible solvent in pump or column	5. Wash system with commonly miscible solvent, e.g., propanol
	6. Selective evaporation of volatile mobile-phase component	6. Cap solvent reservoirs
	L. Baseline noise	
	1. Bubbles in detector cell	1. Degas solvents, install restrictor on detector outlet
	2. Leakage of packing material from column	2. Replace outlet terminator frit
	3. Dirty or misaligned detector cell	3. Clean or realign cell according to manufacturer's instructions
	4. Failing detector lamp	4. Replace lamp
	5. Defective pulse damper	5. Replace damper
	6. Poor instrument grounding, faulty recorder connections	6. Provide adequate grounding; clean or replace connections

unswept volume between injector and column can be diagnosed by thoroughly flushing (in the worst cases, backflushing) transfer lines and injector; improperly seated tubing and ferrules on injector loops or transfer lines are often the cause. Occasionally, proteins that fail to elute quantitatively by the end of a gradient will appear as ghost peaks at the same elution position in subsequent blank gradients, with peak height decreasing progressively. In such instances the column can be washed between runs by injecting a volume (up to several milliliters) of a strong solvent between analyses.

Tailing peaks and split peaks occur frequently with biological materials in HPLC. Peak tailing can arise from sample overload, a poorly packed or degraded column, buildup of adsorbed contaminants on the stationary phase, or extra-column effects. In reversed-phase chromatography of polypeptides, tailing is often a sign of interaction between residual silanol groups on the stationary phase and basic amino acid side chains. The effect can be minimized by operating at acidic pH to suppress silanol ionization, by adding a competing base such as triethylamine or a tetramethylammonium salt to the mobile phase, and by using a column with high surface coverage that has been end-capped (undergone secondary silanization to reduce the number of residual free silanols.) Split peaks or doublets arise by the formation of multiple flow paths through the column, usually by settling or degradation of the column bed to form voids or channels. Peak doublets can also occur as sample contaminants partially plug the inlet frit or column head, creating partial flow resistance. The effect may be accompanied by a gradual rise in operating pressure and can be cured by frit replacement or, as a last resort, removing and repacking the top millimeter of the column bed.

Column Repair

Careful column operation will prolong column lifetime to a year or longer. However, column performance will degrade as strongly adsorbed sample components accumulate on the stationary phase and as a void is formed by the gradual dissolution of the support. Often column life can be extended for old or abused columns by simple operations such as washing and backflushing, frit replacement, and bed repair.

Frit Cleaning and Replacement

To remove a column frit, tightly secure the column tube and carefully remove the top terminator fitting, being careful not to disturb the column bed. Replaceable frits may sometimes remain in the terminator body; they can usually be dislodged with the tip of a spatula blade or by passing a 20- or 24-gauge wire through the terminator inlet. Although replaceable stainless steel frits may sometimes be cleaned by immersing them in a sonic bath containing 3–6 *N* nitric acid, it is easier simply to install a new frit. When refitting the terminator, be sure the new frit is aligned and that surfaces are free of packing material to permit proper seating. Where the terminator ferrule has been distorted, it may be difficult to achieve a leak-free connection with the new frit; installation of double frits will sometimes allow a tight seal to be formed.

Terminators with pressed-in frits can be cleaned by pumping a nitric acid solution through the terminator. If this procedure is ineffective (as indicated by continued high backpressure during flushing), the terminator can be installed on an empty guard column and pumped in the reverse direction to backflush the frit. The analytical column should be capped with a spare terminator during frit cleaning to prevent drying or disruption of the bed.

Column Backflushing

When a column exhibits excessive operating pressure due to partial plugging of the inlet frit or column head, it is often possible to dislodge the material by reversing the direction of flow and backflushing with the mobile phase or a wash solvent; this is a simple remedy when a column begins to show a gradual rise in pressure, and it avoids the hazards of opening a terminator. However, several precautions should be observed. First, the pressure rise must *not* be accompanied by a loss in efficiency. Efficiency loss is indicative of voids in the bed, and backflushing may disturb the bed so as to prevent repair by topping off. Second, not all manufacturers recommend backflushing of columns; the user should refer to the column installation instructions or contact the manufacturer. Third, the column should be disconnected from the detector during backflushing to prevent contaminants or particulates from entering the flow cell. Occasionally, increasing pressure can arise from blockage of the outlet terminator, particularly when fines are generated by degradation of silica supports during operation at high pH, high temperature, or with cationic ion-pair agents. Under such conditions, backflushing will produce only a partial or transient reduction in pressure, and the outlet terminator frit should be replaced.

Repair of Voids and Bed Irregularities

Symptoms such as peak broadening, peak tailing, or split peaks suggest that a void or channel may have been formed in the column bed. If these are observed with a new column upon receipt or during the first few injections of normal operation, it indicates a damaged or poorly packed column, which should be returned. Voids that develop with extended use can be repaired by topping off; original performance will probably not be recovered, but the improvement in resolution is often acceptable, particularly in gradient elution.

After removal of the inlet terminator, the column bed should be level and flush with the face of the column tube. A light gray or brown discoloration on the bed surface is normal after extended use, but dark discoloration indicates a contaminated bed, which should be removed and repacked. Even small depressions in the bed can result in significant band broadening and should be filled; large voids of a centimeter or more probably cannot be repaired successfully. Before filling the void, the bed should be leveled using a square blade; a small flat-tip laboratory spatula trimmed to the column internal diameter works well. The void can be filled with glass beads, a microparticulate packing, or a pellicular guard column packing. Glass beads are not recommended because, unless they have been completely silanized, they will adsorb proteins. If a microparticulate material is used, it should be prepared as a slurry in an organic solvent and added dropwise to the column, allowing excess solvent to permeate the bed. Since microparticulate packings are high capacity, the stationary phase should be the same as that of the column to prevent selectivity changes. In most cases, a pellicular packing will serve as the best repair material since loss in efficiency due to a poorly repacked void is not as significant with a low-capacity packing. Once the void is filled, the bed surface should be leveled flush with the face of the tubing, excess packing material removed from the tubing surface, and the terminator replaced. After the repaired column has been tested, the terminator should be removed and the bed checked for settling. If settling has occurred, the top-off procedure should be repeated.

Column Washing

With extended use, the buildup of strongly retained material such as lipids or basic and hydrophobic proteins will cause increased operating pressure and altered chromatographic behavior of an HPLC column. Often column performance can be recovered by stripping adsorbed material with one or a series of strong solvents. The key in rejuvenating a contaminated column lies in knowing the nature of the contaminants and an appropriate strong solvent. Lipids can be removed by washing with nonpolar solvents such as methanol, acetonitrile, or tetrahydrofuran. There is no solvent or series of solvents that will universally strip all adsorbed proteins from HPLC stationary phases, but Table II lists several strong eluents or solubilizing agents that have been used in specific instances for stripping proteins from HPLC columns. Some columns, particularly those based on organic polymer supports rather than silica, are not compatible with these solvents, and the user should contact the manufacturer regarding solvent compatibility. In most cases, neat organic solvents such as acetonitrile or methanol are not strong eluents in protein chromatography and, therefore, are not effective stripping solvents; 50 : 50 mixtures of organic solvents with a buffer, organic acid, or ion-pairing agent serve as better stripping agents for reversed-phase and size-exclusion columns. It has been observed[8] that repeated gradients followed by retrogradients

[8] F. E. Regnier, this series, Vol. 91, p. 165. See also this volume [8].

TABLE II
WASH SOLVENTS FOR HPLC COLUMNS[a]

Solvent	Composition
Reversed-phase and size-exclusion columns	
Acetic acid	1% in water
Trifluoroacetic acid	1% in water
0.1% Aqueous trifluoroacetic acid/propanol[b]	40:60
TEAP[c]/propanol[b]	40:60
Aqueous urea or guanidine	5–8 *M*
Aqueous sodium chloride, sodium phosphate, sodium sulfate	0.5–1 *M*
DMSO–water[a]	50:50
Ion-exchange columns	
Acetic acid	1% in water
Phosphoric acid	1% in water
Aqueous sodium chloride, sodium phosphate, or sodium sulfate	1–2 *M*

[a] Consult manufacturer for column compatibility.
[b] High viscosity solvent; pump at reduced flow rate to prevent overpressure.
[c] Triethylamine-phosphoric acid; adjust 0.25 *N* phosphoric acid to pH 2.5 with triethylamine.

between aqueous trifluoroacetic acid (TFA) and TFA–propanol can be effective in regenerating contaminated HPLC columns. Although detergents such as sodium dodecyl sulfate and Triton are good protein solvents, they tend to be strongly retained on HPLC stationary phases and may irreversibly change column characteristics; detergent cleanup should be used as a last resort and be followed by extensive washing with water and methanol.

Several precautions should be observed during column cleanup. If the initial wash solvent is not compatible or miscible with the mobile phase, the column should be flushed with an intermediate solvent. Similarly, sets of wash solvents used in series must be compatible or, if not, interspersed with a mutually compatible flushing solvent. For example, after washing with high salt, urea, or guanidine, the column *must* be purged with 5–10 volumes of water before the introduction of any organic solvent. Similarly, if nonpolar solvents such as hexane or methylene chloride are used to strip lipids from a reversed-phase column, a propanol purge is necessary before the introduction of aqueous solvents. To avoid shocking the column bed, it is advisable to introduce wash solvents with gradients over 5–10 column volumes. Viscous solvents such as dimethyl sulfoxide

(DMSO)–water, methanol–water, and propanol–water mixtures should be pumped at reduced flow rates to prevent high pressures.

Successful regeneration of a contaminated column can be a time-consuming process. With microprocessor-controlled HPLC systems, multistep washing sequences can be programmed and run automatically overnight. Alternatively, column regeneration can be carried out off-line with an inexpensive low-pressure pump at reduced flow rates.

[7] High-Performance Size-Exclusion Chromatography

By KLAUS UNGER

Since 1980 the classical gel filtration technique employing soft and semirigid organic gels for protein characterization and purification has received progressively greater competition from high-performance size-exclusion chromatography (HPSEC). The breakthrough of HPSEC is associated with the development of highly efficient buffer-compatible columns operating at elevated back pressures. The columns are packed with microparticulate organic-based or silica-based particles of tailor-made graduated pore size and hydrophilic surface composition. High-resolution separation of proteins on these columns is attained by adjusting appropriate operating conditions. The proteins elute in the sequence of decreasing molecular weight and size. In its simplest version, the retention of the proteins in HPSEC is considered to be a selective permeation of biopolymeric solutes through the pores of the particles of the column bed; smaller proteins should penetrate a larger portion of the pore volume of the packing and hence will be retarded longer than larger proteins. Two limiting cases arise: (1) proteins so large that they are excluded from the pores will be eluted with a volume equal to the interparticle volume of the column, V_0; (2) small proteins that totally permeate the specific pore volume of the packing, V_i, will elute with $V_e = V_0 + V_i$. The total accessible volume is then equal to $V_t = V_0 + V_i$. The intraparticle volume of the column is termed V_i. Between these two limits, a linear relationship is established between the logarithm of molecular weight (M) of the protein and its elution volume, V_e. This retention mechanism is distinctly different from other modes of column liquid chromatography, such as adsorption and ion exchange, and it entails several advantages. Proteins are eluted quickly with relatively narrow bands. A predictable volume inter-

ISBN 0-12-182004-1

val for the elution operates when V_0 and V_t of the column are established. Calibration of the column using the plot of log M against V_e for standard proteins enables the molecular weight of an unknown solute to be estimated by its elution volume. Apart from the selection of a proper column, no intensive method development is required. The eluents used are buffers of appropriate pH and ionic strength, and the necessary instrumentation is also simple. These features more than compensate for the disadvantages involved, such as the limited peak capacity of HPSEC (intrinsic to the method) and restrictions in the pH range of the eluent when applying silica-based packings.

Fundamental Aspects

Although the area of application of HPSEC is very broad, two general aims may be distinguished. In analytical separation and identification, small sample sizes are injected and the primary objective is to gain optimum resolution of components during a reasonable period of time. In preparative separation, the purpose is to isolate large amounts per unit time at a given purity and at reasonable cost.

Under analytical conditions, the elution volume of the protein on an HPSEC column obeys Eq. (1), provided size exclusion is the only separation mechanism.

$$V_e = V_0 + K_{SEC} V_i \quad (1)$$

where K_{SEC} denotes the distribution coefficient of the protein and expresses its degree of permeation. K_{SEC} is defined by Eq. (2), which equals zero for a totally excluded protein, and unity for a totally permeating solute. Appropriate markers for V_0 of HPSEC columns are horse apoferritin (467,000 daltons),[1] pig thyroglobulin[1] (670,000), bovine glutamic dehydrogenase[1] (998,000), and calf thymus DNA.[2] Dextran Blue of 2×10^6 daltons was found to be less suitable owing to its tailing peak.[1,2] As V_t markers, sodium azide and DNP-alanine[1] are usually preferred. Other solutes have also been recommended as V_t markers to check the extent of non-size-exclusion effects on retention.[2,3]

$$K_{SEC} = (V_e - V_0)/(V_t - V_0) \quad (2)$$

To construct a calibration curve, size-exclusion data are arranged by plotting log M vs V_e or K_{SEC} (Fig. 1).

[1] M. E. Himmel and P. G. Squire, *Int. J. Pept. Protein Res.* **17,** 365 (1981).

[2] E. Pfankoch, K. C. Lu, F. E. Regnier, and H. G. Barth, *J. Chromatogr. Sci.* **18,** 430 (1980).

[3] P. Roumeliotis and K. K. Unger, *J. Chromatogr.* **218,** 535 (1981).

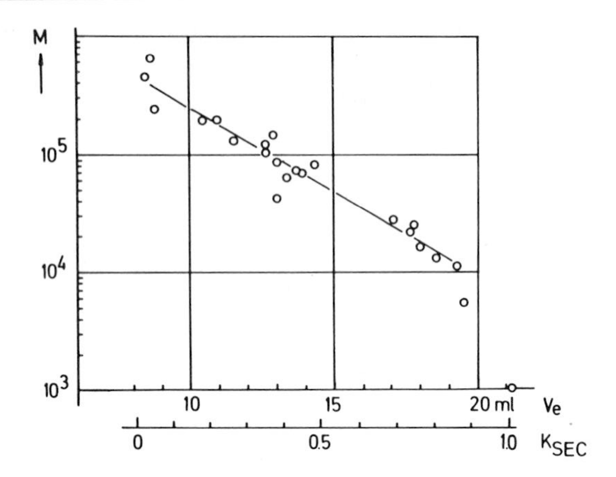

FIG. 1. Calibration curve of native proteins on a TSK 3000 SW column (7.5 × 600 mm). Eluent consists of 10 m*M* potassium phosphate at pH 7.0, 100 m*M* KCl, and 0.02% NaN_3. Data are calculated from Table I of Himmel and Squire[1] with permission of the publisher.

The interesting part of the calibration curve is the linear intermediate portion, i.e., the working range of the column. Above the exclusion limit all excluded proteins coelute with the same V_e. The same is valid at the lower end for the totally permeating solutes. Typically, an HPSEC column with a microparticulate organic-based or silica-based packing elutes proteins of about 1.5–2 decades of molecular weight across the linear range. This means that a total of two or three columns, each with an appropriate calibration curve, is required to cover the whole molecular weight range of about 4 decades from 1000 to 10 million.

A large number of theoretical models have been developed to relate the molecular size and shape of polymeric solutes to the pore size of the packing and to predict the experimental calibration curve with sufficient accuracy.[4] As a rule of thumb, two silica-based HPSEC packings having a mean pore diameter of 10 and 80 nm will span the whole molecular weight range for protein separation.

Since the elution volume of protein is affected by both size and shape, the calibration curve log M vs V_e is valid only for proteins of the same shape, e.g., globular. This aspect has been thoroughly examined for native[1] and denatured[5] proteins on TSK-SW columns. Himmel and Squire[1] stated: "When using a calibration curve F_v vs $M^{1/3}$ and a well-calibrated

[4] W. W. Yau, J. J. Kirkland, and D. D. Bly, "Modern Size Exclusion Chromatography," p. 27. Wiley, New York, 1979.

[5] N. Ui, *Anal. Biochem.* **97,** 65 (1979).

column, the estimated error in calculating molecular weight expressed as standard deviation is 14%." F_v is given by Eq. (3).

$$F_v = (V_e^{1/3} - V_o^{1/3})/(V_t^{1/3} - V_o^{1/3}) \quad (3)$$

Deviations from linearity (i.e., much larger errors) are often caused by proteins being involved in non-size-exclusion interactions on HPSEC columns. Ionic interactions between proteins and charged surface sites of the packing result either in a decrease or increase of V_e, depending on whether repulsion or attraction forces operate. With eluents of high ionic strength, hydrophobic interactions between the protein and hydrophobic surface groups of the packing increase the elution volume. These secondary effects can be suppressed or minimized by appropriate manipulation of the eluent composition.[6] On the other hand, they may be utilized in some cases to generate a superior selectivity than is expected from pure size exclusion alone.[3]

The selectivity of an HPSEC column is expressed by the slope of the linear part of the calibration curve D_2; i.e., the smaller the slope, the better is the discrimination between proteins of different molecular weight. The primary interest in HPSEC is aimed at gaining resolution, R_s. For two solutes of similar peak width and height, R_s is defined by Eq. (4)

$$R_s = \frac{V_{e(2)} - V_{e(1)}}{2(\sigma_1 + \sigma_2)} \qquad V_{e(2)} > V_{e(1)} \quad (4)$$

where σ_1 and σ_2 are the standard deviation of peak 1 and peak 2, respectively. Furthermore,

$$M = D_1 e^{-D_2 V_e} \quad (5)$$

where D_1 is the intercept of the calibration curve. Combining Eq. (4) and Eq. (5) yields Eq. (6).

$$R_s = \frac{\ln (M_2/M_1)}{2D_2(\sigma_1 + \sigma_2)} \quad (6)$$

High resolution is achieved on a column with a small value of $D_2(\sigma_1 + \sigma_2)$. D_2 depends on a proper choice of column packing material and is proportional to the reciprocal of the column length, L. σ is proportional to the square root of the column length.[7] Thus increase in column length increases the resolution.

A special situation, compared to other modes of HPLC, is met in HPSEC with regard to peak dispersion. Peak dispersion is expressed by the standard deviation of the peak (σ_{V_i} in volume units), given by Eq. (7)

[6] H. G. Barth, *J. Chromatogr. Sci.* **18,** 409 (1980).

[7] W. W. Yau, J. J. Kirkland, and D. D. Bly, "Modern Size Exclusion Chromatography," p. 102. Wiley, New York, 1979.

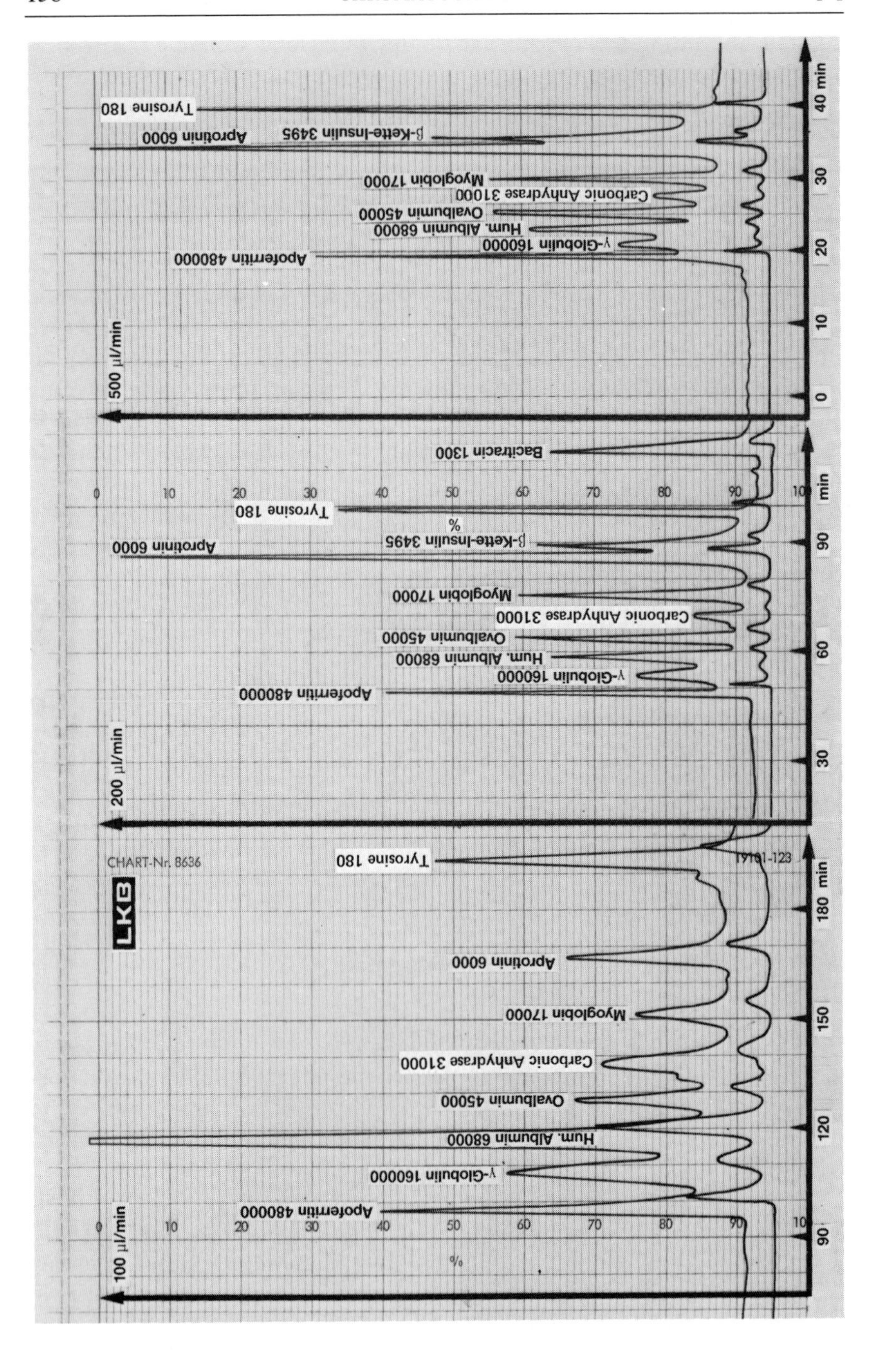
Tyrosine 180
Aprotinin 6000
β-Kette-Insulin 3495
Myoglobin 17000
Carbonic Anhydrase 31000
Ovalbumin 45000
Hum. Albumin 68000
γ-Globulin 160000
Apoferritin 480000
500 μl/min
Bacitracin 1300
200 μl/min
100 μl/min
CHART-Nr. 8636
LKB
%
min

$$\sigma_{V_i} = V_{e(i)}/N_i^{1/2} \quad (7)$$

where N is the number of theoretical plates per column length. The aim is to keep σ_{V_i} to a minimum or to generate a large plate number.

Totally permeating solutes exhibit the highest plate number per column length, whereas N usually diminishes with decreasing elution volume of the proteins at a constant flow rate of the eluent. The drop in N with increasing molecular weight of the solute is due to the inverse relationship between the size of the molecule and the diffusion coefficient. Low diffusivity, however, is associated with slow mass transfer and enhanced peak broadening. In order to compensate for this undesirable effect, the flow rate must be reduced by a factor of about 100, compared to normal HPLC of monomeric solutes. Since the resolution R_s is proportional to $N^{1/2}$, [combining Eqs. (6) and (7) and assuming $\sigma_{V1} = \sigma_{V2}$ and $N_1 = N_2$], R_s is improved with reduced flow rate. This is shown in Fig. 2.

The relation between the elution volume (V_e), the elution time (t_e), and the flow rate (f_v) is given by Eq. (8).

$$V_e = f_v t_e \quad (8)$$

Since the elution volume remains constant, a reduction in flow rate causes an increase in t_e. This is exemplified by the data in Fig. 2. The required resolution must be weighed against an acceptable analysis time.

Isolation of proteins by means of HPSEC involves an optimization governed by the following quantities: resolution, sample throughput per unit time, and purity of protein collected. When the required sample mass and its purity are fixed, the problem reduces to how much sample size can be tolerated in column load while maintaining the resolution attained under analytical conditions. Assuming that the elution volume remains unaffected by increasing sample size, a loss in resolution is expected to occur only through a decrease in the plate number of the HPSEC column. Defining a 20% decrease of the plate number as a loadability limit, the mass load at constant injection volume was found to be 0.1 mg of protein per gram of column packing material[8] for LiChrosorb Diol.

A TSK 3000 SW column of 600-mm length and 7.5-mm i.d. contains about 10 g (for 21.5-mm i.d., about 80 g) of silica support. Mass loadability then calculates to 1 mg (or 8 mg) under these conditions. Increasing the

[8] P. Roumeliotis and K. K. Unger, *J. Chromatogr.* **185,** 445 (1979).

FIG. 2. Effect of eluent flow rate, f_v, on the resolution of proteins. Elution with 0.1 *M* sodium phosphate–0.1 *M* NaCl at pH 6.8 with a TSK 3000 SW column (7.5 × 600 mm) and monitored at 206 nm. The chromatogram was kindly supplied by Dr. Földi, LKB Instrument GmbH, Karlsruhe, FRG.

sample concentration also increases the viscosity. As an approximate guide, the relative viscosity of the sample (compared to the eluent) should be less than 2, corresponding to about 70 mg/ml for proteins in dilute buffers.[9] Higher concentrations may also change the hydrodynamic volume of biopolymeric solutes and thus affect the elution volume. Another means for circumventing mass overload is to inject a large volume of a dilute sample. Again, however, a limit exists that is given by the volume loadability of the column. This is the sample volume, at a given mass of solute, that causes a 20% decrease of the column plate number. This value was found to be about 1–2% of the column bed volume[8]; for a column of 600 × 7.5 mm, the volume amounts to 270–540 μl. On the other hand, a high dilution of the solute is unfavorable when isolation of the protein is considered.

In practice, one can overload the HPSEC column to some extent and accept a certain loss in resolution. Then the central portion of the peaks is collected and the fractions are recycled for repetitive separation. This is the preferred technique for the isolation of 10- to 100-mg quantities of proteins from complex mixtures. Ultimately, however, this approach is costly, because expensive columns are employed.

When analytical conditions no longer apply and real column overload is apparent, the resolution equation defined above loses its original meaning. The elution volume V_E is then defined as the volume at the half-height of the ascending part of the plug of solute emerging from the column. The separation volume, V_{sep}, is then given by $V_{E(2)} - V_{E(1)} (V_{E(2)} > V_{E(1)})$ (see Pharmacia[10]). Typical cases for this approach are group separation and desalting processes, where a high resolution exists under analytical conditions. Sample volumes that correspond to up to 30% of the column bed volume are applied in such processes.[10] Under these circumstances microparticulate packings no longer offer any advantage and large-bore columns packed with coarse particles of about a 50-μm mean diameter are preferred. As a consequence of increasing particle diameter the column back pressure is drastically reduced and much simpler instrumentation is required. Typical packings for this purpose are the Fractogel TSK types marketed by E. Merck, Darmstadt, FRG.[11]

Equipment

HPSEC is carried out with conventional HPLC equipment composed of the following basic components: solvent reservoir, high-pressure

[9] H. G. Barth, *J. Chromatogr. Sci.* **18,** 409 (1980).

[10] Pharmacia Fine Chemicals, "Gel Filtration, Theory and Practice." Uppsala, Sweden.

[11] E. Merck, Fractogel TSK, Reagenzien Merck, Darmstadt, FRG.

pump, injection system, column, detector, recorder, and appropriate data-handling systems.[12] A fraction collector connected to the detector cell outlet is employed when isolation is considered. To check the purity of fractions separated, gel electrophoresis is employed or, alternatively, HPLC using a reversed-phase or ion-exchange column. The equipment can be either modular (i.e., assembled of single components) or a completely integrated apparatus. Since the eluents are buffers, often of high salt concentration, corrosion of stainless steel parts becomes noticeable on long-term use. For that reason pumps have been developed with ceramic-lined cylinders and sapphire pistons and valve seals, e.g., the LKB 2150 HPLC pump. Similarly, columns made of high-pressure-resistance glass have been developed, e.g., those manufactured by Laboratorni Pristoge, Prague, Czechoslovakia. However, a complete apparatus of inert equipment is not yet commercially available. Among other decisive design criteria controlling analytical accuracy, speed of analysis, and separation versatility are the following: a precise and stable flow rate, precise sampling, efficient columns with graduated exclusion limits and sufficient lifetime, low dead volume connections, and good temperature control.

Commercially available pumps are of three types: reciprocating, positive displacement, and constant pressure. A detailed comparison of the different types under the aspects of resettability, drift, short-term precision, accuracy, and cost has been presented.[12] There are two important aspects to consider when employing pumps in HPSEC. As the working volume of HPSEC columns is rather small compared to other HPLC columns, a high precision of the flow rate produced is required. For instance, a TSK 3000 SW column of 600 × 7.5 mm possesses a working volume equal to the intraparticle volume of 13 ml. Within that volume, proteins are eluted spanning a molecular weight of 40,000 to 400,000. This means that the elution volume, measured with a precision of about ±1%, generates a precision of about 10% in molecular weight.[1] Furthermore, for full utilization of the performance of HPSEC columns for high-resolution separation of proteins, flow rates of the order of 10–100 μl/min are necessary for 6-mm bore columns. Most of the currently available HPLC pumps produce flow rates in the range of 0.1–10 ml/min; i.e., columns are operated under far from ideal conditions.

To maintain the efficiency of the column (see this volume [6]), samples are injected by means of an injector as a sharp plug with a reproducibility of better than ±2%. Sample volume varies between 10 and a few hundred microliters. Of the two injection systems, the syringe-septumless device

[12] W. W. Yau, J. J. Kirkland, and D. D. Bly, "Modern Size Exclusion Chromatography," p. 123. Wiley, New York, 1979.

and the microsampling detector valve, the latter offers higher reproducibility and flexibility. By interchanging the loops, up to a few millimeters can be introduced onto the column; automatic versions of the sampling device are available.

Columns made of stainless steel and closed by appropriate low dead-volume end fittings are usually 4–25 mm in bore and 300–600 mm in length. These are packed with particles of a mean diameter of 5–15 μm, mostly 10 μm, and narrow size distribution of about 10–20% standard deviation from the mean. Packings are either organic- or silica-based; in both cases the pore size and size distribution are the most significant packing properties. These are controlled in the case of organic gels by the degree of cross-linking and in the case of porous silica, by the method of synthesis and the nature of the aftertreatment. To become suitable as HPSEC packings for proteins, the surface of these particles must be hydrophilic, free of charged surface sites, and chemically stable toward the buffered eluents. Hydroxyl, ether, and amide functional groups anchored by a hydrocarbon spacer to the surface of the matrix are preferred. The exact bulk and surface composition is not indicated in most of the commercial packings. Detailed examinations have been carried out on diol and amide-modified silica packings.[13,14] The problems arising in the modification of silicas lie in attaining a chemically stable and dense surface layer that essentially protects the residual silanols.

Silanol groups cause adsorption and denaturation of proteins and act as weak cation-exchange sites at pH > 7, giving rise to ionic interactions. Owing to the chemical stability of the Si—O—Si—C link and to the solubility of silica itself, the packings are employed in a pH range of 2 to 8.5. The major aspect in the synthesis of organic-based packings is to overcome the swelling property by appropriate cross-linking. The organic packings also may carry functional groups that undergo protonation or deprotonation, e.g., amino and carboxyl groups. Thus organic-based types of HPSEC packings are applicable in the wider pH range of 2 through 13, although stability problems remain. For instance, the TSK-type PW packings, being a semirigid hydrophilic cross-linked gel containing —$CH_2CHOHCH_2O$— groups as main constituent, do not tolerate salt concentrations higher than 0.5 *M*.[15]

As most proteins absorb in the UV, photometers, preferably of variable wavelength, are usually used as detectors. The latest developments

[13] D. P. Herman and L. R. Field, *J. Chromatogr. Sci.* **19,** 470 (1981).

[14] H. Engelhardt and D. Mathes, *J. Chromatogr.* **185,** 305 (1979).

[15] C. T. Wehr, T. V. Alfredson, and L. Tallman, "Varian, Liquid Chromatography," No. LC-134. Varian Associates, Walnut Creek, CA, 1982.

in this field are the diode array spectrophotometers that scan the whole spectrum (190 to 700 nm) of a solute emerging from the column in 10 nsec.

Flow fluorometers are also applied in HPSEC, whereby UV-activated solutes emit a fraction of absorbed light as fluorescence. In this way a noise-level concentration of 10^{-11} g/ml can be achieved under appropriate conditions. Solutes that do not fluoresce can be converted into fluorescent components by means of an appropriate postcolumn reaction although difficulties arise in ensuring a sufficient conversion (adjustment of the residence time of reactants) and in avoiding large dead volumes.[16,17]

The principle of one-line postcolumn reaction has also been used in developing a direct or coupled enzyme assay using photometric detectors.[18]

Column Selection and Maintenance

The choice of an appropriate HPSEC column and its operation under optimum conditions is the key to a successful separation of proteins (also see this volume [6]). The table lists the commercially available organic-based and silica-based packings and columns. The majority of HPSEC separations are carried out on the TSK columns of Toyo Soda. Thus most of the work on column characterization and use will refer to this type.

The type of column chosen should exhibit a fractionation range that covers the molecular weight range of the proteins to be resolved. This column property is independent of the particle size of the packing and the column length and is solely a function of the pore size distribution for silica-based packings or the degree of cross-linking for organic-based packings.

The most frequent range in which size-exclusion separations of proteins are carried out covers 10,000 to 500,000 daltons. This is almost achieved by a single column: the TSK 3000 SW has a fractionation range of 40,000 to 400,000. The mean pore diameter of this packing is estimated to be about 30 nm. With materials of large pore size, e.g., 100-nm pore diameter, the exclusion limit is extended to above 1 million daltons. Small proteins of $<$40,000 daltons require packings of about 5-nm mean pore diameter. Special effects arise in the application of small and large pore size packings, associated with their structure. Since the mean pore diameter of a rigid porous packing such as silica is inversely proportional to the specific surface area, 5-nm pore size packings offer high surface areas of

[16] R. W. Frei and A. H. M. T. Scholten, *J. Chromatogr. Sci.* **17,** 152 (1979).

[17] J. F. K. Huber, K. M. Jonker, and H. Poppe, *Anal. Chem.* **52,** 2 (1980).

[18] T. D. Schlabach and F. E. Regnier, *J. Chromatogr.* **158,** 349 (1978).

COMMERCIALLY AVAILABLE HPSEC PACKINGS AND COLUMNS

Trade name	Composition	Fractionation range for globular proteins (daltons)	Mean particle size (μm)	Column dimensions, length × inner diameter (mm)	pH stability	Supplier[a]
TSK-gel SW type						
Analytical						
G 2000 SW	Spherical porous silica with bonded	5000–60,000	10 ± 2	300 (600) × 7.5	2.5–8.0	1
G 3000 SW	hydrophilic polar groups; exact	1000–300,000	10 ± 2	300 (600) × 7.5	2.5–8.0	1
G 4000 SW	composition not known	5000–1,000,000	13 ± 2	300 (600) × 7.5	2.5–7.5	1
Preparative						
G 2000 SWG	Spherical porous silica with bonded	5000–60,000	13 ± 2	300 (600) × 21.5	2.5–8.0	1
G 3000 SWG	hydrophilic polar groups; exact	1000–300,000	13 ± 2	300 (600) × 21.5	2.5–8.0	1
G 4000 SWG	composition not known	5000–1,000,000	17 ± 2	300 (600) × 21.5	2.5–7.5	1
TSK-gel PW type						
Analytical						
G 1000 PW	Hydroxylated polyether; exact	NG[b]	10 ± 2	300 (600) × 7.5	2–12	1
G 2000 PW	composition not known	NG	10 ± 2	300 (600) × 7.5	2–12	1
G 3000 PW		NG	13 ± 2	300 (600) × 7.5	2–12	1
G 4000 PW		NG	13 ± 2	300 (600) × 7.5	2–12	1
G 5000 PW		NG	17 ± 2	300 (600) × 7.5	2–12	1
G 6000 PW		NG				
Preparative						
G 1000 PWG	Hydroxylated polyether; exact	NG	NG	300 (600) × 21.5	2–12	1
G 2000 PWG	composition not known	NG	NG	300 (600) × 21.5	2–12	1
G 3000 PWG		NG	NG	300 (600) × 21.5	2–12	1
G 4000 PWG		NG	NG	300 (600) × 21.5	2–12	1
G 5000 PWG		NG	NG	300 (600) × 21.5	2–12	1
G 6000 PWG		NG	NG	300 (600) × 21.5	2–12	1
LiChrosorb Diol	Angular porous silica with bonded	10,000–100,000	5, 7, 10	Bulk material	2–8.5	2
	glycerol propyl groups		7	125 (200) × 4	2–8.5	2

LiChrospher						
100 diol/II	Spherical porous silica with bonded glycerol propyl groups	NG	5, 10	250 × 4	2–8.5	2
300 diol/II		NG	10	250 × 4	2–8.5	2
500 diol/II		NG	10	250 × 4	2–8.5	2
1000 diol/II		NG	10	250 × 4	2–8.5	2
4000 diol/II		NG	10	250 × 4	2–8.5	2
Protein I 60	Porous silica with bonded hydrophilic polar groups (exact composition not known; I 60, irregular silica; I 125 and I 250, spherical silica)	1000–20,000	10	300 × 7.8	2–8.0	3
Protein I 125		2000–50,000	10	300 × 7.8	2–8.0	3
Protein I 250		10,000–500,000	10	300 × 7.8	2–8.0	3
Shodex OH pak	Spherical methacrylate glycerol copolymer	NG	10	NG	4–12	4
Synchropak						
100	Spherical porous silica with bonded glycerol propyl groups	NG	10	NG	2–8.5	5
300		NG	10	NG	2–8.5	5
500		NG	10	NG	2–8.5	5
1000		NG	10	NG	2–8.5	5
4000		NG	10	NG	2–8.5	5
Separon HEMA						
40	Spherical poly(2-hydroxyethyl-methacrylate)	40,000[c]	10 ± 2	Bulk material	Not indicated	6
100		100,000[c]	10 ± 2	Bulk material	Not indicated	6
300		300,000[c]	15 ± 2			
1000		1,000,000[c]	15 ± 2			
Separon HEMA						
300 Glc	Spherical Separon HEMA with bonded glucose	300,000[c]	10 ± 2	Bulk material	Not indicated	6
1000 Glc		1,000,000[c]	15 ± 2	Bulk material	Not indicated	6

[a] *List of suppliers:* (1) Toyo Soda Manufacturing Co. Ltd., Tokyo, Japan; columns are also marketed by LKB Producter AB, Bromma, Sweden and by Varian Associates, Walnut Creek Instruments Division, Walnut Creek, CA; (2) E. Merck, Darmstadt, FRG; (3) Waters Associates, Milford, MA; (4) Showa Denko K. K., Tokyo, Japan; (5) Synchrom Inc., Linden, IN; packings of this type are also marketed by Brownlee Laboratories, Inc., Santa Clara, CA; (6) Lachema, Brno, Czechoslovakia.

[b] NG, Not given.

[c] Exclusion limit for globular proteins.

the order of 400 to 500 m^2/g, possibly slightly less after modification. This gives rise to adsorption effects in protein separation by HPSEC that can be decreased to some extent by manipulation of the composition of the eluent. For large-pore silicas of 100-nm pore size, problems of mechanical stability of the particles bring about limitation in the pressure and flow rate used in the slurry packing method.

Commercial columns differ in their working volume at a given fractionation range, i.e., differences in the intraparticle volume of the columns. For comparison, V_i is related to the V_o of the column. The ratio V_i/V_o, the "phase ratio," varies between 0.6 and 1.70, the latter being the highest value attainable for silica packings.[1,2] As the intraparticle volume determines both separation capacity and column selectivity, these can be improved by coupling two separate columns of the same type or by applying a column of 600-mm length, instead of 300 mm. It will be evident that by doubling the column length the slope of the calibration curve D_2 (in milliliters of V_e per decade of molecular weight) becomes smaller by a factor of 2 while the range of separation remains the same; resolution is improved as expected from Eq. (6).

To characterize the proteins by molecular weight, a calibration curve of log M vs V_e (Fig. 2) is needed. It is advisable to gauge the calibration plot of the column employed by measuring the elution volume of standard proteins at the given eluent composition, rather than relying on literature data. This procedure should be repeated from time to time to monitor possible changes in the column or packing structure. Although the commercial HPSEC columns show high recoveries for native proteins,[2] the data are measured for a given eluent composition and may change under new conditions.

When required, the fractionation range can be extended by coupling columns of different and descending pore sizes. The whole separation range is spanned with three columns of 5-, 30-, and 100-nm pore size. Yau *et al.*[19] have shown that the individual calibration curves should be adjacent but not overlapping, in order to obtain the broadest range and maximum linearity of the calibration curve. The intraparticle volume of each column should be equal, since sigmoidal calibration curves would be obtained otherwise.

Since the pressure Δp across the column is proportional to column length, Δp increases when columns are assembled. Longer columns also enhance the analysis compared to short columns. The relationship be-

[19] W. W. Yau, J. J. Kirkland, and D. D. Bly, "Modern Size Exclusion Chromatography," p. 267. Wiley, New York, 1979.

tween these parameters is given by Eq. (9), where η is the viscosity of the eluent, ϕ' is the flow resistance factor of the column (about 500–1000), d_p is the mean particle diameter of the packing, and t_m is the elution time of a totally permeating solute.

$$\Delta p = [(\eta\phi')/t_m](L/d_p)^2 \tag{9}$$

To keep an HPSEC column functioning properly for a long time, a series of precautions have to be considered (see this volume [6]). Primary attention has to be paid to the eluent composition. Prepacked HPSEC columns of TSK types are filled with a 0.05% sodium azide in aqueous solution on delivery. Solvent replacement with an appropriate buffer is carried out at a low flow rate, e.g., 0.5 ml/min. Column lifetime is limited by both the chemical resistance of the packing and the stability of stainless steel column toward corrosive reagents. The pH range available for the TSK PW type of packings is much wider than that of the TSK SW type owing to the enhanced solubility of silica above pH 8.0 and the sensitivity of the Si—O—Si—C link in alkaline media. High salt concentrations are often applied to suppress ionic interactions between the support and the proteins. To avoid a rise in viscosity and precipitation of salt due to temperature changes, the concentration should be kept below 0.5 *M*. Column life is drastically reduced when proteins are separated in their denatured form by applying sodium dodecyl sulfate (SDS), highly concentrated guanidinium hydrochloride, or urea. When changing to such denaturing conditions, it should be noted that the elution volume of the proteins on the same column becomes shorter owing to the unfolded state of the protein. In principle, HPSEC columns can be operated above room temperature to reduce the viscosity of the eluent, to increase the plate count by enhanced diffusivity of proteins, and to prevent possible adsorption.

For TSK PW and TSK SW columns, a maximum flow rate of 1.2 ml/min (analytical column) and of 8.0 ml/min (preparative column) is tolerable. In routine work it is advisable to run the column overnight at a low flow rate of about 0.5 ml/min. On longer storage the buffer must be replaced by rinsing the column with dilute buffer and finally conditioning with an aqueous solution of 0.05% sodium azide.

Column life is also dependent on the purity of the samples that are introduced. Solutions to be injected should be free of fines and colloidal constituents. Samples must be dissolved in the eluent. Degrading of the column is prevented to some extent by inserting a short guard column between the injector and the column itself. Guard columns are packed with the same material; since their capacity is limited, they must be replaced from time to time.

Operating Conditions

Having decided on the type and dimension of column, an eluent is chosen that is composed of a buffer, a salt, and, if needed, a stabilizer. Typical buffers of pH 5–8 are phosphate, Tris acetate, citrate, and acetate. The ionic strength is adjusted by adding sodium chloride, sodium sulfate, ammonium acetate, and ammonium formate. The addition of salts suppresses ionic interactions between the charged surface sites of the HPSEC packing and the protein. Depending on the pH and the isoelectric point of the protein, the elution volume increases or decreases with increasing ionic strength.[20,21] The charge and type of the ions added also affect the hydration of the protein according to the chaotropic series, which may lead to changes in the elution volume.[3]

For separation of denatured proteins, 0.1% aqueous SDS, 6 *M* guanidine hydrochloride, or 6 *M* urea solutions are applied as eluent. SDS is preferred for purification of membrane proteins, which do not dissolve in common buffers. Kato *et al.*[22–24] established the calibration plots of TSK-gel SW columns for a series of proteins under denaturing conditions. In eluents containing SDS, the sodium phosphate concentration was shown to have a significant effect on the slope of the calibration curve.

Optimum performance of HPLC column, i.e., minimum peak dispersion and highest plate count N, is known to be attained at values of the reduced linear velocity (ν) between 1 and 10 for monomeric solutes.[25] Equation (10) defines ν, where

$$\nu = (ud_p)/D_{im} \tag{10}$$

u is the linear velocity of the eluent, d_p is the mean particle diameter of packing, and D_{im} is the diffusion coefficient of solute in the eluent. With $d_p = 10\ \mu m$ and $D_{im} = 1 \times 10^{-11}\ m^2/sec$ for a protein[26] the linear velocity of the eluent calculates to 0.0001 cm/sec ($\nu = 1$) and 0.001 cm/sec ($\nu = 10$). This corresponds to a volume flow rate (f_v) of 2–20 μl/min for a 7.5-mm i.d. column. As most HPSEC separations are carried out at flow rates of about 0.5–1.0 ml/min for a 7.5-mm i.d. column, these columns are operated far from their optimum. It must be said, however, that such low flow

[20] D. E. Schmidt, R. W. Giese, D. Conron, and B. L. Karger, *Anal. Chem.* **52,** 177 (1980).

[21] P. Roumeliotis, K. K. Unger, J. Kinkel, G. Brunner, R. Wieser, and G. Tschank, *in* "High Pressure Liquid Chromatography in Protein and Peptide Chemistry" (F. Lottspeich, A. Henschen, and K. Hupe, eds.), p. 71. de Gruyter, Berlin, 1981.

[22] Y. Kato, K. Komiya, H. Sasaki, and T. Hashimoto, *J. Chromatogr.* **193,** 29 (1980).

[23] Y. Kato, K. Komiya, H. Sasaki, and T. Hashimoto, *J. Chromatogr.* **193,** 458 (1980).

[24] Y. Kato, K. Komiya, H. Sasaki, and T. Hashimoto, *J. Chromatogr.* **190,** 297 (1980).

[25] P. A. Bristow and J. H. Knox, *Chromatographia* **10,** 279 (1977).

[26] J. M. Schurr, *CRC Crit. Rev. Biochem.* **4,** 371 (1977).

rates would produce long analysis times; assuming an elution volume of 15 ml for a protein on a 600 × 7.5 mm HPSEC column at $f_v = 20$ μl/min, the analysis time calculates [Eq. (8)] to 12.5 hr. In order to complete an HPSEC chromatogram in a more reasonable time (a few hours), the flow rate is increased to about 100 μl/min. This has to be paid for with a loss in plate count and loss in resolution. On a TSK-gel SW 3000, 600 × 7.5 mm column, the following plate counts were achieved for human albumin in a phosphate buffer of pH 6.8: 3300 (1 ml/min), 4100 (0.8 ml/min), 4900 (0.4 ml/min), 8500 (0.2 ml/min), 11,100 (0.1 ml/min) (data kindly supplied by LKB Instruments GmbH, Munich, FRG). Roumeliotis and Unger[8] reported values for N of 4000 and 9000 for chymotrypsinogen A on LiChrosorb Diol columns of 250 × 6 mm and 250 × 23.5 mm, d_p 5 μm, at flow rates of 2 and 21 ml/min. When plotting the plate number of proteins on a given column against the flow rate, N is seen to decrease linearly with increasing f_v.[2] The decrease is more pronounced for proteins of high molecular weight.

Since a gain in resolution is the primary objective, the dependence of R_s on the molecular weight of proteins at a given column is of interest. Data are available on the TSK-gel SW type of columns.[24] The resolution in this case was calculated by

$$R_{sp} = (V_{e(2)} - V_{e(1)})/[2(\sigma_1 + \sigma_2)(\log M_1/M_2)] \quad (11)$$

Eq. (11), which yields values up to 7 for a pair of proteins on G 2000 SW and G 3000 SW columns but drops to 3 on a G 4000 SW column. For each column, the function R_{sp} of proteins passes through a characteristic maximum on plotting against M. The value of M at $R_{sp(max)}$ corresponds to the value in the middle of the linear part of the calibration plot. In conclusion, proteins that differ in molecular weight by a factor of 2 may be resolved.

Applications

HPSEC columns are applied in four areas of protein characterization and purification: prefractionation, analytical separation, molecular weight determination, and preparative isolation. For prefractionation, collected fractions are further separated on ion-exchange or reversed-phase columns. Analytical separations are carried out, for example, in the monitoring of the time course of enzymic reactions. Molecular weight determination of proteins by means of a calibration curve requires carefully purified standards and precise measurement of V_e or t_e. The most frequent use of HPSEC is the isolation of milligram quantities of proteins for further examination. Examples of such use are available.[27]

[27] C. T. Wehr, R. L. Cunico, G. S. Ott, and V. G. Shore, *Anal. Biochem.* **125,** 386 (1982).

[8] High-Performance Ion-Exchange Chromatography

By FRED E. REGNIER

Liquid chromatography of proteins is a separation process based on differential rates of molecular migration through a bed of particles. When purifying a single substance by this process, the objective is to choose conditions and materials that maximize the difference between the migration of this substance and all others in the sample. The difference in migration between the substance being purified and that of any other component in the sample depends on a number of variables, including the resolving power of the system. Resolution (R_s) of a particular component from any other is expressed by Eq. (1)

$$R_s = 2(V_{e_2} - V_{e_1})/(\Delta V_{e_1} + \Delta V_{e_2}) \quad (1)$$

in which V_{e_1} and V_{e_2} refer to the elution volumes of the first and second components, respectively, to elute from the column. Peak width for each of these components is designated, with the appropriate subscript, as ΔV_e. The discussion here will deal with the rationale used in column selection, operation of ion-exchange columns, and maximization of resolution through manipulation of V_e and ΔV_e values.

Obtaining successful ion-exchange separations of proteins from a series of different samples has two requirements: a good column and an understanding of the ion-exchange process. The first requirement is generally satisfied by the purchase of a commercial column, but the second requires an intellectual commitment from the investigator. Actually, many individuals fail to see the need for the second requirement since they view chromatography as a procedure that can be accomplished by a recipe. Although much can be done by recipe, that approach is as restrictive as an attempt to paint a landscape in monochrome. Each protein is unique and, therefore, offers unique opportunities for its purification. The chromatographer must be aware of the means for finding and exploiting these opportunities.

Retention

Protein retention on an ion-exchange column is obviously the result of electrostatic interactions in which retention increases in proportion to the charge density on both the ion-exchange matrix and the protein. Since proteins are amphoteric, their net charge and charge density will vary

METHODS IN ENZYMOLOGY, VOL. 104

ISBN 0-12-182004-1

with the pH of the solution. Two additional variables that may influence retention are the nature of the ion-exchange support and the distribution of charge in the protein. Some ion-exchange materials have broad titration curves that will cause ion-exchange ligand density to vary with the solution pH. The ionic properties of the ion-exchange material and of proteins may vary independently with pH. That all the charges on the surface of a protein may not be able to interact with the ion-exchange matrix simultaneously is an important consideration. That is, two proteins of identical charge can interact differently with an ion-exchange support because of differences in the distribution of charges rather than the net charge alone.

In order to control the retention process in ion-exchange chromatography and to manipulate elution volumes of the various protein components, it is necessary to examine the ionic properties of proteins. Proteins are amphoteric species that bear a net positive charge under acidic conditions and a net negative charge under basic conditions and are isoelectric at a specific intermediate pH designated as the isoelectric point (p*I*). Titration curves indicate that the amount of charge on proteins increases steadily as the pH of the enveloping solution deviates from the p*I* and that each protein has a unique titration curve. This implies that a pH could probably be found at which the difference in charge between the protein being purified and all other proteins will be at a maximum. That pH would obviously be the one at which the protein should be chromatographed on an ion-exchange column. Two factors complicate this rationale: all the charged groups within a protein will not be able to interact simultaneously with the ion-exchange matrix, and charge distribution on the surface of all proteins may not be uniform. Ion-exchange characteristics are based only on those groups of the protein that are capable of interacting with the ion-exchange matrix. Ion-exchange surfaces can recognize the difference between two proteins that are identical in every respect except that one has uniform charge distribution and the other is asymmetric. (Significantly, electrophoretic systems will not distinguish between these two species because both have the same net charge and p*I*.) In contrast, two molecules having a different p*I* and net charge could possibly cochromatograph if they have the same surface characteristics. From these examples, it can be seen that any comparison between electrophoresis and ion-exchange chromatography has the inherent flaw that the two techniques are effective by different mechanisms using charge differences as the basis for separation.

The difference in separation mechanism between ion-exchange and electrophoretic techniques is most prominent at the p*I* where the net charge of a protein is zero and where it does not migrate in an electric

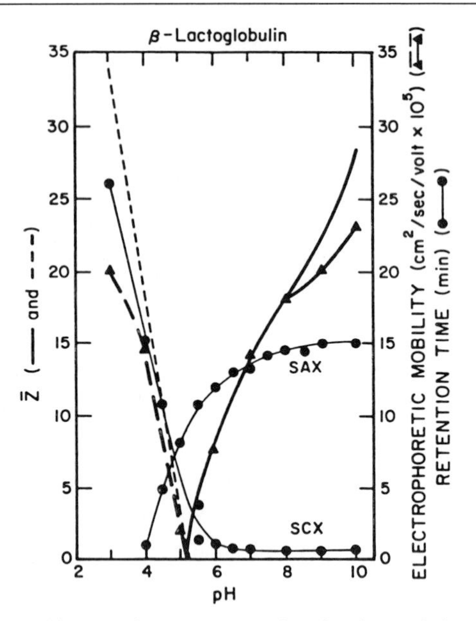

FIG. 1. Chromatographic retention as compared to titration and electrophoretic mobility curves for β-lactoglobulin, p*I* = 5.1. SAX refers to the Pharmacia Mono Q column; SCX, to the Pharmacia Mono S column. A 40-min linear gradient to 1.0 *M* NaCl was used at eluent pH 3.0 on the SCX column.

field. The absence of electrophoretic migration does not mean that the protein is devoid of charge. There will still be cationic and anionic groups within the molecule at its p*I*, and if their distribution is sufficiently asymmetric, i.e., if some of the charges are clustered, they will cause ionic interactions with an ionic surface. Thus, it is not surprising that 75% of the proteins examined by Kopaciewicz[1] were retained on both anion- and cation-exchange columns at their p*I*.

The term "ion-exchange retention," as used here, refers to either the retention time or the elution volume of a protein from a column that has been eluted with a gradient. Two major modes of gradient elution are used in protein separations: ionic strength gradients, and pH gradients. Ionic strength gradients are by far the most common and the easiest to generate. Chromatofocusing provides a form of pH gradient elution and will be discussed separately. The reason for separating most proteins by gradient elution will also be discussed. To examine the difference between electrophoretic and ion-exchange separations, Kopaciewicz[1] compared the

[1] W. Kopaciewicz, M. A. Rounds, J. Fausnaugh, and F. E. Regnier, *J. Chromatogr.* **266,** 3 (1983).

charge, electrophoretic mobility, and chromatographic retention of β-lactoglobulin at a number of pH values. The results, shown in Fig. 1, indicate that retention on both strong cation- and anion-exchange columns occurs at the p*I* at which net charge and electrophoretic mobility are zero. This would indicate that β-lactoglobulin represents a protein in which the charge distribution is asymmetric. The figure also shows that chromatographic retention begins to plateau several pH units above the p*I* of β-lactoglobulin whereas its charge and electrophoretic mobility continue to increase. Further analysis of the titration and retention data indicates that the protein had a total of 17 negatively charged residues at pH 10 but that only 6 were involved in the ion-exchange process.

Plots of chromatographic retention versus pH are referred to as *retention maps*[1] (see also this volume [11]). Comparison of the retention maps (Fig. 2) of several proteins indicates that many proteins are retained on both strong anion- and cation-exchange columns near their p*I*; that retention on both types of columns increases as mobile-phase pH is moved away from the p*I*; that each protein has a unique retention map, just as it has a unique titration curve; and that there is an optimum pH that produces the largest difference in retention of a specific component from others in a mixture.

Questions relating to properties of the ion-exchange support and operating parameters to resolution are discussed separately.

Macromolecules have been separated chromatographically on the basis of charge, hydrophobicity, size, and bioaffinity. Obviously, charge characteristics are the most important in ion-exchange separations, but hydrophobicity and molecular size also play a role. For example, molecules that are of limited water solubility may interact with an ionic matrix by a hydrophobic mechanism in addition to coulombic forces. In extreme cases, a protein may be so hydrophobic that it lacks water solubility. Separation of proteins of limited water solubility, such as membrane proteins, will also be discussed (see also this volume [16] and [18]). The contributions of molecular size to ion-exchange chromatography, in contrast to hydrophobic effects, are usually evident in the capacity of an ion-exchange medium. The pore diameter of a support must be matched to the molecular size to obtain maximum ion-exchange capacity. The role of these variables should become clearer from the discussion that follows.

Column Selection

The most important question confronting the neophyte chromatographer is that of the nature of the column to be used. For a rational choice, the chromatographer should have some knowledge of the protein

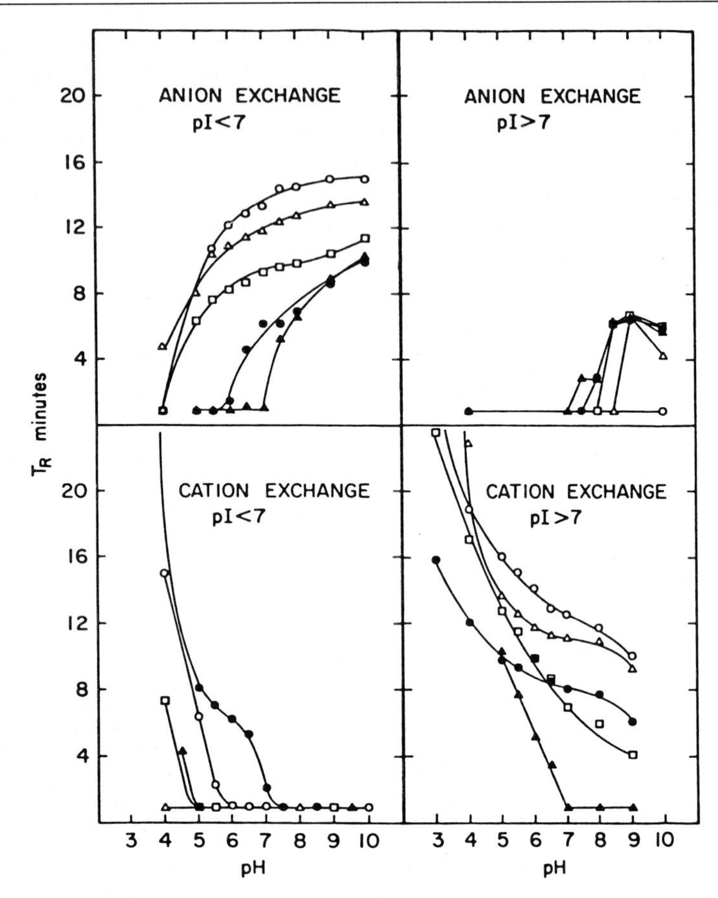

FIG. 2. Retention maps of some common proteins. Five acidic and five basic proteins were chromatographed on both a strong anion (Mono Q) and strong cation (Mono S) exchange column. The symbols represent proteins within the vertical column. A suitable buffering ion (10 m*M*) was chosen for each pH. Proteins were eluted with a 20-min linear gradient from 0 to $I = 0.5$ *M* NaCl at a flow rate of 1 ml/min. Anion exchange: ○, lysozyme; △, cytochrome *c*; □, ribonuclease; ●, chymotrypsin; ▲, carbonic anhydrase; cation exchange: ○, β-lactoglobulin; △, soybean trypsin inhibitor; □, ovalbumin; ▲, α-amylase; ●, conalbumin.

that is to be separated in terms of its p*I*, size, and even hydrophobicity. The retention maps of proteins in Fig. 2 indicate that greatest retention on cation-exchange columns occurs below the isoelectric point, where the protein has a net positive charge, whereas the largest retention on anion-exchange columns occurs above the isoelectric point, where proteins are negatively charged. For example, a protein with a p*I* of 7 would be retained on a cation-exchange column below pH 7 and on an anion-exchange column above pH 7. Less than 5% of the proteins examined are an exception to this statement.[1] In the event that the chromatographer does not know the p*I* of the protein he is trying to separate, the information may be obtained either chromatographically or electrophoretically. A simple technique for approximating p*I* is to chromatograph the protein on both anion- and cation-exchange columns over a range of 3 pH units at near-physiological conditions. Retention properties of the protein may be mapped and the most suitable column selected by this simple process. A superior method for determining p*I* is the isoelectric focusing titration curve of Rhigetti.[2] In addition to giving the isoelectric points of all proteins in a sample, information on the charge characteristics of all sample proteins at any pH from 3 to 10 is also provided. Although there is not a direct correlation between the charge characteristics as observed in electrophoresis and chromatographic retention, electrophoresis data suggest the most useful pH at which to begin ion-exchange separations. A number of instances have been observed in which electrophoresis was used to predict optimum pH for ion-exchange chromatography.[3] Once it has been established that either an anion- or cation-exchange column is the more likely to provide the desired separation, the experimenter may proceed with column selection.

Four basic types of high-performance ion-exchange chromatography (HPIEC) columns are available: (1) weak anion-exchange (WAX), (2) strong anion-exchange (SAX), (3) weak cation-exchange (WCX), and (4) strong cation-exchange (SCX) materials. Unfortunately, the terms "strong" and "weak" are often misunderstood in reference to ion-exchange chromatography. They are not intended to suggest the strength with which a substance is retained on the ion-exchange matrix, as the name would imply. Rather, they indicate the degree of stationary phase ionization at various extremes of pH. Strong cation- and anion-exchange materials are usually sulfonates and quaternary amines, respectively; both remain fully ionized within the normal operating range of high-performance ion-exchange chromatography. In contrast, weak cation- and anion-exchange supports usually contain carboxyl groups or primary and secondary amines, respectively, as the ionizable species. The pK_a of the average weak cation-exchange support is roughly 4, and that of a weak anion-exchange material is in the range of 8 to 10. Some ion-exchange materials even have very broad titration curves that span 5 pH units.[4]

[2] P. G. Righetti and J. W. Drysdale, *in* "Isoelectric Focusing," p. 341. North-Holland Publ., Amsterdam, 1976.

[3] Technical Bulletin, "The Pharmacia HPLC System." Pharmacia Fine Chemicals AB, Uppsala, Sweden.

[4] A. J. Alpert and F. E. Regnier, *J. Chromatogr.* **185,** 375 (1978).

Obviously, as the operating pH of a column approaches within a few pH units of the pK_a, charge on the ion exchange matrix decreases along with ligand density. Since the charge density of both the support and the protein vary with the pH of the mobile phase in an unrelated manner, retention characteristics of proteins on weak ion-exchange columns are less predictable than with strong ion exchangers. There are instances in which the retention of a protein will be greatest at an intermediate pH and will diminish toward the pH extremes as ionization of the WAX or WCX support collapses. As a general rule, it is best to carry out the initial separations of a protein on a strong ion-exchange material, although this is not meant to suggest that the strong ion-exchange support will necessarily be superior to a weak ion-exchange column in resolution. The subject of resolution is discussed below.

A number of ion-exchange columns are now available (Table I). There are two broad classes of ion-exchange supports in terms of support matrices: those totally organic and those that are inorganic with an organic surface coating. The inorganic supports are available in a series of pore diameters ranging from 100 to 4000 Å and particle sizes from 3 to 30 μm. The only microparticulate organic-based support currently available for proteins is the Pharmacia MonoBead material; it is available in a strong cation- and anion-exchange form in addition to a weak anion-exchange material for chromatofocusing (see also this volume [11]). Support pore diameter in the MonoBeads is reported[3] to be approximately 800 Å, with the matrix chemically stable between the limits of pH 2 and 12. This is considerably broader than the pH 2 to 8 limit of silica-based materials. Present silica supports have an upper pH limit of 8 or 9 because of the increased erosion of the silica matrix at basic pH.

An important question centers about the pH to be used in protein separations. Chromatographic band broadening, the tendency of enzymes to denature, and the amount of retained material at the inlet of ion-exchange columns increase substantially when columns are operated more than 2 pH units from physiological pH. This does not mean that ion-exchange separations are limited to the limited pH 5 to 9 range, but rather that most separations can be achieved within this range.

Pressure limits and expected column life are generally not available from manufacturers (see also this volume [6]). Since most separations may be achieved at less than 30 atmospheres with properly designed columns, mechanical stability is probably not a problem with any of the current semirigid and rigid packing materials. Expected column life, in contrast, is more difficult to quantitate since it depends on operating conditions, type of sample applied, and column care. There is now evidence that many columns will survive the analysis of 500 or more serum samples at pH 7 if routinely cleaned when resolution begins to fail.

TABLE I
HIGH-PERFORMANCE ION-EXCHANGE SUPPORTS FOR PROTEINS

Name	Manufacturer	Support material	Bonded phase	Particle diameter (μm)	Pore diameter (Å)	Surface area (m^2/g)	Ion-exchange capacity	Type of exchanger
Pharmacia	Pharmacia Fine Chemicals							
Mono Q		Organic	Quaternary amine	10	NA[a]	NA	0.3[b]	SAX[c]
Mono S		Organic	Sulfonic acid	10	NA	NA	0.15[b]	SCX[c]
Polyanion Si		Silica	Polyamine	7, 16	340	NA	1	WAX[c]
SynChropak	SynChrom, Inc.							
AX 100		Silica	Polyamine	5, 10	100	120	64[d]	WAX
AX 300		Silica	Polyamine	5, 10	300	NA	93	WAX
AX 500		Silica	Polyamine	5, 10	500	50	59	WAX
AX 1000		Silica	Polyamine	5, 10	1000	20	NA	WAX
CM 300		Silica	Carboxymethyl	5, 10	300	75	NA	WCX[c]
QX 300		Silica	Quaternary amine	5, 10	300	NA	47	SAX
Toyo Soda	Toyo Soda Corporation							
IEX 540, DEAE Sil		Silica	NA	5	130	NA	>0.3[e]	WAX
IEX 545, DEAE Sil		Silica	NA	10	240	NA	>0.3[e]	WAX
IEX 530, CM Sil		Silica	NA	5	130	NA	>0.3[e]	WCX
IEX 535, CM Sil		Silica	NA	10	240	NA	>0.3[e]	WCX

[a] NA, Not available.
[b] Milliequivalents per milliliter.
[c] WAX and WCX designate weak anion and cation exchangers, respectively; SAX and SCX represent the corresponding strong anion and cation exchanger.
[d] Ion-exchange capacity is expressed in milligrams of bovine serum albumin bound per gram of support.
[e] Milliequivalents per gram.

Loading Capacity

An additional parameter in column selection is that of the amount of protein that must be separated. In current practice, a 4.6 × 250 mm column is considered to be one for analytical use. Most ion-exchange columns of this size will have a loading capacity ranging up to 10 mg of protein without loss of resolution. As sample loading is increased, resolution decreases steadily up to loads of 40 or 50 mg, at which point protein often saturates most of the length of the column and normally retained proteins begin to break through with other nonretained material. The relative degree of separation between the proteins of interest determines to a large extent whether such a large sample may be accommodated. Loads of 50 mg may give very acceptable separation of proteins that separated widely in an analytical sample, whereas a 15-mg load may produce unacceptable resolution between peaks that are immediately adjacent. The loading capacity of any support is due to a combination of variables that includes the ratio of support pore diameter to solute diameter, support surface area, and ligand density. The loading capacities presented here are in terms of what might be expected with a protein (M_r = 50,000) on a 300-Å pore diameter support. Proteins of higher molecular weight would be expected to exhibit lower loading because they cannot penetrate the support matrix. Smaller proteins might load more heavily because they have greater access to the support surface inside pores.

At submicrogram loadings, sample recovery may decline because the surface area-to-solute mass ratio in the column is very large; the column is underloaded. The small number of imperfections in a support that irreversibly adsorb or denature protein can dominate the separation in an underloaded column. It is better to use a smaller column for micropreparative work. Submicrogram samples also begin to exceed the detection limits of the system unless such sensitive indicators as radioactivity, enzyme activity, or fluorescence are being monitored.

Columns of 50-mm length or less are becoming common in analytical separations because column length contributes little in the resolution of proteins. The loading capacity of a 4.2 × 50 mm column is in the range of 1–2 mg of protein. The principal advantages of short columns are that they are less expensive, mechanical stress on the packing bed is reduced, they may be eluted more quickly, they are easier to pack, and solutes may be eluted in smaller volumes of mobile phase thereby decreasing sample dilution and increasing detector sensitivity. When the intent is to separate very large quantities of protein, the chromatographer must go to the true preparative-scale column. It is to be expected that the loading capacity of ion-exchange columns will increase with the square of the column radius. Thus, by increasing column dimensions from 4 mm (analytical dimensions) to 10 mm (preparative dimensions), loading capacity is expected to increase sixfold. The design and application of preparative ion-exchange columns to protein separations remains in an early stage, but there is sufficient knowledge available to begin a discussion of the subject.

Two types of ion-exchange users are anticipated for the future: those scaling up an analytical procedure to produce a few grams or less of protein, and the industrial user who intends to produce tens of grams to kilograms of protein per hour. Their needs and their approaches will be completely different. In the first case, simply scaling up the analytical separation by using the same packing material and a two- to fourfold larger column would probably provide the best solution. The need for redesigning a separation system and elution protocol is thereby avoided for occasional use. The cost per unit mass of protein isolated may be much more expensive by this approach than with the large, preparative system required in the commercial production of proteins, but it requires much less development.

With the advent of genetic engineering, the ability to prepare kilogram quantities of protein is in demand. The preparative systems used in the purification of these proteins will be designed for specific separations. Even the support materials may be optimized for loading capacity and the resolution of a single protein species.

Elution

A major initial concern of a chromatographer attempting a separation on a new type of column is "how to get a peak." Since ion-exchange chromatography of proteins has been in use for the past 25 years and is quite similar to HPIEC, a large body of information is available on ion-exchange elution protocols. As a first suggestion, it is recommended that the chromatographer draw upon his own experience or search the literature for separations similar to the one he is undertaking. Conditions similar to those reported with conventional gel-type ion-exchange columns usually work well on HPIEC columns. The most widely used technique for eluting ion-exchange columns over the last 25 years has been with ionic-strength gradients at a fixed pH. Although pH gradients have been successful, they are more difficult to produce and generally give poorer resolution than ionic-strength gradients. The possible exception would be in chromatofocusing columns.

The initial step in gradient elution of increasing the ionic strength of an ion-exchange column is to load the sample on the column in a buffer at a low ionic strength and at a pH appropriate for retention to occur. Buffer concentrations of 10–20 m*M* are generally used, although up to 50 m*M* buffer is appropriate when proteins are strongly retained. Desorption and elution of proteins is achieved by gradually increasing the ionic strength of the mobile phase. Proteins are generally displaced from the column in the order of their increasing charge, although the actual number of charges in contact with the surface are more important in determining retention than is net charge.

The next consideration is the selection of a mobile phase consisting of a buffer and displacing ions. Since the ionic strength of the buffer is low, the pK_a of the buffer chosen should be within 1 pH unit of the operating pH of the system. Manufacturer literature supplied with the column should also be consulted, since specific buffers are occasionally recommended.

If there is no information on which to base a selection of the displacing ion or salt, it is best to start with simple salts such as sodium chloride, sodium acetate, or, possibly, sodium bromide. Sodium chloride has been so widely successful in conventional gel-type columns that its use is recommended in HPIEC as well. However, it should be recognized that halide ions erode stainless steel surfaces of pumping systems unless the metal is passivated. The problems of metal erosion and passivation techniques have been discussed in this series.[5] Selection and use of a variety of other salts will be discussed below in the section dealing with optimization. The strong or displacing solvent (solvent B) usually consists of 0.5–1.0 *M* displacing ion added to the initial buffer (solvent A). Seldom will a protein be so strongly adsorbed to a column that more than 1.0 *M* salt is required for elution; when a case of strong retention is encountered, bringing the operating pH of the column closer to the p*I* of the protein will diminish retention. In the event that all protein is eluted from the column by the time the gradient has reached 50% solvent B, it would be desirable to dilute solvent B until the most strongly retained proteins elute near 90% B in gradient elution. Although step gradients are commonly used in conventional gel-type columns and, on occasion, in HPIEC columns, complex mixtures are usually best resolved by a continuous gradient. Analytical columns may be eluted at 1 ml/min with a linear gradient ranging from 0 to 100% B in 20–30 min. A rationale for using different mobile-phase velocities and gradient slopes will be presented in the discussion of optimization.

[5] This series, Vol. 91, p. 137.

A note of caution should be added with regard to particulate matter in buffers, since both the column frits and the column bed have the properties of a filtration medium. Any particulate matter in the mobile phases will accumulate in the column and eventually plug it. Proper care and preparation is discussed in detail in this volume [6].

Chromatofocusing

As noted, charged proteins may be adsorbed to an ion-exchange support and eluted with either a pH or salt gradient. Gradients are normally generated externally and fed into the ion-exchange column to effect elution. Sluyterman and his co-workers[6–8] have developed a new technique for generating a pH gradient: chromatofocusing. A pH gradient is formed by pumping a buffer with a large number of different charged species into a column that has a natural buffering capacity. For example, if the pH of a weak anion-exchange column is adjusted to 8 and polybuffer—a term introduced by Pharmacia Fine Chemicals to indicate a buffer of many charged components—of pH 5 is pumped into the inlet of the weak anion-exchange column, the most acidic components of the buffer will be adsorbed and other components will migrate farther down the column before being adsorbed. The net effect of this process on pH is that the ion-exchange groups at the head of the column will be titrated and the pH of the medium will gradually be raised to that of the inlet buffer. This titration process will be repeated in all segments of the column until the whole length of the column is, in the example, at pH 5. Because the titrating buffer enters the head of the column, titration will occur sequentially in column segments and a pH gradient will be established across the length of the column. If a protein is introduced into the column during this titration process, it will migrate along the pH gradient until it reaches a point in the column where the pH is sufficiently high that it will become negatively charged. This pH will be near, but not necessarily at, the p*I* of the protein because some proteins have intensely charged areas on their surface that bind to an ionic surface even at their p*I*. As the pH gradient moves down the column, a protein is alternately adsorbed and desorbed in such a manner as to have a focusing or concentrating effect on sample components. Proteins will move along the column at or near their p*I*.

As in other types of ion-exchange chromatography, resolution may be influenced by mobile-phase velocity and gradient slope. Figure 3 shows

[6] L. A. E. Sluyterman and J. Wijdenes, *J. Chromatogr.* **150,** 31 (1978).
[7] L. A. E. Sluyterman and J. Wijdenes, *J. Chromatogr.* **206,** 429 (1981).
[8] L. A. E. Sluyterman and J. Wijdenes, *J. Chromatogr.* **206,** 441 (1981).

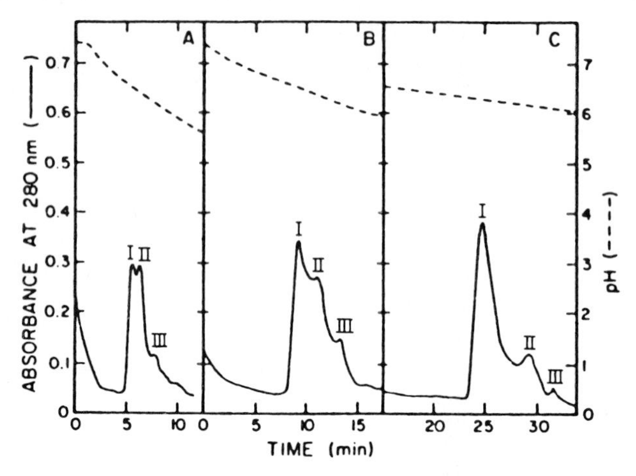

FIG. 3. Chromatofocusing profiles showing the effect of flow rate on resolution of peaks of conalbumin. A 4% suspension of conalbumin (20 μl) was chromatographed at (A) 2, (B) 1, and (C) 0.5 ml/min. The pH gradient (---) was formed using a 1 : 10 dilution of PB 96 polybuffer, pH 6.0, after the column was equilibrated in 0.25 *M* imidazole–acetic acid at pH 7.4. I, II, and III represent the major components of conalbumin.

the relationship between mobile-phase velocity and resolution of components in a conalbumin sample at the same gradient slope (percentage of change per unit volume of mobile phase). Flow rates even lower than 0.5 ml/min are useful if the pumping system retains its accuracy at these low velocities.

Gradient slope is controlled in chromatofocusing by the concentration of polybuffer being pumped into the column. As the buffer becomes more dilute, a greater volume will be required to titrate the support and the pH change of the eluent per unit volume of buffer added to the column will be smaller. It is apparent that the amount of buffer required to titrate the column will be a function of the charge density of the support. Since charge density varies considerably among commercial supports, the column manufacturer's literature should be consulted for guidance, or several dilutions of the polybuffer examined to achieve the desired gradient slope.

Only two types of columns have currently been reported as useful: Pharmacia Mono P and the Synchrom AX series. Both are weak anion-exchange supports. Mono P is a proprietary material of undisclosed pore diameter; presumably, the pore diameter is 800 Å as in the Mono Q support. The Synchropak AX materials have a polyethylenimine phase bonded to silica and are available on 100-, 300-, 500-, or 1000-Å diameter porosity. It has been reported[8] that polyethylenimine is useful in chromatofocusing because it has a very broad titration curve and that the pore diameter contributes to resolution in chromatofocusing. Substantially better resolution of estrogen receptor proteins were obtained on 500-Å pore diameter supports than 300 Å.[9]

Although it is possible to prepare polybuffers in the laboratory, they are generally inferior to the products of Pharmacia.[10] Pharmacia recommends that only a polybuffer be used over an interval of 3 pH units. To cover the full range of pH within which one must work, they offer three polybuffers: polybuffer 74 for the pH 7 to 4 range; polybuffer 96 for the pH 9 to 6 range; and Pharmalyte for the pH 10.5 to 8 range. Pharmacia polybuffers are supplied as sterile filtered solutions at 0.075 mmol per pH unit per milliliter.

Before reusing a chromatofocusing column, all residual components from the previous sample should be removed by adjusting to pH 3 or less, using 1 *M* salt, or by a combination of low pH and high ionic strength. Two to five column volumes should be sufficient for recycling, depending on the nature of the sample. After this step, the column may be brought back to the initial pH required for chromatofocusing.

The advantages of chromatofocusing are in provision of another method of ion-exchange chromatography that has a selectivity different from that of ionic strength gradient elution and requires only a single pump to form a gradient. The principal disadvantage is that fractionation is achieved near the isoelectric point of a protein at low ionic strength, both of which favor denaturation. As yet, there are relatively few reports of high-performance chromatofocusing. The true utility of the technique compared to ionic strength gradient elution remains largely undetermined.

Optimization

One of the questions regarding any separation is whether it is the best that can be achieved. This is particularly pertinent when trying to purify a protein with the minimum number of steps or when a coeluting substance prevents analytical quantitation. The answer to the question requires a more detailed discussion of resolution. Unfortunately, there is no single means by which it may be determined that a column is operating at maximum resolution or to optimize resolution. Two sets of variables control resolution in a chromatographic system: those that are inherent to the column and those that are subject to experimental control. The emphasis

[9] T. W. Hutchens, R. D. Wiehle, N. A. Shahabi, J. V. Evan, and J. L. Wittliff, *J. Chromatogr.* **266,** 115 (1983).

[10] G. Wagner and F. E. Regnier, *Anal. Biochem.* **126,** 37 (1982).

here is with the latter set of variables, among which the most important are mobile-phase velocity, ionic strength, pH of the mobile phase, and the displacing ions.

From the resolution equation [Eq. (1)] it may be seen that resolution can be maximized by decreasing the peak width (ΔV_e) or by maximizing the difference in elution volume ($V_{e_2} - V_{e_1}$) between peaks. These two variables of a separation are often referred to as band spreading and selectivity, respectively. The tendency of chromatographic columns to spread solute bands (increase ΔV_e) is well known and easily explained. The operational variables that contribute strongly to band spreading are mobile-phase velocity (flow rate), diffusion coefficient of the protein, and viscosity of the mobile phase. Mobile-phase velocities of 7.2 ml/min per square centimeter of cross-sectional area in an ion-exchange column are average. Columns operated at this velocity with a 3 to 6% per minute gradient slope will give good separations in 15–30 min. Decreasing the mobile-phase velocity to 3.6 ml/min per square centimeter (0.5 ml/min in a 4.2 × 50 mm column) has been shown to increase resolution 1.5-fold when a 1.66%/min gradient slope was used. Total separation time in this case would be 60 min. Decreasing mobile phase velocity still more to 1.8 ml/min per square centimeter (0.25 ml/min in the 4.2 × 50 mm column), and using a 0.75%/min gradient slope with a separation time of 3 hr produced a separation only 12% better than that at 1 hr. Available information suggests that the small additional increase in resolution that may be gained beyond separation times of 60 min does not warrant that investment of time.

Obviously, the enhancement of resolution gained by decreasing mobile-phase velocity was accomplished by allowing solute molecules longer for equilibration between the stationary and mobile phase. Anything that hinders this equilibration process will diminish resolution. Subambient column operation is one such means. As column temperature is lowered, mobile-phase viscosity increases, with a concomitant decrease in solute diffusion coefficient. Unless there is compelling evidence that enzyme activity is lost in transit through the HPLC column at ambient temperature, subambient column operation is not recommended.

Reference has been made to a relationship between gradient slope and flow rate in resolution, i.e., the percentage of solvent change per minute relative to the flow rate. It is important that these two variables be balanced in a manner that allows elution of one substance from the column before another is desorbed. Gradient slope relative to mobile-phase velocity determines the number of column volumes of mobile phase that will be pumped through the column to effect elution of adsorbed material. The examples given in this section used a gradient slope relative to velocity that produced a 30 column-volume gradient. This means that there was a 3.3% salt gradient across the column. In a 25-cm column, the gradient would be 0.13%/cm; and in a 5-cm column, a 0.66%/cm gradient. Elution volumes equal to 20–40 column-volumes are routine with 5-cm columns. The number of elution volumes used to develop a column is easily calculated by dividing the total elution volume by the liquid volume of the column. In general, the liquid volume of an anion-exchange column is 65–70% of the volume of the empty column.

Selectivity

Selectivity is the word usually used to describe the differential affinity of a column for two solutes. As the difference in retention between two solutes increases, the term $V_{e_2} - V_{e_1}$ in Eq. (1) will increase, with a concomitant increase in resolution. To vary the differential affinity of a column for two solutes, the investigator must find some property in their interaction with the support that may be made different. The two variables in HPIEC of proteins that have been found to be the most useful in altering selectivity are the pH of the mobile phase and the nature of displacing ions. As shown in the retention maps (Fig. 2), the difference in retention between solute species varies with the pH of the mobile phase, which can be traced to unique differences in the pH titration curves of proteins. Although ion-exchange retention is not directly proportional to net charge, the titration curve indicates that differential ionization of ionic groups is taking place. By changing the pH of the mobile phase, the possibility of differentially altering the number of sites at which two proteins are adsorbed to a surface may be explored. It will be recalled from the section on retention that retention on strong anion-exchange columns is increased as the pH of the mobile phase is elevated, whereas retention on strong cation-exchange columns increases with a decrease in mobile-phase pH. The opportunity exists for achieving multiple purification steps on a single column by operating it with several mobile phases of different pH. Impurities coeluting from an ion-exchange column operated at one pH have a reasonable possibility of being separated from the desired component at another pH.

The displacing salt present in the strong buffer is equal in importance to pH in controlling retention and selectivity.[11] For a variable that is so important, selection of a displacing ion is, unfortunately, complex. This is illustrated in Tables II and III, in which the retention time of a number of proteins is examined on anion- and cation-exchange columns. It is seen

[11] W. Kopaciewicz and F. E. Regnier, *Anal. Biochem.* **133,** 251 (1983).

TABLE II
INFLUENCE OF VARIOUS ANIONS ON THE RETENTION AND RESOLUTION OF SELECTED PROTEINS[a]

Anions[b] (sodium salt)	Relative displacing power[c]					R_s Ova/STI	R_s Cyt *c*/Lys
	Ova[d]	STI[d]	Chy[e]	Cyt *c*[e]	Lys[e]		
Bromide	0.70	1.00	0.93	0.77	1.0	4.1	2.2
Phosphate	0.59	0.74	1.00	1.00	0.90	ND[f]	3.8
Citrate	1.00	0.80	0.64	0.44	0.51	5.3	3.6
Chloride	0.60	0.88	0.87	0.66	0.80	5.4	3.6
Perchlorate	0.81	0.88	ND	ND	ND	4.3	ND
Fluoride	0.45	ND	0.74	0.56	0.58	ND	6.7
Bicarbonate	0.61	0.79	ND	ND	ND	4.1	ND
Tartrate	0.70	0.69	0.64	0.56	0.51	4.5	4.4
Sulfate	0.73	0.70	0.70	0.60	0.61	6.1	3.0
Formate	0.54	0.74	ND	ND	ND	3.5	ND
Acetate	0.54	0.59	0.75	0.73	0.70	5.7	4.1
Propionate	0.54	0.56	ND	ND	ND	4.2	ND

[a] The following abbreviations are used for proteins: Ova, ovalbumin; STI, soybean trypsin inhibitor; Chy, chymotrypsin; Cyt *c*, cytochrome *c*; and Lys, lysozyme.
[b] Chromatography was performed at pH 8. The ionic strength of buffer B was 0.5.
[c] Unity refers to the strongest displacing salt for that protein; 0 the least.
[d] Determined on the Q-300 silica-based strong anion-exchange column.
[e] Determined on a Pharmacia Mono S strong cation-exchange column.
[f] ND, Not determined.

that different ions affect the chromatographic behavior uniquely. For soybean trypsin inhibitor, bromide was the most powerful displacing anion in contrast to citrate for ovalbumin. The relative order of displacing power was also seen to vary between anions from a low of 3.5 with formate to a high of 6.1 with sulfate, using an anion-exchange column and the same pair of proteins. In contrast, sulfate produced the lowest selectivity and fluoride the highest with the cytochrome *c*/lysozyme pair on a strong cation-exchange column. The ability to alter selectivity by changing displacing ions is of obvious utility in purification. For cations, similar phenomena were observed. It may be generally stated that magnesium and bromide ions are among the most powerful displacing agents as contrasted to lithium and fluoride ions, which are among the weakest. When salts are used at the same concentration, strong displacing agents decrease retention and compress the chromatogram, whereas weak displacing agents expand the chromatogram. Compression or expansion of elution profiles has minimal influence on resolution unless there is a change in the retention of one protein relative to another.

TABLE III
INFLUENCE OF VARIOUS CATIONS ON THE RETENTION AND RESOLUTION OF FIVE SELECTED PROTEINS[a]

Cations[c]	Relative displacing power[b]					R_s	R_s
	Ova[d]	STI[d]	Chy[e]	Cyt *c*[e]	Lys[e]	Ova/STI	Cyt *c*/Lys
Chloride salt						STI	Lys
Lithium	0.68	0.57	0.60	0.56	0.54	5.2	3.1
Sodium	0.82	0.72	0.93	0.97	0.98	5.4	3.6
Potassium	0.80	0.69	0.93	0.98	0.98	5.0	4.0
Ammonium	0.83	0.74	0.86	0.94	0.93	5.5	3.7
Magnesium	1.0	0.94	1.00	1.00	1.00	3.1	2.8
Calcium	1.0	1.0	ND[f]	ND	ND	2.8	ND
Acetate salt							
Sodium	0.67	0.47	0.81	0.88	0.85	5.7	4.1
Magnesium	0.83	0.57	ND	ND	ND	4.2	ND

[a] See footnote *a* of Table II.
[b] Unity refers to the strongest displacing salt for that protein; 0, the least.
[c] Chromatography was performed at pH 8. The ionic strength of buffer B was 0.5.
[d] Determined on the Q-300 strong anion-exchange support.
[e] Determined on the Pharmacia Mono S strong cation-exchange column.
[f] ND, not determined.

Hydrophobic Proteins

The principal problems with hydrophobic proteins are their limited solubility in water and their tendency to aggregate. In size-exclusion chromatography[12] (see also this volume [7]), a number of denaturing agents and various organic acid–alcohol combinations have been used successfully. The choice of a mobile phase is more dependent on the unique requirements of the protein for dissociation and solubilization than on the chromatography system. In contrast, ion-exchange systems are much more restrictive. Mobile-phase additives that modify the solubility of a protein cannot be ionic; large concentrations of ionic solubilizing agents would swamp the requisite ionic character of ion-exchange columns and prevent electrostatic interaction of the protein with the column. This problem has been circumvented in part by the use of organic solvents or nonionic detergents as solubilizing agents (see this volume [16]).

Despite the hydrophobicity of the M_r 8000 subunit of the chloroplast

[12] J. D. Pearson, E. Pfankoch, and F. E. Regnier, "Food Constituents and Food Residues: Their Chromatographic Determination." Dekker, New York, 1983.

coupling complex, Tandy *et al.*[13] have reported that ion-exchange separations could be obtained in high concentrations of chloroform and methanol. Proteins were applied to a weak anion-exchange column (0.41 × 10 cm SynChropak AX 300) in $CHCl_3$–CH_3OH, and the column was eluted according to the following protocol: 15 min of isocratic elution with $CHCl_3$–CH_3OH (2 : 1); 10 min of isocratic elution with $CHCl_3$–CH_3OH (1 : 1); and, finally, gradient elution in 25 min from $CHCl_3$–CH_3OH–H_2O (3 : 3 : 1) to the same eluent containing 20 m*M* ammonium acetate. Elution from the column during the gradient was on the basis of charge.

In the presence of 0.1% Triton X-100, envelope proteins from Sendai virus have been fractionated on a strong anion-exchange column (0.5 × 5 cm, Mono Q) and eluted with a gradient ranging from 0.02 *M* sodium phosphate at pH 7.2, containing 0.15 *M* sodium chloride and 0.1% Triton X-100, to the same solution except for the addition of 1.5 *M* sodium chloride.[14] A similar type of separation of bovine viral diarrhea proteins has been reported with Berol as the detergent.[15]

Six-molar urea has been used to dissociate alkaline phosphate complexes.[16] Although this reagent works well and has little influence on the ion-exchange process, the high concentration of urea results in high viscosity and, therefore, large column back pressure. Urea is recommended only when other techniques are ineffective.

Cleaning Columns

During the course of many analyses, residual protein will accumulate on the column. This may be the result of denaturation, low solubility in the mobile phase, or very strong interaction with the support surface. If residual protein is not periodically removed, the column loses ion-exchange capacity and resolution and can eventually plug. When performance tests indicate deterioration of resolution, residual materials should be eluted from ion-exchange columns by extremes in pH, high ionic strength, or organic solvents (see also this volume [6]). A combination of these techniques using gradients of 0.1 to 0.5% trifluoroacetic acid (TFA) to 60% propanol in 0.1 to 0.5% TFA may be preferred. This solvent system is widely used in reversed-phase chromatography and appears to be able to solubilize both the most ionic and hydrophobic proteins. Five to

[13] N. Tandy, F. E. Regnier, and R. Dilly, *J. Chromatogr.* (in press).

[14] G. W. Welling, G. Kroen, and S. Welling-Webster, *J. Chromatogr.* (in press).

[15] P. Karsncs, J. Moreno-Lopez, and T. Kristiansen, *J. Chromatogr.* (in press).

[16] T. D. Schlabach, "High Speed Isoenzyme Profiling by HPLC." Ph.D. Thesis, Purdue University, Lafayette, Indiana, 1978.

10 repetitive gradient cycles with the TFA–propanol mixture is usually sufficient for cleaning. In the event that some proteins remain adsorbed, substitution of 60% formic acid for TFA will probably remove them. However, it is advisable to consult the manufacturer before using such severe solvents because some support matrices are destroyed under very acidic conditions.

Sample Preparation

The amount of sample preparation required in HPIEC depends on the origin and history of the sample. Obviously, all particulate matter must be eliminated; thus, all cell extracts require preliminary treatment.

When the protein has been previously subjected to column, salt, or solvent fractionation, the ionic strength and pH of the sample must be adjusted to allow electrostatic association with the column. Although solvent adjustment of this sort may be accomplished in many ways, dialysis seems to be the most common. If the sample is very dilute and ionic strength has been reduced to less than 20 m*M* of buffer by dialysis, large volumes of sample may be applied to a column. For example, 300–500 ml of a solution containing 0.2 mg of protein per milliliter can be pumped onto a 1×25 cm polyanion silica column and gradient eluted to produce a normal chromatogram.[17]

Column Switching

Purification of proteins is usually a labor-intensive, multiple-column process. In addition to the time required for chromatography, considerable additional time is consumed in collecting and preparing samples for the next chromatographic step. Part of this expenditure of labor may be circumvented if the sample can be directly transferred from one column to another. The coupling of two different types of columns with the aid of switching valves has been shown[18] to be applicable in the transfer of a protein directly from a size-exclusion to an ion-exchange column without an intermediate collection step. Direct transfer from ion-exchange to reversed-phase columns was equally successful. Although not widely used currently, apparatus for column switching is being introduced commercially and should become more common in the near future.

[17] W. Kopaciewicz and F. E. Regnier, *Anal. Biochem.*, submitted for publication.

[18] W. Kopaciewicz and F. E. Regnier, *Anal. Biochem.* **129,** 472 (1983).

[9] Reversed-Phase High-Performance Liquid Chromatography

By MILTON T. W. HEARN

In recent years, reversed-phase high-performance liquid chromatography with porous, microparticulate, chemically bonded alkylsilicas has emerged as a very rapid and selective method for peptide and protein purification.[1,2] Although the role of a variety of mobile-phase parameters, such as the concentration of the organic solvent modifier, the pH, and the buffer ion composition on peptide resolution in general is now well documented, the purification of proteins by similar reversed-phase techniques still requires careful optimization on a case-by-case basis. Investigations with a variety of silica-based hydrocarbonaceous supports with nominal pore diameters encompassing the range of 6 to 50 nm have allowed several crucial stationary phase characteristics to be identified[3-5] for a number of specific structural features of proteins based on studies using selected protein standards. For peptides and proteins separated on adequately bonded alkylsilicas, retention correlates well with the topographic surface polarity of these molecules. As a consequence, chromatographic behavior for a group of peptides under a particular set of elution conditions can largely be described in terms of the concepts of linear elution development with selectivity following regular free-energy relationships, which can often be directly estimated from consideration of the amino acid composition and sequence differences. Although these empirical relationships have enabled, in a variety of selected peptidic cases, the prediction of elution orders from topological indices,[1] the current data base does not yet adequately take into account amino acid residue positional effects or changes in the unique secondary or tertiary structural features of a peptide or protein that may arise with different elution conditions.

Recognition of the synergistic features of mobile-phase and stationary-phase parameters and their adequate control such that the chromato-

[1] M. T. W. Hearn, *Adv. Chromatogr.* **20,** 1 (1982).

[2] M. T. W. Hearn, F. E. Regnier, and C. T. Wehr, *Am. Lab.* (*Boston*) **14,** 18 (1982).

[3] J. D. Pearson, N. T. Lin, and F. E. Regnier, *in* "High Performance Liquid Chromatography of Proteins and Peptides" (M. T. W. Hearn, F. E. Regnier, and C. T. Wehr, eds.), p. 81. Academic Press, New York, 1983.

[4] M. T. W. Hearn and B. Grego, *J. Chromatogr.* **288** (in press).

[5] M. J. O'Hare, M. W. Capp, E. C. Nice, N. H. C. Cooke, and B. G. Archer, *in* "High Performance Liquid Chromatography of Proteins and Peptides" (M. T. W. Hearn, F. E. Regnier, and C. T. Wehr, eds.), p. 161. Academic Press, New York, 1983.

ISBN 0-12-182004-1

graphic distribution ideally involves a single (conformationally stable) molecular species is of the utmost importance for the separation of large polypeptides and proteins by the reversed-phase techniques. The great versatility of peptide separation by reversed-phase HPLC arises from the ease with which mobile-phase elutropicity can be manipulated. However, the present popularity of several elution conditions that were developed initially with peptide separations in mind, and their application largely in an empirical manner, has tended to obscure the requirements for optimal resolution and recovery of larger polypeptides and proteins with alkylsilica supports. In this chapter, focus is on the selection of mobile-phase and stationary-phase parameters that lead to improved strategies for the purification of both native and denatured proteins by reversed-phase methods. General reviews on the use of HPLC in protein chemistry have been published[6,7] and one is extensively recorded in this volume [6]–[8]. Shorter discussions of procedures for resolution optimization, including pairing-ion modulation of retention behavior of polypeptides and proteins on alkylsilicas, have also appeared in the scientific literature.[8–10]

Background

There are basically two types of interactive retention mechanisms that can be proposed for protein separation by reversed-phase HPLC: partition and adsorption. The concept of liquid–liquid partition implies that a homogeneous solution of a protein in the nonpolar phase occurs, and further, the concept requires that the intercept (C) of the linear plot (with slope equal to unity) of the logarithmic capacity factor versus the logarithmic partition coefficient is less than the phase ratio, Φ, of the chromatographic system. Experimentally, the observed C values for polypeptides and even for amino acids and small peptides separated on alkylsilicas are much larger than the Φ values. As a consequence, the retention process should be more appropriately characterized as involving adsorption processes and alkylsilicas, based on bonded monolayers of a nonpolar phase, considered from a formal point of view as diffuse solid adsorbents. In contrast to classical liquid–solid chromatography of proteins on neat silica or glass surfaces, where the interactions between solute and adsorb-

[6] S. Stein, *in* "The Peptides" (E. Gross and J. Meienhofer, eds.), Vol. 4, p. 73. Academic Press, New York, 1981.

[7] M. T. W. Hearn, *in* "High-Performance Liquid Chromatography: Advances and Perspectives" (C. Horvath, ed.), p. 87. Academic Press, New York, 1983.

[8] M. T. W. Hearn and B. Grego, *J. Chromatogr.* **255,** 125 (1983).

[9] R. A. Barford, B. L. Sliwinski, and H. L. Rothbart, *J. Chromatogr.* **235,** 281 (1982).

[10] M. T. W. Hearn and B. Grego, *J. Chromatogr.* **218,** 497 (1981).

ent involve strong polar forces, protein–stationary-phase interactions in reversed-phase HPLC are predominantly due to weak, nonspecific van der Waals forces. However, with most bonded alkylsilicas currently available, mixed retention mechanisms involving nonpolar and polar contributions operate. The intensity of the polar contributions in the retention process has been found to be directly related to the extent of alkyl ligand coverage and the chemical nature of the mobile-phase components that can adsorb to the surface of the nonpolar support.

Protein Separation by Reversed-Phase Chromatography

To date, most protein separations on hydrocarbonaceous silica-based surfaces have been phenomenological in nature with the methodology applied largely in an empirical manner. Recent experience has shown that the silica-based supports with pore diameters in the range 6–10 nm are far from optimal for the chromatography of proteins with regard to resolution and high recovery. Further, the separation of proteins on alkylsilicas with water–organic solvent combinations often involves much more complex phenomena than is evident with peptides, e.g., adsorption–desorption kinetics which may be slow compared to the chromatographic time scale. Interpretation of the structure-retention relationships of proteins separated on alkylsilicas, in contrast to the reversed-phase HPLC of peptides, is thus complicated by multiple chromatographic and nonchromatographic variables that may affect either protein structure in solution or protein structure at a polar liquid–hydrocarbonaceous surface interface. Proteins exhibit great structural diversity, with the native three-dimensional structure very sensitive to changes in the microenvironment of solvation. Contact with organic solvents, adsorption at air–liquid or liquid–solid interfaces, high ionic strengths of dissociating salt species, or extremes of pH may cause denaturation and loss of biological activity. As a consequence, several practical requirements must be met for successful reversed-phase HPLC separations of proteins.

The first, and most important, requirement when preservation of biological function is a necessary goal of the chromatographic separation is the issue of irreversible conformational changes induced by mobile-phase composition or the characteristics of the stationary phase. Analysis of the effects of mobile-phase composition on biological activity in preliminary incubation experiments aids the selection of suitable eluents but does not guarantee optimal resolution or recovery.

A second requirement that must be satisfied is associated with the phenomenon of multisite attachment of protein solutes to alkylsilica stationary phases. The concept of cooperative multisite interactions between proteins, such as glycolytic enzymes and *n*-alkylated agaroses, has been well developed by Jennissen,[11,12] and similar behavior is clearly evident with large-pore diameter alkylsilicas. This behavior invariably results in a pronounced sensitivity of protein elution on the composition of the mobile phase and is particularly noticeable when isocratic behavior and gradient retention behavior for a group of proteins are compared. With water–organic solvent eluents, this behavior is manifested by severe dependencies of the logarithmic capacity factor on the volume percentage of the organic solvent modifier. Furthermore, owing to heterogeneous protein–stationary-phase interactions in the multisite attachment mode, broadened or multiple peak shapes can occur.

A third requirement arises with complex mixtures where, from practical necessities, it is important to ensure that chromatographic retention of a specific protein is essentially mediated by a single molecular species. Variations in protein shape, extent of ionization, and conformation may be induced by either the eluent composition or stationary-phase characteristics. Size-exclusion phenomena as well as distorted peak shapes may result. Because proteins under these conditions may exhibit unique populations of reversible or irreversible states involving native, partially unfolded, and fully unfolded conformers in solution, different manifestations of these complex surface-directed interactive phenomena can occur under a given set of chromatographic conditions. Differences in interactive affinities, attendant on deformational changes induced by eluent or stationary-phase effects, which may reveal additional sequestered hydrophobic or polar domains of a polypeptide or protein, will affect chromatographic performance. As a consequence, optimization of chromatographic performance with porous, microparticulate alkylsilicas requires the same level of tactical consideration as, for example, is used to define the best strategy for approaching the amino acid sequence determination of a given protein. The availability of the starting material, the end use of the purified protein, the number of discrete high-resolution separation steps that may be required to achieve the best purification factors, will all determine the most appropriate choice of chromatographic conditions.

Specific clues guiding the rational selection of chromatographic conditions for optimal resolution and recovery come from physical measurements of solubility, exploratory analytical elution experiments, evaluation of selectivity matrices (and, in particular, the evaluation of τ plots[4,7]), and attention to the susceptibility of some proteins to respond to pairing-ion stabilization of their secondary and tertiary structure, which may thus

[11] H. P. Jennissen, *Biochemistry* **15,** 5683 (1976).
[12] H. P. Jennissen, *J. Chromatogr.* **159,** 71 (1978).

facilitate their separation. Attention must also be given to methods for eluent preparation as well as procedures for handling samples at the pre- and postcolumn stages. In a study with several different protein hormones,[13,14] improved methods of handling the chromatographic fractions greatly facilitated the recovery of the proteins in biologically active form, particularly in circumstances where the abundance of the desired components was low and their initial dilution high, i.e., $<10\ \mu g/ml$. Practically, it is obviously expedient to optimize selectivity with volatile buffers where possible. Although such choice of eluents should be contemplated at an exploratory stage of an investigation, their use may not necessarily be desirable with respect to the highest resolution or recovery.

The most efficient strategy for successful reversed-phase HPLC purifications of proteins is one that continuously redefines the practical limitations of the elution conditions as the molecular features of the protein of interest become more evident. Recovery for many proteins from alkylsilicas can be improved by the addition of stabilizing cations or cofactors to the eluent. For example, the addition of 2 mM $CaCl_2$ to water–organic solvent mixtures enhanced the recovery of both trypsin and parvalbumin from reversed-phase silicas.[15–17] Reappraisal of strategy may necessitate the use of alternative columns of different dimensions, chosen according to rational resolution optimization requirements and a battery of different alkylsilica phases, characterized in terms of ligand coverage and porosity at different stages of the purification scheme. Rigorous adherence to a single elution condition or stationary phase cannot be expected to generate efficient high-resolution separations except for the most straightforward problems, typified by the separation of denatured polypeptides where it may not be essential to recover all components.

Consideration of Retention Behavior

Protein selectivity in this method is dependent on the different distribution coefficients established by the several protein species present between a polar, water-rich mobile phase and a nonpolar stationary phase solvated by extractable mobile-phase components. On the assumption that protein retention involves the distribution of the solute between an

[13] M. Dobos, H. K. Burger, M. T. W. Hearn, and F. J. Morgan, *J. Mol. Cell. Endocrinol.* **31,** 187 (1983).

[14] G. S. Baldwin, B. Grego, M. T. W. Hearn, J. A. Knesel, F. J. Morgan, and R. J. Simpson, *Proc. Natl. Acad. Sci. U.S.A.* **80,** 5276 (1983).

[15] K. Titani, J. Sasagawa, K. Resing, and K. A. Walsh, *Anal. Biochem.* **123,** 408 (1982).

[16] W. S. Hancock, C. A. Bishop, R. L. Prestidge, and M. T. W. Hearn, *Anal. Biochem.* **89,** 203 (1978).

[17] M. W. Berchtold, K. J. Wilson, and C. W. Heizmann, *Biochemistry* **21,** 6552 (1982).

interfacial monolayer (m/s) and a mobile eluent (m), a capacity factor, k_i', for a sparingly soluble protein solute, i, can be described by Eq. (1).

$$\ln k_i' = \ln(C_{m/s,i}/C_{m,i}) = \ln(u_{i,m}^{\infty}/C_{m/s}) - \ln(u_{i,m/s}^{\infty}/C_m) \quad (1)$$

$C_{m/s,i}$ and $C_{m,i}$ are the number of moles of the solute in the stagnant interfacial phase and the mobile phase; $u_{i,m}^{\infty}$ and $u_{i,m/s}^{\infty}$ are the solute activity coefficients in the mobile phase and the bulk solvated stationary phase; and $n_{m,s}$ and n_m are the number of moles of adsorbed and mobile eluent components. If ΔA_i is the area occupied by a protein molecule on the nonpolar surface, $u_{i,m/s}^{\infty}$ is the corresponding solute activity coefficient at the interface of a solute monolayer, and $\gamma_{m/s}$ and $\gamma_{i,s}$ are the surface tensions of the adsorbent–eluent and adsorbent–solute interfaces, then retention can be given by Eq. (2).

$$\ln k_i' = \ln(n_{m/s}/n_m) + (\ln u_{i,m}^{\infty} - \ln u_{i,m/s}^{\infty}) + \Delta A_i N(\gamma_{m/s} - \gamma_{i,s})/RT \quad (2)$$

Several corollaries arise on the basis of this displacement–adsorption model. First, the contact area, ΔA_i, for a protein will usually be considerably larger than that occupied by a desorbing agent, such as an organic solvent molecule. As a consequence, the binding of a polypeptide or protein to the nonpolar surface will displace multiple molecules of the eluent, e.g., an alcohol or acetonitrile. As multisite binding becomes more heterogeneous owing to unfolding transitions in the bulk mobile phase or at the interface, the assumption that all the molecules of a particular protein interact in an identical manner with the same number of sites on the surface—a basic tenet of linear elution development of low-molecular-weight solutes—will clearly no longer be valid. Furthermore, desorption of these heterogeneously bound species will require a mobile phase of higher eluotropic strength even when a single molecular conformer is the dominant binding species. This "unzippering" requirement underlies the well-documented observation that proteins can usually be successfully chromatographed on alkylsilicas only under gradient elution conditions. For example, Lewis *et al.*[18] noted that bovine serum albumin eluted from a C_8 column in about 13 min with an isocratic mobile phase of 0.5 *M* formic acid–0.4 *M* pyridine at pH 4.0, containing 34% 1-propanol, but did not elute when the mobile phase contained 32% 1-propanol. In fact, such behavior is typical of globular proteins eluted from alkylsilicas. Typical examples of the pronounced biomodal dependencies of retention on organic solvent content have been reported for growth hormone, collagen

[18] R. V. Lewis, A. Fallon, S. Stein, K. D. Gibson, and S. Udenfriend, *Anal. Biochem.* **104,** 153 (1980).

chains, ribonuclease, and phosphorylase *a*.[19–21] Where bimodal dependencies exist, inappropriate changes in solvent composition may account for the resolution of native and unfolded forms of the same protein under certain gradient elution conditions.

Second, protein retention is strongly dependent on the extent to which mobile-phase components can be extracted onto the stationary phase [the first term of Eq. (2)]. Retention modulation will occur as a consequence of the extraction of buffer ion species, which may engage in ion-pair or dynamic ion-exchange phenomena, as well as owing to the extraction of organic solvent molecules. The influence of these secondary adsorption isotherms on polypeptide retention has been extensively documented experimentally.[1] Changes in retention of fully permeating polypeptide or protein solutes can thus be equated with changes in the extracted solvent density and differences in solvent eluotropic strengths. The pronounced effect on stationary phase porosity and matrix field effects due to eluent component isotherms has been accommodated[4,19] into the so-called softball model of protein retention with alkylsilicas.

Finally, retention of proteins on alkylsilicas is anticipated from a theoretical analysis of activity coefficient dependencies, as well as from more empirical treatments of experimental data, to be the sum of the retention components due to size-exclusion and interactive processes that involve not only the nonpolar topographic features of the protein molecule itself, but also composite polar electrostatic and hydrogen-bonding characteristics. The general empirical expression for protein retention to alkylsilicas has the form of Eq. (3).

$$k' = \rho_r k'_r + \rho_e k'_e + \rho_h k'_h + \rho_s k'_s \tag{3}$$

The subscripts r, e, h, and s refer to the hydrophobic, electrostatic, hydrogen-bonding, and size-exclusion components of retention, and ρ_r, ρ_e, ρ_h, and ρ_s are the weighted mole fractions of the solute in each retention mode. Despite the inherent difficulties in evaluating the k'_e and k'_h terms in Eq. (3) due to the complexity of protein structure in solution and of the multiple participating variables, including ionic strength, pH, dielectric constant, etc., which influence the electrostatic and hydrogen bonding terms, a beginning has been made by several investigators[19,20,22] in including these effects in the displacement model for protein separation on alkylsilicas. For a specified column, temperature, and buffer composition,

[19] M. T. W. Hearn and B. Grego, *J. Chromatogr.* (in press).

[20] M. T. W. Hearn and B. Grego, *J. Chromatogr.* **266,** 75 (1983).

[21] K. Cohen, B. L. Karger, K. Schellenberg, B. Grego, and M. T. W. Hearn, submitted for publication.

[22] X. Geng and F. E. Regnier, *Proc. Am. Pept. Assoc.* (in press).

the linearized form of the dependency of retention on solvent composition has been shown[20] to follow Eq. (4).

$$k'_{T,i} = k'_{w,i}e^{-S_i\psi} + (k'_{o,i}B_i\psi)^{-1} + k'_{s,i}f(\psi) \quad (4)$$

where k'_w, k'_o, and k'_s are the capacity factors in neat water (related to the solubility parameter δ_w), in the final organic solvent modifier (an extrapolated value related to k'_e and k'_h) and due to size exclusion, respectively. The variables, S, B, and f are solute and condition dependent. Since the volume fraction of organic modifier, ψ_s, is directly related to the bulk surface tension of the eluent, γ, through the modified Fowkes' equation, then protein retention on alkylsilicas may also be evaluated in terms of changes in intrinsic surface tensions associated with interaction at the liquid–solid interface.

Mobile Phase Dependencies

Although literature values of the microscopic surface tensions for proteins under reversed-phase conditions are generally not readily available, they can be derived from plots of the logarithmic selectivity parameter ($\ln \alpha = \tau$) versus the eluent surface tension at fixed temperature with a defined stationary phase. Alternatively, they may be estimated from contact angle measurements. A major interpretative concern that such approaches raise has been the desire to express solute activity coefficients predominantly in terms of mobile-phase parameters. In this treatment, protein retention is equated with the composite free-energy changes associated with the formation of a cavity in the eluent of dimensions sufficient to accommodate the protein molecule, with the van der Waals interactions between the solute and eluent and with the electrostatic interactions between the eluent and solute. Retention in this model may then be viewed as a consequence of the propensity of a protein solute to be expelled from the mobile phase with the stationary phase playing a passive role as a protein acceptor–displacer surface. Composite strong (electrostatic, hydrogen bonding) and weak (hydrophobic) forces that occur in the mobile phase are manifested by intrinsic differences in interfacial surface tension and dielectric constant. Although it was originally assumed that only weak dispersive forces were associated with the stationary phase, recent detailed studies[3,19,20] with well-characterized stationary phases have demonstrated that secondary effects, due to the stationary phase ligand coverage and matrix preparation and pretreatment, can significantly influence protein selectivity. Undoubtedly, the most successful theoretical approach to date that accounts for solvent-mediated interactions and the effect of salts on the macroscopic properties of the eluent has been the solvophobic model proposed by Horvath and co-workers.[23] According to this model, selectivity for pairs of solutes eluted under isocratic conditions may be given by Eq. (5).

$$\tau_{i,j} = [\Delta(\Delta G_{vdw})_{i,j} + \gamma N(\Delta A_j - \Delta A_i)]/RT \tag{5}$$

Relative retention of a protein with isocratic eluents y and z on a given column can be given by Eq. (6).

$$\tau_{i,y/z} = C + \Delta A_i(\kappa_y^e \delta_y - \kappa_z^e \delta_z) \tag{6}$$

The parameter κ^e corrects for the effect of curvature of the cavity surface on the macroscopic surface tension of the eluent and is both solute and eluent dependent. Evaluation of protein selectivity on alkylsilicas in terms of the cavity factor, κ^e, and interfacial contact area, ΔA, from plots of τ versus γ with slope given by the ω factor[19,20] permits the evaluation of difference in van der Waals energies of interaction, $\Delta(\Delta G_{vdw})$, as well as the free energies of association (ΔG_{assoc}) for different polypeptide or protein solutes.

When isocratic retention for polypeptides and proteins over an appropriate ψ-value range obeys Eq. (7)

$$\ln k_i' = \ln k_{i,w}' - S\psi \tag{7}$$

(as is often observed with polypeptides and globular proteins for the range $1 < k' < 10$ under optimized low pH conditions), then the value of capacity factor at the column inlet for linear solvent strength gradient elution starting from neat aqueous conditions is given by Eq. 8.

$$\ln k_i' = \ln k_{i,w}' - b(t/t_o) \tag{8}$$

and the gradient retention time is given by Eq. (9)[20]

$$t_{g,i} = (t_o/b) \log[2.3bk_{i,w}(t_{sec}/t_o)] + 1 + t_{sec} + t_L \tag{9}$$

where $b = SV_m(d\psi/dt)F^{-1}$, F = volumetric flow rate, S = solvent desorption parameter of a specific protein, $t_o = V_oF^{-1}$, t_{sec} = retention time for size exclusion alone, t_L = lag time for the gradient system, V_i = mobile-phase pore volume, V_o = mobile-phase interstitial volume, and V_m = total mobile-phase volume (equal to $V_o + V_i$).

From Eqs. (7)–(9), it can be seen that for polypeptides and proteins with large k_w and S values, very little elution development is expected. The observed sensitivity of protein elution to mobile-phase composition, particularly the very narrow operational range of solvent strength conditions that can be exploited under appropriate circumstances of isocratic

[23] C. Horvath, W. Melander, and I. Molnar, *J. Chromatogr.* **125,** 129 (1976).

and batch elution, is in accord with these predictions. Linear time-based gradient elution is not anticipated to yield optimal resolution for complex mixtures of proteins. Rather, nonlinear gradients that accommodate s-value differences are required. Since k'_w values are usually very large [e.g., k'_w (lysozyme) > 1000] the solute band will not migrate significantly during passage of the lag volume of mobile phase through the band center. This means that the protein zone is subjected to microscopic solvent strength increases during gradient elution, which results in band compression factors substantially different from those observed for small molecules (e.g., $M_r \cong 1000$). The consequence of this behavior is a pronounced dependence of peak height (and recovery) on flow rate. Other corollaries that emerge directly from Eqs. (7)–(9) relate to optimal column dimensions from the S values versus V_m analysis, optimal flow rate for different gradient $d\psi/dt$ rates, and solvent strength limits for particular t_g values. Experimental data obtained with large polypeptides and globular proteins separated on alkylsilicas under denaturing, low pH elution conditions have been in accord with these theoretical predictions.[4,20,21,24]

Selection of Chromatographic Parameters

The success of polypeptide and protein separation on alkylsilica supports depends to a very large extent on the choice of mobile-phase conditions. For gradient elution with organic solvent modifiers, appropriate selection of the initial and final compositions is important if optimal resolution and recovery are to be achieved. Relative retention is dependent on the surface area of the nonpolar adsorbent within the column and the chemical nature of the bonded surface. For most investigators without access to "in-house" silica-bonding technology, control over the different, and frequently complex, distribution coefficients can be achieved only through manipulation of the eluent composition. This limitation immediately preempts the question of the purpose of the separation. Many hydrocarbonaceous silica-based supports exhibit chemical instability at high pH values that effectively limits their use to below pH 7.5. With mixed solvent combinations under acidic conditions, many proteins are denatured. It is often feasible, however, to reconstitute proteins isolated by reversed-phase HPLC into functional proteins either by the addition of stabilizing cofactors, cations, or other reagents to the eluent or by careful preferential removal of the organic solvent modifier under nitrogen, or by pH and buffer ion changes immediately after chromatographic separation. Examples of the use of one or other of these manipulations can be found

[24] N. H. C. Cooke, B. G. Archer, M. J. O'Hare, E. C. Nice, and M. Capp, *J. Chromatogr.* **255,** 115 (1983).

in studies with purification of enzymes such as trypsin[15] or papain,[25] ribosomal proteins,[26] calcium-binding proteins,[17] and protein hormones.[14]

Not infrequently, impaired recovery of biological activity can be attributed to handling procedures subsequent to chromatographic fractionation, such as adsorption to glass or plastic ware following lyophilization. Difficulties in the purification of human prostatic acid phosphatase[27] and follicular inhibins[13] on reversed-phase silicas were circumvented by appropriate handling procedures. A characteristic of protein chromatography on alkylsilicas for which poor mass and activity recovery are found is the appearance of ghost peaks in a blank gradient run immediately after the initial protein separation has been completed. Often this carryover effect will give rise to multiple peaks, the later eluting peaks corresponding to irreversibly denatured protein. The presence of ghost peaks can be usually traced back to the use of stationary phases with bonded nonpolar surface areas inappropriate for the specific separation.[20,21] Stationary phases with either lower surface areas or, alternatively, decreased ligand densities frequently circumvent this difficulty without a major adjustment in mobile-phase composition being necessary.

Influence of Organic Solvent Composition

All the water-miscible organic solvents have found use for protein separation in reversed-phase HPLC. Their efficacy in terms of eluotropicity can be related directly to the nature of the relationship between the volume fraction (usually expressed as the percentage of solvent) and the bulk eluent surface tension. Comparative studies with methanol, acetonitrile, and 1- or 2-propanol have confirmed that the slope of the retention dependency, log k' versus ψ, follows the eluotropic strength, ε^0, of the organic solvent; i.e., the slope is larger for 1-propanol than, for example, methanol, the relative retention decreasing as ε^0 decreases. Although relative retention for a particular polypeptide decreases in the order methanol < ethanol < acetonitrile < 1-propanol or 2-propanol, it does not follow that resolution will improve with more effective desorbing solvents such as the propanols. In fact, ternary mixtures containing water and two organic solvents, e.g., acetonitrile–1-propanol or ethanol–1-propanol, at compositions chosen to maximize the ln k' versus γ slope of Eq. (7), will often result in improved selectivity; proteins are thereby eluted with mo-

[25] S. A. Cohen, S. Dong, K. Benedek, and B. L. Karger, *in* "Affinity Chromatography and Biological Recognition" (I. Chaiken, M. Wilchek, and I. Parikh, eds.). Academic Press, New York (in press).

[26] A. R. Kerlavage, C. J. Weitzmann, T. Hasan, and B. S. Cooperman, *J. Chromatogr.* **266,** 225 (1983).

[27] M. P. Strickler, J. Kintzios, and M. J. Gemski, *J. Liq. Chromatogr.* **5,** 1921 (1982).

bile phases of higher water content than can be achieved with either individual solvent. Such approaches have been successfully applied in a number of studies on the isolation of membrane proteins.[28]

Because of the magnitude of the S and k'_w values for large polypeptides and proteins, gradient elution provides the most powerful routine method for their separation on alkylsilicas. Based on exploratory gradient experiments, however, optimized batch elution conditions can be employed for larger scale preparative separations. In agreement with theory, resolution is dependent on the gradient steepness parameter, b. Peak capacity follows an inverse asymptotic dependence on b [Eq. (8)]. With b values up to about 1.5, peak capacity gradually decreases; above about 2.0, significant loss in resolution occurs. Since the b term is related to the flow rate as well as to the rate of change in percentage of the organic modifier, comparable resolution can be achieved when $(d\psi/dt)F^{-1}$ is held constant. This can be achieved under conditions of long separation times, i.e., a shallow gradient and low flow rate, or short separation times with a high flow rate and steep gradient. Practical constraints will apply in individual cases, but flow rates in excess of 50 ml/min can be utilized over suitable $d\psi/dt$ ranges for preparative separations, particularly where inappropriate dwell times for a protein mixture on the column may lead to impaired recoveries associated with either dissociation or denaturation. A further corollary arises insofar as solute detectability is also asymptotically dependent on the gradient steepness parameter. Irrespective of whether spectrophotometric or biological (immunological or biological assay procedures) detection methods are used, advantage should be taken of the optimal b value compatible with resolution so that the band width of the eluting protein solute experiences the maximum band compression. Under such conditions, recovery is favored provided shear degradation or irreversible changes in conformation do not occur. To a large extent, the commonly used gradient conditions (1%/min and 1 ml/min for a 25 × 0.4 cm column) should be viewed as representing only a starting point for exploratory experiments designed to establish the correct limits of mobile-phase gradient composition. Once established, significantly reduced separation times can then be achieved[19,24] by rigorous optimization procedures.

When applied in strictly aqueous solutions, most polypeptides and proteins are strongly adsorpted to most modern microparticulate reversed-phase packing materials. The magnitude of the binding effect is dependent on the number and accessibility of the hydrophobic domains in these biopolymers. Most water-soluble proteins contain a fairly large

[28] S. Welling-Wester, T. Boer, G. W. Welling, and J. B. Wilterdink, *in* "Proceedings of the Second International Symposium on HPLC of Proteins, Peptides and Polynucleotides" (M. T. W. Hearn, F. E. Regnier, C. J. Wehr, eds.). Elsevier, Amsterdam, 1983. In press.

number of sites of weak affinity for hydrocarbons whereas some proteins, e.g., β-lactoglobulin and the serum lipoproteins, have hydrophobic binding sites of high affinity. When high-adsorption isotherms are observed, as with very hydrophobic polypeptides and proteins, special care must be given to the parameters that influence the sorption–desorption events. Included would be the initial volume fraction or the rate of change of volume fraction of the organic solvent modifier in isocratic or gradient elution, respectively, and pH and buffer ion effects. If these parameters are not controlled with precision, poor resolution, associated with large and erratic k' values, will occur as a symptom of "irreversible" binding. Although the magnitude of the solute–ligand interactions is expected to decrease with increasing organic solvent modifier in the mobile phase, the low solubility parameters of polypeptides and proteins (and of some buffer components) in mobile phases containing high levels of the common organic solvents, i.e., where $\psi < 0.75$, limits the secondary solvent composition. The problems associated with limited solubility of polypeptides and proteins in hydroorganic eluents can be partially circumvented by choosing a different modifier combination of lower eluotropic properties. In this regard, the use of mixed 1- or 2-propanol-based eluents with detergents has proved to be efficacious for some macroglobulin separations.[29] With labile proteins, optimal elution conditions for analytical separations may be incompatible with larger scale preparative isolations or with subsequent assay methods; a compromise between resolution and recovery must be found. Illustrative of this approach is the purification of ovine thyrotrophin from a pituitary extract on octyl- and octadecylsilica supports.[30] On a 50-nm pore diameter, end-capped octylsilica, this glycoprotein hormone could be readily fractionated using a triethylammonium formate–acetonitrile gradient eluent at pH 4.0 to pH 6.5, whereas subunit dissociation appears to occur in triethylammonium phosphate at pH 3.0. Similar results have been obtained with luteotropin and human chorionic gonadotropin.[31]

Influence of Buffer and Ionic Strength

The application of solvophobic theory and pairing-ion concepts[32] has enabled much of the retention behavior, on reversed-phase columns, of polypeptides and proteins in the presence of different ionic buffers and

[29] R. A. Barford, B. J. Sliwinski, A. C. Breyer, and H. L. Rothbart, *J. Chromatogr.* (in press).

[30] M. T. W. Hearn, P. G. Stanton, and B. Grego, *J. Chromatogr.* (in press).

[31] G. J. Putterman, M. B. Spear, K. S. Meade-Cobun, M. Widra, and C. V. Hixson, *J. Liq. Chromatogr.* **5,** 715 (1982).

[32] M. T. W. Hearn, *Adv. Chromatogr.* **18,** 1 (1980).

salts to be rationalized and judicious choice made for analytical and micropreparative elution conditions. The addition of millimolar concentrations of many inorganic salts to the mobile phase tends to decrease relative retention and improve peak shape, presumably owing to suppression of polar interactive equilibria. At high salt concentrations, e.g., >100 mM, salting-out effects become evident. This has been exploited with butyl- and phenyl-bonded TSK-G3000SW, a silica-based packing material with a 250-Å mean pore diameter.[33] With this support, it was possible to separate proteins with high efficiency under mild eluting conditions similar to those conventionally used in hydrophobic interaction chromatography. Linear gradient elution has been possible with ammonium sulfate concentration decreasing from 2 to 0 M in 100 mM phosphate buffer at pH 6.0. With organic acids and bases (and their salts) relative retention of proteins on alkylsilicas can be increased or decreased depending on the chemical nature and polarity of the ionic substance. Considerable flexibility is thereby introduced into the chromatographic separation by the use of suitable buffer co- and counterions. With appropriate choice of the polarity and concentration of the added buffer co- and counterions, specific electrostatic and solution field interactions can be optimized that lead to apparent increases or decreases in protein polarity and hence changes in relative retention. Since they are directly lyophilizable, the volatile organic acids (formic, acetic, trifluoroacetic, and heptafluorobutyric) and their alkylammonium salts have been widely utilized. The most appropriate concentration for optimal resolution can be evaluated from pairing ion considerations and τ plots.[4,19] The nonvolatile inorganic salts, such as phosphate and perchlorate, as well as the nonvolatile organics, such as dodecyl sulfate and heptane sulfonate, also exhibit similar relationships (asymptotic retention versus concentration) with proteins.

Because of the ease of desalting samples by reversed-phase systems, removal of inorganic salts from recovered fractions is usually straightforward. A large variety of ionic and neutral species, which at pH $\leq$ 7.0 modify the retention characteristics of polypeptides and proteins, have been described.[1,7] Table I lists a selection of co- and counterions that have been used successfully in reversed-phase separations of proteins (see footnotes 7 and 32 for more extensive reviews of applications and conditions). Because of the surfactant properties of many of the more hydrophobic anionic and cationic pairing ions, disaggregations of protein–protein noncovalent complexes can be achieved, although this may be at the expense of protein denaturation. However, a dramatic improvement in peak shape and recovery for polypeptides and proteins is often seen[20,34]

[33] Y. Kato, T. Kitamura, and T. Hashimoto, *J. Chromatogr.* **266,** 49 (1983).

[34] J. Rivier, *J. Liq. Chromatogr.* **1,** 343 (1978).

TABLE I
ANIONIC AND CATIONIC[a] CO- AND COUNTERIONS USEFUL IN RP-HPLC SEPARATIONS OF PROTEINS

Anion	Cation
Formate	R_4N^+
Acetate	R_3N^+H
Propionate	$R_2N^+H_2$
Trifluoroacetate	RN^+H_3
Heptafluoroacetate	Inorganic M^+, M^{2+}
C_3–C_{12} sulfonate	N^+H_4, pyridine
C_3–C_{12} sulfate	
Phosphate	
Perchlorate	
Bicarbonate	

[a] The chain length for the cationic reagents usually ranges from C_1 to C_7 (see footnotes 1, 4, and 32 for a compendium of applications).

with these reagents. Above critical concentrations with nonionic and ionic surfactants, micellar chromatography of proteins occurs on alkylsilicas.[20,29] Of the organic acids, trifluoroacetic acid in particular has been found to induce desirably reversed-phase selectivity for protein separation owing to its effectiveness as a solubilizing ion-pairing reagent. Similarly, alkylammonium phosphates such as triethylammonium phosphate, typically at 0.1 *M*, have provided alternative isolative capabilities, particularly in the range pH 3–5.

Influence of pH

With proteins, as with peptides, pH values in the range of 2–3 tend to give better resolution on alkylsilicas. The possibility that a specific protein will be insoluble at or near its p*I* value must be taken into account when a pH is chosen. Similarly, dissociation of multisubunit proteins or the unfolding of the tertiary structure may occur at low pH. The adverse effects of reduced solubility, dissociation, or unfolding arising from low pH mobile phases can be readily assessed in static experiments. A suitable choice of pH, permitting local resolution and optimum recovery, is rarely a practical problem. Although low pH values favor suppression of ionization of carboxyl groups (as well as silanol groups on the stationary phase), careful handling is required when proteins are exposed to aqueous organic solvent mixtures at low pH. For example, both human growth hormone and insulin deamidate[35,36] at a moderate rate at pH 2.5 and residue cleavage at acid-labile sites may also occur, as appears to be the case with apolipoproteins.[37] Significant broadening of the eluted peak can occur with only modest increases in pH, e.g., with $ACTH_{1-39}$,[38] whereas other proteins require very low pH values for elution, e.g., β-microglobulin.[39] Pyridine–acetic acid or formic acid systems, as used for human fibroblast interferon[40] in the range of pH 3.5–5.0, provide comparable results to alkylammonium phosphate of lower pH value, with the added benefit of volatility. For proteins that are unstable at a low pH, these systems offer considerable flexibility, particularly when on-line fluorescent detection is available. When combinations of pH and pairing-ion modulation under gradient elution conditions are used, it is possible to separate very closely related molecules. For example, Terabe *et al.*[41] have used butane sulfonate to resolve Thr_{30}-bovine insulin, bovine insulin, human insulin, and porcine insulin, the last two differing by only 1 out of 51 amino acid residues. Related techniques have been used for the isolation of insulins from animal pancreas,[42] corticotropin analogs,[43] relaxins,[44] the neurophysins,[45] the bovine pancreatic trypsin inhibitor aprotinin,[46] ferritin,[47] collagen α-chains,[18] thyroglobulins,[48] cytochrome *c*,[34,47] albumins,[38,47] and acyl carrier proteins and analogs.[49]

Column Material

Several important practical benefits arise as a result of the high affinity of polypeptides and proteins for alkylsilicas. First, the solute mixture can be loaded onto a column in a large volume of a very dilute solution and

[35] B. Grego, F. Lambrou, and M. T. W. Hearn, *J. Chromatogr.* **266,** 89 (1983).
[36] M. E. F. Beimond, W. A. Sipman, and J. Olivie, *J. Liq. Chromatogr.* **2,** 1407 (1979).
[37] W. S. Hancock, C. A. Bishop, A. M. Gotto, D. R. K. Harding, S. M. Lamplugh, and J. T. Sparrow, *J. Lipid Res.* **16,** 250 (1981).
[38] M. J. O'Hare and E. C. Nice, *J. Chromatogr.* **171,** 209 (1979).
[39] V. L. Alvarez, C. A. Roitsch, and O. Henriksen, *Anal. Biochem.* **115,** 353 (1981).
[40] S. Stein, C. Kenny, H. J. Friesen, J. Shively, U. Del Valle, and S. Pestka, *Proc. Natl. Acad. Sci. U.S.A.* **77,** 5716 (1980).
[41] S. Terabe, R. Konaka, and K. Inouye, *J. Chromatogr.* **172,** 163 (1979).
[42] M. T. W. Hearn, B. Grego, J. Cutfield, and M. McInnes, in preparation.
[43] H. P. J. Bennett, *J. Chromatogr.* **266** (in press).
[44] J. R. Walsh and H. D. Niall, *Endocrinology* **107,** 1258 (1980).
[45] J. A. Glasel, *J. Chromatogr.* **145,** 469 (1978).
[46] K. Krummen, *J. Liq. Chromatogr.* **3,** 1243 (1980).
[47] W. Monch and W. Dehnen, *J. Chromatogr.* **147,** 415 (1978).
[48] A. J. Paterson and M. T. W. Hearn, *J. Protein Chem.,* submitted for publication.
[49] W. S. Hancock, C. A. Bishop, R. L. Prestidge, D. R. K. Harding, and M. T. W. Hearn, *Science* **200,** 1168 (1978).

concentrated *in situ*. Second, high loadings for some solutes can be achieved; i.e., a loading up to 50 mg of ribonuclease did not noticeably affect the resolution on an analytical column (25 × 0.4 m, Si 100 A, octadecyl, dp 5 μm).[50] Third, the sample capacity of large proteins is greater on macroporous than on mesoporous reversed phases, although at comparable loadings similar elution times are observed with otherwise identical chromatographc conditions.

Studies by Lewis *et al.*,[18] Regnier *et al.*,[3] and Karger and Hearn[20,21] have delineated some of the options for high-capacity supports for the reversed-phase HPLC of large proteins. High coverage, 25- to 50-nm nominal pore diameter, silica bonded with butyl or octyl phases and end-capped, in particular, appear to be very useful supports for the chromatography of semipreparative quantities of proteins exhibiting good recoveries, high sample capacities, and good resolution. For stationary phases bonded to the same alkyl chain density, the chain length has been found to have only a small influence on relative retention but does affect recoveries. Several studies have implied that the resolution and recoveries are higher with short-chain alkyl ligands. For example, Rubinstein[50] has favored LiChrosorb RP-8 over RP-18 or RP-2 for the fractionation of proteins of $M_r > 30{,}000$.

The changes in elution behavior noted in these and other early studies can now be more readily accommodated in terms of the extent of carbon coverage of the stationary phase. What is apparent is that selectivity is influenced by the chemical characteristics of the parent silica matrix and the bound ligand: cyanopropyl-, octyl-, and octadecylsilicas, for example, show noticeable selectivity differences to the diphenyl- or cyclohexylsilicas. Sample loading also affects relative retention of proteins with alkylsilicas, but, interestingly, overloading does not result in total loss of resolution as may occur in conventional size-exclusion or ion-exchange chromatography. Similarly, column length does not appear to be important for protein separation. A 10-fold increase in column length increased the gradient elution resolution of human growth hormone ($M_r = 22{,}000$) and the human growth hormone ($M_r = 20{,}000$)[19] by only 11%, and similar results have been observed[3] for the separation of bovine serum albumin and ovalbumin. In contrast, relative retention of proteins on alkylsilicas is directly dependent on pore size, and, in particular, on the accessible bonded surface area of the stationary phase. As a consequence, those alkylsilicas that currently afford the greatest utility for protein separation all exhibit surface areas in the range 60–100 m^2/g with ligand densities of 2.7–2.5 μmol of alkyl chain/m^2. For many commercial reversed phases

[50] M. Rubinstein, *Anal. Biochem.* **98,** 1 (1979).

with nominal pore diameters of 80–100 Å, it is probable that many protein molecules are entropically excluded from the pore interior, with consequent impairment of sample capacity.[4] Protein solutes that do interact strongly with small-pore silicas may not be recovered in good yield. It is also apparent from two studies[3,19] that wide-pore (200–500 Å) silicas are not necessarily suitable for every case of protein separation.

Data obtained[3,4,19] with polypeptide hormones and protein standards on reversed phases of different pore diameters, but identical alkyl chain length, have indicated that the resolving power of the system decreases as hydrophobicity of the ligand increases, but that the retention dependence on solvent content is essentially independent of pore size at low sample loadings, i.e., that the solute S values are constant. These results stress the importance of choosing elution conditions that generate adequate selectivity factors [$\Delta\alpha/\alpha$; cf. Eq. (5)] if resolution between two proteins of similar hydrophobicity is to be achieved. For these reasons, attention must be given to the potential available for optimization of resolution arising from the manipulation of secondary chemical equilibria, the temperature, the flow rate, the eluotropic properties of the organic solvent modifier, and the extent of nonpolar ligand coverage on the stationary phase.

Influence of Flow Rate and Temperature

Because of the slower rates of diffusion of macroglobulins, lowering the mobile-phase velocity generally improves the resolution. The flow rate and eluent composition had a significant influence on the column efficiency for bovine serum albumin, for example, on an Ultrasphere-octyl (d_p 5 μm, 10 nm) column.[51] Three solvent systems (15 mM H_3PO_4, 0.1% HCOOH, and 500 mM HCOOH–400 mM pyridine) were compared in this study, using a 1-propanol gradient. Under all flow rate conditions the formic acid–pyridine buffer showed the highest efficiency, the phosphoric acid system the lowest. These results are in accord with the expected differences in the molarity of the buffer. More interesting, however, was the finding that the differences in plate number (N) were accentuated as the flow rate was decreased; e.g., the plate numbers for phosphoric acid system showed little affect when the flow rate was decreased from 75 to 15 ml/hr, whereas the formic acid–pyridine system showed a threefold increase.

Although column efficiency for polypeptides and proteins appears to increase at higher temperature,[21,25] the lability of many biologically func-

[51] B. N. Jones, R. J. Lewis, S. Paabo, K. Kojima, S. Kimura, and S. Stein, *J. Liq. Chromatogr.* **3,** 1373 (1980).

TABLE II
SELECTED EXAMPLES OF POLYPEPTIDES AND PROTEINS PURIFIED BY REVERSED-PHASE HPLC[a]

Polypeptide or protein	Column	Mobile phase	Reference
Growth hormone	μ-C_{18}	8% Pyr, 2.5% AcOH, 0–100% CH_3CN : *i*-PiOH; 1 : 1, 4 ml/min	52
Growth hormone	LS-C_4	100 m*M* ammonium bicarbonate, 0–50% CH_3CN, 2 ml/min, 90 min	14
Thyrotropin and α- and β-subunits	B-C_{18}	100 m*M* ammonium bicarbonate, 0–50% CH_3CN, 2 ml/min, 90 min	30
β-Lipotropin/ β-endorphin	LS-C_{18}	1 *M* Pyr-AcOH, 0–40% *i*-PrOH 0.5 m/min, 2 hr	53
Relaxin	LS-C_8	200 m*M* ammonium bicarbonate, 0–40% CH_3CN, 1 ml/min, 15 min	54
Relaxin	μ-AP	100 m*M* ammonium acetate, 0–50% CH_3CN–0.1% TFA, 1.2 ml/min, 90 min	54
Human chorionic gonadotropin and subunits	μ-AP	100 m*M* ammonium acetate, 0–60% CH_3CN–0.1% TFA, 1.2 ml/min, 90 min	55
Human chorionic gonadotropin	μ-C_{18}	0.1% TFA, 0–60% CH_3CN, 1.0 ml/min, 109 min	31
Human fibroblast interferon	LS-C_8	0.8 *M* Pyr–1 *M* formic acid, 0–60% *n*-PrOH, 22 ml/hr	40
ACTH-related polypeptide	μ-C_{18}	0.1% TFA or 0.13% HFBA, 0–50% CH_3CN, 1.5 ml/min, 60 min	56
Neurophysins	μ-C_{18}	10 m*M* NaH_2PO_4, pH 6.0, 40–50% MeOH, 2.5 ml/min, 90 min	57
Calcitonin-related	H-C_{18}	0.155 *M* NaCl, pH 2.1, 0–60% CH_3CN, 1 ml/min, 60 min	58
Apolipoproteins A-I, A-II, C-I, C-II, C-III	μ-AP	1% TEAP, pH 3.2, 0–80% CH_3CN, 1.5 ml/min, 30 min	59
β_2-Microglobulin	μ-C_{18}	0.012 *M* HCl, 35–80% EtOH, 0.8 ml/min, 30 min	39
Thyroglobulin	LS-C_{18}	15 m*M* H_3PO_4, 0–60% CH_3CN, 1.0 ml/min, 90 min	48
Collagen types I, II, III	B-AP	0.1% TFA, 0–50% CH_3CN, 1.0 ml/min, 90 min	60

Collagens I, II, III	μ-CN	1.5 *M* Pyr-AcOH, 0–40% *n*-PrOH, 0.3 ml/min, 80 min	61
α-, β-, γ-, δ-Hemoglobin chains	H-C_{18}	0.1% TFA, 0–50% CH_3CN, 1.0 ml/min, 60 min	62
Cytochrome *c*	μ-CN	0.1 *M* NaH_2PO_4, 0.05 *M* Na_2SO_4, 27.5% CH_3CN, 2 ml/min	63
Trypsin	μ-C_{18}, SC-RPP, μ-CN	0.1% TFA, 2 m*M* $CaCl_2$, 30–65% CH_3CN, 2 ml/min, 30 min	15
Ela, Elb adenovirus protein	A-8, SC-RPP	0.5 *M* Pyr-formate, 0–60% 1-propanol, 0.33 ml/min, 100 min	64
MSV-transformed growth factors	μ-C_{18}	0.1% TFA, 20–40% CH_3CN, 0.8 ml/min	65
Phosphorylase *b*	A-3	0.155 NaCl, pH 2.1, 0–50% 1-propanol, 1 ml/min, 50 min	5
Trypanosoma glycoproteins	μ-C_{18}	0.05% TFA, 30–45% CH_3CN, 1 ml/min, 180 min	66
Amunine	V-C_{18}	TEAP, pH 2.25, 0–60% CH_3CN, 2 ml/min, 20 min	67
Chymotrypsinogen	TSK-3000SWP	0–2 *M* ammonium sulfate in 0.1 *M* NaH_2PO_4, pH 6.0, 1 ml/min, 25 min	33
Cytochrome *c* oxidase subunits	μC_{18}	0.05% TEA–TFA, 0–70% CH_3CN, 0.5 ml/min, 115 min	68
Sendai virus protein	SU-LC318, N-C_{18}	12 m*M* HCl, 0–60% EtOH–*n*-BuOH (80 : 20, v/v), 1 ml/min, 24 min	28
Leukocyte interferon	W-P	0.05 *M* KH_2PO_4, 20–80% 2-methoxyethanol–propanol (5 : 95), 1 ml/min, 45 min	69
Ribosomal proteins	SC-RPP	0.1% TFA, 15–45% CH_3CN, 0.7 ml/min, 120 min	26
Parvalbumin	LS-C_{18}	50 m*M* Tris-HCl, 0.1 m*M* $CaCl_2$, 0–70% *i*-PrOH, 1.0 ml/min, 75 min	17
Secretin	N-C_{18}	0.1% TFA, 15% CH_3OH, 1 ml/min, 70 min	70
Histones	H-C_{18}	0.1 *M* $NaClO_4$–0.1% H_3PO_4, 0–60% CH_3CN, 1 ml/min, 70 min	71
Prostatic acid phosphatase	μ-C_{18}	0.1% TFA, 12–70% CH_3CN, 1.5 ml/min, 30 min	27
Thymosin β-4	A-C_{18}	0.2 *M* Pyr–1 *M* formic acid, 0–40% *n*-PrOH, 0.53 ml/min, 100 min	72
M_r 24,000 T-cell factor	LS-C_8	1 *M* Pyr–0.5 *M* acetic acid, 0–50% *n*-PrOH step	73
M_r 26,000 Hepatocyte cell factor	A-C_{18}	0.05% TFA, 0–20% CH_3CN, 1.0 ml/min, 30 min	74

(*continued*)

TABLE II (*continued*)

Polypeptide or protein	Column	Mobile phase	Reference
Calmodulin	A-RP300	0.1% H_3PO_4, 10 m*M* $NaClO_4$, 0–60% CH_3CN, 1.0 ml/min, 60 min	75
Tyrosinase	LS-C_8	0.2 *M* Pyr–1 *M* formic acid, 0–40% *n*-ProH, 0.5 ml/min, 100 min	18
Ia antigen membrane proteins	μ-C_{18}	0.1% TEA, 0.2% Triton X-100–TFA, 60–100% CH_3CN, 1 ml/min, 60 min	76
Monoclonal M8 IgG	A-C_3	0.1% TFA, 0–50% *n*-PrOH, 1 ml/min, 50 min	5
Ribonuclease	V-C_8	0.1% TFA, 0–80% *i*-PrOH, 0.7 ml/min, 40 min	3
Papain	LS-C_4	10 m*M* H_3PO_4, *n*-PrOH, 1 ml/min, 15 min	25
Neurotoxin III	H-C_{18}	100 m*M* NaH_2PO_4–H_3PO_4, pH 2.1, 0–50% CH_3CN, 1 ml/min, 60 min	38
POMC fragments	μ-C_{18}	0.02 *M* TEAP, 5–90% CH_3CN, 1 ml/min, 90 min	77
Ferritin	N-C_{18}	0.05 *M* NaH_2PO_4, pH 2.0, 10–50% methoxyethanol, 2 ml/min, 30 min	47
Carbonic anhydrase	A-3	0.155 *M* NaCl, pH 2.1, 0–75% CH_3CN, 1 ml/min, 90 min	23
Human TGF	μ-C_{18}	0.035% TFA, 0–20% *n*-PrOH, 1 ml/min, 60 min	77
Maclurin	A-C_3	0.1% TFA, 0–50% CH_3CN, 2 ml/min, 30 min	78
Bacteriorhodopsin fragments	μ-C_{18}	5% Formic acid, 40–80% EtOH, 6 ml/min, 60 min	79

[a] The following abbreviations have been employed. Columns: μ-C_{18}, μ-Bondapak C_{18}; LS-C_4, LiChrospher 500-*n*-butyl; B-C_{18}, Bakerbond C_{18}; LS-C_{18}, LiChrosorb C_{18}; LS-C_8, LiChrospher 500-*n*-octyl; μ-AP, μ-Bondapak alkylphenyl; H-C_{18}, Hypersil ODS; B-AP, Bakerbond diphenyl; μ-CN, μ-Bondapak CN; A-C_{18}, Ultrasphere ODS; SD-RPP, Synchropak RPP; A-8, Ultrasphere-octyl; A-3, Ultrasphere-propyl; V-C_{18}, Vydac TP-C_{18}; TSK-3000SWP, Toyasoda TSK3000SW phenyl; Su-LC318, Supelco LC_3-C_{18}; N-C_{18}, Nucleosil-C_{18}; V-C_8, Vydac TP-C_8; W-P, Whatman Proteosil 300 diphenyl. Column sizes varied between 5 and 25 × 0.4 cm. Pyr, pyridine; TFA, trifluoroacetic acid; TEAP, triethylammonium phosphate; HFBA, heptofluorobutyric acid.

tional proteins may limit the utility of this approach for improving resolution to temperatures below 50°. In some cases, low temperatures, e.g., 5°, may be essential for preservation of biological activity. Although elevated temperatures are normally thought to reduce ionic interactions between the solutes and the stationary phase, temperature-induced changes in secondary equilibria and protein dynamics in the mobile phase will also have a significant effect on retention and resolution.

Comments

Some of the potential of reversed-phase HPLC in the purification of polypeptides and proteins can be demonstrated from the selected examples summarized in Table II.[51–79] It should be emphasized that most of the

[52] J. Meienhofer, T. F. Gabriel, J. Michalewsky, and C. H. Li, *in* "Peptides 78," p. 243. Wroclaw Univ. Press, Wroclaw, Poland, 1979.

[53] M. Rubinstein, S. Stein, L. D. Gerber, and S. Udenfriend, *Proc. Natl. Acad. Sci. U.S.A.* **74,** 3052 (1977).

[54] M. T. W. Hearn and B. Grego, *J. Liq. Chromatogr.* (in press).

[55] M. T. W. Hearn and B. Grego, submitted.

[56] H. P. J. Bennett, C. A. Browne, and S. Solomon, *Proc. Natl. Acad. Sci. U.S.A.* **78,** 4713 (1981).

[57] W. Richter and P. Schwandt, *J. Neurochem.* **36,** 1279 (1981).

[58] E. C. Nice, M. Capp, and M. J. O'Hare, *J. Chromatogr.* **147,** 413 (1979).

[59] W. S. Hancock, H. J. Pownall, A. M. Gotto, and J. T. Sparrow, *J. Chromatogr.* **216,** 285 (1981).

[60] S. J. Skinner, B. Grego, M. T. W. Hearn, and C. G. Liggins, *J. Chromatogr.* (in press).

[61] A. Fallon, R. A. Lewis, and K. D. Gibson, *Anal. Biochem.* **110,** 318 (1981).

[62] W. A. Schroeder, J. B. Shelton, J. R. Shelton, and D. Powers, *J. Chromatogr.* **174,** 385 (1979).

[63] S. Terabe, H. Nishi, and T. Ando, *J. Chromatogr.* **212,** 293 (1981).

[64] M. Green and K. H. Brackmann, *Anal. Biochem.* **124,** 209 (1982).

[65] M. A. Anzano, A. B. Roberts, J. M. Smith, L. C. Lamb, and M. B. Sporn, *Anal. Biochem.* **125,** 217 (1982).

[66] J. J. L'Italien and J. L. Strickler, *Anal. Biochem.* **127,** 198 (1982).

[67] J. Spiess, J. Rivier, C. Rivier, and M. Vale, *Proc. Natl. Acad. Sci. U.S.A.* **78,** 6517 (1981).

[68] S. D. Power, M. A. Lochrie, and R. O. Poyton, *J. Chromatogr.* **266,** 585 (1983).

[69] S. W. Herring and R. K. Enns, *J. Chromatogr.* **266,** 249 (1983).

[70] D. Voskamp, C. Olieman, and H. C. Bayerman, *Recl. Trav. Chim. Pays-Bas* **99,** 105 (1980).

[71] U. Certa and G. J. Ehrenstein, *Anal. Biochem.* **118,** 147 (1981).

[72] G. J. Xu, E. Hannappel, J. Morgan, J. Hempstead, and B. L. Horecker, *Proc. Natl. Acad. Sci. U.S.A.* **79,** 4006 (1982).

[73] K. Krupen, B. A. Araneo, L. Brink, J. A. Kapp, S. Stein, K. J. Weider, and D. R. Webb, *Proc. Natl. Acad. Sci. U.S.A.* **79,** 1254 (1982).

[74] J. B. McManon, J. G. Farrelly, and P. T. Iype, *Proc. Natl. Acad. Sci. U.S.A.* **79,** 456 (1982).

polypeptides or proteins successfully purified by this technique have relatively low molecular weights and conformation and aggregation states that are not strongly dependent on intramolecular interactions induced by the elution conditions. However, the usefulness of reversed-phase HPLC as an integral part of the purification of proteins in their native form with unaltered physiological properties is increasingly being established. Because of the versatility of the technique, membrane and transmembrane proteins can be resolved in the presence and in the absence of detergents (see also this volume [18]). At this stage, the greatest strength that these procedures currently offer to the protein chemist is in the micropreparative isolation of proteins or protein fragments required for structural elucidation. As a consequence, reversed-phase methods are increasingly replacing conventional means as an efficient prelude to structural studies.

Acknowledgment

The support of the National Health and Medical Research Council of Australia is gratefully acknowledged.

[75] K. J. Wilson, M. W. Berchtold, P. Zumskin, S. Klause, and G. J. Hughes, *in* "Methods in Protein Sequence Analysis" (M. Elizinga, ed.).
[76] J. C. Chan, N. G. Seidah, C. Gianoulakis, A. Belanger, and M. Chretien, *J. Clin. Endocrin. Metab.* **51,** 364 (1980).
[77] A. Marquardt and G. J. Todaro, *J. Biol. Chem.* **257,** 5220 (1982).
[78] M. T. W. Hearn, P. A. Smith, and A. K. Mallia, *Biosci. Rep.* **2,** 247 (1982).
[79] G. E. Gerber, R. J. Anderegg, W. C. Herlihy, C. P. Gray, K. Biemann, and H. G. Khorana, *Proc. Natl. Acad. Sci. U.S.A.* **76,** 227 (1979).

[10] High-Performance Liquid Affinity Chromatography

By PER-OLOF LARSSON

High-performance liquid affinity chromatography combines with the remarkable specificity of (bio)affinity techniques with the efficiency, sensitivity, and speed of operation of HPLC techniques.

Preparation of Adsorbents

The affinity adsorbents described herein are based on porous silica, which has excellent mechanical properties. Unfortunately, silica is unstable toward alkaline conditions and therefore should be used only for brief

METHODS IN ENZYMOLOGY, VOL. 104

ISBN 0-12-182004-1

periods with buffers of pH 8 or above.[1] An alternative packing material with lower pressure resistance but better stability toward hydrolysis is exemplified by cross-linked hydroxymethyl methacrylate.[2]

Porous silica is commercially available in several pore sizes (60–4000 Å). Ordinarily, HPLC material is based on silica with 60- or 100-Å pores, a size sufficient for separating small molecules. In affinity chromatography larger pores are needed since the separated molecules or the silica-bound molecules are of high molecular weight. In order to achieve unhindered diffusion in the pores, the pore size must be considerably larger than the chromatographed molecules[3,4]; a pore size of 300–1000 Å should be adequate for most applications. When separating very large entities such as immune complexes, the 4000-Å pore size would be advantageous.

Native silica contains acidic silanol groups, among other surface groups, that cause strong, often irreversible, adsorption of proteins.[5] Therefore an important task when preparing affinity adsorbents from silica is to mask such binding sites that would otherwise interfere with the reversible binding process. In the present case this is accomplished by coating the silica with glycerylpropyl groups (diol–silica; Fig. 1), thereby producing a hydrophilic and nonionic surface.[3,5]

Coupling to epoxy-silica is the simplest of the three routes outlined in Fig. 1. The poor reactivity makes it best suited for strong nucleophiles or for stable molecules that will withstand rather drastic conditions. The tresyl chloride method is efficient and suitable for coupling of all types of molecules containing primary amino groups. The aldehyde method involves several steps but generally gives good results, although reduction with $NaBH_4$ may be deleterious. Both the tresyl chloride method and the aldehyde method may use commercially available diol–silica as a starting material.[6]

Preparation of Epoxy-Silica (Fig. 1)

Silica (10 g) with 1000-Å pores[7] is placed in a 500-ml three-necked flask. The flask is connected to a vacuum line and heated to 150° for 4 hr to

[1] K. K. Unger, "Porous Silica: Its Properties and Use as Support in Column Liquid Chromatography." Elsevier, Amsterdam, 1979.

[2] J. Turková, K. Blahá, and K. Adamamová, *J. Chromatogr.* **236,** 375 (1982).

[3] P.-O. Larsson, M. Glad, L. Hansson, M.-O. Månsson, S. Ohlson, and K. Mosbach, *in* "Advances in Chromatography" (J. C. Giddings, E. Grushka, J. Cazes, and P. R. Brown, eds.), p. 41. Dekker, New York, 1983.

[4] R. R. Walters, *J. Chromatogr.* **249,** 19 (1982).

[5] S. H. Chang, K. M. Gooding, and F. E. Regnier, *J. Chromatogr.* **120,** 321 (1976).

[6] For example, LiChrosorb Diol (100-Å pore size), manufactured by E. Merck, Darmstadt, FRG.

[7] LiChrospher Si 1000 (10-μm particles; surface area 20 m^2/g) from E. Merck, Darmstadt, FRG.

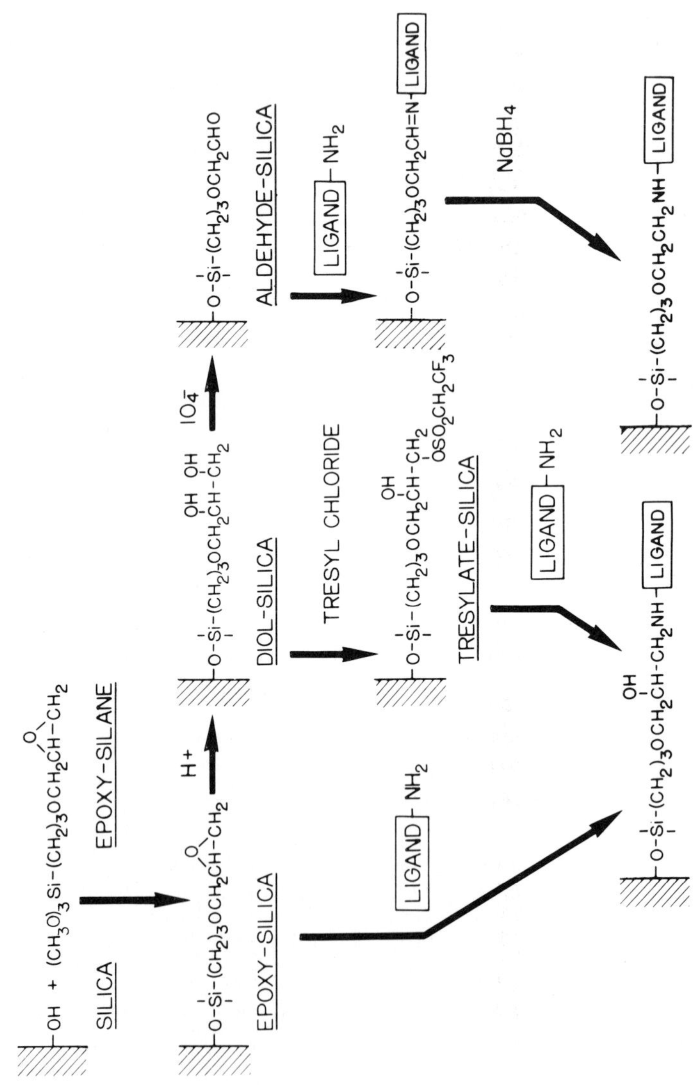

FIG. 1. Three routes for preparing ligand-substituted silica. Reproduced from Larsson *et al.*,[3] with permission.

remove adsorbed moisture. After cooling to 50–100°, the vacuum is disconnected and the flask is immediately fitted with a reflux condenser and is provided with a $CaCl_2$ drying tube and stirrer. Sodium-dried or molecular sieve-dried toluene (150 ml) is added, followed by 2 ml of γ-glycidoxypropyltrimethoxysilane[8] and 0.05 ml of triethylamine.[9] The mixture is refluxed with stirring for 16 hr. The epoxy-silica obtained is washed on a glass filter with toluene, acetone, and ether and dried under reduced pressure.

Well-coated silica should contain about 2.5 μmol of epoxy groups per square meter. The epoxy group content is determined by titration of hydroxyl ions released by the reaction of epoxy groups with thiosulfate[3]: Epoxy-silica (25–100 mg) is carefully slurried in 2 ml of water, the pH is adjusted to 7.0, and 1 ml of 3 *M* sodium thiosulfate at pH 7 is added to start the reaction; 100 m*M* HCl is provided, preferably with an automatic titrator, to maintain the pH at 7.0 for 1 hr. The consumption of HCl is a direct measure of the epoxy group content.

Preparation of Glycerylpropyl-Silica, Diol-Silica (Fig. 1)

Epoxy-silica (5 g) is suspended in 500 ml of 10 m*M* H_2SO_4 and heated to 90° for 1 hr. The glycerylpropyl-silica (diol-silica) obtained is filtered; washed sequentially with water, ethanol, and ether, and dried under reduced pressure.

Preparation of Aldehyde-Silica (Fig. 1)

Diol-silica (5 g) is suspended in 100 ml of acetic acid–water (90 : 10), 5 g of sodium periodate is added, and the suspension is stirred for 2 hr at room temperature. The resultant aldehyde-silica is washed sequentially with water, ethanol, and ether and dried under reduced pressure.

Preparation of Tresyl-Silica (Fig. 1)

Tresyl chloride (2,2,2-trifluoroethanesulfonyl chloride), which is used in the activation process, is readily hydrolyzed by water; dry solvents (molecular sieve 4 Å) and a minimum of exposure to humid conditions are imperative. The activated silica, on the other hand, is stable in aqueous solutions at low pH (pH 3).[10]

[8] Dow Corning Z-6040, Midland, MI.

[9] When activating silica with larger surface area, e.g., LiChrospher Si 300 and LiChrospher Si 100 (area ≈ 250 m^2/g), five times the amount of silane and amine was used.

[10] The tresyl chloride activation method is described by Nilsson and Mosbach in this volume [2].

Diol-silica (2 g; 1000-Å pores) is washed three times with 50 ml of dry acetone on a glass filter. The wet silica (5 g) is immediately transferred to a vessel containing 2.5 ml of dry acetone and 200 μl of dry pyridine. The suspension is cooled to 0°, and 65 μl of tresyl chloride[11] is added with vigorous stirring. As soon as the addition is made, the stirring speed is slowed to avoid generation of "fines." After 20 min the silica is washed on a glass filter with acetone, acetone–5 m*M* HCl (1 : 1), 5 m*M* HCl, and, finally, acetone followed by drying under reduced pressure. If the activated silica is to be used within a day, the last acetone wash and the drying process are unnecessary. The degree of activation may be determined by elemental analysis of sulfur.

Coupling to Epoxy-Silica (Fig. 1)

Coupling to epoxy-silica is a comparatively slow process. As high a concentration as possible of the ligand should therefore be attempted. The method is exemplified here by coupling of a spacer-provided NAD, N^6-[*N*-(6-aminohexyl)carbamoylmethyl]-NAD.[12] Epoxy-silica (100-Å pores; 1 g) is suspended in 3 ml of 0.1 *M* sodium pyrophosphate at pH 8, containing 20 mg of the NAD analog. The suspension is gently shaken for 5 days at room temperature. The NAD-silica obtained is filtered, washed with water, and suspended in 100 ml of dilute H_2SO_4 (pH 2) and heated to 50° for 4 hr to hydrolyze excess epoxy groups to the diols. The NAD-silica is finally washed with water, pyrophosphate buffer, water, ethanol, and ether and dried under reduced pressure. The NAD-analog content may be determined from UV spectra to be 1 μmol per gram of silica.

Coupling of Alcohol Dehydrogenase to Tresyl-Silica (Fig. 1)

Tresyl-silica (1000-Å pores, 10 μm, 0.7 g dry weight) is suspended in 2 ml of 0.4 *M* sodium phosphate at pH 7.0, containing 0.2 *M* isobutyramide and 2 m*M* NADH. The suspension is deaerated for 5 min. Horse liver alcohol dehydrogenase,[13] 15 mg in 2 ml, is added, and the coupling is allowed to proceed for 20 hr at room temperature. Remaining tresyl groups are subsequently removed by treatment with 0.2 *M* Tris-HCl containing 1 m*M* dithioerythritol at pH 8 for 1 hr. The gel is then washed extensively with 0.1 *M* sodium phosphate at pH 7.5, containing 0.5 *M*

[11] Obtained from Fluka, Buchs, Switzerland.

[12] K. Mosbach, P.-O. Larsson, and C. R. Lowe, this series, Vol. 44, p. 859. The compound is available from Sigma, St Louis, MO.

[13] Boehringer, Mannheim, FRG. The commercial preparation was dialyzed against 0.075 *M* sodium phosphate at pH 7.9, and cleared by centrifugation.

NaCl and 1 mM dithioerythritol, followed by washing with the same solution but without NaCl. The alcohol dehydrogenase-silica may be stored at 4° until used. Enzyme content can be determined by amino acid analysis or from UV spectra and is generally of the order of 20 mg per gram of dry silica, i.e., 100% coupling yield.

Coupling of Concanavalin A to Aldehyde-Silica (Fig. 1)

Aldehyde-silica (1000-Å pores; 10 μm; 2 g dry weight) is suspended in 10 ml containing 1 mM $CaCl_2$, 1 mM $MnCl_2$, and 0.1 M sodium phosphate at pH 6.0. The suspension is deaerated under reduced pressure for 5 min. Concanavalin A (150 mg) in 20 ml of the above solution is added, followed by 75 mg of $NaCNBH_3$ (to reduce the Schiff base formed to a secondary amine). The suspension is stirred very gently at 4° overnight.[14] The solution is adjusted to pH 8.0, and 50 mg of $NaBH_4$ is added in portions during 30 min (to reduce remaining aldehyde groups to their alcohols). After 1 hr the concanavalin A-silica is washed on a glass filter with the above buffered solution supplemented with 0.5 M NaCl. The concanavalin A-silica (about 60 mg per gram of silica; determined spectrophotometrically) may be stored at 4° until used.

Determination of Silica-Bound Ligands

Ligand density often may be determined spectrophotometrically by one or both of the following methods:

Method 1. Ligand-substituted silica (10–200 mg) is carefully suspended in 2.5 ml of saturated aqueous sucrose solution in a spectrophotometer cell. The cell is repeatedly evacuated to ensure complete filling of the pores. Spectra are recorded using a proper reference, and the ligand concentration is calculated. The sucrose solution has approximately the same refractive index as the silica. Light scattering from silica particles suspended in saturated sucrose is therefore negligible, allowing meaningful spectra to be obtained.[3]

Method 2. Ligand-substituted silica (1–100 mg) is heated to 60° for 30 min in 1 ml of 1 M NaOH. To the solubilized silica gel are added 4 ml of 0.1 M sodium phosphate at pH 7, 1 ml of 1 M HCl, and water to a final volume of 10 ml. A reference cuvette is prepared by parallel treatment of a blank silica. The spectrum is recorded, and the ligand density is calculated. The neutralization step provides a defined pH and avoids alkaline attack on the spectrophotometer cells.

[14] Coupling at room temperature for 2 hr will give approximately the same result.

Packing of Columns

Stainless steel columns are used that are 50 or 100 mm in length, $\frac{1}{4}$-in. o.d., 5-mm i.d., and provided with compression fittings.[15] A 50-mm column has an inner volume of 1 ml; when packed, it contains 0.44 g of silica (1000-Å pores) and has a liquid-phase volume of 0.80 ml (pore volume + interstitial volume).

Packing of 50-mm Column. Derivatized silica (0.7–1 g) is suspended in 5 ml of buffer containing 50% sucrose. The suspension is treated for 5 min in an ultrasonic bath and then immediately poured into the packing device. The packing device is prepared from a 250-mm, $\frac{1}{4}$-in. stainless steel tube, connected at one end to the column to be packed via a bored-through union. After filling, the other end of the packing device is connected to a standard HPLC pump, and the pumping is started immediately. The pump should deliver buffer at a pressure of about 4000 psi. After 30 min of pumping, packing is considered to be complete and the column may be disconnected and fitted with the remaining end piece.

Chromatographic Procedures

Normal HPLC equipment may be used, generally at room temperature.

The flow rate is usually maintained within 0.1–3 ml/min for a standard 5-mm i.d. column. A typical value is 1 ml/min, which gives a pressure drop over a 50-mm column loaded with 5-μm particles of about 500 psi.

Applications of High-Performance Liquid Affinity Chromatography

Table I gives a summary of applications with high-performance liquid affinity chromatography.

Separation of Dehydrogenases on Silica-Bound AMP

AMP is an inhibitor, competitive with NAD for many dehydrogenases, and silica-bound AMP might, therefore, be expected to resolve this class of enzymes. AMP-silica has been prepared by coupling a spacer-provided AMP analog, N^6-(6-aminohexyl)-AMP, to tresyl-activated silica.[16] When a mixture of serum albumin, liver alcohol dehydrogenase, and lactate dehydrogenase was injected into a column prepared with this analog, albumin eluted unretarded, whereas both dehydro-

[15] hetp, Macclesfield, Cheshire, U.K.

[16] P.-O. Larsson and K. Mosbach, *Biochem. Soc. Trans.* **9,** 285 (1981).

TABLE I
HIGH-PERFORMANCE LIQUID AFFINITY CHROMATOGRAPHY WITH SILICA-BOUND LIGANDS

Silica-bound ligands	Interacting molecules
Nucleotides	
AMP[a]	Alcohol dehydrogenase
NAD[b]	Lactate dehydrogenase
Dyes	
Cibachron Blue F3G-A[c]	Dehydrogenases, kinases, others
Procion Blue MX-R[d]	Lactate dehydrogenase
Procion Green MX-5BR[d]	Hexokinase
Procion Brown MX-5BR[d]	Tryptophanyl-tRNA synthetase
Procion Yellow H-A[d]	Carboxypeptidase
Procion Red H-8BN[d]	Alkaline phosphatase
Antibodies	
Antihuman serum albumin[a]	Albumin
Antibovine insulin[e] (monoclonal)	Insulin
Proteins	
Soybean trypsin inhibitor[f]	Chymotrypsin
Bovine serum albumin[g]	DL-Amino acids
Alcohol dehydrogenase[h]	Nucleosides, nucleotide dyes
Concanavalin A[i]	Carbohydrates, glycoenzymes
Protein A[j]	Immunoglobulins
Miscellaneous	
Boronic acid[k]	Nucleotides, carbohydrates
Glucoseamine[l]	Concanavalin A
L-Phe-D-Phe-OCH_3[m] (methacrylate support)	Pepsin
Thymine[n] (methacrylate support)	Nucleic acid derivatives

[a] S. Ohlson, L. Hansson, P.-O. Larsson, and K. Mosbach, *FEBS Lett.* **93,** 5 (1978).
[b] Larsson *et al.*[3]
[c] C. R. Lowe, M. Glad, P.-O. Larsson, S. Ohlson, D. A. P. Small, T. Atkinson, and K. Mosbach, *J. Chromatogr.* **215,** 303 (1981).
[d] D. A. P. Small, T. Atkinson, and C. R. Lowe, *J. Chromatogr.* **216,** 175 (1981).
[e] J. R. Sportsman and G. Wilson, *Anal. Chem.* **52,** 2013 (1980).
[f] V. Kasche, K. Buchholz, and B. Galunsky, *J. Chromatogr.* **216,** 169 (1981).
[g] S. Allenmark, *Chem. Scr.* **20,** 5 (1982).
[h] K. Nilsson and P.-O. Larsson.[19]
[i] A. Borchert *et al.*[17]
[j] S. Ohlson and U. Niss, Swedish Patent application, 8104876-1 (1981).
[k] M. Glad, S. Ohlson, L. Hansson, M.-O. Månsson, and K. Mosbach, *J. Chromatogr.* **200,** 254 (1980).
[l] R. R. Walters.[4]
[m] J. Turková *et al.*[2]
[n] Y. Kato, T. Seita, T. Hashimoto, and A. Shimizu, *J. Chromatogr.* **134,** 204 (1977).

genases were strongly adsorbed. A specific elution method, ternary complex formation, released the dehydrogenases. Addition of NAD and oxalate eluted lactate dehydrogenase, whereas addition of NAD and pyrazole eluted alcohol dehydrogenase. The NAD concentration used, 0.1 m*M*, was in itself too weak to effect desorption.

Separations with Antibody–Silica

The specific interaction between an antibody and its antigen allows very effective separations. In illustration, a mixture of two serum albumins, human and bovine, was injected in a 5 × 50 mm antihuman serum albumin–silica column (particle size, 10 μm; pore size, 60 Å). The bovine albumin was eluted unretarded, whereas human albumin was strongly adsorbed. To elute the human albumin, drastic conditions in the form of a pulse of 0.2 *M* glycine–HCl at pH 2.2 were required. By the pulse elution technique, the separation process could be managed within 5 min. The acid eluent had no apparent adverse effects, since the procedure could be repeated more than 20 times without any observable decrease in performance of the system.

Separations on Concanavalin A–Silica

The lectin concanavalin A was coupled to silica (60 mg of lectin per gram of silica), and its specificity for α-D-mannose and α-D-glucose was used for the separation and purification of carbohydrates and glycoproteins.[17] A sample of commercial horseradish peroxidase (4.1 mg in 4.1 ml of buffer) was injected in a concanavalin A column (5 × 50 mm; particle size, 10 μm; pore size, 1000 Å) operated at room temperature at a flow rate of 1 ml/min. Approximately 50% of the total protein, but less than 2% of the peroxidase, passed the column unretarded.

A pulse of the competitively acting α-methylglucoside was subsequently used to elute pure peroxidase. The separation illustrates that even a small analytical column can be used for preparative purposes, at least in the final stages of a purification process. Aside from a twofold purification in 20 min, the enzyme became concentrated in the process, 90% of the enzyme being collected at four times its original concentration.

Reversed Affinity Chromatography with Silica-Bound Alcohol Dehydrogenase

Enzymes are seldom used as adsorbents for compounds of low molecular weight in conventional affinity chromatography.[18] Although several

[17] A. Borchert, P.-O. Larsson, and K. Mosbach, *J. Chromatogr.* **244,** 49 (1982).
[18] K. Das, P. Dunnill, and M. D. Lilly, *Biochim. Biophys. Acta* **397,** 277 (1975).

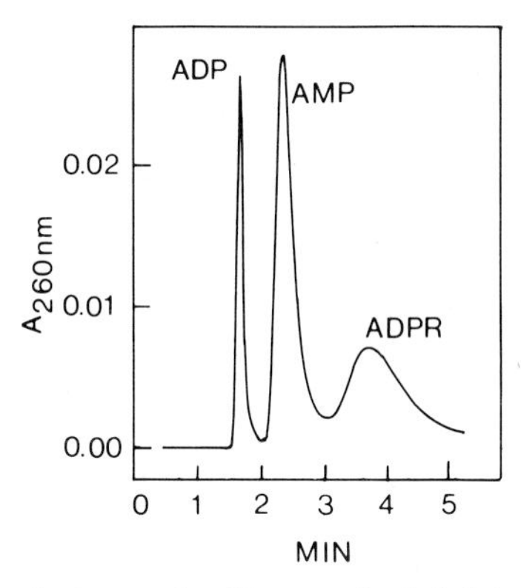

FIG. 2. Separation of adenine nucleotides on alcohol dehydrogenase-silica. Column: 50 × 5 mm. Content: Silica (1000 Å) with 12 mg of horse liver alcohol dehydrogenase per gram of silica coupled after tresyl chloride activation. Binding site concentration: 120 μM. Sample: 0.5 nmol of AMP, 0.2 nmol of ADP and 1 nmol of ADP-ribose in 15 μl of sodium phosphate at pH 7.5. Mobile phase: 0.25 M sodium phosphate at pH 7.5, containing 1 μM $ZnSO_4$. Flow rate: 1.0 ml min^{-1}. Reproduced from Nilsson and Larsson,[19] with permission.

difficulties apply, the basic problem is that the capacity, at least on a weight basis, is very limited in such reversed affinity systems. This obviously has consequences in preparative applications although problems may also arise in an analytical context. For example, poor resolution due to overloading may occur. However, in combination with the HPLC technique, with its emphasis on resolution and sensitivity, the reversed affinity chromatography concept could be useful.

An example is given in Fig. 2 of a mixture of adenine nucleotides chromatographed on a horse liver alcohol dehydrogenase silica column.[19] The separation pattern obviously reflects the strength of the interaction between the silica-bound enzyme and the chromatographed nucleotide. A quantitative relationship between the retention (k') of a substance, the dissociation constant, K_d, for the binary complex involved, and the ligand density, [HLADH], has been verified for alcohol dehydrogenase-silica[19]: $k' = [\mathrm{HLADH}]/K_d$. The capacity factor, k', is commonly used in HPLC contexts and refers to the retention of a substance in column volumes. It may be experimentally calculated from $k' = (V_e - V_0)/V_0$, where V_e is the elution volume for the compound in question and V_0 is the elution volume for a nonadsorbed, nonexcluded compound. If the ligand density and the dissociation constant are known, the chromatographic behavior could be predicted from the equation. Conversely, knowledge of ligand density and chromatographic behavior allow calculation of the K_d constant. Interest-

[19] K. Nilsson and P.-O. Larsson, *Anal. Biochem.* **133** (1983).

TABLE II
CHROMATOGRAPHIC DATA FOR ADENINE NUCLEOTIDES ON ALCOHOL DEHYDROGENASE-SILICA[a]

Compound	Spacer composition	k'-value	K_d (μM)	
			Calculated	Literature
AMP		2.15	134	70
N^6-(6-Aminohexyl)-AMP	$H_2N(CH_2)_6$—	5.6	51	32
N^6-(2-Aminoethyl)-AMP	$H_2N(CH_2)_2$—	0.28	1030	—
N^6-Carboxymethyl-AMP	$HOOCCH_2$—	2.7	107	—
ADP		0.19	1500	390
ATP		0.06	5000	—
ADP-ribose		7.5	38	18
NAD		2.4	120	96
N^6-[N-(6-Aminohexyl)carbamoylmethyl]-NAD	$H_2N(CH_2)_6NHCOCH_2$—	1.56	185	—
N^6-[N-(2-Aminoethyl)carbamoylmethyl]-NAD	$H_2N(CH_2)_2NHCOCH_2$—	0.40	720	—

[a] The adsorbent contained 21 mg of horse liver alcohol dehydrogenase per gram of silica (1000 Å; 10 μm).

ingly, it has been shown that alcohol dehydrogenase immobilized to silica by the tresyl chloride method preserves many of its native properties.[10] Thus, dissociation constants referring to silica-bound enzyme may be valid also for the free enzyme. This obviously opens up the possibility of rapid screening of a large number of compounds in order to determine their K_d. Table II exemplifies this idea: the k' values, the calculated dissociation constants, and the corresponding literature data are presented for several adenine nucleotides of potential interest as affinity ligands. The chromatographically derived dissociation constants are generally higher than those presented previously. These observations may be due to chromatography being carried out with high ionic strength buffers, whereas data from the literature refer mainly to weak buffers. Table II also indicates that K_d values for very weakly retained substances are less reliable, probably because of the greater imprecision in the determination of k' for these compounds. Nevertheless, Table II reveals interesting affinity differences for the several analogs—differences that may be explained by the attraction or the repulsion between the N^6 substituents and certain amino acid residues on the enzyme surface.

Comments

High-performance liquid affinity chromatography as described here allows very rapid separation, making the technique suitable for analytical as well as micropreparative purposes. Moderate scaling up should give preparative systems with high capacity. The technique provides a convenient means of obtaining quantitative information about biological complexes.

Acknowledgment

Discussions with S. Birnbaum, A. Borchert, M. Glad, and L. Hansson are gratefully acknowledged.

[11] Optimal pH Conditions for Ion Exchangers on Macroporous Supports

By JAMES S. RICHEY

Two relatively recent techniques are available for determining the most suitable pH conditions for performing ion-exchange chromatography with macroporous media: the electrophoretic titration curve approach and the retention mapping method. The ready determination of this parameter should provide optimal conditions for truly high-performance ion-exchange chromatography of biologically active macromolecules. However, both methods are known to be effective only with macroporous media; whether these approaches are reliable or have predictive value for other HPLC or conventional ion-exchange materials is not tested (see also [8]).

Macroporous Ion-Exchange Matrix

Recently, a major change in the polymer chemistry industry has allowed the production of a rigid chromatographic support media with very large pore structure. The macroporous structure of these beads allows rapid movement of both fluid and large molecules through packed beds with very low resistance. Another factor, the monodispersity or very

ISBN 0-12-182004-1

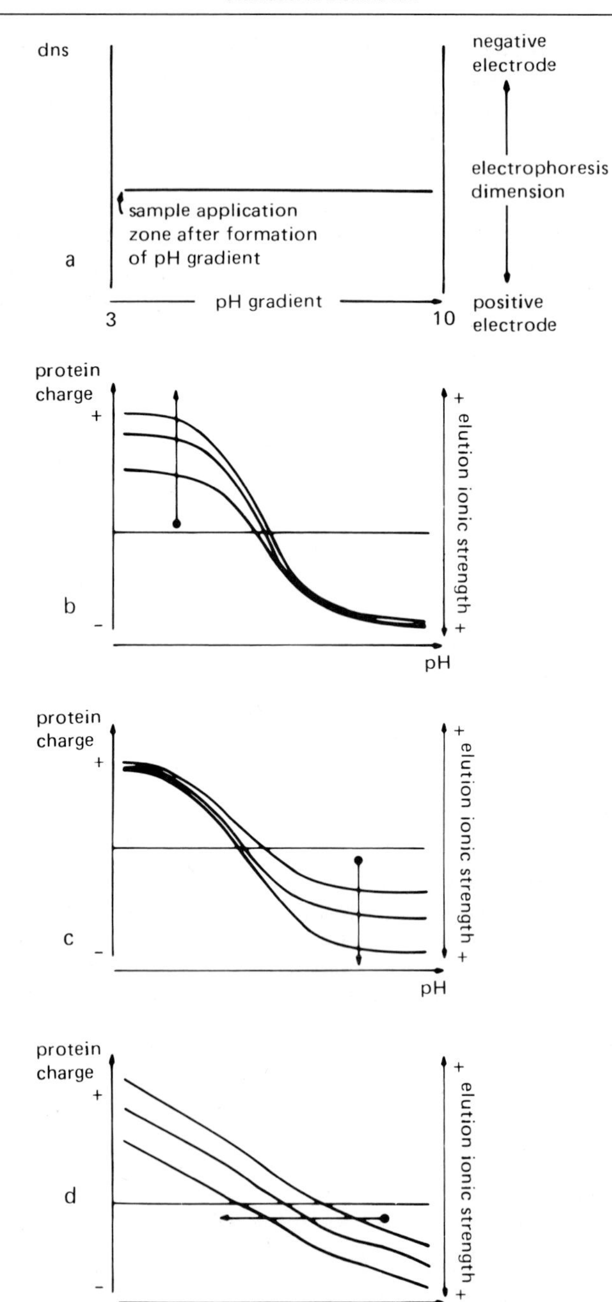
dns
negative electrode
electrophoresis dimension
sample application zone after formation of pH gradient
a
pH gradient
3
10
positive electrode
protein charge
+
-
b
pH
+ elution ionic strength +
protein charge
+
-
c
pH
+ elution ionic strength +
protein charge
+
-
d
pH
+ elution ionic strength +

narrow particle size distribution of this rigid material, is responsible for drastically reduced back pressures and allows elution equipment more compatible with recovery of enzymic activity than the stainless steel normally used with very high-pressure pumping systems. A review of the properties of these macroporous exchange materials is available.[1] The large pore structure and monodisperse particle size distribution are thought to be the prime factors responsible for the ability of the two techniques described here to function in a predictive and reliable manner to aid in the determination of optimal separation conditions on macroporous ion exchangers for macromolecules.[2]

Electrophoretic Titration Curves

This method, first described in 1977,[3] provides a relatively simple determination of the pH-dependent mobility of components within a protein mixture. The objective, of course, is the determination of the most suitable pH at which to perform a chromatographic separation. The principal steps of this two-dimensional electrophoretic technique are outlined in Fig. 1. Briefly, the first dimension involves the creation of a pH gradient by isoelectric focusing with carrier ampholytes in a polyacrylamide or agarose gel matrix. Once the pH gradient is established, sample is applied in a trough across the gradient and electrophoresis is performed at right angles. Righetti and Gianazza[4] recognized the potential of the technique in predicting conditions for charge-dependent separations and showed in 1979 that the sigmoidal curves generated by this method were propor-

[1] J. Richey, *Am. Lab.* **14,** 104 (1982).

[2] Commercial availability of these supports is presently limited to the MonoBead column series from Pharmacia Fine Chemicals, Piscataway, New Jersey. Mono Q is a strong anion exchanger, Mono S is a strong cation exchanger, and Mono P is a chromatofocusing medium.

[3] A. Rosengren, B. Bjellqvist, and V. Gasparic, "Electrofocusing and Isotachophoresis." de Gruyter, Berlin, 1977.

[4] P. G. Righetti and E. Gianazza, *in* "Electrophoresis '79," p. 23. de Gruyter, Berlin, 1980.

FIG. 1. Generation of titration curves by two-dimensional electrophoresis is useful for prediction of chromatographic pH-dependent charge mobility. (a) A pH gradient is formed with carrier ampholytes in an electric field in the horizontal dimension, followed by sample application in a central trough. Electrophoresis is then performed in the vertical dimension. (b) A titration curve of this mixture, indicating that optimal separation should take place at lower than neutral pH (pH 4–5) where proteins are positively charged, i.e., a cation exchanger should be used. (c) In this case, optimal separation will be at a pH greater than 7.0 and an anion exchanger should be used to resolve the net negatively charged proteins. (d) Where titration curves are essentially parallel, separation near neutral charge, i.e., near the p*I*, is optimal.

tional to the theoretical titration curve of the protein. Proportionality, however, is not sufficient for prediction. In order to exploit the method fully, two factors must be resolved: sieving effects caused by the electrophoresis gel matrix should be eliminated, and the chromatographic media must separate on the basis of net protein charge (not localized charge) and without extraneous sieving or hydrophobic interaction.

The polyacrylamide electrophoretic gel matrix described here has a pore structure that allows free mobility of proteins up to 1.5×10^5 daltons.[5] The agarose matrix described here is recommended for mixtures containing larger proteins up to about 6.6×10^5 daltons. By selecting and using the two matrices, sieving effects within the electrophoresis matrix are minimized.

Electrophoretic Titration Curve Method

Materials

All focusing and electrophoreses are performed on a Pharmacia Flatbed Apparatus FBE-3000 with an ECPS 3000/150 power supply and VH-1 volt-hour integrator. A circulating bath is required for temperature control and cooling. Gels have been cast in a custom-made casting frame, which forms a trough for sample application.

Gel Preparation

Polyacrylamide Gels. Prepared from a stock solution (10% acrylamide, 3% bisacrylamide); the stock solution is deionized by mixing 1 g of Amberlite MB-3 per 100 ml of solution for 1 hr. Solution may be stored refrigerated, over MB-3, for up to 1 week. A solution sufficient for two gels contained 22.5 ml acrylamide stock, 12 ml glycerol (50%), and 3 ml of Pharmalyte 3-10. The solution is diluted to 45 ml, filtered through Whatman #1 paper, and degassed. One hundred microliters of a 60 mg/ml ammonium peroxydisulfate solution is added to the mixture and the solution is quickly applied to the casting frame after insertion of Silane 174 (Pharmacia)-treated glass plates. Polymerization is complete and stable after 90 min at room temperature. Gels may be stored in a humidity chamber at 4°.

Agarose Gels. Agarose–Sephadex gels are prepared from a solution containing 0.45 g Agarose-IEF (Pharmacia), 0.75 g Sephadex G-200 Superfine, and 4.5 g D-sorbitol. The solution is diluted to 45 ml with hot

[5] A. B. Bosisio, C. Loeherlein, R. S. Snyder, and P. G. Righetti, *J. Chromatogr.* **189,** 317 (1980).

water, boiled, and cooled to 70° before 3 ml of Pharmalyte 3-10 is added. Gel Bond (Marine Colloids), cut to the size of the casting mold, is allowed to adhere to the frame with the hydrophobic side to the frame. The agarose–Sephadex solution is then injected into the casting frame. After hardening at room temperature for 45 min, the gels may be stored in a moist environment at 4°.

Running and General Staining Conditions

Both types of gels are electrofocused on the flatbed apparatus at 12°, with the sample trough perpendicular to the electrodes. The anode electrode strip is soaked in 1 *M* phosphoric acid for agarose gels or 40 m*M* aspartic acid for polyacrylamide gels. Cathode strips are always prepared by immersing in 1 *M* sodium hydroxide solution.

Focusing is complete for agarose gels (size: 80 × 80 mm) at 7 W constant power for 750 V-hr whereas polyacrylamide gels are run at 15 W for 750 V-hr. Second-dimension conditions start with sample application and rotation of the gel a quarter of a turn, such that the sample trough is parallel to the electrode wicks; the wicks are freshly soaked again in their respective solutions. Agarose gels are electrophoresed at 1000 V for 100 V-hr and polyacrylamide gels require 1000 V for 150 V-hr.

Both types of gels may then be fixed for 30 min in 10% trichloroacetic acid and 5% sulfosalicylic acid. Agarose gels are dried by blotting for 30 min, followed by drying in a hot air stream until dry to the touch. Both types of gels may be stained in either 0.2% Coomassie Brilliant Blue R-250 or 0.2% Page Blue 83 in 35% methanol and 10% acetic acid. Protein material in agarose gels binds stain rapidly, in about 10 min. Polyacrylamide gels require much longer and are normally left overnight (6–18 hr). Both types of gels may then be destained in 35% methanol and 10% acetic acid.

In practice, specific detection methods such as zymogram staining or fluorescent-labeled substrate-binding techniques are used to locate the patterns of specific enzymes.[6] Comparison of a nonspecifically stained gel to one stained for enzyme activity presents the full picture in terms of pH-dependent mobility in relation to the other proteins in the mixture.

Retention Mapping

Most easily described as "running the sample at a number of different pH values," the special properties of the strong anion and cation ex-

[6] M. J. Heeb and O. Gabriel, this volume [28]. See also this series, Vol. 22 [40].

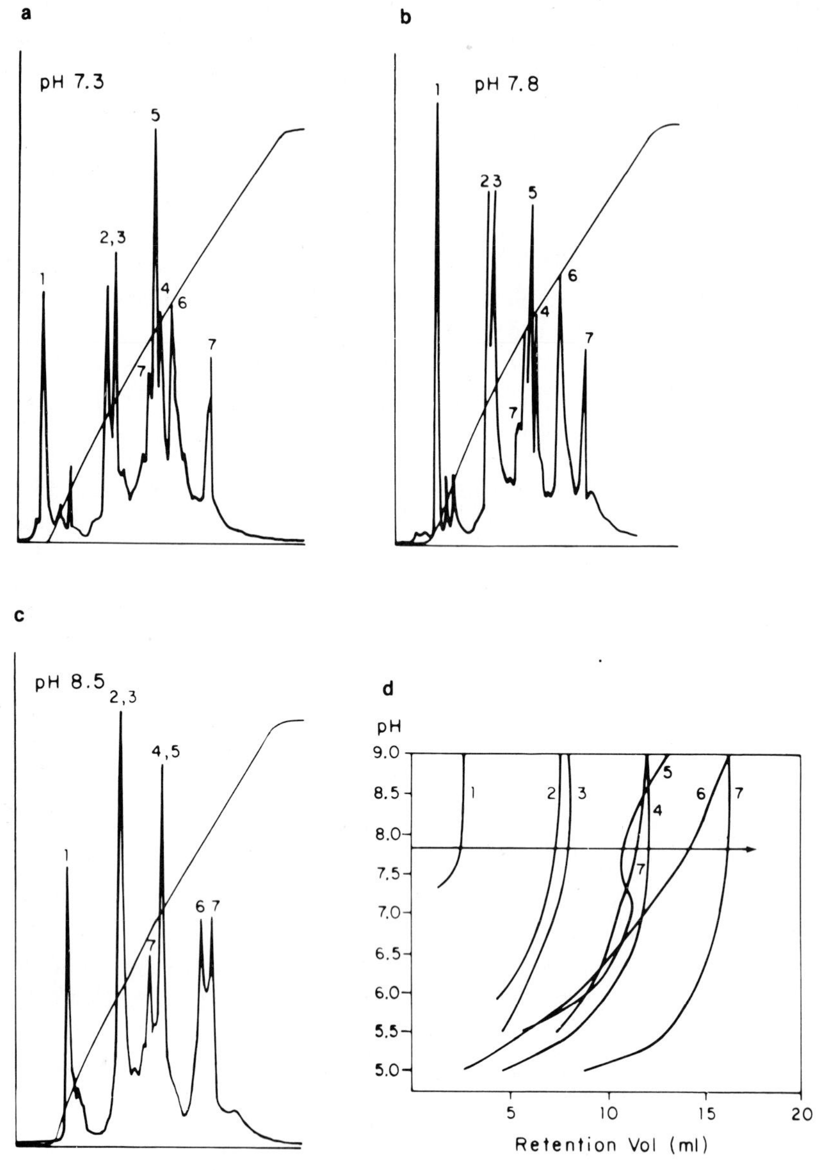

FIG. 2. Retention mapping method for chromatographic optimization using a Pharmacia FPLC System with the fast anion exchanger, Mono Q, at a flow rate of 1 ml per minute. Protein: (1) carbonic anhydrase (carbonate dehydratase), 1 mg/ml; (2) conalbumin type 2, 2 mg/ml; (3) transferrin, 1 mg/ml; (4) ovalbumin, 3 mg/ml; (5) α-lactalbumin, 1 mg/ml; (6) bovine serum albumin, 5 mg/ml; (7) trypsin inhibitor, 4 mg/ml. (a) Buffer A: 20 mM triethanolamine

changers' rapid separation times make this approach practical. All buffers for the various pH values have been tested by the manufacturer and their recommendations are supplied with the product. Preparation of four or five of these buffers for one or both ion exchangers is sufficient for obtaining practical retention map information. Macroporous ion exchangers provide consistent, interpretable results within a time frame that makes the method very attractive (10–20 min per trial). Figure 2a, b, and c provides examples of chromatograms run at different pH values, and Fig. 2d represents the retention volume as a function of pH. It is a simple matter to examine the curves and select the pH most appropriate for optimal resolution of the components of interest.

Retention Mapping and Electrophoretic Titration Curves: An Example

Haff *et al.*[7] have reported a study of the behavior of a number of samples with the macroporous ion exchangers and the value of both of these techniques in predicting conditions for optimal separation. One system, the lactate dehydrogenase isoenzymes from beef heart and muscle prepared by a quick-freeze, slow-thaw method,[8] was chosen because of the very high degree of physical similarity among the isomeric subunits and the possible resultant isoenzymes. Freezing and thawing produces hybrids containing subunits from both native sources [beef heart (H) and muscle (M)] in all the tetrameric conformations, H_3M, H_2M_2, HM_3, H_4, and M_4. The investigators reasoned that chromatographic behavior would not be complicated by size or shape differences, but would be influenced primarily by surface charge. The lactate dehydrogenase, then, should serve as a model system for comparing retention mapping and electrophoretic titration curves.

The agarose–Sephadex electrophoretic titration curve (Fig. 3) reveals the five major bands to be fairly well resolved in the anionic side of the gel. Although polyacrylamide gels were reported to show slightly sharper bands, splitting of the M_3H and M_2H_2 bands could just barely be discerned

[7] L. A. Haff, L. G. Fagerstam, and A. R. Barry, *J. Chromatogr.*, **266**, 409 (1983).

[8] A. Stolzenbach, *in* "Methods in Enzymology: Carbohydrate Metabolism" (Willis A. Wood, ed.), p. 287. Academic Press, New York, 1966.

chloride, pH 7.3; buffer B: buffer A + 0.35 *M* NaCl; gradient elution: 0–20 ml (0–100% B). (b) pH 7.8; all other conditions as in (a). (c) Buffer A: 20 m*M* diethanolamine chloride, pH 8.5; buffer B: buffer A + 0.35 *M* NaCl; all other conditions as in (a). (d) Elution volume plotted as a function of pH. Each curve is representative of the charge-dependent mobility of the protein it represents.

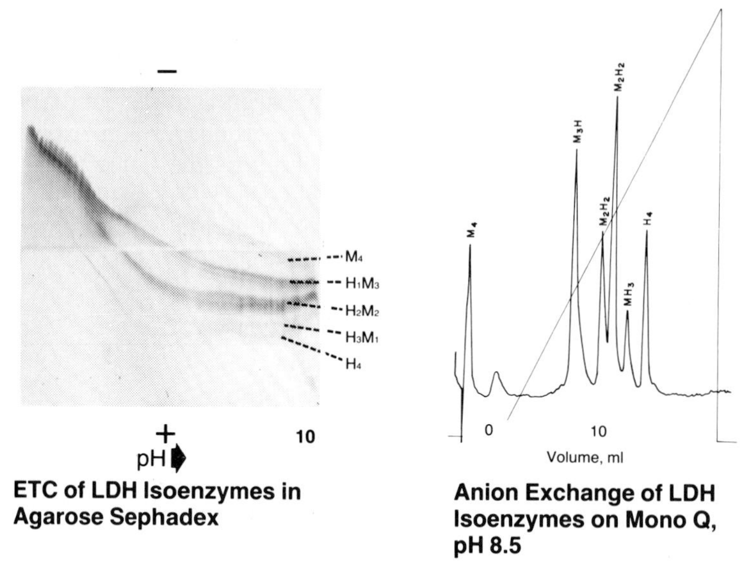

FIG. 3. Electrophoresis titration curve of lactate dehydrogenase isoenzymes in an agarose–Sephadex gel matrix compared to anion exchange chromatography at pH 8.5 on Mono Q, a fast anion-exchange material.

whereas they are plainly seen here. Notice the smearing of protein material in the cationic portion. This is valuable information to the chromatographer because it immediately suggests that the mixture is unstable or insoluble in this pH region. Chromatographic manipulations of this enzyme in the acidic regions must therefore be avoided because little or no sample recovery could be expected. Using this information, a logical choice for performing ion-exchange chromatography would be an anion matrix, i.e., one charged positively, developed under basic conditions. The results of four trials at pH 8.5 through 10.0 are presented in Fig. 4. Overall resolution is clearly superior in the separation at pH 8.5, as predicted from the electrophoretic curve.

Retention mapping was performed in the range of pH 5.5 to 10 with retention volume, in milliliters, plotted against the pH of the separation for M_3H, MH_3, and H_4 (Fig. 5a) and the two bands of M_2H_2 (Fig. 5b). That the M_2H_2 band splits in this manner is not as curious as it might seem at first. Although the subunit structures are precisely identical, two different conformational states may exist that result in a different net surface charge, hence a different retention volume.

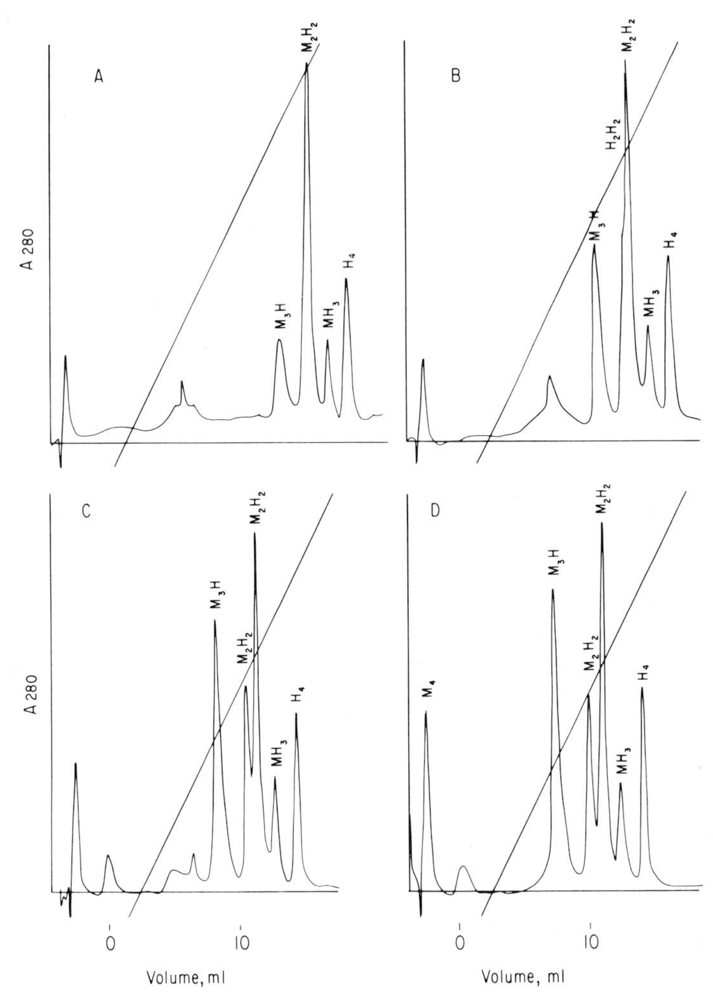

FIG. 4. High-performance anion exchange chromatography on Mono Q of lactate dehydrogenase isoenzymes at several pH values. Buffers: (A) 20 mM 1,3-diaminopropane, pH 10.0; (B) 20 mM piperazine, pH 9.5; (C) 20 mM ethanolamine, pH 9.0; (D) 20 mM diethanolamine, pH 8.5. Flow rate, 1 ml/min. All were developed in their respective buffers with a linear salt gradient to 0.35 M in NaCl.

Conclusions

Obtaining electrophoretic titration curves provides "advanced" information regarding stability and resolution of proteins in a complex mixture. Specific detection methods applicable to polyacrylamide or agarose

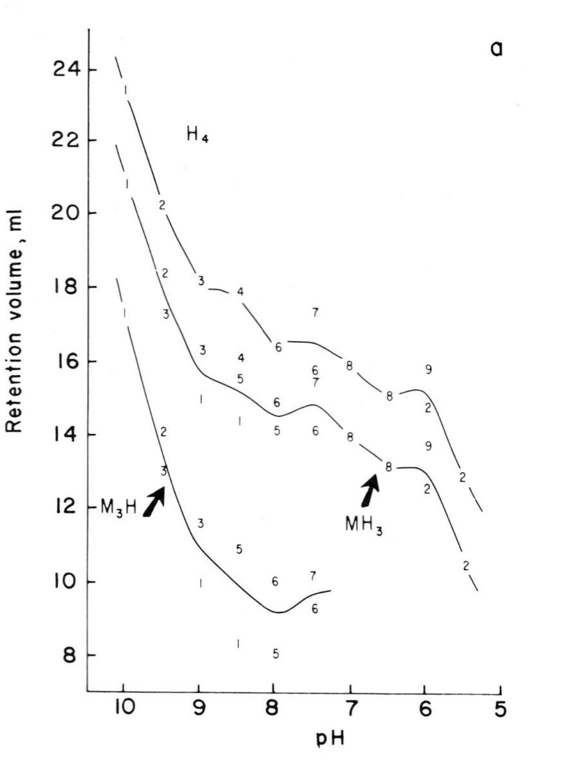

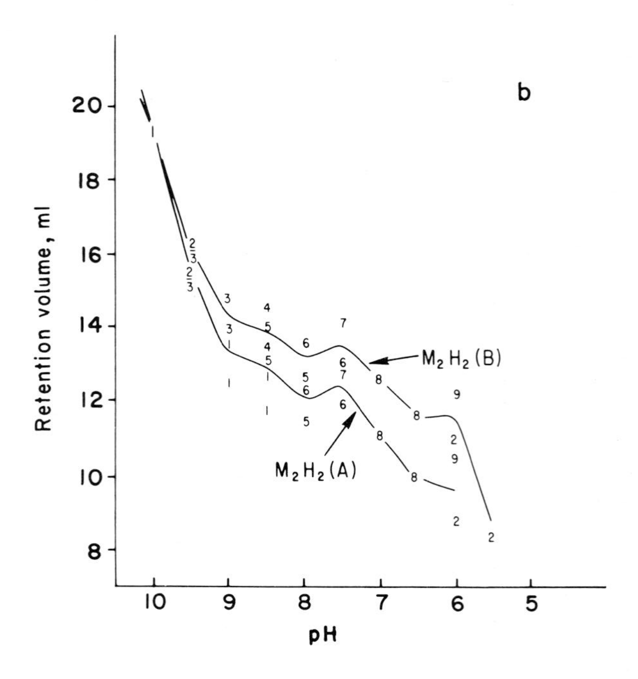

FIG. 5. Retention maps of lactate dehydrogenase isoenzymes chromatographed at several pH values on Mono Q. The numbers within the graphs refer to the buffers: (1) 20 m*M* 1,3-diaminopropane; (2) 20 m*M* piperazine; (3) 20 m*M* ethanolamine; (4) 20 m*M* diethanolamine; (5) 20 m*M* *N*-methyldiethanolamine; (6) 20 m*M* tris(hydroxylmethyl)methane; (7) 20 m*M* triethanolamine–HCl; (8) 20 m*M* bis-tris-propane; (9) 20 m*M* bis-tris. (a) Retention values for H_4, M_3H, and MH_3 species. M_4 failed to bind under any of the conditions tested. (b) Retention value for the two M_2H_2 species.

may be used for electrophoretic titration curves. However, if such techniques are not available, retention mapping provides another, also empirical, alternative with actual fractions of recovered material available for specific assay.

Although published uses of these techniques are as yet infrequent, the producer of the macroporous exchangers reports numerous communications of helpful applications and has published on the subject.[1,7]

Section II

Electrophoresis

[12] Systems for Polyacrylamide Gel Electrophoresis

By PERRY J. BLACKSHEAR

Separation of proteins in complex mixtures by polyacrylamide gel electrophoresis remains a powerful and versatile analytical technique. Advances in the use of sensitive protein staining procedures (see this volume [28–32]) and autoradiography (see this volume [34]) have greatly increased the sensitivity of the primary technique. Perhaps equally important are extensions of gel electrophoresis for specialized purposes, including two-dimensional electrophoresis, partial proteolytic digestion within the stacking gel, electrophoretic transfer to paper for immunoblotting procedures, and others.

Presented here are some of the polyacrylamide gel electrophoretic separation methods that we have found to be useful for a variety of purposes. No attempt is made to review the many electrophoretic procedures in use today, although some of them are recorded elsewhere in this volume. Most of the procedures presented here deal with electrophoresis in slab gels, but virtually all the comments are applicable to electrophoresis with rod gels.

General Principles

Most gel systems widely used today make use of the strategy for zonal or discontinuous electrophoresis in polyacrylamide gels developed by Ornstein[1] and Davis[2] and Raymond and Weintraub.[3] The advantages inherent in the use of a stacking gel with multiphasic buffer systems have been described in detail by Chrambach and Rodbard[4] and include the ability to study very dilute samples; the ability to concentrate proteins into very thin starting zones of high local protein concentration, resulting in improved resolution; and the provision of a dye front, which allows calculation of relative mobilities of proteins, as well as a means of following the progress of a given electrophoretic run.

The principle behind the use of a stacking gel is derived from the so-called Kohlrausch regulating function. In brief, at the pH of the stacking gel, the sample ion (in most cases, a mixture of proteins or their detergent

[1] L. Ornstein, *Ann. N.Y. Acad. Sci.* **121,** 321 (1964).
[2] B. J. Davis, *Ann. N.Y. Acad. Sci.* **121,** 404 (1964).
[3] S. Raymond and L. Weintraub, *Science* **130,** 711 (1959).
[4] A. Chrambach and D. Rodbard, *Science* **172,** 440 (1971).

ISBN 0-12-182004-1

derivatives) is introduced near the boundary of two ions of the same sign as the sample ion at that pH. One ion migrates faster than, and the other slower than, the sample ion when a current is passed through the mixture, and the sample ion, being of intermediate mobility, is sandwiched between them, forming a narrow zone or band of high concentration. When the sample ion migrates into the running or separating gel, under conditions of different pH and/or pore size, the trailing ion continuously overtakes and passes the sample ion, "establishing a comparatively uniform voltage gradient in which electrophoretic separation of samples occurs."[2]

A further refinement that has attained widespread use is the introduction of electrophoresis of sodium dodecyl sulfate (SDS) derivatives of proteins by Shapiro, Viñuela, and Maizel in 1967[5] and subsequent descriptions of the utility of this system as a means of separating proteins by their subunit molecular weight, largely independent of tertiary conformation, amino acid composition, or isoelectric point.[6] Denaturation and binding of SDS to proteins and peptides generally result in a relatively uniform negative charge, since most peptides bind about 1.4 mg of SDS per milligram of protein; furthermore, changes in electrophoretic behavior resulting from different tertiary protein structures are largely obviated because of the uncoiling introduced by SDS binding. Thus, when combined with discontinuous or zonal electrophoresis, SDS electrophoresis in polyacrylamide gels combines the advantages of a stacking system with those of the combined gel filtration and electrophoretic properties of the separating gel, resulting in sharp separation of proteins based largely on their subunit molecular weights. The gel systems described by Laemmli,[7] who added SDS to the Tris–glycine buffer system of Ornstein and Davis, and by Neville,[8] using an SDS–Tris–borate buffer, are probably the most versatile and widely used types of denaturing, discontinuous electrophoresis.

Other systems have since been developed for more specialized purposes. For example, linear or curvilinear gradient gels, in which the acrylamide concentration increases linearly in the separating gel (often from 3 to 30%) have become useful for maximizing separation of protein bands in a complex mixture of proteins of widely varying molecular weights. Other buffer systems employing SDS–urea, SDS–agarose, or other combinations have been useful in the study of small proteins and peptides. Acidic detergents such as tetradecyltrimethylammonium bromide (TDAB) or

[5] A. L. Shapiro, E. Viñuela, and J. V. Maizel, *Biochem. Biophys. Res. Commun.* **28,** 815 (1967).

[6] K. Weber and M. Osborn, *J. Biol. Chem.* **244,** 4406 (1969).

[7] U. K. Laemmli, *Nature (London)* **277,** 680 (1970).

[8] D. M. Neville, *J. Biol. Chem.* **246,** 6328 (1971).

hexadecylpyridinium chloride (HDPC) have found acceptance for the separation of membrane acylphosphate intermediates in a discontinuous system. Careful use of disulfide reducing agents with and without SDS has been successful in determination of the subunit structure of a number of complex proteins. Finally, two-dimensional electrophoresis in which proteins are separated by isoelectric focusing and then by SDS-polyacrylamide gel electrophoresis has revolutionized the separation of proteins in crude mixtures.

The emphasis here is on discontinuous or zonal electrophoresis in which a stacking gel is used. These systems have so many advantages over continuous electrophoresis that there are few situations in which the latter is necessary. However, Hames[9] has noted that the concentration of a protein that occurs during stacking can occasionally lead to protein aggregation or precipitation in nondenaturing solutions. This leads to failure of the protein to enter the gel, as well as to artifacts such as streaking in the path of the sample lane. Where a continuous electrophoresis system seems advantageous, almost any buffer can be used, provided the proteins of interest retain either a positive or a negative charge; this can be accomplished using one of the three buffers described below for nondenaturing gels. Such systems require that the smallest possible volume of sample be applied and that the sample buffers be of low ionic strength to ensure maximal band sharpness despite the absence of a stacking gel.

Nondenaturing Gels

When it is necessary to separate intact proteins, especially oligomeric proteins, by a nondestructive means for later assessment of biological activity, gels must be prepared under nondenaturing conditions. Since denaturation of protein is not involved, protein separation depends on a combination of differences in molecular size and shape as well as charge. Separation by size is accomplished by varying the pore size of the acrylamide polymer as a function of both the concentration of the acrylamide (range about 3–30%, w/v) and the amount of cross-linker used. In general, the higher the acrylamide concentration, the smaller the proteins that remain in the gel; this can be counteracted by decreasing the amount of cross-linker used, which in turn increases the degree of gel swelling during standard staining and washing procedures. Separation by charge in nondenaturing gel systems is permitted because the protein separation can be performed at any pH between 3 and 11, to allow for maximal charge differences between neighboring protein species. For these rea-

[9] B. D. Hames, *in* "Gel Electrophoresis of Proteins: A Practical Approach" (B. D. Hames and D. Rickwood, eds.), p. 1. IRL Press, Oxford and Washington D.C., 1981.

sons, Gabriel[10] has recommended that several properties of an enzyme be established before it is subjected to electrophoresis to determine, for example, its purity in a heterogeneous solution. These include the molecular weight of the protein, its isoelectric point, and, if enzymic activity is to be retained, its stability in the range of pH 4 to 9. Once these parameters are known, optimal resolution can be obtained by varying certain components within a single gel system. For example, the initial choice of a gel system for separation of an acidic protein of molecular weight 20,000 might be a system operating at an alkaline pH (i.e., 9) with an acrylamide concentration of 12–15%.

In this series, Gabriel[10] has described in detail the gel formulations and buffer systems to be used in the construction of three gel systems for discontinuous, nondenaturing electrophoresis at both acid and alkaline pH. Detailed formulations for these gel systems, summarized from previously published methods, have been compiled in a more convenient format by Hames,[9] and these details are described below. Many hundreds of possible buffer combinations can be generated by computer analysis of discontinuous gel systems,[4] and suitable combinations of pH, buffer components, and gel constituents are obtainable for almost every requirement. It should be noted that biologically active proteins such as enzymes often require special electrophoretic conditions in order to retain activity after separation. For example, the heat lability of some enzymes requires that electrophoresis be conducted most conveniently in a refrigerated room or, if necessary, using circulating water of 0–4° in the cooling jacket of the slab gel apparatus. In addition, ammonium persulfate, an agent commonly used in the gel polymerization reaction, often interferes with enzyme activity after elution from the gel. For this reason, the tables include the use of the photoactivated polymerizing agent riboflavin, as well as ammonium persulfate, for use in the preparation of the stacking gel for the various systems; if ammonium persulfate affects enzyme activity, then preelectrophoresis can be used to remove it from the gel if a continuous electrophoresis system is used. Dithiothreitol (40–100 m*M*) or 2-mercaptoethanol (up to 1 *M*) may be included in the sample buffer to reduce some disulfide linkages, although detergent denaturation is often necessary for complete disulfide reduction.

Using the compilation of Hames,[9] the stock solutions in Table I should be prepared. These can be combined with the appropriate buffers to form the solutions for the three types of discontinuous, nondenaturing electrophoresis described in Table II. Depending on the desired acrylamide concentration, the solutions can be combined to form the gels and buffers in Table III. Details of solution preparation are included in the tables in most

[10] O. Gabriel, this series, Vol. 22, p. 565.

TABLE I
STOCK SOLUTIONS FOR NONDENATURING DISCONTINUOUS ELECTROPHORESIS[a]

Solution	Formula
1. Acrylamide–bisacrylamide, 30:0.8	30 g of acrylamide, 0.8 g of bisacrylamide per 100 ml; store at room temperature
2. TEMED (*N,N,N',N'*-tetramethylethylenediamine)	Use as supplied; store in the cold in dark bottle
3. Ammonium persulfate, 1.5%, w/v	1.5 g in 100 ml; make up fresh on day of electrophoresis
4. Riboflavin, 0.004%, w/v	4 mg in 100 ml
5. SDS, 10%, w/v	10 g in 100 ml; filter through Whatman No. 1 paper before use; store at room temperature

[a] From Hames.[9]

cases; details of solution mixing, gel pouring, sample loading, etc., are described in greater detail in the following section on electrophoresis under denaturing conditions.

Denaturing Gel Systems

Protein separation for analytical purposes by denaturing, discontinuous electrophoresis has achieved widespread popularity. It is a relatively rapid, inexpensive, and reproducible means of separating hundreds of proteins in large numbers of heterogeneous samples, largely on the basis of their molecular weights. The gels can then be analyzed for protein staining by several techniques of differing sensitivity and specificity for determination, for example, of protein purity during a purification procedure. Selected protein bands can be cut out of the gel and used for the determination of radioactivity from a variety of *in vivo* and *in vitro* labeling techniques, amino acid compositions, and sometimes sequences, for preparation of antibody, and, occasionally, for assay of retained enzyme activity. Dried gels can be used for autoradiography of proteins labeled by similar techniques.

Among the advantages of denaturing gel electrophoresis using sodium dodecyl sulfate (SDS) is that most proteins in crude mixtures are soluble in SDS and bind it avidly; even the most basic proteins are converted to their acidic SDS derivatives and thereby migrate toward the anode at pH 7–9. Since SDS derivatives are largely unfolded, they migrate closer to their true subunit molecular weight and are more susceptible to reduction by disulfide-reducing agents, so that protein subunit structure can be more accurately assessed. The addition of hot SDS to an enzymic reaction or to mixtures of proteins is usually sufficient to stop enzyme activity, thereby

TABLE II
BUFFERS FOR NONDISSOCIATING DISCONTINUOUS SYSTEMS[a]

Buffer	Preparation
High pH discontinuous[b]: stacks at pH 8.3, separates at pH 9.5	
Stacking gel buffer (Tris-HCl at pH 6.8) 6.8)	6.0 g of Tris in 40 ml titrated to pH 6.8 with 1 *M* HCl; bring to 100 ml
Resolving gel buffer (Tris-HCl at pH 8.8)	36.3 g of Tris and 48.0 ml of 1 *M* HCl brought to 100 ml; titrate to pH 8.8 with HCl, if necessary
Reservoir buffer (Tris-glycine at pH 8.3)	At the correct concentration for use; 3 g of Tris and 14.4 g of glycine in 1 liter
Neutral pH discontinuous[c]: stacks at pH 7.0, separates at pH 8.0	
Stacking gel buffer (Tris-phosphate at pH 5.5)	4.95 g of Tris in 40 ml titrated to pH 5.5 with 1 *M* phosphoric acid; bring to 100 ml
Resolving gel buffer (Tris-HCl at pH 7.5)	6.85 g of Tris in 40 ml of water titrated to pH 7.5 with 1 *M* HCl; bring to 100 ml
Reservoir buffer (Tris-diethylbarbiturate at pH 7.0)	5.52 g of diethylbarbituric acid and 1.0 g of Tris to 1 liter
Low pH discontinuous[d]: Stacks at pH 5.0, separates at pH 3.8	
Stacking gel buffer (acetic acid–KOH at pH 6.8)	48.0 ml of 1 *M* KOH and 2.9 ml of glacial acetic acid; bring to 100 ml
Resolving gel buffer (acetic acid–KOH at pH 4.3)	48.0 ml of 1 *M* KOH and 17.2 ml of glacial acetic acid are mixed and diluted to 100 ml
Reservoir buffer (acetic acid–β-alanine at pH 4.5)	31.2 g of β-alanine and 8.0 ml of glacial acetic acid to 1 liter

[a] Reprinted from B. D. Hames, *in* "Gel Electrophoresis of Proteins: A Practical Approach" (B. D. Hames and D. Rickwood, eds.), p. 30. IRL Press, Oxford and Washington D.C., 1981, with permission.
[b] Davis.[2]
[c] D. E. Williams and R. A. Reisfeld, *Ann. N.Y. Acad. Sci.* **121,** 373 (1964).
[d] R. A. Reisfeld, V. J. Lewis, and D. E. Williams, *Nature (London)* **195,** 281 (1962).

halting the effects of proteases, protein kinases, and protein phosphatases.

The main drawback of the technique is that denaturation with SDS and similar detergents is usually irreversible, and most proteins cannot be

TABLE III
RECIPE FOR GEL PREPARATION USING NONDISSOCIATING DISCONTINUOUS BUFFER SYSTEMS[a]

Stock solution	Stacking gel (riboflavin as catalyst)	Final acrylamide concentration in resolving gel (%)[b]							Reservoir buffer[c]
		20.0	17.5	15.0	12.5	10.0	7.5	5.0	
Acrylamide–bisacrylamide, 30:0.8	2.5	20.0	17.5	15.0	12.5	10.0	7.5	5.0	—
Stacking gel buffer stock[d]	5.0	—	—	—	—	—	—	—	—
Resolving gel buffer stock[d]	—	3.75	3.75	3.75	3.75	3.75	3.75	3.75	—
Reservoir buffer stock[d]	—	—	—	—	—	—	—	—	1000 (i.e., undiluted)
Ammonium persulfate, 1.5%[e]	—	1.5	1.5	1.5	1.5	1.5	1.5	1.5	—
Riboflavin, 0.004%	2.5	—	—	—	—	—	—	—	—
Water	10.0	4.75	7.25	9.75	12.25	14.75	17.25	19.75	—
TEMED[e]	0.015	0.015	0.015	0.015	0.015	0.015	0.015	0.015	—

[a] Reprinted from B. D. Hames, *in* "Gel Electrophoresis of Proteins: A Practical Approach" (B. D. Hames and D. Rickwood, eds.), p. 31. IRL Press, Oxford and Washington D.C., 1981, with permission.

[b] The columns represent volumes (ml) of the various reagents required to make 30 ml of gel mixture.

[c] Volumes (ml) of reagents required to make 1 liter of reservoir buffer.

[d] Stock solution prepared as described in Table II.

[e] When the low pH discontinuous buffer system is used with ammonium persulfate as catalyst, the volume of TEMED should be increased to 0.15 ml for the resolving gel and the water volume adjusted accordingly. Riboflavin is usually more effective than ammonium persulfate–TEMED at low pH, whereas the latter is more effective at high pH.

recovered with intact biological activities, although several exceptions have been recorded. Variable binding of SDS to some proteins, particularly low-molecular-weight acidic and basic proteins, has been described; it often results in inaccurate molecular weight determinations. A number of posttranslational modifications, including glycosylation and phosphorylation, can affect electrophoretic mobility; the latter modification affects mobility of certain proteins, but not all, probably by affecting the net charge of the SDS derivative. Some of these factors can be used to advantage; others can be obviated by using another type of denaturing electrophoresis.

One SDS–polyacrylamide gel system that is in use in our laboratory, and some of its variations, is presented here in detail. Subsequently, several other types of denaturing electrophoretic systems, those useful in specialized applications, are also described.

SDS-Discontinuous Electrophoresis

As noted, the anionic detergent SDS denatures proteins by binding avidly to them at about 1.4 mg of SDS per milligram of protein, resulting in strongly negatively charged proteins at neutral pH that are of approximately uniform charge per unit of protein. Proteins are exposed to SDS, usually 1–2% (w/v), for 3 min at 100° to effect complete denaturation. This is performed with or without reducing agents, generally dithiothreitol (40–100 m*M*) or 2-mercaptoethanol (up to 1 *M*), to reduce disulfide bonds as needed. A substance of high specific gravity [10% (w/v) glycerol or 0.25 *M* sucrose] is added to the protein sample to allow it to sink to the bottom of the sample well in the stacking gel, and a tracking dye (pyronine Y or bromophenol blue) is added to allow visualization of the fastest migrating components in the mixture. The negatively charged SDS–protein derivatives are then subjected to an electric current at pH 7–9, causing them to migrate toward the anode. The pore size of the polyacrylamide matrix allows components of lower molecular weight to migrate faster, thus separating the protein components in the mixture according to the apparent molecular weights of their SDS derivatives. Apparent molecular weights of sample proteins can be determined by comparison with protein standards of known molecular weight.

At the end of the run, when the dye front has reached a predetermined point near the end of the gel, the current is turned off, the separating gel is separated from the stacking gel, and the former is subjected to protein fixation and staining or to whatever further analytical techniques are desired. Fixation and staining of proteins in polyacrylamide gels are discussed in this volume [29–32].

TABLE IV
STOCK SOLUTIONS FOR TRIS–GLYCINE (LAEMMLI) DENATURING (SDS) DISCONTINUOUS ELECTROPHORESIS[a]

Solution	Preparation
1. Acrylamide–bisacrylamide, 37.5 : 1.0 or 37.5 : 0.5	37.5 g of acrylamide and 1 g or 0.5 g of bisacrylamide to 100 ml; store at 20°
2. Lower gel buffer, 1.5 *M* Tris-HCl, pH 8.8	Tris, 181.5 g. Bring to pH 8.8 and to 1 liter; store at 4°
3. Upper gel buffer, 0.5 *M* Tris-HCl, pH 6.8	Tris, 60.5 g. Bring to pH 6.8 with concentrated HCl and to 1 liter; store at 4°
4. SDS, 20% (w/v)	SDS, 100 g, in 500 ml. Filter through Whatman No. 1 before use; store at 20°
5. TEMED, 0.5% (v/v)	TEMED, 0.5 ml in 100 ml; store in dark bottle at 4°
6. Ammonium persulfate, 1.5%	Ammonium persulfate, 1.5 g in 100 ml; make up fresh
7. Running buffer	Tris, 12 g, and glycine, 57.6 g, in water; add 10 ml of 20% SDS and bring to 2 liters
8. Gel overlay, 0.1% SDS	SDS, 100 mg in 100 ml; dispense from spray bottle
9. Sample "quench"	20% (w/v) SDS, 6.0 ml; 1 *M* dithiothreitol, 4.8 ml; 0.1% (w/v) pyronine Y, 1.2 ml; and 50% (w/v) sucrose, 8.0 ml. Bring to 20 ml
10. Stain	Glacial acetic acid, 100 ml; methanol, 500 ml; H_2O, 400 ml; Coomassie Brilliant Blue, 1–2 g. Use about 500 ml per gel; stain for 45–60 min at 20°
11. Destain	10% (v/v) acetic acid; use about 500 ml per gel, change as needed

[a] Modified slightly from U. K. Laemmli, *Nature (London)* **277,** 680 (1970).

Specific Details of Procedure

Preparation of the Gel. The electrophoresis system described in detail here is a minor modification of the SDS–Tris–glycine system of Laemmli[7] (Tables IV and V). Similar procedures can be used with the SDS–Tris–borate system[8]; stock solutions and gel formulas for this system are described in Tables VI and VII. Commercially available preparations of acrylamide and *N,N'*-methylene bisacrylamide (bisacrylamide), as well as other gel components, are of an adequate level of purity without recrystallization. Standard slab gel apparatus, available from Bio-Rad, Hoeffer, or other suppliers, in which the separating and stacking gels are successively

TABLE V
FORMULAS FOR TRIS–GLYCINE (LAEMMLI) DENATURING (SDS) DISCONTINUOUS ELECTROPHORESIS

	Final percentage of acrylamide desired				
	7.2%	9.0%	12.0%	15.0%	20.0%
A. Lower (separating) gel, for one slab of average size					
Mix first					
1. Acrylamide–bisacrylamide[a] (ml)	5.75	7.2	9.6	12.0	16.0
2. Lower gel buffer (ml)	7.5	7.5	7.5	7.5	7.5
3. SDS, 20% (ml)	0.15	0.15	0.15	0.15	0.15
4. Degassed water (ml)	13.45	12.0	9.6	7.2	3.2
Add to polymerize					
5. Ammonium persulfate, 1.5% (ml)	0.75	0.75	0.75	0.75	0.75
6. TEMED, 0.5% (ml)	2.4	2.4	2.4	2.4	2.4
B. Upper (stacking) gel, for one slab of average size					
Mix first					
1. Acrylamide–bisacrylamide (ml)		0.8			
2. Upper gel buffer (ml)		2.5			
3. SDS, 20% (ml)		0.05			
4. Degassed water (ml)		5.15			
Add to polymerize					
5. Ammonium persulfate, 1.5% (ml)		0.7			
6. TEMED, 0.5% (ml)		0.8			

[a] For 15 and 20% gels, 37.5 : 0.5 or 37.5 : 0.25 may be needed.

poured between a mold formed by two glass plates, separated by spacers of 0.75- to 3-mm thickness, are all adequate. The thin gels (0.75 mm) are necessary if the sensitive silver stain of Merril *et al.*[11] is to be used; 3-mm gels are valuable for large sample volumes or for preparative purposes. Once the gel apparatus has been arranged with extremely clean and grease-free glass plates, the separating gel is poured. Table IV lists the stock solutions, and Table V lists the components of these gels as prepared in our laboratory; it should be stressed that the higher ratios of acrylamide–bisacrylamide are used when the final acrylamide concentration is greater than about 12%. At high acrylamide concentrations, the gels become stiffer and more brittle and require more cooling during the run; in addition, at high ratios of acrylamide to bisacrylamide, gels tend to destain more quickly and swell considerably during destaining in 7–10%

[11] C. R. Merril, D. Goldman, S. A. Sedman, and M. H. Ebert, *Science* **211,** 1437 (1981). See also this volume [30].

acetic acid. The deionized water used should be degassed prior to use; the ammonium persulfate solution should be prepared on the day of use.

To prepare the gel, mix the appropriate components together in a small flask, swirl rapidly without causing bubble formation or aeration, and pipette rapidly a suitable volume of the final solution into the mold formed by the glass plates. No bubbles or leaks should be present. To ensure a flat junction between separating and stacking gels, we routinely spray the inside of one plate with a gel overlay solution consisting of 0.1% (w/v) SDS, which will maintain a flat meniscus on top of the separating gel during polymerization. For most gels, polymerization usually is complete within 30–60 min at room temperature; this can be checked by noting a discrete line of separation between the gel and the overlay or by tilting the mold to make sure that the gel is polymerized. If it is covered by a reasonable volume of overlay to prevent drying out, the separating gel can be stored at 4° for up to a week without evident ill effects.

After polymerization of the separating gel is complete, the gel overlay is poured off and the stacking gel mixture is pipetted into the mold. If several samples are to be evaluated "combs" providing a template for 10–

TABLE VI
STOCK SOLUTIONS FOR TRIS–BORATE (NEVILLE) DENATURING (SDS) DISCONTINUOUS ELECTROPHORESIS[a]

Solution	Preparation
1. Acrylamide–bisacrylamide, 40 : 1.5	40 g of acrylamide and 1.5 g of bisacrylamide in 100 ml of water; store at room temperature
2. Lower (separating) gel buffer	318.5 g of Tris in water. Adjust to pH 9.81 with 1 *N* HCl and bring to 1 liter; store in cold
3. Upper (stacking) gel buffer	270.6 g of Tris brought to pH 6.1 with 1 *N* H_2SO_4 and diluted to 1 liter; store in cold
4. Upper tray (cathode) stock (10×)	99.3 g of Tris and 49.5 g of boric acid brought to pH 8.64 with 1 *N* NaOH. Add 100 ml of 20% SDS and dilute to 2 liters; store at room temperature. Dilute 1 : 10 for use
5. Lower tray (anode) buffer	Dilute lower gel buffer 1 : 10 for use
6. SDS, 20% (w/v)	See Table IV
7. TEMED, 0.5% (v/v)	See Table IV
8. Gel overlay	See Table IV
9. Sample "quench"	See Table IV
10. Stain	See Table IV
11. Destain	See Table IV

[a] Modified slightly from D. M. Neville, *J. Biol. Chem.* **246,** 6328 (1971).

TABLE VII
FORMULAS FOR TRIS–BORATE (NEVILLE) DENATURING (SDS) DISCONTINUOUS ELECTROPHORESIS

	Final percentage of acrylamide desired		
	7.2	9.0	12.0
A. Lower (separating) gel, for one slab of average size			
Mix first			
1. Acrylamide–bisacrylamide (ml)	5.4	6.75	9.0
2. Lower gel buffer (ml)	15.0	15.0	15.0
3. Degassed water (ml)	6.6	5.25	3.0
Add to polymerize			
4. Ammonium persulfate, 1.5% (ml)	1.5	1.5	1.5
5. TEMED, 0.5% (ml)	1.5	1.5	1.5
B. Upper (stacking) gel, for one slab of average size			
Mix first			
1. Acrylamide–bisacrylamide (ml)		1.15	
2. Upper gel buffer (ml)		1.5	
3. Degassed water (ml)		10.1	
Add to polymerize			
4. Ammonium persulfate, 1.5% (ml)		1.2	
5. TEMED, 0.5% (ml)		1.05	

20 sample wells can be inserted into the liquid stacking gel solution before polymerization begins. If a flat surface is desired for, for example, a first-dimension isoelectric focusing gel, then gel overlay solution is again sprayed on top of the running gel before polymerization to ensure a smooth, flat meniscus. A similar procedure can be followed if the gel is to be used for preparative electrophoresis of larger volumes. For this purpose, the stacking gel is poured and covered with overlay, leaving sufficient space between the plates for application of the sample in a volume of up to 3 ml. Once polymerization of the upper or stacking gel is complete, usually taking longer than the lower gel but still less than an hour at room temperature, the template or comb is removed, the upper and lower aspects of the gel are placed in running buffer, and sample application can proceed.

Preparation of the Sample and Sample Loading. When preparing crude protein mixtures or solutions for electrophoresis, the various samples are routinely diluted until they are of the same protein concentration or trichloroacetic acid-precipitable radioactivity, and then one-fifth volume of a "quench" solution (Table IV) is added to each. Generally, the ratio of SDS to protein in the final sample should be at least 3 : 1 (by weight). The final dithiothreitol concentration should be at least 40 mM to ensure complete reduction of both interchain and intrachain disulfide bonds. The samples are normally placed in a boiling water bath for 3–5 min, allowed to cool, and then loaded onto the gel lanes after centrifugation of the samples to remove any insoluble components. In general, not more than 250 μg of a complex mixture of protein in a volume of 50 μl or less should be loaded into 4-mm wells in a 1.5-mm-thick gel; 1 mg of protein, or more, in less than 200 μl can usually be loaded into a 1-cm well in a 1.5-mm gel provided the stacking gel is sufficiently long. The high specific gravity of the sucrose makes the samples sink to the bottom of the well, so that sample gels are not necessary. We have found that capillary pipettes with a hand-held applicator are the best means for loading samples, since they fit nicely between the plates separated by 1.5-mm spacers, and they can be inserted directly into the sample wells. Again, air bubbles should be avoided during sample loading.

Running the Gel. After samples have been loaded, the entire gel apparatus is placed into a box so that both the upper and lower aspects of the gel are exposed to running buffer (Table IV). The lower buffer chamber is connected to the anode, so that negatively charged SDS derivatives migrate downward when the current is turned on. Air bubbles should be removed from the lower aspect of the gel, using a piece of rubber tubing or a bent Pasteur pipette. The current is then turned on; the procedure is adequate at 25 mA per side during the stacking phase and 30–50 mA per side during the separating phase of the run, providing the voltage remains below 200 V. Higher voltages are associated with excessive heat production and sometimes blurring of protein bands. At high acrylamide concentrations (greater than 20%, w/v), it is often necessary to run the gel at 4° with a cooling jacket to prevent excessive heating of the gel. In most cases, a single gel run takes 3–4 hr; if necessary, gels can be run overnight at low levels of current without apparent loss of resolution.

After the completion of the electrophoretic run, the gels are removed from the plates, stained, and destained. For routine use, gels are immersed for about 1 hr at room temperature in a staining solution consisting of 50% (v/v) methanol, 10% (v/v) acetic acid, 40% (v/v) water, and 0.1% Coomassie Brilliant Blue, with moderate shaking. Destaining is carried out overnight in 10% acetic acid in water. Other treatment of the destained gels can also be carried out, including removal of phosphorylated nucleic acids by treatment of the gel with 5% trichloroacetic acid at 90° for 30 min,[12] hydrolysis of phosphoserine by immersion in 1 N KOH at

[12] S. Auerbach and T. Pederson, *Biochem. Biophys. Res. Commun.* **63,** 149 (1975).

55° for 2 hr,[13] or even radioiodination of proteins within the gel.[14] The destained gels may then be analyzed by densitometry, or the stained bands may be cut out and analyzed or subjected to liquid scintillation counting. The gels may be dried, subjected to direct or indirect autoradiography or fluorography, or analyzed in a number of other ways as described in Section IV of this volume.

Molecular Weight Determinations of Unknown Proteins. SDS–polyacrylamide gel electrophoresis is useful for determining molecular weights of proteins and their subunits, although some proteins behave anomalously. In general, proteins are separated on a slab gel in which one or more lanes have been devoted to the electrophoresis of protein standards, i.e., proteins of relative purity of known subunit molecular weights. A number of suitable proteins and their subunit molecular weights (after disulfide reduction) are listed in Table VIII; for routine use, both Sigma and Bio-Rad provide kits containing five or six proteins of well-defined molecular weights in the region of interest.

After completion of electrophoresis, staining, and destaining, relative mobilities are calculated for each of the standard proteins and the unknown proteins of interest. Relative mobility (R_f) is defined as the distance migrated by protein divided by the distance migrated by tracking dye, where distance refers to the distance from the junction of the stacking and separating gels. If the $\log_{10}$ of the molecular weights of the standard proteins is plotted as a function of their R_f, a straight line is usually formed that encompasses about the middle 80% of the area of the separating gel. The apparent molecular weights of the sample proteins can then be determined by matching their R_f value with the appropriate point on the standard curves. Hames[9] has estimated that linear standard curves can be obtained within the following useful ranges at three different acrylamide concentrations: 5% for molecular weights 60,000 to 212,000; 10% for 18,000 to 75,000; and 15% for 15,000 to 45,000. Increasing the acrylamide-to-bisacrylamide ratio from the standard ratio results in an increase in the size of proteins for which a linear relationship between molecular weight and R_f obtains, for any given acrylamide concentration.

This general method, using uniform separating gels (rather than gradient gels) is not very useful for proteins of less than about 10,000 molecular weight. This problem has been approached by Swank and Munkries,[15] who used SDS–polyacrylamide gel electrophoresis in the presence of 8 *M* urea in gels composed of 12.5% acrylamide and 1.25% bisacrylamide. An

[13] J. A. Cooper and T. Hunter, *Mol. Cell Biol.* **1,** 165 (1981).

[14] J. H. Elder, R. A. Pickett, J. Hampton, and R. A. Lerner, *J. Biol. Chem.* **252,** 6510 (1977).

[15] R. T. Swank and K. D. Munkries, *Anal. Biochem.* **39,** 462 (1971).

TABLE VIII
MOLECULAR WEIGHT (M_r) OF POLYPEPTIDE STANDARDS[a]

Polypeptide	M_r
Myosin (rabbit muscle) heavy chain	212,000
RNA polymerase (*E. coli*) β'-subunit	165,000
β-subunit	155,000
β-Galactosidase (*E. coli*)	130,000
Phosphorylase *a* (rabbit muscle)	92,000
Bovine serum albumin	68,000
Catalase (bovine liver)	57,500
Pyruvate kinase (rabbit muscle)	57,200
Glutamate dehydrogenase (bovine liver)	53,000
Fumarase (pig liver)	48,500
Ovalbumin	43,000
Enolase (rabbit muscle)	42,000
Alcohol dehydrogenase (horse liver)	41,000
Aldolase (rabbit muscle)	40,000
RNA polymerase (*E. coli*) α-subunit	39,000
Glyceraldehyde-3-phosphate dehydrogenase (rabbit muscle)	36,000
Lactate dehydrogenase (pig heart)	36,000
Carbonic anhydrase	29,000
Chymotrypsinogen A	25,700
Trypsin inhibitor (soybean)	20,100
Myoglobin (horse heart)	16,950
α-Lactalbumin (bovine milk)	14,400
Lysozyme (egg white)	14,300
Cytochrome *c*	11,700

[a] The data for M_r are in the presence of excess thiol reagent. Reprinted from B. D. Hames, *in* "Gel Electrophoresis of Proteins: A Practical Approach" (B. D. Hames and D. Rickwood, eds.), p. 39. IRL Press, Oxford and Washington D.C., 1981, with permission.

essentially linear relationship was observed between the log of the molecular weight and R_f of oligopeptides of molecular weights between 1500 and 15,000 using this system.

In addition to low-molecular-weight peptides, anomalous migration behavior can be exhibited by very basic proteins, very acidic proteins, glycoproteins, membrane proteins, phosphoproteins, and proteolipids. For glycoproteins, one approach to the true molecular weight has been the use of many different acrylamide concentrations as described by Segrest and Jackson.[16] Some of the methods in use for molecular weight determination in instances of anomalous migration behavior will be de-

[16] J. P. Segrest and R. L. Jackson, this series, Vol. 28, p. 54.

scribed further below. Use of gradient gels for this application is discussed in the following section.

Gradient Gels. Gradient gels, in which the concentration of acrylamide increases regularly from top to bottom of the separating gel, have gained popularity for several reasons. First, protein molecular weights can be estimated for a much broader range of proteins in a single gel than can be measured in a uniform gel. Second, the separation of protein bands from neighboring bands is better, and each band is sharper than in nonoptimal zones of uniform gels. Finally, some anomalous migration behavior may be decreased on gradient gels. However, band separation in the optimum area of a uniform gel remains superior to that achieved in gradient gels. Thus, gradient gels are most useful when performing initial studies of protein mixtures encompassing a large range of molecular weights.

In practice, linear gradients are formed by adding equal volumes of a final separating gel mixture (i.e., after addition of both ammonium persulfate and TEMED) to each side of a common gradient-forming device. A common recipe for a single slab gel of 20 ml would involve 10 ml of a 3% acrylamide solution and 10 ml of a 30% solution. As the mixed solution is withdrawn by gravity or through the use of a peristaltic pump, it is infused into a slab gel apparatus in such a way as to prevent aeration and bubble formation. When the entire gel is poured, the gel overlay and stacking gel are poured in the usual manner. Obviously, the gradient mixer should be arranged so that the 30% acrylamide solution reaches the bottom of the gel first, rather than the other way around.

Any of the gel recipes described in the foregoing sections can be used for gradient gels, although the discontinuous SDS systems have probably been used most widely. It is wise to decrease the amount of TEMED used in the gel recipes so that more time is available for pouring the gel before it polymerizes. Some authors recommend dissolving sucrose in the gel solution with the higher concentration of acrylamide, to form a density gradient with the acrylamide gradient that will prevent mixing with added solution during pouring of the gel. We have not found this to be necessary.

For determination of molecular weights in a gradient gel, the log of the molecular weights of standard proteins run in a neighboring track on a slab gel is plotted vs the log of acrylamide concentration of the gel, assuming a linear gradient. The molecular weights of unknown proteins are determined by measuring the R_f for each protein, assigning it to the appropriate value of acrylamide concentration, and reading the apparent molecular weight off the standard curve. Hames[9] has listed the approximate ranges of molecular weights that can be determined for different acrylamide gradient gels. These include: 7 to 25% acrylamide for molecular

TABLE IX
STOCK SOLUTIONS FOR HDPC (CATIONIC DETERGENT) DENATURING DISCONTINUOUS ELECTROPHORESIS[a]

Solution	Preparation
1. Acrylamide–bisacrylamide, 37.5 : 0.5	37.5 g of acrylamide, 0.5 g of bisacrylamide in 100 ml; store at 20°
2. Phosphate, 1.5 *M*, pH 2	Adjust pH of phosphoric acid (85% = 14.7 *M*) to 2.0 with KOH; bring to 1.5 *M*
3. Ascorbic acid, 1 (w/v)	1.0 g of ascorbic acid in 100 ml
4. Ferrous sulfate (0.003, w/v, $FeSO_4 \cdot 7\ H_2O$)	3 mg $FeSO_4 \cdot 7\ H_2O$ in 100 ml
5. Hexadecylpyridinium chloride, 0.175 *M*	595 mg in 10 ml
6. Hydrogen peroxide, 0.03 (w/v), H_2O_2	Dilute 1 ml of 30% H_2O_2 to 1 liter before use
7. Acrylamide–bisacrylamide, 30 : 2.5	30 g of acrylamide and 2.5 g of bisacrylamide in 100 ml; store at 20°
8. Phosphate buffer, 0.5 *M*, pH 4	Adjust pH of phosphoric acid (85% = 14.7 *M*) to 4.0 with KOH; dilute to 0.5 *M*
9. Running buffer	Hexadecylpyridinium chloride, 2.5 g, and glycine, 11.25 g, to pH 3.0 with H_3PO_4; bring to 2 liters
10. Sample buffer	0.78 ml of 2-mercaptoethanol, 4.28 g of sucrose, 10 ml of hexadecylpyridinium chloride (0.175 *M*), 0.5 g of pyronine Y, 10 ml of phosphate buffer at pH 4 (No. 8). Bring to 50 ml and adjust pH to 4
11. Separating gel overlay	1.26 ml of phosphate buffer at pH 4 (No. 8) and 0.1 ml of hexadecylpyridinium chloride (0.175 *M*); bring to 5 ml

[a] Modified from a similar system described by A. Amory, F. Foury, and A. Goffeau, *J. Biol. Chem.* **255,** 9353 (1980), largely according to the modification of M. D. Resh, *J. Biol. Chem.* **257,** 6978 (1982).

weight 14,000 to 330,000; 5 to 20% for 14,000 to 210,000; 3 to 30% for 13,000 to 950,000.

Discontinuous Denaturing Electrophoresis with Cationic Detergents

Occasionally, a denaturing electrophoresis system is required for the separation or determination of the molecular weight of a protein or its subunits that exhibit anomalous behavior on SDS gels. This can occur because of incomplete SDS binding due to extremes of isoelectric point of the protein, either acidic or basic, protein glycosylation, membrane asso-

TABLE X
FORMULAS FOR HDPC (CATIONIC DETERGENT) DENATURING DISCONTINUOUS ELECTROPHORESIS

	Final percentage of acrylamide desired				
	5%	7.5%	10%	15%	20%
A. Lower (separating) gel, for one slab of average size					
Mix first					
1. Acrylamide–bisacrylamide, 37.5:0.5 (ml)	4	6	8	12	16
2. Degassed water (ml)	15.2	13.2	11.2	7.2	3.2
3. Phosphate buffer, pH 2 (ml)	1.5	1.5	1.5	1.5	1.5
4. Ascorbic acid, 1% (ml)	2.3	2.3	2.3	2.3	2.3
5. Ferrous sulfate, 0.003% (ml)	2.3	2.3	2.3	2.3	2.3
Degas × 2 min					
6. Hexadecylpyridinium chloride, 0.175 *M* (ml)	0.4	0.4	0.4	0.4	0.4
Degas × 2 min; then add to polymerize					
7. Hydrogen peroxide, 0.03% (ml)	2.3	2.3	2.3	2.3	2.3
Degas × 30 sec, then pour separating gel					
B. Upper (stacking) gel, for one slab of average size (4% acrylamide)					
Mix first					
1. Acrylamide, 30:2.5 (ml)	0.67				
2. Degassed water (ml)	1.83				
3. Phosphate buffer, pH 4 (ml)	1.26				
4. Ascorbic acid, 1% (ml)	0.38				
5. Ferrous sulfate, 0.003% (ml)	0.38				
Degas for 2 min					
6. Hexadecylpyridinium chloride, 0.175 *M* (ml)	0.10				
Degas for 2 min; then add to polymerize					
7. Hydrogen peroxide, 0.03% (ml)	0.38				
Degas for 30 sec; then pour stacking gel					

ciation, or other factors. Because of such difficulties, denaturing systems have been devised for discontinuous electrophoresis using the cationic detergents tetradecyltrimethylammonium bromide[17] or cetylthiomethylammonium bromide[18] for the separation of the phosphorylated form of yeast plasma membrane ATPase and histones, respectively. We have used a modification of the former system using the cationic detergent hexadecylpyridinium chloride (HDPC); because of errors in the printing of the method of gel preparation of Amory *et al.*,[17] details of our modification of this method are shown in Tables IX and X. Amory *et al.*[17] have

[17] A. Amory, F. Foury, and A. Goffeau, *J. Biol. Chem.* **255,** 9353 (1980).
[18] V. V. Schmatchenko and A. J. Varshavskey, *Anal. Biochem.* **85,** 42 (1978).

provided good evidence that in this system, as with SDS–polyacrylamide gel electrophoresis, proteins are denatured by the detergent and migrate according to the molecular weights of their HDPC derivatives; the R_f of proteins subjected to electrophoresis using this system appear to be related in a linear way to the $\log_{10}$ of the true molecular weights of the proteins. Obviously, when a cationic detergent in used, the polarity of the electrophoresis chamber is reversed compared to SDS electrophoresis; i.e., the HDPC-proteins migrate toward the cathode instead of the anode. Otherwise, gel preparation, sample loading, and the like are as described above for SDS electrophoresis.

Other Discontinuous Denaturing Systems

Several other discontinuous systems have been described for special applications of detergent electrophoresis. As mentioned above, Swank and Munkries[15] have used a combination of SDS with 8 *M* urea to separate and determine the molecular weights of peptides of molecular weights less than 10,000. For some proteins, a urea–glycerol system has apparently been useful in maximizing the separation of native and phosphorylated forms, e.g., with myosin light chains.[19] For separation of histones, a number of techniques have been used successfully, as reviewed by Hardison and Chalkley[20]; these include acetic acid–urea and Triton–acetic acid–urea systems. For some membrane proteins, nonionic detergents such as Triton X-100 and sodium deoxycholate can be used to both solubilize the proteins and substitute for SDS in the electrophoresis buffers.[21,22] Details of these and related techniques can be found in the original references and several excellent general reviews.[9,10,23,24]

[19] W. T. Perrie and S. V. Perry, *Biochem. J.* **19,** 31 (1970).

[20] R. Hardison and R. Chalkley, *Methods Cell Biol.* **17,** 235 (1978).

[21] B. Dewald, J. T. Dulaney, and O. Touster, this series, Vol. 32, p. 82.

[22] A. C. Newby and A. Chrambach, *Biochem. J.* **177,** 623 (1979).

[23] A. H. Gordon, "Electrophoresis of Proteins in Polyacrylamide and Starch Gels," Vol. 1, Pt. 1. North-Holland Publ., Amsterdam, 1975.

[24] R. C. Allen and H. R. Maurer (eds.). "Electrophoresis and Isoelectric Focusing in Polyacrylamide Gel." de Gruyter, Berlin, 1974.

[13] High-Resolution Preparative Isoelectric Focusing

By BERTOLD J. RADOLA

Isoelectric focusing presents an ingenious addition to modern separation methods for the fractionation and isolation of enzymes. Among charge fractionation methods preparative isoelectric focusing is particularly attractive owing to high resolution and load capacity. The method concentrates and separates amphoteric substances in a stable pH gradient according to differences in isoelectric points (p*I*).[1] Isoelectric focusing is a steady-state method that offers two distinct advantages over the more usual kinetic methods of electrophoresis. (1) After the steady state has been attained, the process is time-independent so that zone definition does not deteriorate with further passage of time. In practice some minor deterioration becomes apparent after extended focusing periods, particularly in gel-stabilized systems, owing to the operation of certain nonideal effects.[2] In most experiments this deterioration is almost negligible. (2) Since solutes migrate from all positions of the separation system toward the final steady-state position, no definite starting zone is required, and the initial solute mixture may indeed occupy the entire volume of the separation system. Minor components, sometimes accompanied by major contaminants, are concentrated at their p*I* and may be isolated. Isoelectric focusing is the predominant charge fractionation method, mainly because of high resolution. Proteins differing in their p*I* values by 0.001 to 0.01 in pH are resolved by variations of this method.[3–6] The number of protein zones found in one dimension can be of the order of 100–120. High-resolution preparative isoelectric focusing reaches the indicated resolution in terms of resolvable p*I* differences and number of zones under conditions of high total load.[7,8]

[1] H. Rilbe, *in* "Isoelectric Focusing" (N. Catsimpoolas, ed.), p. 13. Academic Press, New York, 1976.

[2] H. Rilbe, *in* "Electrofocusing and Isotachophoresis" (B. J. Radola and D. Graesslin, eds.), p. 35. de Gruyter, Berlin, 1977.

[3] O. Vesterberg and H. Svensson, *Acta Chem. Scand.* **20,** 820 (1966).

[4] R. C. Allen, R. A. Hasley, and R. C. Talamo, *Am. J. Clin. Pathol.* **62,** 732 (1974).

[5] R. Charlionet, J. P. Martin, R. Sesboué, P. J. Madec, and F. Lefebvre, *J. Chromatogr.* **176,** 89 (1979).

[6] B. Bjellqvist, K. Ek, P. G. Righetti, E. Gianazza, A. Görg, R. Westermeier, and W. Postel, *J. Biochem. Biophys. Methods* **6,** 317 (1982).

[7] M. D. Frey and B. J. Radola, *Electrophoresis* **3,** 216 (1982).

[8] M. Flieger, M. D. Frey, and B. J. Radola, *Electrophoresis* **5** (in press).

ISBN 0-12-182004-1

The potential of isoelectric focusing for preparative separations was recognized early,[9,10] but applications have remained modest. Depending on the scale of fractionation, all preparative isoelectric focusing techniques can be classified into two categories: (1) techniques for laboratory-scale fractionation of milligram quantities with 1 g as an upper limit; in only a few applications has 0.5–1 g of protein actually been separated[11–17]; (2) techniques for large-scale fractionation of gram amounts (and more) with the potential of industrial applications.[18–20] Both groups of techniques comprise a plethora of diverse systems, but most do not seem to have been applied owing to unsolved practical problems. The two most widely used methods of preparative isoelectric focusing use density-gradient columns[10] and layers of granulated gels.[14,21] The former technique has remained unchanged over the past years and so have its shortcomings.[cf. 14,22]

Although granulated gels of the Sephadex and BioGel series are also imperfect, they afford the most challenging potential for preparative isoelectric focusing. High-resolution isoelectric focusing in layers of granulated gels has evolved from the previously described technique by the following modifications. Resolution is improved by using high field strengths (100–300 V/cm) in thin (0.2- to 1-mm) instead of the thick (2- to 12-mm) gel layers employed thus far. Loading is limited to 1 g because of inherent limitations in heat dissipation in the thicker gel layers that are required for larger amounts. Increased flexibility is achieved by using dry,

[9] H. Rilbe, *Protides Biol. Fluids* **17,** 369 (1970).

[10] O. Vesterberg, this series, Vol. 22, p. 559.

[11] J. S. Fawcett, *in* "Isoelectric Focusing" (J. P. Arbuthnott and J. A. Beeley, eds.), p. 23. Butterworth, London, 1975.

[12] H. Rilbe and S. Pettersson, *in* "Isoelectric Focusing" (J. P. Arbuthnott and J. A. Beeley, eds.), p. 44. Butterworth, London, 1975.

[13] J. P. Arbuthnott, A. C. McNiven, and C. J. Smyth, *in* "Isoelectric Focusing" (J. P. Arbuthnott and J. A. Beeley, eds.), p. 212. Butterworth, London, 1975.

[14] B. J. Radola, *in* "Isoelectric Focusing" (N. Catsimpoolas, ed.), p. 119. Academic Press, New York, 1976.

[15] J. Bours, M. Grabers, and O. Hockwin, *in* "Electrophoresis '79" (B. J. Radola, ed.), p. 529. de Gruyter, Berlin, 1980.

[16] R. van Driel, *J. Mol. Biol.* **138,** 27 (1980).

[17] J. P. Liberty and M. S. Miller, *J. Biol. Chem.* **255,** 1023 (1980).

[18] J. S. Fawcett, *in* "Isoelectric Focusing" (N. Catsimpoolas, ed.), p. 173. Academic Press, New York, 1976.

[19] M. Bier, N. B. Egen, T. T. Allgyer, G. E. Twitty, and R. A. Mosher, *in* "Peptides: Structure and Biological Function" (E. Gross and J. Meinenhofer, eds.), p. 79. Pierce Chemical, Rockford, Illinois, 1979.

[20] M. Jonsson and H. Rilbe, *Electrophoresis* **1,** 3 (1980).

[21] B. J. Radola, *Biochim. Biophys. Acta* **386,** 181 (1975).

[22] M. Jonsson, J. Ståhlberg, and S. Fredriksson, *Electrophoresis* **1,** 113 (1980).

rehydratable gels[8] on polyester films instead of wet gel layers on glass plates or in troughs. Rapid focusing with resultant short residence times of the separated samples in the gel layer is possible using shorter separation distances, extended prefocusing and cascades, or a combination of these approaches. Detection of proteins and enzymes is improved by new print techniques utilizing cellulose acetate membranes,[7] trichloroacetic acid-impregnated paper strips,[8] or high-resolution enzyme visualization techniques.[23] The time for detection by these techniques is shortened to a few minutes.

Apparatus

Most horizontal systems for analytical flatbed isoelectric focusing can be used without adaptation or with only minor adaptation for high-resolution preparative isoelectric focusing. The main requirements for preparative work are high efficiency of the cooling system; adequate loading capacity provided by the surface of the cooling plate; movable electrodes for increased flexibility in the choice of separation distance; and a power supply capable of yielding 3000–6000 V. The horizontal flatbed apparatus excels over systems with a cylindrical geometry for either density gradients[10] or polyacrylamide gels[24–26] in more efficient heat dissipation and versatility. Cooling is most favorable with thin layers owing to a high ratio of cooling surface to total volume. Some features of the most widely used advanced systems for isoelectric focusing are summarized in Table I. The maximum voltage of 3000 V provided by most commercial systems may prove to be too low for high-resolution preparative isoelectric focusing. An option is to combine the flatbed apparatus with power supplies manufactured locally or available for other electrophoretic techniques, e.g., high-voltage paper electrophoresis or nucleic acid sequencing. In working with voltages of 6000–8000 V, the safety features of the commercially available flatbed apparatus should be carefully reevaluated. Troughs and fractionation grids are obsolete in high-resolution preparative isoelectric focusing. The 0.2- to 1-mm layers are prepared more conveniently on plastic supports[8] or thin glass plates[7] than in troughs. The grid is incompatible with high-resolution preparative isoelectric focusing because it divides the layer arbitrarily into a number of segments with inevitable remixing of components separated *in situ*.

[23] A. Kinzkofer and B. J. Radola, *Electrophoresis* **4** (in press).

[24] A. Chrambach, T. M. Jovin, P. J. Svendsen, and D. Rodbard, *in* "Methods of Protein Separation" (N. Catsimpoolas, ed.), Vol. 2, p. 27. Plenum, New York, 1976.

[25] B. An der Lan and A. Chrambach, *in* "Gel Electrophoresis of Proteins: A Practical Approach" (B. D. Hames and D. Rickwood, eds.), p. 157. IRL Press, London, 1981.

[26] N. Y. Nguyen and A. Chrambach, *J. Biochem. Biophys. Methods* **1**, 171 (1979).

TABLE I
FLATBED APPARATUS FOR ISOELECTRIC FOCUSING

Company	Power supply[a] and maximum voltage	Focusing unit	Cooling plate	Maximum loading[b] (mg protein)
Bio-Rad Laboratories, Richmond, CA	3000/300, 3000 V	1415	12.5 × 22 cm	270
		1415	12.5 × 43 cm (Glass and plastic)	540
Desaga, Heidelberg, West Germany	Desatronic, 6000 V	Mediphor	12.5 × 26.5 cm	330
		Desaphor	12.5 × 26.5 cm	—
			20 × 26.5 cm	530
			16.5 × 30 cm	800
			26.5 × 40 cm (Aluminum coated with epoxide)	1060
LKB, Bromma, Sweden	2197[c], 2500 V	Ultrophor	12.6 × 26.1 cm (Aluminum insulated with glass)	330
Pharmacia, Uppsala, Sweden	ECPS[d], 3000/150, 3000 V	FBE 3000	25 × 25 cm (Aluminum with replaceable Teflon coating)	625

[a] All power supplies have adequate power outputs (100–300 W) and operate in three modes: constant power, constant voltage, or constant current.

[b] Calculated for high-resolution isoelectric focusing at a loading capacity of 10 mg of protein per milliliter of focusing gel volume in a 1-mm gel layer. All systems can be operated at 5–10 times higher total loading in 5–10-mm gel layers at lower final field strengths and inferior resolution. High resolution and increased total loading can be achieved by cascade focusing combining in two steps, isoelectric focusing in thick and thin gel layers (for details see text).

[c] Macrodrive power supply for 5000 V is available for nucleic acid sequencing.

[d] Optional extra: the VH-1 integrates volts with time and records volt-hours. This enables running conditions to be controlled accurately, particularly when the voltage changes with time.

Generation of pH Gradients

The generation of stable pH gradients is the key problem in isoelectric focusing. Only three approaches (1–3 below) are of practical value at present.

1. pH gradients are usually formed with the aid of mixtures of synthetic carrier ampholytes.[3] Several commercial products [Ampholine (LKB), Servalyt (Serva), and Pharmalyte (Pharmacia)] are available, differing with respect to synthesis[27–29] and physicochemical properties.[30] In addition to wide-range carrier ampholytes for the generation of steep pH gradients covering 5–6 pH units (e.g., the range pH 3–10), restricted pH ranges are offered covering 2–2.5 pH units, 1 pH unit, and even only 0.5 pH unit (Serva). Servalyt T carrier ampholytes, which are much cheaper than other products, appear to be particularly well suited for preparative isoelectric focusing. Small amounts of colored material present in Servalyt T, as well in most laboratory-made preparations,[31–34] do not interfere with preparative isoelectric focusing, with the rare exception of protein detection at 280 nm *in situ*.[21] Carrier ampholytes of any restricted pH range can be easily prepared from the commercial products, or even a blend of different products, for an increased number of carrier ampholyte species,[35] by preparative isoelectric focusing.[21,36]

Most of the desirable properties of carrier ampholytes defined for analytical work[37] apply also to preparative isoelectric focusing. A critical property is the molecular weight distribution of carrier ampholytes in those applications in which they have to be removed from the focused proteins. In contrast to many manufacturers' assertions, such removal is not as simple a procedure as is suggested. Nonstandard batches, uncontrolled changes on storage, and the presence of species of high molecular weight may seriously hamper the removal of carrier ampholytes from focused proteins.[38–41] If high-molecular-weight species are detected, e.g.,

[27] O. Vesterberg, *Acta Chem. Scand.* **23,** 2653 (1969).

[28] N. Grubhofer and C. Borja, *in* "Electrofocusing and Isotachophoresis" (B. J. Radola and D. Graesslin, eds.), p. 111. de Gruyter, Berlin, 1977.

[29] K. W. Williams and L. Söderberg, *Int. Lab.* **1,** 45 (1979).

[30] W. J. Gelsema, C. L. de Ligny, and N. G. van der Veen, *J. Chromatogr.* **173,** 33 (1979).

[31] S. N. Vinogrador, S. Lowenkron, H. R. Andonian, H. R. Baghshaw, J. Felgenhauer, and J. Pak, *Biochem. Biophys. Res. Commun.* **54,** 501 (1973).

[32] R. Charlionet, J. P. Martin, R. Sesboüé, P. J. Madec, and F. Lefebvre, *J. Chromatogr.* **176,** 89 (1979).

[33] W. W. Just, *Anal. Biochem.* **102,** 134 (1980).

[34] S. B. Binion and L. S. Rodkey, *Anal. Biochem.* **112,** 362 (1981).

[35] B. J. Thompson, M. J. Dunn, A. H. M. Burghes, and V. Dubowitz, *Electrophoresis* **3,** 307 (1982).

[36] A. Kinzkofer and B. J. Radola, *Electrophoresis* **2,** 174 (1981).

[37] O. Vesterberg, *in* "Isoelectric Focusing" (N. Catsimpoolas, ed.), p. 53. Academic Press, New York, 1976.

[38] G. Baumann and A. Chrambach, *Anal. Biochem.* **64,** 530 (1975).

[39] W. Otavsky and J. W. Drysdale, *Anal. Biochem.* **65,** 533 (1975).

[40] K. Goerth and B. J. Radola, *in* "Electrophoresis '79" (B. J. Radola, ed.), p. 955. de Gruyter, Berlin, 1980.

with the aid of thin-layer gel chromatography,[40,41] their removal is recommended; gel chromatography or, preferably, ultrafiltration through membranes with appropriate retention characteristics[40] are suggested.

2. Buffer electrofocusing uses mixtures of amphoteric or nonamphoteric buffers as carrier constituents.[25,42] With mixtures containing 2–14 constituents, narrow pH ranges are generated, and with a 47-component buffer mixture, gradients between pH 3 and 10 can be obtained.[43] Most of the work with buffer electrofocusing is confined to analytical separations, but their utility for preparative work has also been demonstrated.[26,44] There appears to be no limit to the choice of the desired pH range in buffer electrofocusing. Addition of acidic constituents to the buffer mixture causes a shift in the acidic direction, whereas basic constituents shift toward the basic direction. A single constituent, added in large amounts compared to the concentration of other constituents, flattens the pH gradient in the vicinity of its steady-state position.[45]

Buffer electrofocusing offers a number of advantages over isoelectric focusing utilizing synthetic carrier ampholytes.[43] The average molecular weight of the buffer constituents is 150, so that these components can be easily removed. There is no evidence that the buffer constituents bind to the proteins under electrofocusing conditions, although interactions between buffer constituents cannot be excluded. Buffer electrofocusing provides a higher degree of reproducibility due to the defined composition of the buffer mixtures.

3. Immobilized pH gradients are prepared with the aid of acidic and basic acryloyl derivatives with defined p*K* values.[6] These derivatives are used to generate two buffer solutions, which are linearly mixed and copolymerized with acrylamide and *N,N'*-methylene bisacrylamide. When an electric field is applied to the immobilized pH gradient, amphoteric substances are focused within their p*I* ranges while other charged solutes collect at the electrodes. Immobilized pH gradients are claimed to overcome problems associated with pH drift and irregularities in zone formation. Their major advantage is increased resolution in extremely flat pH gradients. Their disadvantages include prolonged focusing time, high electroendoosmosis due to H^+ and OH^- imparting a net charge to the medium at below pH 5 and above pH 9, and longer time for sample entry into the gel with the risk of protein precipitation and re-

[41] B. J. Radola, *Electrophoresis* **1,** 43 (1980).

[42] A. Chrambach, L. Hjelmeland, and N. Y. Nguyen, *in* "Electrophoresis '79" (B. J. Radola, ed.), p. 3. de Gruyter, Berlin, 1980.

[43] C. B. Cuono and G. A. Chapo, *Electrophoresis* **3,** 65 (1982).

[44] R. L. Prestidge and M. T. W. Hearn, *Anal. Biochem.* **97,** 95 (1979).

[45] M. L. Caspers, Y. Posey, and R. K. Brown, *Anal. Biochem.* **79,** 166 (1977).

stricted flexibility with respect to sample application and gel matrix selection. These disadvantages limit the utility of immobilized pH gradients for preparative work. An attractive feature is that neither carrier ampholytes nor buffer constituents need be removed from the focused proteins. However, elution of proteins may be expected to be difficult, and soluble nonproteinaceous impurities derived from the polyacrylamide gel are likely to contaminate the eluate.[46]

Gel Matrices

Isoelectric focusing requires a nonrestrictive anticonvective gel. Nonrestrictive gels are imperative because molecular sieving will retard migration of proteins, resulting in prolonged focusing periods. The migration velocities decrease as proteins approach their p*I*s, and any molecular sieving effect not only will retard the migration of the proteins, but may appear to halt it entirely, giving rise to fallacious p*I* values.[25] However, a steady state need not be attained because many proteins are sufficiently separated under nonequilibrium conditions. The following gels, arranged according to decreasing restrictiveness, are suitable for anticonvective stabilization: polyacrylamide gels for proteins with molecular weights up to 500,000; agarose for molecules up to several millions and particles 30–80 nm in radius,[47] and granulated gels, which may be used for all molecules but not for cells.

Polyacrylamide Gel

Continuously polymerized polyacrylamide gels dominate in analytical applications of isoelectric focusing, but they have failed to attract much interest in preparative separations. Cylindrical gels have a poor geometry for heat dissipation and are also afflicted with problems of wall adherence that can disturb the operation of preparative columns.[24,25] Flatbed polyacrylamide gels were only occasionally used for preparative separations.[48,49] Gels of rather low total monomer and cross-linking concentrations are usually used in analytical isoelectric focusing, and a gel composed of 5% T and 3% C_{Bis} (cross-linked with *N*,*N*′-methylene bisacrylamide[50]) has become particularly popular.[51] A drawback of polyacrylamide gels is the difficult recovery of separated proteins.[46] Proteins

[46] N. Y. Nguyen, J. DiFonzo, and A. Chrambach, *Anal. Biochem.* **106,** 78 (1980).
[47] P. Serwer and S. J. Hayes, *Electrophoresis* **3,** 80 (1982).
[48] D. H. Leaback and A. C. Rutter, *Biochem. Biophys. Res. Commun.* **32,** 447 (1968).
[49] D. Graesslin, H. C. Weise, and M. Rick, *Anal. Biochem.* **71,** 492 (1976).
[50] S. Hjertén, *Arch. Biochem. Biophys. Suppl.* **1,** 147 (1962).
[51] O. Vesterberg, *Biochim. Biophys. Acta* **257,** 11 (1972).

that are electrophoretically extracted from polyacrylamide gels may be contaminated with nonproteinaceous impurities.

Agarose

Agarose with low electroendoosmosis has been proposed as an alternative to polyacrylamide gels in analytical isoelectric focusing of high-molecular-weight proteins.[52,53] In a few reports, agarose has also been used for preparative isoelectric focusing.[54–56] High resolution, easy handling, nontoxicity, and absence of molecular sieving for high-molecular-weight proteins are some of its advantages. Although the purified or charge-balanced agaroses are claimed to fulfill many requirements of a good anticonvective support, there is evidence for certain disadvantages. A severe degree of surface flooding at the cathode, water transport to both electrodes resulting in distorted pH gradients, protein loss into the water accumulated on the gel surface, and protein trailing at the edges of the gel are some of the shortcomings.[35,54,57] To overcome these drawbacks, a composite agarose–Sephadex matrix was developed.[57] Higher field strength than in agarose could be used for improved resolution. In analytical experiments, photopolymerized composite agarose (0.5%)–polyacrylamide gels (2.5%) proved to be superior to each of the single gels for the analysis of crude tissue extracts containing a wide molecular range.[58] Preparative isoelectric focusing of immunoglobulins was improved by adding 0.5% non-cross-linked polyacrylamide to 1% agarose.[59] A drawback of all agarose-containing gels is the unsatisfactory protein recovery from macerated gels, which may have to be digested for extended periods with a mixture of agarase–hemicellulase.[56]

Granulated Gels

Horizontal layers of granulated gels of the Sephadex or BioGel type were introduced for anticonvective stabilization of the pH gradient with the intent of overcoming some of the limitations of both the density gradi-

[52] A. Rosén, K. Ek, and P. Åman, *J. Immunol. Methods* **28,** 1 (1979).

[53] C. A. Saravis, M. O'Brien, and N. Zamcheck, *J. Immunol. Methods* **29,** 91, 97 (1979).

[54] G. C. Ebers, G. P. Rice, and H. Armstrong, *J. Immunol. Methods* **37,** 315 (1980).

[55] C. Chapuis-Cellier and P. Arnaud, *Anal. Biochem.* **113,** 325 (1981).

[56] W. D. Cantarow, C. A. Saravis, D. V. Ives, and N. Zamcheck, *Electrophoresis* **3,** 84 (1982).

[57] A. Manrique and M. Lasky, *Electrophoresis* **2,** 315 (1981).

[58] E. A. Quindlen, P. E. McKeever, and P. L. Kornblith, *in* "Electrophoresis '81" (R. C. Allen and P. Arnaud, eds.), p. 539. de Gruyter, Berlin, 1981.

[59] R. McLachlan and F. N. Cornell, *in* "Electrophoresis '82" (D. Stathakos, ed.), p. 697. de Gruyter, Berlin, 1983.

ent technique and continuously polymerized polyacrylamide gels.[21,60,61] Granulated gels have also been used for preparative isoelectric focusing in columns[62] and in continuous-flow configurations,[18,63] but the horizontal systems offer distinct advantages over the vertical, closed systems. Granulated gels excel over other gel matrices in a number of properties: high load capacity, quantitative elution of the focused proteins from the gel, simple handling, absence of molecular sieving for high-molecular-weight proteins (making these gel particularly suitable for isoelectric focusing of molecules >500,000), and availability of granulated gels ready for use, partly in a prewashed form, with well-defined chemical and physical properties. Their inertness toward biopolymers under a wide range of conditions is well established owing to their widespread use in gel chromatography. The focused proteins and enzymes can be conveniently and rapidly located in the gel with the print technique. Drawbacks to granulated gels have been reported. Preparation of a gel bed with optimum consistency has been considered difficult[55,57,64] or laborious.[65] Mixtures of Sephadex and Pevikon[64] [a copolymer of poly(vinyl chloride) and poly(vinyl acetate)[66]] and Pevikon alone[65] were suggested as possible supports. Inferior resolution and bad printing properties are shortcomings of Pevikon-containing layers. Loss of resolution due to diffusion on protein detection with the paper print technique was a drawback of granulated gels in a comparative study of different gel matrices.[57] With the new printing techniques, there is essentially no loss in resolution.[7,8]

Sephadex G-200 is the gel of choice for most applications. It exhibits the best load capacity and handling properties and the highest water regain and is most economical.[7,8] Sephadex G-200 was not practical owing to unsatisfactory printing properties,[21] but the new printing techniques[7,8] have overcome this limitation. Sephacryl S-200, prepared by covalently cross-linking allyl dextran with *N*,*N*′-methylene bisacrylamide, can be handled as conveniently as Sephadex G-200 and also has good printing properties but inferior resolution. The higher G-numbered Sephadex gels may contain as much as 10% free dextran, which would contaminate the eluates. The enzymically resistant polyacrylamide gel BioGel P-60 is potentially useful, superior to Sephadex in work with crude preparations of

[60] B. J. Radola, *Biochim. Biophys. Acta* **295,** 412 (1973).

[61] B. J. Radola, *Ann. N.Y. Acad. Sci.* **209,** 127 (1973).

[62] T. J. O'Brien, H. H. Liebke, H. S. Cheung, and L. K. Johnson, *Anal. Biochem.* **72,** 38 (1976).

[63] G. Hedenskog, *J. Chromatogr.* **107,** 91 (1975).

[64] W. I. Otavsky, T. Bell, C. Saravis, and J. W. Drysdale, *Anal. Biochem.* **78,** 301 (1977).

[65] B. M. Harpel and F. Kueppers, *Anal. Biochem.* **104,** 173 (1980).

[66] H. J. Müller-Eberhard, *Scand. J. Clin. Lab. Invest.* **12,** 33 (1960).

cellulases and hemicellulases.[7] All granulated gels must be extensively washed with distilled water before use to remove charged solutes interfering with the formation of the pH gradient.[67] Optimum results are obtained with gels with a dry bead diameter of 10–40 μm ("Superfine" or minus 400 mesh).

Rehydratable Gels

Until recently, granulated gels were prepared as wet layers on a glass plate or in a trough[21]; they could not be stored. Preparation of rehydratable layers is simple and allows storage. After spreading the gel suspension of the correct consistency over a support, the gel is dried in air. The dry gel firmly adheres to the support, is mechanically stable, and can be preserved indefinitely.

Instead of glass plates or troughs, the rehydratable gel layers are preferably prepared on a plastic film. Best results are obtained with 100-μm polyester films (Mylar D, Du Pont) treated with alkali to impart hydrophilic properties to the film.[41] Two commercially available supports (GelBond for agarose, from Marine Colloids; and Gel-Fix, from Serva) are also suitable.

Rehydratable gels can be prepared with carrier ampholytes, which, owing to their hygroscopic properties, ensure the residual moisture necessary for storage.

Even greater versatility is provided by preparing "empty" gels, gels without added carrier ampholytes, but supplemented with 1–2% glycerol. Before use, the rehydratable gels, containing carrier ampholytes, are sprayed with an amount of water, calculated from the surface and thickness of the gel layer; empty gels are sprayed with a 2–3% solution of carrier ampholytes. Any formulation of carrier ampholytes, supplemented if necessary with such additives as urea, can be used for rehydration.

Load Capacity

In order to compare isoelectric focusing in systems employing a different geometry, pH gradient, or other forms of anticonvective stabilization, load capacity is defined as the amount of protein (in milligrams) per milliliter of focusing volume.[21] Load capacity is calculated by dividing the total load by total volume of the gel layer. The appearance of straight zones is used as the criterion for determining the highest permissible protein load capacity. Overloading results in irregular zones, which, with additional

[67] A. Winter, *in* "Electrofocusing and Isotachophoresis" (B. J. Radola and D. Graesslin, eds.), p. 433. de Gruyter, Berlin, 1977.

protein, cause the major zones to decay into droplets. The irregularities of the major zones in some parts of the gel layer usually have no detrimental effect on zone definition of minor components in other parts of the gel. Thus, load capacity for total protein is much higher when minor components are to be separated from an excess of major components rather than when all protein zones have to be well defined.[21]

Load capacity has been determined for natural and artificial mixtures of proteins as well as for single protein and carrier ampholytes of different pH ranges using 0.3- to 1-mm layers of Sephadex G-200 and BioGel P-60. The highest loads are attained for protein mixtures with a uniform distribution of protein zones over a wide pH range, e.g., crude tissue extracts. The decisive parameter for preparative systems is the amount of material to be fractionated, which, depending on the load capacity of the system, requires a specific focusing volume. In early work with granulated gels, thick (2-mm) layers were used,[68] and subsequent separations also employed thick layers of up to 12 mm.[14,21,61] The notion that preparative isoelectric focusing requires thick layers became established. More recent work appears to have made the thick layer obsolete.[7,8] In high-resolution preparative isoelectric focusing, thin layers afford the following advantages: higher field strengths due to more efficient heat dissipation; better resolution as the result of higher field strength and absence of skew zones; shorter focusing time; easier and more rapid gel preparation; better moisture control; reduced cost due to lower consumption of carrier ampholytes, buffer, and gels; improved recovery and higher concentrations of the recovered proteins; and reduced risk of inactivation of labile substances, e.g., as a result of chelating activity of carrier ampholytes. High-resolution preparative isoelectric focusing should be carried out at as high a load level as possible. The optimum capacity is 3–5 mg protein per milliliter of gel bed volume, 10–20 mg/ml remaining well tolerated although this is dependent on the specific properties of the separated material. An increase of total load to more than 1 g can be achieved by selecting an apparatus with larger surface of the cooling plate (see Table II) or, preferably, by applying a two-step cascade with prefractionation of up to several grams in the first step, followed by high-resolution isoelectric focusing of selected parts of the gel in the second step.

Resolution

The excellent resolution of analytical isoelectric focusing is a challenge for any preparative focusing method. Resolution in isoelectric fo-

[68] B. J. Radola, *Biochim. Biophys. Acta* **194,** 335 (1969).

cusing depends on several factors: design of apparatus, anticonvective stabilization of the pH gradient, parameters of the separation process (field strength and shallowness of the pH gradient), and properties of the separated material (diffusion coefficient and mobility slope in the vicinity of p*I*).[1] Apparatus in which a continuous pH gradient is built up, e.g., all gel or density gradient systems, afford superior resolution to instruments with a segmented design such as zone convection isoelectric focusing[69–71] and multicompartment electrolysis apparatus.[20,72] The popularity of analytical gel-stabilized systems stems from an operational advantage of gel matrices, namely, rapid fixation of the focusing pattern by conversion of the diffusible species into insoluble precipitates.[73] The main drawback of the segmented design is mixing of the zones within a single compartment and retarded equilibration; the latter may result from the use of membranes or other devices.

Resolution can be influenced by the field strength and the shallowness of the pH gradient.[1] Whereas high field strengths are being increasingly applied in analytical work,[36,74,75] preparative isoelectric focusing has been carried out at more moderate field strengths, mainly because of difficult heat dissipation. The added advantage of high field strength is a shorter focusing time; long focusing times repeatedly have been held as a shortcoming of preparative isoelectric focusing.[18,22,56] Extended residence times of labile substances at extreme pH values or close to their p*I* incurs the risk of inactivation. By reducing the thickness of the gel layer, field strengths of 100–500 V/cm can be applied in preparative isoelectric focusing with resultant improved resolution over a drastically shortened period of time.[7,8] In prefocused gels over a 10-cm separation distance, the residence time of the sample is decreased to only 30–40 min under steady-state conditions for granulated gels. The most conspicuous effect of high field strength is improved resolution. Proteins differing by only 0.01 to 0.15 pH are resolved on 40-cm gels using a pH 4 to 6 gradient.[7] This resolution had been achieved previously only with the analytical system. In preparative experiments, two components should not only be visibly resolved but also be amenable to elution from the gel layer by a simple

[69] E. Valmet, *Sci. Tools* **15,** 8 (1969).

[70] J. Bours, *in* "Isoelectric Focusing" (N. Catsimpoolas, ed.), p. 209. Academic Press, New York, 1976.

[71] R. Quast, *in* "Electrokinetic Separation Methods" (P. G. Righetti, C. J. van Oss, and J. W. Vanderhoff, eds.), p. 221. Elsevier/North-Holland, Amsterdam, 1979.

[72] M. Jonsson and S. Fredriksson, *Electrophoresis* **2,** 193 (1981).

[73] M. D. Frey and B. J. Radola, *Electrophoresis* **3,** 27 (1982).

[74] R. C. Allen, *Electrophoresis* **1,** 32 (1980).

[75] T. Låås, I. Olsson, and L. Söderberg, *Anal. Biochem.* **101,** 449 (1980).

slicing technique. The main argument for using longer separation distances is that zones can be handled more easily on elution.

There are several approaches toward improved resolution by flattening the pH gradient.

1. Selection of narrow-range carrier ampholytes. These are either commercially available or can be prepared by fractionation of the commercial products by preparative isoelectric focusing.[21,36]

2. Increased separation distance. For longer separation distances the gradient is flattened linearly and resolution is improved if focusing is carried out at the same field strength. For a 40-cm separation distance, pH gradients are flattened from 0.15 pH/cm to 0.025 pH/cm for wide range and 0.5 to 1 pH range carrier ampholytes, respectively.

3. Cascade focusing combines in a two-step or multistep procedure, prefractionation of the sample with a fractionation of the carrier ampholytes. In the first step the sample is focused at a high load and with lower resolution in a steep pH gradient. In subsequent steps, parts of the gel layer, enriched with the components of interest, are transferred to a prefocused narrow-range pH gradient. The carrier ampholytes, transferred with the sample, flatten the pH gradient and greatly improve resolution. In an experiment with a crude fungal enzyme resolution was improved to as little as 0.0013–0.005 pH.[8]

4. Addition of separators. Single or multiple amphoteric substances added in large amounts (5–50 mg/ml) to carrier ampholytes flatten the pH gradient in the vicinity of their steady-state positions. At present, manipulation of the pH gradient has to be conducted in a systematic but empirical manner because our understanding of the mechanism of gradient formation remains inadequate for predicting the gradient from the p*K* values of the separators.[25]

5. Buffer isoelectric focusing. With some buffer mixtures in cylindrical polyacrylamide gels, using a 14-cm separation distance, the pH gradient was flattened to 0.02–0.04 pH/cm.[26,76] By addition or deletion of buffer constituents, the course of the pH gradient may be manipulated.

6. Local increase of gel volume[77] or concentration[78] of carrier ampholytes. Both approaches have been described for analytical isoelectric focusing, but with 0.2- to 0.3-mm layers they could be useful also in preparative separations.

7. Continuous displacement and pH of the anolyte. The pH gradient

[76] N. Y. Nguyen and A. Chrambach, *Electrophoresis* **1,** 14 (1980).

[77] K. Altland and M. Kaempfer, *Electrophoresis* **1,** 57 (1980).

[78] T. Låås and I. Olsson, *Anal. Biochem.* **114,** 167 (1981).

can be flattened by suitable choice of anolyte and catholyte, both chosen so that they fall within the pH range of the gradient.[79-81]

8. Immobilized pH gradients. With the aid of Immobilines (LKB) the most shallow pH gradients can be created, thereby improving resolution and increasing the distance between separated zones. This will facilitate isolation of zones without contamination by adjacent zones. Although potentially interesting, limitations to the use of immobilized pH gradients exist (see section on granulated gels, above).

The advantages of increased field strength and flat pH gradients are documented for analytical isoelectric focusing.[36,74,75] The few data available indicate that the approaches outlined should be equally successful for preparative separation. The degree to which gradient shallowness is desirable remains unclear.[25] Retarded migration of proteins with flat titration curves near their p*I*, impeded entrance of sample into the gel, increased band width, and prolonged focusing time are all negative factors of isoelectric focusing in very flat pH gradients. While annoying in the most restrictive gel matrix, i.e., polyacrylamide gels of standard composition, some of these limitations may be alleviated by optimizing the separation with respect to selection of gel matrix, field strength, and mode of generation of the pH gradient. Even more important is to disregard the notion that preparative isoelectric focusing has to attain the steady state. By adopting the strategy outlined in this section, good separation may be expected for most samples also under nonequilibrium conditions.

Detection of Proteins

Prior to recovery, the focused proteins and enzymes must be located in the gel layer by detection techniques that should be rapid, simple, and preferably nondestructive. Speed is important because keeping the gels without voltage or at reduced voltage will broaden the zones as a result of diffusion, an effect less conspicuous with long separation distances. The gel layer may be rapidly frozen if this is compatible with the separated material. There are several approaches to locating proteins and enzymes in horizontal gel layers (see also this volume [28]–[32]).

1. Transparent zones. At high protein loading the major components are visible in the gel layer after focusing as transparent zones owing to

[79] A. G. McCornick, L. E. M. Miles, and A. Chrambach, *Anal. Biochem.* **75,** 314 (1976).
[80] A. G. McCornick, H. Wachslicht, and A. Chrambach, *Anal. Biochem.* **85,** 209 (1978).
[81] B. An der Lan and A. Chrambach, *Electrophoresis* **1,** 23 (1980).

changes in refraction relative to the surrounding gel.[7,8,21] This permits rapid visual identification, p*I* determination, and direct isolation by gel slicing.

2. Membrane and paper printing is the most versatile technique for protein location. Originally, paper prints were obtained with chromatographic papers. The drawbacks of chromatographic paper for printing are diffuse zones resulting from the coarse structure of paper relative to the gel matrix, limited applicability to some gel matrices, e.g., Sephadex G-200, and long visualization time. These shortcomings are overcome by cellulose acetate membranes[7] or trichloroacetic acid-impregnated paper.[8] Ponceau S, rather than the more sensitive triphenylmethane dyes,[41,73] is preferred for staining at high protein load. With membrane printing, the total time required for fixation, staining, and destaining is 2–3 min. Destaining depends on the chemical properties of the carrier ampholytes[41] and is most rapid for Servalyt. Narrow (1–2 cm) strips are sufficient for printing, only small amounts of proteins being removed. Membrane printing is nondestructive but the trichloroacetic acid from the paper exerts a fixative effect on the proteins in the gel layer. The stained prints are a convenient document that can be preserved easily and evaluated densitometrically. Location of radioactivity in the print by a strip scanner has been reported.[82]

3. Ultraviolet densitometry is more sensitive than zone transparency, but less sensitive than staining of a print, and requires an expensive instrument.

4. A topographic method is based on the fluorescence of Servalyt carrier ampholytes in a paper print.[83–85]

5. Fluorescence. Without printing, proteins can be visualized in the gel layer with 8-anilino-1-naphthalenesulfonic acid by spraying a water solution of the reagent on the gel surface.[86]

6. Enzyme visualization. Substrate-impregnated papers,[61,87] dimensionally stable polyamide membranes,[23] or 100- to 200-μm ultrathin agarose layers[23] containing a high concentration of the substrate and coupling dyes can be used for enzyme location.

[82] A. J. MacGillivray and D. Rickwood, *in* "Isoelectric Focusing" (J. P. Arbuthnott and J. A. Beeley, eds.), p. 254. Butterworth, London, 1975.

[83] J. Bonitati, *J. Biochem. Biophys. Methods* **2,** 341 (1980).

[84] J. Bonitati, *J. Biochem. Biophys. Methods* **4,** 49 (1981).

[85] J. Bonitati, B. Sabatino, and J. B. Van Liew, *Electrophoresis* **3,** 326 (1982).

[86] W. E. Merz, U. Hilgenfeldt, M. Dörner, and R. Brossmer, *Hoppe-Seyler's Z. Physiol. Chem.* **355,** 1035 (1975).

[87] H. Delincée and B. J. Radola, *Anal. Biochem.* **48,** 536 (1972).

7. Activity determination in eluates. In those cases in which visualization reactions are not available, the enzyme activity has to be determined in eluates of gel segments, with some unavoidable zone remixing within a single segment.

Recovery

Recovery in preparative isoelectric focusing will depend on a number of factors that are related either to the proper separation, including elution from the gel, or to additional steps that may be necessary for removal of the carrier ampholytes or concentration of the isolated fractions. Elution from granulated gels is simple, rapid, and quantitative. A loss of recovery at this step is negligible in comparison with elution from compact polyacrylamide gels.[46] The recovery of isoelectrically homogeneous proteins was studied in preparative refocusing experiments for which protein recovery of 85–92% was found.[61] For crude protein mixtures recoveries of 80–90% were determined by eluting all proteins simultaneously from gel strips removed lengthwise from the layer[21]; this approach gives a more reliable estimate than procedures in which protein recovery is calculated by summation of the protein content of individual isolated fractions.[88] The total activity in a portion of the gel layer is therefore a means of checking inactivation inherent to the separation process. Recovery may depend strongly on load capacity[89]; for Pronase E at loads from 0.5 to 10 mg per milliliter of gel suspension, recovery of activity increases with increasing load from 14 to 80%.

The chelating properties[90] of the carrier ampholytes have been implicated as resulting in the dependence of enzyme recovery on the ratio of enzyme to the carrier ampholytes.[89]

Since extreme pH values during focusing may cause denaturation, the sample is applied at a sufficient distance from the electrodes. Components that are focused at extreme pH values should be protected by efficient temperature control and a short focusing period. The risk of denaturation can be lowered somewhat by establishing a pH gradient, prefocusing in the absence of the sample. The residence time of the sample is thereby reduced, in some instances at the expense of not reaching steady-state conditions.

[88] H. Delincée and B. J. Radola, *Eur. J. Biochem.* **52,** 321 (1975).

[89] B. J. Radola, *in* "Isoelectric Focusing" (J. P. Arbuthnott and J. A. Beeley, eds.), p. 182. Butterworth, London, 1975.

[90] H. Davies, *Protides Biol. Fluids* **17,** 389 (1970).

Dialysis,[11,39,91] electrodialysis,[92] ultrafiltration,[11] salting out,[93] gel chromatography,[94] ion-exchange chromatography,[38] hydrophobic interaction chromatography,[95] and two-phase extraction with *n*-pentanol[96] have been suggested for the removal of carrier ampholytes. A recently described technique is based on electrophoresis of carrier ampholytes through a dialysis membrane into a filter paper sheet soaked with buffer.[7] The proteins are retained by the membrane and can be recovered with nearly 100% yield. The technique is simple, flexible with respect to gel volume and processing of multiple samples. The technique can be easily used with most commercially available equipment for flatbed isoelectric focusing with buffer vessels of sufficient capacity.

Technique[7,8]

Preparation of Gel Suspension. Sephadex G-200 (Superfine) or BioGel P-60 (minus 400 mesh) is suspended in distilled water, and the swollen gel is washed on a sintered-glass funnel with 20 volumes of deionized water. To avoid mechanical damage of the gel beads, water is added without stirring the gel. After washing, the thick gel suspension is dehydrated with several changes of methanol and dried in a vacuum oven at 35°. From the dry, washed gel suspensions for coating are prepared according to Table II.

Preparation of Rehydratable Gels. Mylar D polyester films (100 μm) are treated with 6 *N* NaOH for 15 min to render the surface hydrophilic.[8,41] The films are washed with tap water, followed by deionized water, and are dried at 80°. Gel-Bond films for agarose (Marine Colloids) and Gel-Fix films (Serva) may be also used. The required amount of a deaerated (water pump) gel suspension is spread with the aid of a glass rod over the plastic film. The gels are dried with the aid of a hot fan for 1 hr, or without a fan at room temperature overnight. Gels without carrier ampholytes should be supplemented with 1–2% glycerol. The dry gel can be preserved without noticeable changes for at least several months when stored at room temperature. Prior to use, the gels containing carrier ampholytes are sprayed in a zigzag course with a water spray. Gels without ampholytes are sprayed with a 2% (w/v) solution of carrier ampholytes. A small excess of the solvent will produce a shiny surface. After drying for a

[91] J. F. Gierthy, K. A. O. Ellem, and J. R. Kongsvik, *Anal. Biochem.* **98,** 27 (1979).
[92] T. G. Bloomster and D. W. Watson, *Anal. Biochem.* **113,** 79 (1981).
[93] P. Nilsson, T. Wadström, and O. Vesterberg, *Biochim. Biophys. Acta* **221,** 146 (1970).
[94] O. Vesterberg, *Sci. Tools* **16,** 24 (1969).
[95] W. J. Gelsema, C. L. De Ligny, and W. M. Blanken, *J. Chromatogr.* **196,** 51 (1980).
[96] H. P. Köst and E. Köst-Reyes, *in* "Electrophoresis '79" (B. J. Radola, ed.), p. 565. de Gruyter, Berlin, 1980.

TABLE II
COMPOSITION OF THE GEL SUSPENSION FOR PREPARATION OF GEL LAYERS OF GRANULATED GELS[a]

Thickness of layer (μm)	Amount of dry gel: BioGel P-60 (mg)	Amount of dry gel: Sephadex G-200 (mg)	Wet gel (ml)	Distilled water[b] (ml)	40% Solution of carrier ampholytes (μl)
300	207	120	3	5.0	150
500	345	200	5	7.5	250
1000	690	400	10	12.5	500

[a] The amounts are for 10 × 10 cm gels.
[b] Different amounts of water are recommended to facilitate spreading of the gel suspension on the plastic support or glass plate.

few minutes, the gel is ready for use. Alternatively, wet gel layers may be used by coating the films or glass plates and drying the layer with a fan until irregular 1- to 3-mm fissures appear at the edges of the gel.[60] Gel layers dried to a consistency defined by this easily recognizable criterion do not flow when inclined to an angle of ≥45°.

Electrode Strips and Solutions. Strips (1 cm in width) of MN 866 paper (Macherey & Nagel) are soaked with the following solutions: 25 m*M* aspartic acid and 25 m*M* glutamic acid (anolyte) and 2 *M* ethylenediamine containing 25 m*M* arginine and 25 m*M* lysine (catholyte). Care should be taken to establish good contact of the strips to the gel layer. Adjustable platinum ribbon or wire electrodes are placed on the electrode strips and weighed to ensure good contact.

Sample Application. The sample is applied as a streak on the gel surface with a microscope slide or a rectangular glass plate of a width 2 cm less than that of the gel layer (as compensation for possible edge effects). Approximately 30 μl of solution per centimeter of applicator is easily deposited by this technique. Greater sample volumes can be applied in intervals of 0.5–1 cm by this means or by using commercially available sample applicators (Bio-Rad, LKB, and Pharmacia). The sample should be applied over a 20–25% distance of the total separation length from the electrodes, preferably in a part of the gel layer in which components of interest are not expected after focusing. Although proteins from dilute samples can be concentrated by preparative isoelectric focusing, the more practical way is to concentrate the dilute samples prior to sample applications, e.g., by ultrafiltration, and to apply 2–10% protein solutions. All samples should be desalted to ≤0.05 *M* salt content.

Isoelectric focusing. A cooling solution at 4° is circulated through the apparatus from a constant-temperature circulating bath. Typical running conditions for different separation distances are (a) 10 cm: prefocusing at 50 V/cm, 20–30 min, final field strength 300 V/cm, 2000 to 3000 V-hr; (b) 20 cm: prefocusing at 20–50 V/cm, final field strength 300 V/cm, 10,000 to 13,000 V-hr; (c) 40 cm: prefocusing at 20–40 V/cm, 16 hr, final field strength 100–200 V/cm, 30,000 to 40,000 V-hr. With adequate cooling, 0.05–0.1 W/cm^2 is tolerated.

Printing. Of the several cellulose acetate membranes tested, Sartorius membranes (catalog No. 11106) have optimal printing properties. Strips, 1–2 cm wide and of appropriate length, are rolled from one end onto the gel layer and gently pressed to ensure uniform wetting of the membrane. Care should be taken not to entrap air between the membrane and the gel. After contact for 1 min, which suffices for uniform wetting, the membrane is removed and placed for 1–2 min in 10% trichloroacetic acid (w/v). Adhering gel particles are removed by washing. The strips are stained for 1 min in 1% Ponceau S in 10% trichloroacetic acid and destained for a few seconds in 5% acetic acid. Prints are dried in air, yielding a white background. Staining intensity decreases on drying, which is of advantage in work at high protein loads.

Removal of Carrier Ampholytes. Carrier ampholytes are removed from the gel electrophoretically with a dialysis membrane. A sheet of filter paper, Whatman No. 3 or Macherey & Nagel MN 827, is wetted with 0.05 *M* Tris-HCl at pH 8.5 and blotted with dry filter paper to remove excess liquid. The wet paper is mounted on a 1-mm thin glass plate on the cooling plate of the focusing chamber and connected with pads of MN 866 paper with the electrode vessels containing the same buffer as the paper. Visking dialysis membrane is placed on the buffered paper and pressed to establish good contact. A rectangular 2-mm silicone rubber strip with an appropriate slit in the middle, about 0.5 to 1 cm × 5 cm, is laid on the dialysis membrane. The gel, liquefied with a small amount of water, is transferred into the slit, and electrophoresis is conducted at 800–1000 V for 40–60 min. The dialysis membrane retains the protein while the carrier ampholytes migrate into the paper. The efficiency of the removal is checked by drying the paper at 120° for 5–10 min and staining with 1% Amido Black 10B in methanol–acetic acid–water (45 : 5 : 25, v/v/v) followed by destaining in the same solvent. With Amido Black 10B approximately 0.5–1 μg of carrier ampholytes are detected per square centimeter. Thus, failure to stain the paper beneath the slit indicates a reduction of carrier ampholytes to $<0.01\%$.

Elution of Proteins. The gel with the focused proteins, directly from the gel layer or after electrophoretic removal of carrier ampholytes, is transferred into a centrifuge tube and suspended in a 1 : 1.5 ratio in distilled water. After centrifugation for 10 min at 35,000 *g* at 2°, the supernatant fluid is collected and the sediment suspended in a fresh amount of water. In the two to three combined supernatant fractions, protein is determined against a blank of carrier ampholytes of the same pH range as that used for the focusing procedure. Alternatively, proteins are eluted in a minicolumn assembly consisting of cotton-plugged Eppendorf pipette tips supported in a test tube. A 1.5 : 2 ratio of eluent to gel is sufficient for protein elution.

Determination of pH Gradient. The pH gradient is determined directly in the layer in a part of the gel containing the focused proteins. Measurement of pH at the edge of the layer or in protein-free parts of the gel may introduce great errors owing to a different distribution of carrier ampholytes.

[14] Affinity Electrophoresis

By VÁCLAV HOŘEJŠÍ

The principle of affinity electrophoresis is simple: a macromolecule migrates electrophoretically in a gel medium containing effectively immobilized ligands ("affinity gel") capable of interaction with the migrating macromolecule. As a result of this interaction, the macromolecule is more or less retarded in the affinity gel as compared to a control gel. Such control gels include media devoid of the immobilized ligand or media containing an immobilized mock ligand incapable of complex formation with the migrating macromolecule; the mobility[1] of other macromolecules contained in the sample is not affected in affinity gels. Thus, affinity electrophoresis is essentially analogous to affinity chromatography, since both methods are based on separation of macromolecules due to their affinity toward a ligand immobilized on a solid-phase carrier. Since reviews on affinity electrophoresis are available,[2–4] this chapter concen-

[1] As used here, *mobility* denotes the distance migrated by the protein during the duration of the experiment under a specific set of conditions; the term is not intended to imply the physicochemical quantity of *electrophoretic mobility* as estimated by free-flow electrophoresis.

[2] V. Hořejší, *Anal. Biochem.* **112,** 1 (1981).

[3] V. Hořejší and J. Kocourek, this series, Vol. 34, p. 178.

[4] T. C. Bøg-Hansen, *in* "Proceedings of the Third International Symposium on Affinity Chromatography and Molecular Interactions" (J. M. Egly, ed.), p. 399. INSERM Symposium Series, Paris, 1979.

ISBN 0-12-182004-1

trates mainly on one of the current experimental modifications of the method, i.e., affinity-chromatography-like technique.

Experimental Modifications of Affinity Electrophoresis

Several variants of affinity electrophoresis exist that differ in the manner of immobilization of the ligand in the gel, in the medium, and in the buffer system used.

Immunoelectrophoresis-Like Modification

This is sometimes called crossed immuno-affino-electrophoresis[4–6] and is closely related to some immunoelectrophoretic techniques, namely, those in which antibody is incorporated into the agarose gel and interacts with the migrating antigen. Usually, lectins are incorporated into the gel instead of antibodies although other interacting proteins would serve. The conditions (pH, electroendoosmosis) are chosen so that the lectin has a very low mobility. Such an affinity gel causes retardation of those glycoproteins that form complexes with the incorporated lectin; control gel is either devoid of the lectin or contains a carbohydrate inhibitor of the lectin. The position of the glycoprotein zones after electrophoresis is most conveniently detected by crossed immunoelectrophoresis, i.e., electrophoresis of the separated proteins in a second perpendicular dimension into the agarose slab gel with incorporated, precipitating polyvalent antiserum against the sample. The lectin concentration may be chosen so that lectin–glycoprotein precipitates are formed during electrophoresis in a manner analogous to antigen–antibody precipitates.[7] Instead of merely incorporating the lectin in agarose gel, a slurry of ligand-substituted beads in melted agarose can be used for preparation of affinity gels.[8]

Affinity Electrophoresis in Polyacrylamide Gel

In contrast to the above-mentioned technique, applicable to the study of interactions between two macromolecules, e.g., lectin and glycoprotein, polyacrylamide gel is usually used as a carrier of various ligands, e.g., enzyme substrates and inhibitors, or haptens interacting with lectins or antibodies. Such affinity gels are employed in the studies of the respec-

[5] T. C. Bøg-Hansen, *Anal. Biochem.* **56,** 480 (1973).

[6] T. C. Bøg-Hansen, *in* "Electrophoresis, A Survey of Techniques and Applications," Part B (Z. Deyl, ed.), p. 219. Elsevier, Amsterdam, 1983.

[7] T. C. Bøg-Hansen, O. J. Bjerrum, and C. H. Brogren, *Anal. Biochem.* **81,** 78 (1977).

[8] M. Raftell, *Immunochemistry* **14,** 787 (1977).

tive protein–ligand interactions. Immobilization of the ligand in the polyacrylamide gel matrix can be achieved by several means.

1. A polymerizable derivative of the ligand can be prepared, e.g., an allyl- or acryloyl derivative, and copolymerized with the monomers normally used during preparation of polyacrylamide gel (acrylamide and bisacrylamide). This procedure has been discussed in detail in this series.[3]

2. A soluble macromolecular derivative of the ligand can be prepared, which is then added to the polymerization mixture normally used for preparation of polyacrylamide gels. After completion of polymerization, the macromolecular carrier, if sufficiently large, remains entrapped within the polyacrylamide gel network and is effectively immobilized. Several types of macromolecular carriers have been used. Polysaccharides can be used directly without modification when proteins bind to them.[9–12] Alternatively, a polysaccharide or synthetic polymer can be substituted with the ligand,[13–15] or a soluble copolymer can be prepared by copolymerization of acrylamide with a suitable derivative such as allyl or acryloyl.[16–19]

3. Instead of soluble macromolecular derivatives of the ligand, beads substituted with the ligand can be entrapped in the polyacrylamide gel matrix. A simple procedure ensures homogeneously tight packing of the beads within the gel and eliminates potential problems due to irregular sedimentation of the beads before polymerization occurs.[20]

Affinity Electrophoresis in Agarose or Agarose–Polyacrylamide Gels with Covalently Bound Ligand[20]

For the purposes of affinity chromatography, the ligands are usually coupled to suitable beaded gels, agarose gel beads generally being the material of choice. However, most of the reactions used for "activation" of agarose result in partially cross-linked gels that do not melt upon heating. If periodate oxidation is used for activation with subsequent reduc-

[9] H. Stegemann, *Hoppe-Seyler's Z. Physiol. Chem.* **348,** 951 (1967).
[10] K. Takeo and S. Nakamura, *Arch. Biochem. Biophys.* **153,** 1 (1972).
[11] K. Takeo and E. A. Kabat, *J. Immunol.* **121,** 2305 (1978).
[12] C. Borrebaeck and M. E. Etzler, *FEBS Lett.* **117,** 237 (1980).
[13] K. Čeřovský, M. Tichá, V. Hořejší, and J. Kocourek, *J. Biochem. Biophys. Methods* **3,** 163 (1980).
[14] V. Čeřovský, M. Tichá, J. Turková, and J. Labský, *J. Chromatogr.* **194,** 175 (1980).
[15] M. Tichá, V. Hořejší, and J. Barthová, *Biochim. Biophys. Acta* **534,** 58 (1978).
[16] V. Hořejší, P. Smolek, and J. Kocourek, *Biochim. Biophys. Acta* **538,** 293 (1978).
[17] K. Nakamura, A. Kuwahara, H. Ogata, and K. Takeo, *J. Chromatogr.* **192,** 351 (1980).
[18] P. Masson, A. Privat de Garilhe, and P. Burnat, *Biochim. Biophys. Acta* **701,** 269 (1982).
[19] Jang-Lin Chen and H. Morawetz, *J. Biol. Chem.* **256,** 9221 (1981).
[20] V. Hořejší, M. Tichá, P. Tichý, and A. Holý, *Anal. Biochem.* **125,** 358 (1982).

tive amination by cyanoborohydride for binding of an amino derivative of the ligand to the "activated" aldehydic derivative of agarose beads, non-cross-linked, normally melting ligand-substituted beads are obtained that can be used conveniently in affinity electrophoresis. They can be simply melted on a water bath, mixed in a suitable ratio with unsubstituted agarose, and, after cooling, they result in homogeneous affinity gels. Alternatively, these agarose gel derivatives can be used for preparing mixed agarose–polyacrylamide gels. The last method of ligand immobilization and affinity gel preparation, described in greater detail below, seems to be easy and generally applicable.

Affinity Isoelectric Focusing

Immobilization of a ligand in a gel medium containing carrier ampholytes yields affinity gels applicable to affinity isoelectric focusing.[21] In this method, a combination of gel isoelectric focusing and affinity interaction, the ligand-binding protein present in a complex mixture is again "captured" in the gel near the start. The noninteracting proteins are sharply resolved and focus normally at the positions corresponding to their isoelectric points. The ligand may be immobilized by some of the methods mentioned above, although care must be taken to avoid gel media with increased electroendoosmosis, the latter being incompatible with the isoelectric focusing procedure.[20]

Applicability and Limitations of Affinity Electrophoresis

The phenomenon of retardation of a specific ligand-binding protein under the conditions of affinity electrophoresis can be exploited either qualitatively or quantitatively.[22]

Qualitative Applications

Affinity electrophoresis, or affinity isoelectric focusing, can be used for detection and identification of a ligand-binding protein in a complex protein mixture; for detection of the presence of inactive admixtures in purified preparations of the ligand-binding proteins, and for estimation of their ligand-binding heterogeneity; and for checking the results of chemical modification reactions affecting, presumably, the ligand-binding site. The method is also useful in testing materials to be used for affinity chromatography. In all these applications, the patterns observed on con-

[21] V. Hořejší and M. Tichá, *Anal. Biochem.* **116,** 22 (1981).

[22] For a review with references to more particular examples, see Hořejší.[2]

trol and affinity gels are compared and the zones retarded on the affinity gels are readily identified.

Quantitative Applications

The degree of retardation of a ligand-binding protein in an affinity gel depends primarily on the concentration of the immobilized ligand and on the value of the dissociation constant of the protein–ligand complex. Thus, the dependence of mobility on the concentration of the immobilized ligand can be used for estimation of apparent dissociation constant of the protein-immobilized ligand complexes, as shown first by Gerbrandy and Doorgeest[23] and Takeo and Nakamura.[10] This factor may also be useful for estimation of dissociation constants of protein-free (mobile) ligand complexes, if the free ligand, in addition to the immobilized one, is present in the affinity gel.[24] The basic theoretical background necessary for extraction of quantitative information from affinity electrophoresis experiments performed under various conditions has been developed.[25–27] These quantitative applications of affinity electrophoresis fall beyond the scope of this chapter, and the relevant references can be found elsewhere.[2]

Preparation of Some Macromolecular Carriers of the Ligands

Water-Soluble Copolymers of Acrylamide with an Unsaturated Derivative of the Ligand[15–18]

Acrylamide (0.5 g; an example of a copolymerizable ligand derivative),[16] allylglycoside (1 g), and ammonium persulfate (10 mg) are dissolved in 10 ml of water and heated for 5 min on a boiling water bath. The viscous product is diluted, extensively dialyzed against water, and lyophilized. The copolymer contains approximately one carbohydrate residue per six acrylamide units. The ligand content can be regulated by the amount of the ligand monomer used in the reaction. The molecular weight of the copolymer may be varied by altering the concentration of the polymerization catalyst (persulfate) and probably also by temperature and duration of the polymerization reaction. Studies on the kinetics of acrylamide polymerization may be relevant for optimization of the preparation of copolymers.[28]

[23] S. J. Gerbrandy and A. Doorgeest, *Phytochemistry* **11,** 2403 (1972).
[24] V. Hořejší, M. Tichá, and J. Kocourek, *Biochim. Biophys. Acta* **499,** 290 (1977).
[25] V. Hořejší, *J. Chromatogr.* **178,** 1 (1979).
[26] V. Hořejší and M. Tichá, *J. Chromatogr.* **216,** 43 (1981).
[27] V. Matoušek and V. Hořejší, *J. Chromatogr.* **245,** 271 (1982).
[28] C. Gelfi and P. G. Righetti, *Electrophoresis* **2,** 213 (1981).

Ligand-Substituted Dextrans[13]

A solution of 0.1 *M* $NaIO_4$ and 1% dextran T-500 (Pharmacia) is maintained at 25° for 1 hr and then exhaustively dialyzed against water. A 1% solution of this oxidized dextran, 20 ml, is mixed with 20 ml of a 10–50 m*M* solution of the ligand bearing a primary amino group in 0.2 *M* carbonate-bicarbonate at pH 9.2 and stirred at room temperature for 2 hr. Then 10 mg $NaBH_4$ is added in small portions during 10 min; after another 10 min, 0.1 ml of acetone is added. The reaction mixture is dialyzed exhaustively against distilled water and lyophilized. The ligand content of the product is approximately 10% (w/w). Commercially available Blue Dextran, i.e., Cibracron Blue F3G-substituted dextran (Pharmacia), may be used for affinity electrophoresis of proteins interacting with this dye.[15]

Ligand-Substituted Meltable Derivatives of Agarose Gel[20,29]

Washed Sepharose 4B (5 ml) is added to 5 ml of 0.5 *M* $NaIO_4$. The suspension is agitated for 2 hr at room temperature, after which the oxidized gel is thoroughly washed with water and stored at 4°. The gel (2 ml) is washed with 0.5 *M* potassium or sodium phosphate at pH 6 and added to 2 ml of the same buffer containing 20 m*M* ligand bearing a primary amino group and 20 m*M* $NaCNBH_3$. The mixture is agitated for 24 hr at room temperature. After washing, the gel is stored in 0.1 *M* Tris-HCl at pH 8.2 containing 0.1% NaN_3. Typically, the concentration of immobilized ligand in the gel is 1–2 m*M*.[30]

Triazine dyes may be covalently bound to a nonactivated agarose gel to yield meltable derivatives applicable to affinity electrophoresis.[31]

Procedure for Affinity Electrophoresis

Affinity electrophoresis is performed exactly as is the corresponding "parent" electrophoretic method—i.e., a tube or slab gel arrangement can be used. Glass tubes 2–2.5 mm in diameter are convenient, especially owing to the low consumption of the immobilized ligand derivatives. Some ligands, or soluble high-molecular-weight ligand derivatives, may inhibit acrylamide polymerization. Should this occur, it becomes necessary to increase the concentration of the polymerization catalyst.

[29] I. Parikh, S. March, and P. Cuatrecasas, this series, Vol. 34, p. 77.

[30] If the ligand possesses suitable spectral properties, its concentration in the substituted beads can be estimated most conveniently by spectrophotometry of solubilized gel solutions or directly by spectrophotometry of the agarose gel bead suspension in 50% glycerol solution, using the free ligand solution as a standard.

[31] S. J. Johnson, E. C. Metcalf, and P. D. G. Dean, *Anal. Biochem.* **109,** 63 (1980).

When meltable agarose derivatives are used for immobilization of the ligand in polyacrylamide affinity gels, polymerization is performed at higher temperatures (60–70°) to prevent premature gelling of agarose[20]; under such conditions, 5- to 10-fold lower concentrations of polymerization catalysts must be used. Generally, the catalyst concentration that is optimal for polymerization of polyacrylamide-based affinity gels must be found empirically.

The affinity and control gels to be compared should be run under identical conditions, i.e., simultaneously in the same electrophoretic tank. The properties of the control gel should be as similar as possible to those of the affinity gel. The control gel should contain an immobilized mock ligand very similar to the affinity ligand. The mobility of the interacting protein zone should be related to the mobility of a noninteracting reference protein that is present in all the samples.

[15] Preparative Isotachophoresis

By CHRISTOPHER J. HOLLOWAY and RÜDIGER V. BATTERSBY

Isotachophoresis is the least well known of modern electrophoretic techniques. Although the theoretical groundwork for the method was published in 1897, isotachophoresis has been applied practically to any significant extent only since the early 1960s. It is the exception rather than the rule to find an article devoted to this technique on a preparative scale, particularly with regard to protein separations, since isotachophoresis is most commonly associated with the analytical capillary equipment,[1] in which small ionic species, such as nucleotides,[2] amino acids, or small peptides,[3] rather than macromolecules, present the majority of applications. Nevertheless, there are very good reasons for use of isotachophoresis in protein separation and purification. An attempt has been made here to present the possible advantages and useful areas of application insofar as this is necessary for satisfactory exploitation of the method.

That isotachophoresis is only one of several suggested names for this technique causes some difficulty in literature searches. Other synonymous terms are ionic migration method, displacement electrophoresis,

[1] F. M. Everaerts, J. L. Beckers, T. P. E. M. Verheggen, *J. Chromatogr. Libr.* **6** (1976).
[2] C. J. Holloway and J. Lüstorff, *Electrophoresis* **1,** 129 (1980).
[3] C. J. Holloway and V. Pingoud, *Electrophoresis* **2,** 127 (1981).

ISBN 0-12-182004-1

cons electrophoresis, ionophoresis, omegaphoresis, and steady-state stacking. From the last, it will become apparent that the concentrating step of the well-known Ornstein[4] and Davis[5] disc electrophoresis, i.e., steady-state stacking, is actually an isotachophoretic configuration.

Two reviews on preparative-scale isotachophoresis that contain practical details have been published, but these represent mainly the approach of the individual groups.[6,7] We present here the methodology for the better known preparative procedure in a specially designed apparatus and we also describe a method on a flat gel slab that can easily be performed with standard equipment of most laboratories involved in protein chemistry.

Basic Principles of Isotachophoretic Separations

We briefly outline here the essentials of the principles of isotachophoresis. A more comprehensive theoretical treatment[1] and extensive descriptions of the principles have been presented.[8,9]

Definition of the Isotachophoretic System

In contrast to conventional zonal electrophoresis, in which sample ions migrate through a background electrolyte of uniform composition, isotachophoresis uses a discontinuous electrolyte system. The sample ions are preceded in the migration stack by a leading ion (L) of the same charge quality, but with higher effective mobility than all the sample ions of interest. The displacing electrolyte terminating the stack (T) is a species of lower effective mobility than all the sample ions in the stack. The system is buffered at the required pH by a common counterion. Owing to this arrangement, the samples are forced to migrate in the steady state at the same velocity; hence the name isotachophoresis.

Some Properties of the Isotachophoretic System

Isotachophoresis is normally performed under conditions of constant current. In the fundamental equation governing all electrophoretic separations [Eq. (1)],

[4] L. Ornstein, *Ann. N.Y. Acad. Sci.* **121,** 321 (1964).

[5] B. J. Davis, *Ann. N.Y. Acad. Sci.* **121,** 404 (1964).

[6] N. Y. Nguyen and A. Chrambach, *in* "Gel Electrophoresis of Proteins" (B. D. Hames and D. Rickwood, eds.), p. 145. IRL Press, London, 1981.

[7] P. J. Svendsen, *in* "Electrophoresis, A Survey of Techniques and Applications" (Z. Deyl, ed.), p. 345. Elsevier, Amsterdam, 1979.

[8] C. J. Holloway and I. Trautschold, *Z. Anal. Chem.* **311,** 81 (1982).

[9] S. G. Hjalmarsson and A. Baldesten, *CRC Crit. Rev. Anal. Chem.* **261** (1981).

$$v = \bar{m}E \tag{1}$$

v is the migration velocity, $\bar{m}$ the effective mobility, and E the electric field strength. Since v is constant, the electric field strength is inversely proportional to the effective mobility. Thus, in the stack, the field strength increases stepwise from L to T at each zone boundary.

Since electrophoretic systems are subject to Joule heating effects, in isotachophoresis the temperature pertaining in the zones will increase stepwise with decreasing mobility, i.e., with increasing field strength. For this reason, a vertical apparatus is always arranged such that the hotter terminating electrolyte is uppermost. In this way, convectional disturbances are minimized. An additional property of the isotachophoretic system can be deduced from Ohm's law. Under conditions of constant current, with the stepwise increase in field strength, the conductivity will decrease at each zone boundary from high to low mobility. These properties are exploited in analytical capillary equipment as detectors, but have not yet been incorporated in preparative apparatus.

The Steady State

Conventional zone electrophoresis never reaches a steady state. In isoelectric focusing, the steady state is reached when proteins are focused in a narrow band at a position in the pH gradient equal to their isoelectric points. In isotachophoresis, the steady state is reached when no partially mixed zones are present in the stack. If two sample ions have differing effective mobilities in the mixed state under the conditions of the separation, then they are completely separable. If they have identical effective mobilities in the mixed state, then they are inseparable. "Partial" separation means that the steady state has not been reached. A lower sample load or greater separation volume is then required.

An important consideration in connection with the steady state is the regulation of zone concentrations in isotachophoresis. The so-called "*beharrliche Function*" laid down by Kohlrausch in 1897 defines the conditions at a boundary between two electrolyte solutions with a common counterion migrating in an electric field. As described by Eq. (2),

$$\frac{c_L}{c_X} = \frac{m_L}{(m_L + m_R)} + \frac{(m_X + m_R)}{m_X} \tag{2}$$

c represents the concentration of the two ions L and X, m represents their effective mobilities, and R refers to the common counterion. This form of the equation is a simplification, not taking into account the possible differing electric charges of sample and leading ions.

Having set the concentration of leading ion, the concentration of the following sample zone in the steady state is regulated to a set concentration, which is a function of the several mobilities of the components. In turn, therefore, the concentration of each component of the stack is regulated. In the steady state, the stack migrates in zones with unchanging concentrations and the stack itself has a constant length. This property has been used to determine whether the steady state has been reached.[6]

Practically, the steady state, i.e., the best possible separation, must be reached before sample zones are eluted from the apparatus. Each system has a defined load capacity, which, if exceeded, will not allow the steady state to be reached. The load capacity increases with increasing separation volume. In theory, it does not vary with the applied separation current, since the total coulombs applied up to elution will remain the same. However, increasing the leading electrolyte concentration is a way of increasing the load capacity, since more coulombs are required for complete migration through the apparatus. However, increasing the concentration of leading electrolyte also leads to higher concentrations in the sample zones (from the Kohlrausch function), and solubility problems may be encountered with proteins. The load capacity also depends on the actual differences in mobilities of the sample ions. A large number of species with only slightly differing effective mobilities will require a much greater separation volume than do a smaller number of separands of widely differing mobilities.

The Spacer Principle

In zone electrophoresis, the sample (protein) zones migrate at different velocities and move apart during the separation process. In isotachophoresis, this is not the case. If the sample stack contains only protein, a protein continuum between leading and terminating electrolytes is formed. The zone boundaries cannot be distinguished by staining procedures, and elution of the pure zones becomes impossible. Thus, the "spacer technique" is often applied. This involves adding nonprotein species to the sample with effective mobilities between those of the separands of interest. Amino acids are an example of such spacers. However, finding suitable spacers in preparative systems is a laborious matter of trial and error, although some catalogs of spacers have been published.[10] In principle, carrier ampholytes can be applied to form a more or less continuous spacer mobility gradient. Since these mixtures contain a large number of compounds of closely related mobilities, difficulties in load capacity and separation quality can result.

[10] S. Husmann-Holloway and E. Borriss, *Z. Anal. Chem.* **311,** 465 (1982).

The spacer principle can be employed much more usefully for selective stacking or unstacking. If a separand of primary interest has high mobility and migrates at the front of the stack, the terminating electrolyte can be replaced by another of high mobility, ideally one slightly lower than that of the separand. This separand will migrate in the stack as previously, although many of the other components of the sample of lower mobility will be excluded from the stack; the separand in question is said to be selectively stacked. The separand can be selectively unstacked if it has lower mobility than the other components of the sample mixture.

The Zone "Sharpening" Effect

The main reason for the low resolving power of zone electrophoresis is that no property of the system can counteract diffusion. In isoelectric focusing, diffusion is countered by the continuous migration into a point of zero charge. In isotachophoresis, diffusion is also continuously countered, providing high resolution; the zone "sharpening" effect is responsible. Suppose that a sample ion, X, diffuses into the following zone, Y, where it will find itself in an environment of higher electric field strength than its own zone in the stack. According to Eq. 1, the fundamental migration equation, X will be accelerated over and above the "isotachophoretic" migration velocity and will overtake the X/Y zone boundary before being forced back into its own zone.

Procedures in a Column Apparatus with Polyacrylamide Gel

By far the greatest number of investigations involving preparative isotachophoresis report the use of commercially available instruments for isotachophoresis in columns. Such an apparatus is shown schematically in Fig. 1. The column is filled with the stabilizing medium, e.g., polyacrylamide. The prerequisite here is low electroendoosmosis. In principle, the newer types of agarose with electroendoosmotic values approaching zero could be employed. Some investigators have used granulated dextran-type gels, or density gradients with, for example, sucrose. However, polyacrylamide is the most common gel, whereby rather low concentrations may be selected to minimize molecular sieving.

The gel is prepared in the leading electrolyte system, and the electrode region above the gel is filled with terminating electrolyte. In any vertical apparatus, as explained previously, it is important that the terminator be placed above the leading system, i.e., migration is in a downward direction, since the temperature of the terminating zone is highest. With the terminator above the leading system, therefore, convectional effects are

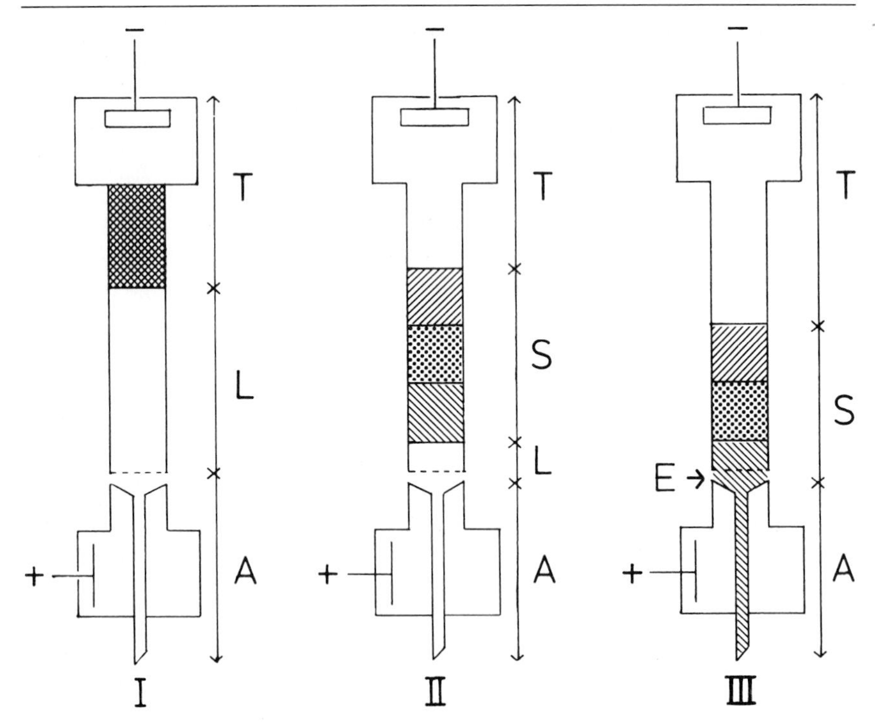

FIG. 1. Schematic representation of preparative isotachophoresis in a column apparatus. Column I shows the initial situation, in which the sample mixture (shaded region) has been introduced into the upper part of the gel cylinder. The upper electrode compartment (in this case, the cathode) is filled with terminating electrolyte (T); the lower gel region is filled with leading electrolyte (L). The lower electrode compartment is filled with anolyte (A), which also serves as elution buffer (E). In column II, the sample species have separated into the stack (S) and migrate down the gel cylinder. The upper gel region is displaced by T. When the separands migrate out of the gel (column III), they are eluted. For simplicity, the cooling mantle of the column has not been shown, although this is an essential component of the equipment.

lower than they would be with the reverse configuration and an upward migration. The sample is applied to the upper gel surface before adding terminating electrolyte. It is preferable to dissolve the sample or to dialyze it against the terminating electrolyte.

Since protein separations are under discussion, some sort of spacer mixture will have to be added to the sample. Although carrier ampholyte mixtures are most frequently employed, problems of load capacity can arise if excessive amounts are applied. As far as the column apparatus is concerned, ampholytes may lead to swelling of polyacrylamide gels. Such effects have led to a movement away from glass columns, since polyacrylamide adheres strongly to glass. Swelling of the gel will give rise to distortion, whereby the expansion takes place through the center of the gel cylinder. The upper surface of the gel becomes convex, and the zones curve. Plastic columns have proved to be more satisfactory, since little if any adherence of the gel is observed; polyacrylamide is free to expand and contract uniformly.

When current is applied, the sample ions migrate down the column and begin to separate. Eventually, they emerge from the gel into an elution chamber of small volume, through which elution buffer is passed at a constant flow rate. The elution tube can be passed through a detector to a fraction collector.

Procedures

Electrolyte Systems. One of the major difficulties associated with isotachophoresis is the selection of suitable leading and terminating electrolytes. Frequently, this is a matter of trial and error, although guidelines are given in Table I. All these systems are designed for anionic separations and may be unsuitable for highly basic proteins. For this latter group a cationic system, such as that presented in a later section on flatbed procedures, can be used.

In principle, the leading electrolyte is simply chosen from the list in Table I, according to the required pH. A system at lower pH will be suitable for selectively collecting acidic proteins; a higher pH yields a broader spectrum of separands. For separation of immunoglobulins, for example, a system at a higher pH is essential. The isoelectric points of the proteins of interest are a useful parameter in selecting the electrolyte. The terminating electrolytes are chosen for selectively stacking or unstacking sample proteins. In order to obtain stacking of as many separands as possible, the EACA system, which has the lowest mobility in the list, is chosen. Clearly, the higher mobility glycine terminator is suitable for selective stacking of the more acidic proteins.

Gel Preparation. The following solutions are required for the polyacrylamide gel: acrylamide, 33% w/v in distilled water; Bis, 1% (w/v) in distilled water (these solutions are stable for up to 10 days at 4°); riboflavin 5′-phosphate, 0.008% (w/v) in distilled water; ammonium persulfate 0.1% (w/v) in distilled water (these solutions must be made up immediately before preparing the gel). Note that photopolymerization is preferable.

The following recipe for the gel is given for a final total volume of 100 ml. Using these proportions, the total quantity can be varied according to the dimensions of the column apparatus.

TABLE I
ELECTROLYTE SYSTEMS FOR ANIONIC SEPARATIONS IN COLUMN ISOTACHOPHORESIS[a]

Approximate pH	Leading ion	Counterion
	A. Leading electrolytes[b]	
4	HAc, 30 ml	Tris, 10 g
6	MES, 73 g	Tris, 15 g
7	1 *M* H_3PO_4, 300 ml	Tris, 60 g
8	1 *M* H_3PO_4, 300 ml	Tris, 120 g
8.5	1 *M* HCl, 600 ml	Tris, 220 g
9	1 *M* HCl, 600 ml	Ammediol, 120 g
	B. Terminating electrolytes[c]	
8.5	Glycine, 15 g	Tris, 3.0 g
8.7	Alanine, 17.5 g	Tris, 2.5 g
8.9	EACA, 30 g	Tris, 1.5 g

[a] Abbreviations: HAc, glacial acetic acid; MES, 2-(*N*-morpholino)ethanesulfonic acid; Tris, tris(hydroxymethyl)-aminomethane; ammediol, 2-amino-2-methylpropane-1,3-diol; EACA, 6-aminohexanoic acid.

[b] The quantities given are for a final stock volume of 1000 ml. Leading and counter species are mixed and diluted to 1000 ml by the addition of distilled water. The exact pH is measured at the running temperature (10°) after diluting a small portion 10-fold, i.e., to the final concentration in the gel.

[c] The quantities given are for a final volume of 1000 ml. The terminating electrolytes thus obtained are used as prepared, i.e., undiluted. The exact pH should be measured at the operating temperature (10°).

Leading electrolyte, stock solution, 10 ml
Acrylamide solution, 10 ml
Bis solution, 10 ml
Riboflavin 5′-phosphate solution, 10 ml
Ammonium persulfate solution, 10 ml
Distilled water, 50 ml

The mixture is poured carefully into the column apparatus without the introduction of air bubbles. This is best achieved by pouring down the inside wall of the column. The lower outlet of the column is suitably blocked until polymerization is complete. It is important to ensure that the upper surface of the gel is flat. The gel mixture is, therefore, overlayed with about 5 ml of leading electrolyte (diluted 10-fold to yield the same concentration as in the gel mixture). Photopolymerization is carried out with the aid of suitable lamps. Although the reaction should be complete within 1 or 2 hr, storage of the gel at 10° overnight is recommended. In order to prevent undue swelling and contracting, the gel should be cast and stored at all times at the final running temperature, e.g., 10°.

Application of Sample. Shortly before use, the overlaying leading electrolyte is removed from above the gel surface, and the sample is carefully applied and allowed to diffuse into the gel matrix. The free space above the gel is then filled with terminating electrolyte.

Separation Procedure. Since migration velocity is dependent on the applied constant current, it is preferable to use as high a driving current as possible, limited only by the cooling capabilities of the apparatus. In practice, the absolute limit is generally 10 mA. Usually, 5–8 mA are applied. Depending on the dimensions of the apparatus and the type of electrolyte system selected, separation requires from 4 to 24 hr. Immediately before commencing the separation, the lower electrolyte region of the equipment must be filled (air-bubble free) with anolyte. This solution is also used as elution buffer and can be varied somewhat to suit the stability of the separands. However, a solution of 30 ml of 1 *M* H_2SO_4 with 8 g of Tris per liter (giving a pH of about 7) has been satisfactory in many cases.

The elution flow rate must be set according to the rate of appearance of the sample species from the gel. This will depend on the driving current applied and the electrolyte system employed. In general, flow rates of the order of 5–30 ml/hr are suitable. It is often useful to mix a small quantity of dyestuff with the sample. This makes visualization of the position of the stack possible. Again, the dyestuff to be used depends on the electrolyte system. However, it is of prime importance that the dye migrate within the stack, not by zone electrophoresis in the leading or terminating electrolytes. During the separation process, the dye should become concentrated and finally retain a fixed zone length during migration. If the dyestuff zone becomes diffuse during the procedure, it is probably not within the stack. In systems of low pH, amaranth red is useful. Dyes with lower anionic mobility are fluorescein, bromophenol blue, and bromocresol green. For cationic systems, Janus green has been useful.

Procedures in a Horizontal Slab of Granulated Gel

The method described in the preceding section requires a specially designed column. Although the schematic representation in the figure appears simple, the actual commercial equipment available requires considerable care and patience for effective use. Since this equipment may not be generally available, we suggest an alternative procedure that is less time-consuming and requires only minor additions to flatbed electrophoresis equipment already present in many laboratories.

Apparatus

The basic requirement is a flatbed electrophoresis chamber with an effective cooling plate of dimensions at least 25 × 10 cm. The electric field is applied across the longer side with electrode tanks accommodating at least 500 ml of anolyte and catholyte, respectively. The cooling plate should be attached to circulating temperature-controlled water supplies, with the option, via Y valves, of 50° for use in the casting procedure and of 10° for cooling during the separation. The power supply should deliver up to 2000 V. A maximum current delivery of 20 mA is adequate.

A gel tray is required for casting the slab. This consists of a glass plate, 1 mm thick, cut to the required dimensions, but no larger than the cooling plate. A silicone rubber gasket of cross section 5 × 5 mm is attached around the rim of this glass plate.

Since the slab is exposed to the atmosphere during the procedure, it may be useful to flush the chamber with carbon dioxide-free air, especially where high pH conditions are employed. The absorbance of carbon dioxide into the system produces a long carbonate zone that can negatively influence the quality of the separation through wastage of load capacity.

Procedures

For these procedures, with a gel tray size of approximately 25 × 10 cm, the final gel slab thickness does not exceed 3 mm. For other dimensions, slab thickness will have to be modified accordingly.

Gel Casting. The types of material suitable for casting granulated gels in isotachophoresis should fulfill the same requirements as for isoelectric focusing (see this volume [13]). In general, Sephadex G-75 SF or G-200 SF (Pharmacia) or Ultrodex (equivalent to G-75 SF, LKB) have been found satisfactory. The gel (4 g) is suspended as a slurry in 100 ml of diluted leading electrolyte. Table II lists some stock electrolytes, which are diluted prior to mixing with the gel material. Of immediate importance is the final leading electrolyte concentration in the gel. However, the gel is initially applied as a slurry, and excess water is evaporated. Thus, the actual dilution of the stock leading electrolyte will depend on the evaporation limit of a particular batch of gel. This information should be quoted by the manufacturers.

TABLE II
ELECTROLYTE SYSTEMS FOR ISOTACHOPHORETIC SEPARATIONS IN GRANULATED SLAB GELS[a]

Approximate pH	Leading ion	Counterion
	A. Leading electrolytes[b]	
4 (anionic)	HAc, 30 ml	Tris, 10 g
6 (anionic)	1 *M* H_3PO_4, 300 ml	Tris, 40 g
8.5 (anionic)	1 *M* HCl, 600 ml	Tris, 140 g
5 (cationic)	1 *M* KOH, 600 ml	HAc, 70 ml
	B. Terminating electrolytes[c]	
8.7 (anionic)	β-Alanine, 18 g	Tris, 2.5 g
8.9 (anionic)	EACA, 30 g	Tris, 1.5 g
4.5 (cationic)	Alanine, 18 g	HAc, 11 ml

[a] Abbreviations as in Table I.
[b] The quantities given are for a final stock volume of 1000 ml. Leading and counter species are mixed and diluted to 1000 ml by the addition of distilled water. The amount of stock electrolyte required for the gel slurry is calculated as explained in the text. The exact pH should be measured in the final dilution in gel slurry at the running temperature (10°). Other leading electrolyte systems can be adapted from Table I.
[c] The quantities given are for a final volume of 1000 ml. The terminating electrolytes are used undiluted. The pH of the terminating electrolytes can be varied by altering the concentration of counterion.

Equation (3) can be used to determine the amount of stock solution, in milliliters, that should be diluted to a final volume of 100 ml with water for addition to 4 g of gel material.

$$\text{Volume of stock solution} = \frac{100 - [1.04 \times \text{evap limit } (\%)]}{10} \quad (3)$$

Filter paper strips 1 cm wide are cut to fit into the gel tray at each end, i.e., strips 10 cm long. These strips are soaked in the diluted leading electrolyte and are stacked to a height of 4 mm. At this stage, the tray with paper strips is weighed. The tray is then placed on the cooling block of the apparatus, ensuring that the latter is exactly level, thus providing an even thickness of gel slab.

The vessel containing the gel slurry is weighed before and after pouring into the tray. The difference gives the initial total weight of slurry applied. In order to assure relatively rapid evaporation of excess water from the slurry, the block is heated to 50° by circulating water while a gentle current of air from a fan is passed over the gel surface. At regular intervals, the gel tray is removed and weighed. The evaporation step is complete when the slurry has been reduced in weight by the evaporation limit, e.g., 36%.

The circulating water supply at 10° is now allowed to pass through the cooling block. The filter strips at the terminating end of the tray are carefully removed and replaced by similar strips soaked in terminating electrolyte. The ends of the gel slab are connected to the electrode vessels by several layers of filter paper cut exactly to the width of the slab. The leading end of the slab is connected by wicks soaked in leading electrolyte at a concentration fourfold that of the final concentration in the gel. The terminating end is connected by wicks soaked in terminating electrolyte. It is advisable to cover the latter wicks with a layer of plastic film to prevent drying, since the terminating region, in particular, can become quite hot during the procedure.

Prerun. Since the sample should be applied in terminating electrolyte, but the gel initially contains only leading electrolyte, a prerun must be performed in order to allow terminating electrolyte to migrate into the gel, a distance of approximately 5 cm. For this purpose, a current of up to 20 mA can be applied. The boundary between leading and terminating electrolytes can be identified by colored (brown) impurities from the strips that concentrate at this point. However, a drop of bromophenol blue may be added to visualize the boundary.

Sample Application. A section of the gel is removed on the terminating side of the zone boundary. This section should be about 2 cm deep across the whole width of the gel. The gel is scraped out with a spatula and is resuspended in a maximum of 3 ml of sample solution (including the volume of any spacers). This suspension is poured back into the cavity in the gel and, owing to its consistency, should yield an even gel surface. For convenience, sample application devices are available commercially.

Running Conditions. The isotachophoretic separation is carried out under conditions of limiting current. Values in the range 4 to 6 mA are suitable. Since the migration velocity is constant at constant current, this latter value can be conveniently set to provide complete migration through the slab overnight. The position of the trailing edge of the leading electrolyte zone can be observed by the trace of dye added at the beginning of the prerun.

Paper Prints. Before cutting the gel into sections for collecting the separands, it is useful to obtain a paper print to establish the position of the zones and estimate the quality of the separation. Since it is important that straight zone boundaries be obtained, it is particularly useful to print the whole surface of the gel. In this volume [13] printing procedures are described that should be satisfactory also for isotachophoresis. Here, we restrict ourselves to two simple and rapid methods of paper printing, which have proved to be satisfactory.

1. Serva Blue W is a water-soluble dye that eliminates the need for organic solvents. The paper print is fixed in a 10% trichloroacetic acid–5% sulfosalicylic acid solution and stained in an aqueous solution (0.5 g per liter of the dye). Destaining is performed with slightly acidified water.
2. Bromophenol blue provides a yellow stain in acid solution. The dye is dissolved in a mixture (9 : 1, v/v) of ethanol and glacial acetic acid (1%, w/v). Destaining is performed with water.

Fractionation of the Gel. The simplest method for fractionation is the use of a grid of at least 30 parallel blades oriented across the width of the gel slab. Such grids are available commercially. However, for isotachophoresis it is useful to have grid widths of rather less than 5 mm (2 mm is ideal), since the fractions are concentrated only in a part of the slab. The gel is first fractionated by using the grid. The pH of each section can then be measured with a surface contact pH electrode. The leading and terminating zones have different pH values, and the protein stack can be detected as that portion of the gel with steadily changing pH values.

After pH measurement, the gel is scraped out of the grid fraction with a spatula and is transferred to small columns fitted with nylon sieves. The fractions are equilibrated with one gel volume of a suitable elution buffer. After the liquid has drained through the sieve, a second volume of buffer is added to the gel. In this way, most of the protein is eluted from the granulated gel. The fractions can then be assayed for total protein and for the desired specific properties, such as enzymic activity.

Alternative Equipment for Preparative Isotachophoresis

In this chapter we have concentrated on the column apparatus as the most widely employed method for preparative isotachophoresis. As can be seen in Table III, compared with Tables IV and V, below, approximately 80% of the applications were performed with this type of equipment. For normal research applications, however, the flatbed technique is more convenient. Although precise details of all the methods presently available cannot be presented here, it is useful to provide a short summary of possible alternatives.

TABLE III
PREPARATIVE-SCALE PROTEIN SEPARATION IN POLYACRYLAMIDE GELS

Separands	Source	Apparatus	Electrolytes[a]	Reference[b]
Transferrin	Human serum	Glass tube	L1, T1	1
Serum proteins	Human serum	Glass tube	L2, T2	2
Histocompatibility antigens	Mouse lymphocytes	Glass tube	L3, T3	3
Fibrinogen breakdown product	Fibrinogen	Column	L2, T1	4
Membrane proteins	Human erythrocytes	Column	L4, T4	5
Carcinoma antigen	Human lung tumors	Column	L2, T4	6
Enterotoxin	*Escherichia coli*	Column	L2, T4	7
HGPT	Human erythrocytes	Column	L2, T2	8
CHE	Human serum	Column	L4, T4	9
Membrane proteins	Human erythrocytes	Column	NG[c]	10
Serum proteins	Human serum	Column	L4, T4	11
Enterotoxin	*E. coli*	Column	L2, T4	12
Serum proteins	Human serum	Column	L2, T4	13
Growth hormone	Human	Glass tube	Various	14
Enterotoxin	*Clostridium perfringens*	Glass tube	L1, T1	15
Antibodies	Rabbit serum	Glass tube	NG	16
Growth hormone	Human	Glass tube	Various	17
Crystallin	Bovine lens	Glass tube	L5, T5	18
Immunoglobulins	Myeloma lines	Flatbed	L2, T4	19
α_1-Macroglobulin	Rat serum	Column	L4, T4	20
Skin test antigens	Blastomycin	Column	L6, T1	21
Enterotoxin	*E. coli*	Column	L4, T4	22
Serum proteins	Human serum	Column	L5, T5	23
Prolactin	Canine pituitaries	Column	NG	24
Histoplasmin	Human *Histoplasma capsulatum*	Column	L6, T1	25
Inorganic pyrophosphatase	Bakers' yeast	Column	L4, T4	26
Enterotoxin	*E. coli*	Column	NG	27
Antibodies	Rabbit serum	Glass tube	NG	28
Serum proteins	Human serum	Flatbed	L2, T4	29
Tumor-specific surface antigen	Mice tumors	Glass tube	NG	30
Erythropoetin	Human urine	Column	L1, T6	31
NAD-dependent formate dehydrogenase	Methylotrophic bacteria	Column	L2, T5	32
Migration inhibition factor	Human lymphocytes	Column	L2, T4	33
Hibernating triggers	Woodchuck plasma	Column	L6, T6	34
Albumin and hemoglobin	Bovine blood	Flatbed	L4, T4	35

TABLE III (*continued*)

Separands	Source	Apparatus	Electrolytes[a]	Reference[b]
CSF proteins	Human CSF	Glass tube	L7, T7	36
CSF proteins	Human CSF	Flatbed	Various	37
LMW kininogen	Human plasma	Column	L8, T8	38
High-density lipoproteins	Human plasma	Column	L9, T4	39

[a] A key to the electrolyte systems is given in Table VI.

[b] Key to references:

1. D. B. Ramsden and L. Lewis, *Protides Biol. Fluids* **19,** 521 (1972).
2. N. Catsimpoolas and J. Kenney, *Biochim. Biophys. Acta* **285,** 287 (1972).
3. M. Hess and D. A. L. Davies, *Eur. J. Biochem* **41,** 1 (1974).
4. I. Clemmensen and P. J. Svensen, *Sci. Tools* **20,** 5 (1973).
5. T. C. Bøg-Hansen, O. J. Bjerrum, and P. J. Svensen, *Sci. Tools* **21,** 33 (1974).
6. M. J. Frost, G. T. Rogers, and K. D. Bagshawe, *Br. J. Cancer* **31,** 379 (1975).
7. R. Möllby, S. G. Hjalmarsson, and T. Wadström, *FEBS Lett.* **56,** 30 (1975).
8. B. Bakay and W. L. Nyhan, *Arch. Biochem. Biophys.* **168,** 26 (1975).
9. C.-H. Brogren, P. J. Svendsen, and T. C. Bøg-Hansen, *in* "Progress in Isoelectric Focusing and Isotachophoresis" (P. G. Righetti, ed.), p. 359. North-Holland Publ., Amsterdam, 1975.
10. T. C. Bøg-Hansen, P. J. Svendsen, O. J. Bjerrum, C. S. Nielsen, and J. Ramlau, *Protides Biol. Fluids* **22,** 679 (1975).
11. T. C. Bøg-Hansen, P. J. Svendsen, and O. J. Bjerrum, "Progress in Isoelectric Focusing and Isotachophoresis" (P. G. Righetti, ed.), p. 347. North-Holland Publ., Amsterdam, 1975.
12. F. Dorner, *J. Biol. Chem.* **250,** 8712 (1975).
13. S.-G. Hjalmarsson, *Sci. Tools* **22,** 35 (1975).
14. A. Chrambach and J. S. Skyler, *Protides of Biol. Fluids* **22,** 701 (1975).
15. W. W. Yotis and N. Catsimpoolas, *J. Appl. Bacteriol.* **39,** 147 (1975).
16. H. Brogren and G. Peltre, *Ann. Immunol.* **126,** 363 (1975).
17. G. Baumann and A. Chrambach, *Proc. Natl. Acad. Sci. U.S.A.* **73,** 732 (1976).
18. F. S. M. van Kleef, M. Peeters, and H. J. Hoenders, *Anal. Biochem.* **77,** 122 (1977).
19. A. Ziegler and G. Köhler, *FEBS Lett.* **71,** 142 (1976).
20. F. Gauthier, N. Gutman, J. P. Muh, and H. Mouray, *Anal. Biochem.* **71,** 181 (1976).
21. M. V. Lancaster and R. F. Sprouse, *Infect. Immun.* **13,** 758 (1976).
22. F. Dorner, H. Jaschke, and W. Stöckl, *J. Infect. Dis.* **133,** 142 (1976).
23. A. Kopwillem, W. G. Merriman, R. M. Cuddeback, A. J. K. Smolka, and M. Bier, *J. Chromatogr.* **118,** 35 (1976).
24. P. J. Knight, M. Gronow, and J. M. Hamilton, *J. Endocrinol.* **69,** 127 (1976).
25. M. V. Lancaster and R. F. Sprouse, *Anal. Biochem.* **77,** 158 (1977).
26. V. N. Kasho and S. M. Avaeva, *Int. J. Biochem.* **9,** 51 (1978).
27. T. Wadström, R. Möllby, B. Olsson, J. Söderholm, and C. J. Smyth, "Electrofocusing and Isotachophoresis" (B. J. Radola and D. Graesslin, eds.), p. 443. de Gruyter, Berlin, 1977.
28. C.-H. Brogren and G. Peltre, *Scand. J. Immunol.* **6,** 685 (1977).
29. C.-H. Brogren, "Electrofocusing and Isotachophoresis" (B. J. Radola and D. Graesslin, eds.), p. 549. de Gruyter, Berlin, 1977.

(*continued*)

References to TABLE III (*continued*)

30. T. Natori, L. W. Law, and E. Appella, *Cancer Res.* **38,** 359 (1978).
31. M. Puschman, W. Thorn, and Y. Yen, *Res. Exp. Med.* **173,** 293 (1978).
32. Y. V. Rodionov, T. V. Avilova, E. V. Zakharova, L. S. Platonenkova, A. M. Egorov, and I. V. Berezin, *Biokhimiya* **42,** 1896 (1977).
33. L. H. Block, H. Jaksche, S. Bamberger, and G. Ruhenstroth-Bauer, *J. Exp. Med.* **147,** 541 (1978).
34. P. R. Oeltgen, L. C. Bergmann, W. A. Spurrier, and S. B. Jones, *Prep. Biochem.* **8,** 171 (1978).
35. F. Hampson and A. J. P. Martin, *J. Chromatogr.* **174,** 61 (1979).
36. K. G. Kjellin and L. Hallander, *J. Neurol.* **221,** 225 (1979).
37. L. Hallander and K. G. Kjellin, *Anal. Chem. Symp. Ser.* **5,** 245 (1980).
38. A. Adam, J. Damas, C. Schots, G. Heynen, and P. Franchimont, *Anal. Chem. Symp. Ser.* **5,** 47 (1980).
39. G. B. Bon, G. Cazzolato, and P. Avogaro, *J. Lipid Res.* **22,** 998 (1981).

[c] NG, Not given.

Small-Scale Preparative Isotachophoresis in Glass Tubes

This type of separation has been described in detail by Chrambach and his group.[6] Isotachophoresis is performed in glass tubes of 5- to 6-mm i.d. containing polyacrylamide gel. Separations can be carried out on most disc electrophoresis equipment, provided that cooling is adequate. The gels are removed as cylinders from the tubes after completion of the migration and sliced into fractions. One disadvantage of this method, apart from the previously mentioned drawbacks of glass, is the limited load capacity. Handling of the gel is rather awkward, as it is with disc electrophoresis in this type of equipment. Possibly, the newer type of vertical slab chambers could be useful.

Hollow Cylinder Technique

This variant, devised by Hampson and Martin,[11] is a logical development of the glass tube method. The configuration of a vertical tube is basically maintained, but with the following modifications: the diameter of the tube is increased to 20 mm, and the gel is cast as a 1- to 2-mm-thick cylinder around the outside wall. Cooling liquid is circulated through the center of the tube. To prevent evaporation, the equipment is immersed in a cooled reservoir of nonaqueous, water-immiscible liquid, e.g., *o*-dichlorobenzene. The terminating electrolyte is floated at the top of the cylin-

[11] F. Hampson, *in* "Electrophoresis '79" (B. J. Radola, ed.), p. 583. de Gruyter, Berlin, 1980.

der. The cooling capability of this apparatus is excellent. The possible drawbacks of the technique, as with much self-built equipment, is complicated handling. The toxicity of the cooling liquid may also be a hindering factor.

Free-Flow Procedures

In the main, two groups have been working on this type of equipment for several years.[12,13] The apparatus is based on free-flow cell electrophoresis, but with a modified chamber arrangement. The separation chamber consists of a flat rectangular cavity (0.5 mm thick) in a cooling block, through which the electrolytes and sample are continuously pumped. An electric field is applied across the cavity, at right angles to the direction of laminar flow. This equipment has a potentially high load capacity, owing to the variability of both current and flow rate. Furthermore, the absence of stabilizing medium eliminates molecular sieving. The major advantage of the technique, however, is the continuous nature of the separation; all other procedures are batch processes. For industrial purposes, therefore, free flow offers the greatest possibilities, and it is expected that commercial apparatus for free-flow isotachophoresis will become available shortly.

Cooling

It will have become apparent that efficient cooling is essential for suitable separation conditions and gel stability. Moreover, when separating heat-labile proteins, cooling is an obvious need. It is reasonable to analyze the cooling systems of the two major types of equipment, column and flatbed, to ascertain whether a major advantage pertains to one or the other system. Although it is difficult to define the precise temperature gradients and heat-exchange characteristics for these systems, the cooling area as a function of gel volume can be compared, realizing that gel volume is the decisive factor in load capacity and separability.

The first assumption here must be that the heat exchange per unit area cooling in the column and flatbed are similar. In order to attain comparable electrical conditions and load capacity, the length of gel slab is defined identical to the length of gel column; i.e., the distance between electrodes is similar, and the total gel volume is the same.

[12] Z. Prusik, J. Stepanek, and V. Kasicka, *in* "Electrophoresis '79" (B. J. Radola, ed.), p. 287. de Gruyter, Berlin, 1980.

[13] H. Wagner and V. Mang, *in* "Analytical Isotachophoresis" (F. M. Everaerts, ed.), p. 41. Elsevier, Amsterdam, 1980.

TABLE IV
PREPARATIVE-SCALE PROTEIN SEPARATION IN AGAROSE GELS

Separands	Source	Apparatus	Electrolytes[a]	Reference[b]
Urinary proteins	Human urine	NG[c]	L1, T1	1
Albumin and transferrin	Human urine and CSF	NG	L1, T1	2
Sweat proteins	Human sweat	Flatbed	L1, T1	3
Sweat and urine proteins	Human sweat and urine	Flatbed	L1, T1	4
Serum proteins	Bovine blood	NG	NG	5
Albumin and hemoglobin	Bovine blood	Hollow cylinder	L4, T9	6
Various proteins	Diverse sources	Glass tube	Various	7

[a] A key to the electrolyte systems is given in Table VI.
[b] Key to references:
1. V. Blaton, K. Uyttendaele, and H. Peeters, *Acta Med. Acad. Sci. Hung.* **31,** 277 (1974).
2. K. Uyttendaele, M. de Groote, V. Blaton, H. Peeters, and F. Alexander, *Protides Biol. Fluids* **22,** 743 (1978).
3. K. Uyttendaele, V. Blaton, F. Alexander, H. Peeters, M. de Groote, N. Vinaimont-Vandecasteele, and J. Chevalier, "Progress in Isoelectric Focusing and Isotachophoresis" (P. G. Righetti, ed.), p. 341. North-Holland Publ., Amsterdam, 1975.
4. K. Uyttendaele, M. de Groote, V. Blaton, H. Peeters, and F. Alexander, *J. Chromatogr.* **132,** 261 (1977).
5. A. J. P. Martin and F. Hampson, Br. UK Patent Appl. 2026546 (1980).
6. F. Hampson, in "Electrophoresis '79" (B. J. Radola, ed.), p. 287. de Gruyter, Berlin, 1980.
7. Z. Buzás and A. Chrambach, *Electrophoresis* **3,** 121 (1982).

[c] NG, Not given.

The flatbed apparatus with base dimensions of 25 × 10 cm, provides a cooling area of 250 cm^2. At a gel thickness of d cm, gel volume would be $250 \times d\ \text{cm}^3$. In the column configuration, the gel volume is also $250 \times d\ \text{cm}^3$, and the length of the column is 25 cm; column diameter is then $(40 \times d/\pi)^{1/2}$. Given that the cooling area of the slab should be at least as great as that of the column, i.e., cooling area of slab ≥ cooling area of column, then, $250\ \text{cm}^2 \geq 25\pi\sqrt{(40d/\pi)}^{1/2}$ and, therefore, $d \leq 0.8$ cm.

Thus, where the gel slab is less than 8 mm thick, the cooling area per unit volume gel is greater than that of the column apparatus. Normally, d is of the order 3–4 mm, so that we can assume realistic cooling efficiency compared with the column apparatus. The apparatus devised by Martin and Hampson[11] offers even better cooling, since both sides of a "cylindrical slab" are exposed to cooling fluid.

TABLE V
PREPARATIVE-SCALE PROTEIN SEPARATION IN GRANULATED GELS

Separands	Source	Apparatus	Electrolytes[a]	Reference[b]
Antibodies	Rabbit serum	Flatbed	L2 or L6 T5 or T4	1
Antibodies	Rabbit serum	Flatbed	L2, T4	2
Plasma protein	Human blood	Column	L5, T10	3
Plasma protein	Human blood	Column	L5, T10	4
Serum proteins	Human serum	Flatbed	L2 or L6 T5 or T4	5
Glutathione *S*-transferases	Cat liver	Flatbed	L10, T4	6

[a] A key to the electrolyte systems is given in Table VI.
[b] Key to references:
1. G. Peltre and C.-H. Brogren, "Electrofocusing and Isotachophoresis" (B. J. Radola and D. Graesslin, eds.), p. 577. de Gruyter, Berlin, 1977.
2. C.-H. Brogren and G. Peltre, "Electrofocusing and Isotachophoresis" (B. J. Radola and D. Graesslin, eds.), p. 587. de Gruyter, Berlin, 1977.
3. M. Bier and A. Kopwillem, "Electrofocusing and Isotachophoresis" (B. J. Radola and D. Graesslin, eds.), p. 567. de Gruyter, Berlin, 1977.
4. M. Bier, R. M. Cuddeback, and A. Kopwillem, *J. Chromatogr.* **132,** 437 (1977).
5. A. Winter, H. Brogren, and T. Dobson, LKB Appl. Note 318 (1980).
6. R. V. Battersby and C. J. Holloway, *Electrophoresis* **3,** 275 (1982).

Applications of Preparative-Scale Isotachophoresis for the Separation and Purification of Proteins

Although the discussion has been limited to two types of configuration of apparatus, many variants of the electrolyte system and modifications of apparatus and running conditions have been described. To provide an overview for general reference, a summary of a range of publications in this field for the three major support media (polyacrylamide, agarose, and granulated gels) is presented in Tables III, IV, and V, respectively. The various electrolyte systems (where reported) are summarized in Table VI.

Predictable Areas of Application of Preparative-Scale Isotachophoresis of Proteins

Compared with preparative isoelectric focusing, and even more so when compared with the chromatographic methods, there is too little published work on isotachophoresis to conclude any specific applications for which the latter technique is ideally suited. Indeed, most of the reported applications do not exploit the special properties of the isotachophoretic principle. Thus, for the guidance of those interested in this

TABLE VI
ELECTROLYTE SYSTEMS EMPLOYED IN THE VARIOUS LITERATURE REFERENCES IN TABLES III, IV, AND V

Electrolyte system	Ion	Counterion	Approximate pH
	Leading		
L1	Phosphate	Tris	5.5
L2	Phosphate	Tris	6.1
L3	Chloride	Ammediol	8.4
L4	Phosphate	Tris	8.1
L5	Cacodylate	Tris	7.0
L6	Acetate	Tris	4.4
L7	TES	Tris	7.5
L8	Chloride	Histidine	6.2
L9	Sulfate	Tris	7.1
L10	Chloride	Tris	7.0
	Terminating		
T1	Glycine	Tris	8.6
T2	β-Alanine	Tris	8.0
T3	Phenolate	Ammediol	7.8
T4	EACA	Tris	8.9
T5	β-Alanine	Tris	8.9
T6	Glycine	Tris	7.2
T7	EACA	$Ba(OH)_2$	9.5
T8	Diethylbarbiturate	Histidine	7.2
T9	2-Aminopropanoic acid	Tris	Not given
T10	β-Alanine	$Ba(OH)_2$	9.2

technique, it is worthwhile to discuss briefly the particular areas, which we believe would be more realistic.

It is not possible to provide definite figures for the capacity of the system described. At best, an extremely complex protein mixture of close mobilities would be limited to less than 100 mg. For the isolation of a specific protein, 1 g or more has been applied. It is, therefore, a matter of the actual sample, i.e., load capacity, which must be determined experimentally.

A comparison with preparative isoelectric focusing is inevitable, since only these two members of the electrophoresis family are worth considering from the point of view of resolution capability for preparative-scale work. In both methods, the sample is concentrated and diffusional effects are actively countered. The slightly higher cost of isoelectric focusing, due to the larger amounts of carrier ampholyte, is a decisive argument in favor of isotachophoresis. Certainly, in the flatbed configuration, the fractionation of the gel is easier for the more widely separated focused zones than for the protein stack in isotachophoresis. However, for those interested in separating and purifying enzymes, isotachophoresis does have one intrinsic advantage over isoelectric focusing: we have frequently observed that enzymes lose considerable activity when exposed for extended periods to an environment at their isoelectric points. In isotachophoresis, it is possible to vary the electrolyte systems widely and thereby provide the conditions under which an enzyme is most stable. Both methods are useful in that the samples are concentrated rather than diluted during the separation process, a factor of importance when dealing with labile proteins.

Section III

Techniques for Membrane Proteins

[16] Solubilization of Functional Membrane Proteins

By LEONARD M. HJELMELAND and ANDREAS CHRAMBACH

Although the solubilization of functional membrane proteins is a central task in modern membrane biochemistry, systematic experimentation directed at achieving this goal has not been successful. Even less success has attended attempts at optimizing procedures tailored to specific systems. There are, however, several approaches to both problems, and these can be illustrated with specific examples.

The material outlined here is intended to present a systematic approach to the solubilization of membrane proteins. Specific details for the initial investigations, directed at achieving solubilization, are outlined. This is not the place for a discussion of the properties of detergents or their many specialized uses in membrane biochemistry; several reviews treat these topics in depth.[1-7] We believe that the initial trials presented here for solubilizing functional membrane proteins are simple and that, although guarantees cannot be given, the outlined approach will be successful in many cases.

Choice of Criteria for Solubility

Before attempting solubilization, it is necessary to consider criteria that can be used for distinguishing between soluble and insoluble membrane proteins after treatment with detergents. Such criteria are by their nature operational and refer to a defined set of conditions that must be controlled throughout. The major criterion is that of retention of function after centrifugation for 1 hr at 105,000 *g*. This process depends both on the density and the temperature of the medium. Thus, a medium containing 50% glycerol (d = 1.11 at 20°) may render some species soluble by reason of nonsedimentability, whereas the same species may sediment in media of lower density. Since density is also a function of temperature, this variable should be kept constant. Another criterion for solubility involves partitioning of species between the void volume and the included volume

[1] A. Helenius and K. Simons, *Biochim. Biophys. Acta* **415,** 29 (1975).
[2] C. Tanford, "The Hydrophobic Effect," 2nd Ed. Wiley, New York, 1980.
[3] C. Tanford and J. A. Reynolds, *Biochim. Biophys. Acta* **457,** 133 (1976).
[4] A. Helenius, D. R. McCaslin, E. Fries, and C. Tanford, this series, Vol. 56, p. 734.
[5] J. C. H. Steele, Jr., C. Tanford, and J. A. Reynolds, this series, Vol. 48, p. 11.
[6] L. M. Hjelmeland and A. Chrambach, *Electrophoresis* **2,** 1 (1981).
[7] A. Tzagoloff and H. S. Penefsky, this series, Vol. 22, p. 219.

METHODS IN ENZYMOLOGY, VOL. 104 ISBN 0-12-182004-1

of a gel filtration column. Again, the choice of gel filtration medium has an obvious effect on how such proteins partition. The usual choice for such a medium is one of the cross-linked agarose preparations, such as Sepharose 6B (Pharmacia). Operationally, proteins that elute in the void volume are considered to be insoluble, whereas proteins with larger elution volumes are considered to be soluble. If uncertainty exists concerning the assignment of a peak to the void volume or the included volume, a different gel filtration medium can be used for clarifying the point. In general, it is of less importance whether one or the other criterion for solubility is adopted than it is to decide on one of them, even though such choice is arbitrary, and to apply the selected criterion consistently.

Assay of Soluble Activity

The other major consideration before beginning a solubilization trial is the choice of a suitable assay for activity. The assay for solubilized activity may be identical to that carried out with a particulate preparation, the probable case for enzyme activity, but binding assays require additional consideration. Since many assays with particulate protein are carried out by centrifugation, as for the separation of free ligand from bound ligand, a method must be devised for separating the two when both are soluble. The most convenient is the precipitation of membrane protein and bound ligand with polyethylene glycol[8] or ammonium sulfate. Alternatively, gel filtration may be used, especially when large differences in molecular weight exist between the ligand and membrane protein. A third protocol for binding consists of a filtration assay in which the protein–ligand complex is bound to glass-fiber filters that are coated with polyethyleneimine.[9] Since most proteins are negatively charged at physiological pH, they bind to a positively charged filter; basic or neutral ligands pass through. Filters and eluate can be assayed to determine bound and free ligand, respectively. Details concerning the assay of solubilized proteins are provided elsewhere.[10]

Choice of a Suitable Detergent

Although a great variety of detergents are commercially available, many possess similar chemical structures. A review on the physical prop-

[8] B. Desbuquois and G. D. Aurbach, *Biochem. J.* **126,** 717 (1972).

[9] R. F. Bruns, K. Lawson, and T. A. Pugsley, *Anal. Biochem.* **132,** 74 (1983).

[10] M. El-Rafai, *in* "Receptor Biochemistry and Methodology" (J. C. Venter and L. Harrison, eds.), Vol. 1. Liss, New York, 1983. In press.

erties of detergents gives an exhaustive list of equivalent trade names (Table II of Helenius *et al.*[4]). In particular, many trade names exist for the popular Triton X-100 and Lubrol PX, but there is little if any evidence for chemically or functionally distinct products. In our experience, a restricted list of eight detergents (Fig. 1 and the table) should allow solubilization of most functional membrane proteins. Except for digitonin all the detergents are homogeneous, and all but deoxy-BIGCHAP[11] are commercially available. Commercial preparations of digitonin contain only approximately 40% digitonin, the remainder being closely related steroidal components.[12] Unfortunately, digitonin cannot be recovered from that mixture by fractional crystallization and separates only by chromatography on cellulose. It cannot be assumed, however, that any specific choice from this list of eight will be successful. In general, digitonin, CHAPS, and octyl glucoside have enjoyed popularity with difficult cases. Instances exist wherein the lesser known members of this list were the only suitable choice for achieving solubilization and preserving function. Zwittergent 3-14, for instance, was the detergent among those in the table most capable of solubilizing 5′-nucleotidase,[13] whereas only deoxy-BIGCHAP was effective in solubilizing cysteine S-conjugate *N*-acetyltransferase from kidney in a form that could be fractionated by ion-exchange chromatography.[14]

Other factors, not directly related to solubilization, may affect the choice of detergent. If detergent must be removed, e.g., then the detergents with high values of the critical micelle concentration (CMC) and low micelle molecular weights that can be dialyzed are suggested.[6] If absorbance at 280 nm is an important parameter, Triton must be excluded. If separation techniques for exploiting predominantly or solely molecular charge differences (charge fractionation) are to be employed, e.g., ion-exchange chromatography or electrophoresis, charged detergents should be avoided. If divalent cations are essential for function of the species in question, sodium cholate with its carboxylic acid polar group should not be used, since it forms insoluble complexes with divalent metals. Finally, if precise physical data are to be obtained, a detergent with precisely evaluated physical parameters, especially the partial specific volume, must be used. The relevant physical properties of the eight detergents listed in Fig. 1 are summarized in the table. Examination of the size of the solubilized species also affects the choice of detergent, since it has been suggested that nonionic detergents, such as Triton X-100 or Lubrol PX,

[11] L. M. Hjelmeland, W. A. Klee, and J. C. Osborne, Jr., *Anal. Biochem.* **130,** 72 (1983).
[12] R. Tschesche and G. Wulff, *Tetrahedron* **19,** 621 (1963).
[13] E. M. Baiyles, J. P. Luzio, and A. C. Newby, *Biochem. Soc. Trans.* **9,** 140 (1981).
[14] M. W. Duffel and W. B. Jakoby, *Mol. Pharmacol.* **21,** 444 (1982).

Structural Formula	Chemical or Trade Name
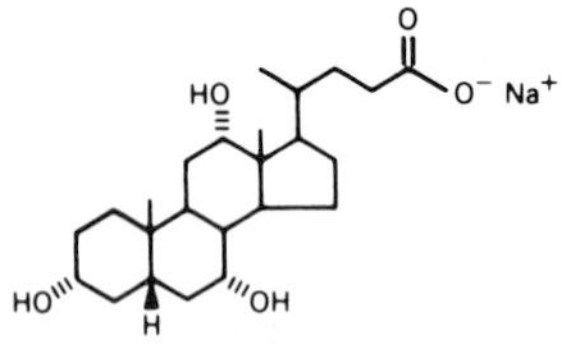	Sodium Cholate
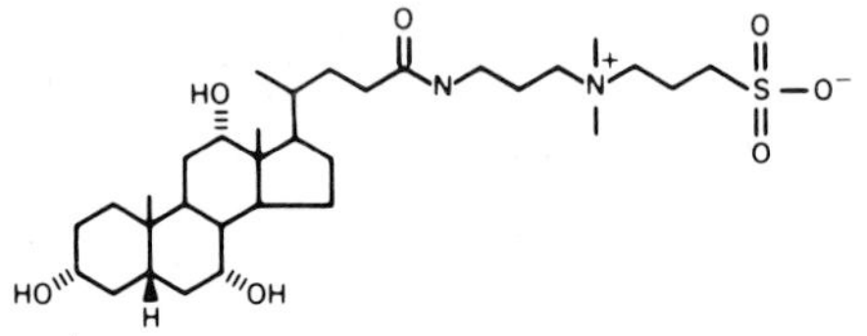	CHAPS
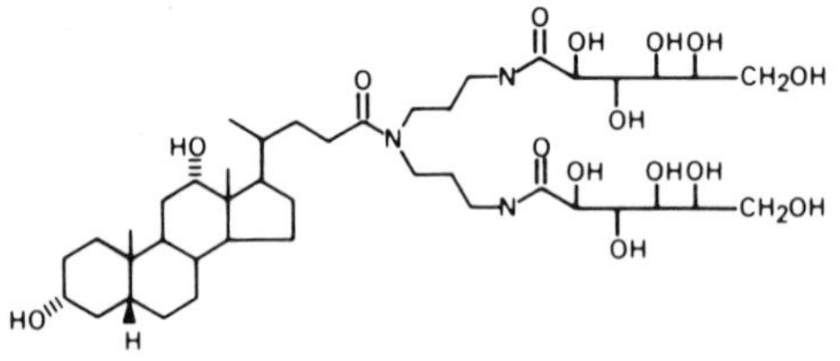	Deoxy-BiGCHAP
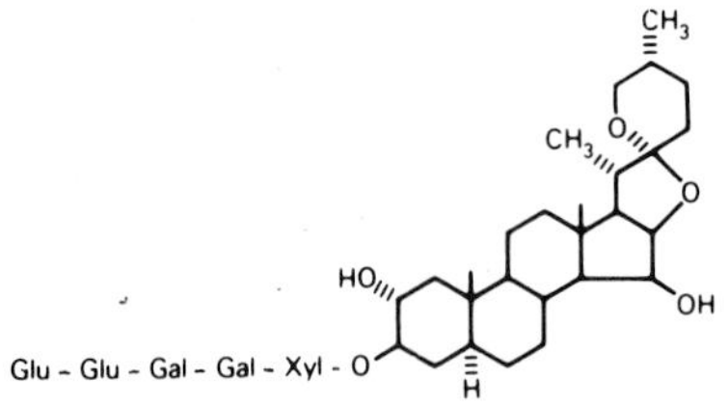 Glu - Glu - Gal - Gal - Xyl - O	Digitonin
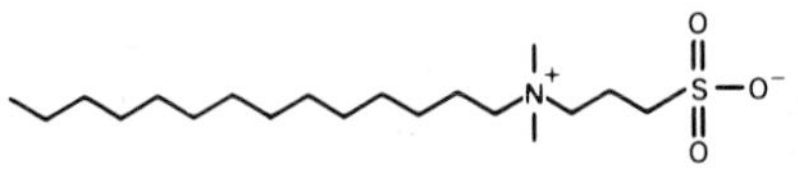	Zwittergent 3-14
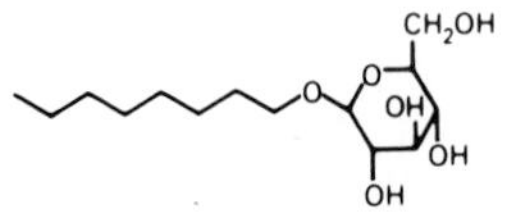	Octyl Glucoside
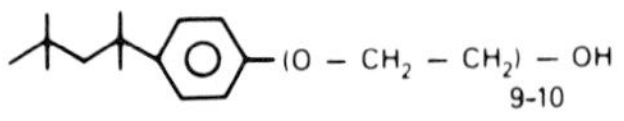 $(O-CH_2-CH_2)_{9-10}-OH$	Triton X-100
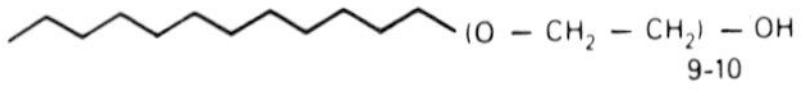 $(O-CH_2-CH_2)_{9-10}-OH$	Lubrol PX

FIG. 1. Structures and conventional names of detergents useful for the solubilization of membrane proteins.

PROPERTIES OF DETERGENTS

Property	Sodium cholate	CHAPS	BIGCHAP	Digitonin	Zwittergent 3-14	Octyl glucoside	Triton X-100	Lubrol PX
Monomer molecular weight	431	615	862	1229[a]	364	292	650[a]	582[a]
Micelle molecular weight	1700	6150	6900, 13,800	70,000	30,000	8,000	90,000	64,000
Critical micelle concentration								
% (w/v)	0.36	0.49	0.12	—	0.011	0.73	0.02	0.006
(m*M*)	8.0	1.4	1.4	—	0.3	25.0	0.3	0.1[b]
Dialyzability	+	+	+	−	−	+	−	−
Suitability for "charge fractionation"	−	+	+	+	+	+	+	+
Binds divalent cations	+	−	−	−	−	−	−	−
Significant A_{280}	−	−	±	−	−	−	+	−
Interference with protein assays	−	−	+		−	−	+	+

[a] Nominal.
[b] Estimate.

do not disaggregate proteins nearly as well as the bile salts and their derivatives (see below). Given those considerations, a promising start for the initial solubilization trial is to use only octyl glucoside and CHAPS.

Choice of Initial Conditions of Buffer and Temperature

Since interaction between membrane-bound macromolecules can be polar as well as nonpolar, the ionic strength of the solubilization medium is a critical consideration. In the absence of a specific requirement for low ionic strength, a high ionic strength should be used for the initial attempts at solubilization. Concentrations between 0.1 and 0.5 *M* KCl are suggested. The specific buffer that is used may also have an important effect. To assure adequate buffer capacity, the concentration of buffer should be at least 25 m*M* and the pH should be close to the p*K*. Finally, specific considerations exist for the choice of specific buffer ions. Borate, for example, is not suggested for use with glycoproteins, or with any system requiring nucleotides, owing to its interactions with the cis-hydroxyl groups of sugars. On the other hand, substitution of phosphate for KCl at concentrations of 0.1–0.2 *M* is often capable of solubilizing many proteins with detergent that cannot be brought into solution with the latter salt.[15] The origins of this peculiar effect of strong phosphate buffers are not well understood although modifications of the structure of water, as well as interaction with divalent cations, have been suggested in explanation. Consideration of the experimental protocols that will follow the solubilization step will influence the choice of buffers. For example, if ion-exchange chromatography is to follow, a buffer should be chosen that does not bind to the ion exchanger. Basic buffers are suitable for anion-exchange and acidic buffers for cation-exchange chromatography. Buffers with low ionic mobility should be chosen when electrophoresis is to follow solubilization; this choice reduces net conductance and thereby reduces Joule heating and increases field strength.

In the absence of specific reasons to the contrary, initial trials should be at 4°. Several "stabilizing" agents may be used in an attempt at preserving the activity of proteins that are unstable under normal storage conditions. This group of compounds will be discussed here, but it is suggested that their use be avoided in the initial trials unless specific reasons exist for their inclusion. Experiments designed to evaluate the need for stabilizing additives correctly belong in the later stages, at which time other parameters are optimized.

[15] A. C. Dey, R. Sheilagh, R. L. Rimsay, and I. R. Senciall, *Anal. Biochem.* **110,** 373 (1981).

The Initial Solubilization Trial

The initial solubilization experiment serves to survey conditions that lead to the maintenance of function of the desired protein. Particulate protein preparations, obtained as either crude membrane fractions or whole cells, are suspended in 50 m*M* buffer containing 0.15 *M* KCl, at a protein concentration of 10 mg/ml and 4°. Detergent stock solutions are made in the same buffer at a concentration of 10% (w/v), except for digitonin, which must be prepared at 4% (w/v) owing to limited solubility. Detergent stock solutions, suspended protein, and buffer are mixed to a final concentration of 5 mg of protein per milliliter, and a series of detergent concentrations including 0.01, 0.03, 0.1, 0.3, 1.0, and 3.0% (w/v) are prepared for each detergent to be tested. If only a restricted number of detergents are to be evaluated initially, it is suggested that octyl glucoside and CHAPS be considered first. As pointed out, however, it will be found useful to examine all the detergents enumerated in Fig. 1.

The individual aliquots are stirred gently with a magnetic stirrer for 1 hr at 4°. It is important to avoid foaming and sonication, both of which lead to protein denaturation. The individual preparations are centrifuged at 105,000 *g* for 1 hr at 4°. The supernatant liquid from each sample is removed from the corresponding pellet, after which the pellet is resuspended in an equal volume of buffer with the identical detergent concentration.

Assays for function and protein concentration are performed both on the supernatant liquid and on the resuspended pellet for each centrifuged sample. Care should be taken to resuspend each pellet completely; a hand-held homogenizer or similar device may be helpful.

The results of this trial should be plotted as percentage of the particulate activity on the ordinate, against detergent concentration on the abscissa, for both the solubilized supernatant liquid and the resuspended pellet. It is also useful to plot the sum of these two curves on the same graph.

Examination of the summed activities directly reveals whether the use of any detergent leads to a progressive loss of the total activity as concentration is increased. The detergent or detergents that yield the highest activity in the supernatant liquids as well as the highest total activity, should be chosen for further work.

Four general cases can be distinguished. In the first, detergents lead to progressive solubilization of activity that is stable at high concentrations of detergent. Figure 2 illustrates this for the solubilization of immunoglobulin E (IgE) receptor with several detergents.[16] In the second case,

[16] B. Rivnay and H. Metzger, *J. Biol. Chem.* **257,** 12800 (1982).

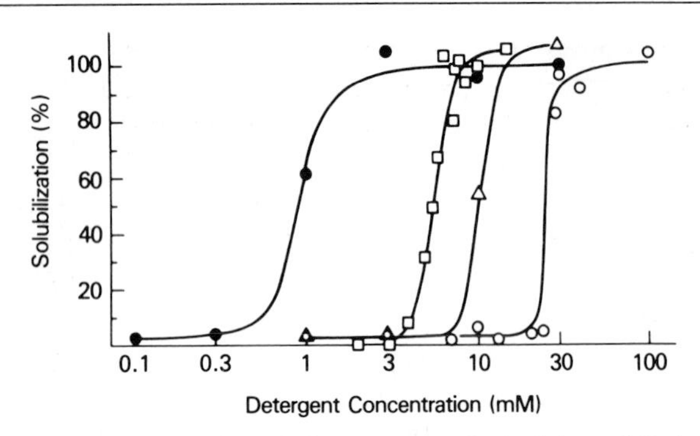

FIG. 2. Solubilization of immunoglobulin E-receptor from rat basophilic leukemia cells as a function of the concentration of several detergents.[16] ●——●, Triton X-100; □——□, CHAPS; △——△, sodium cholate; ○——○, octyl glucoside.

soluble activity is seen to increase and then to decrease as detergent concentration is raised, giving rise to an optimum value of activity at some detergent concentration. This has been shown for the solubilization of opiate receptor by CHAPS[17]; the peak of soluble activity in this case is presumably the result of progressive solubilization and inactivation at higher detergent concentrations. In case 3, most or all of the activity remains in the pellet.

If all eight detergents have proved to be unsuccessful, a number of *mixtures* of detergents should be considered next. Specifically, one of the four steroidal detergents might be mixed with either Triton X-100 or Lubrol PX on a 1 : 1 (w/w) basis. Such an approach has been useful for solubilizing active β-adrenergic receptors (digitonin and Triton X-100)[18] and cytochrome *P*-450 (cholate or CHAPS and Triton N-101).[19]

In case 4, all activity is lost upon treatment with detergent. In this instance, evaluation of the buffer and the protein : detergent ratio is indicated as detailed in the two sections below.

Stabilization

If either very low activity or no activity at all is observed with any of the eight detergents, systematic changes of the buffer with additives known to stabilize solubilized proteins should be considered. Four com-

[17] W. F. Simonds, G. Koski, R. A. Streaty, L. M. Hjelmeland, and W. A. Klee, *Proc. Natl. Acad. Sci. U.S.A.* **77,** 4623 (1980).

[18] W. L. Strauss, G. Gahi, C. M. Frazer, and J. C. Venter, *Arch. Biochem. Biophys.* **196,** 566 (1979).

[19] M. Warner, M. Vella La Marca, and A. H. Neims, *Drug Metab. Dispos.* **6,** 353 (1978).

mon classes of compounds that perform this function are polyols, reducing agents, chelating agents, and protease inhibitors. The following should be considered: 25 and 50% glycerol (v/v); 1 m*M* dithiothreitol or 5 m*M* mercaptoethanol; 1 m*M* EDTA; and phenylmethylsulfonyl fluoride (PMSF) at 75 μg/ml, leupeptin at 20 μg/ml, and pepstatin (at acidic pH) at 20 μg/ml. Cytochrome *P*-450, for example, has an absolute requirement for 20% glycerol for solubilization and is further protected by both dithiothreitol and EDTA.

If no combination of detergents and protective reagents produces any soluble activity, the problem will be a very difficult one. On the other hand, if there is substantial recovery, albeit at low yields, the same measures that are outlined here will be useful in improving recovery in the solubilization procedure.

Optimization of Protein-to-Detergent Ratio

The initial solubilization experiment is directed at finding a detergent and buffer system that will serve as the basis for refinement. Many of the proteins that have been solubilized appear to have well-defined detergent-to-protein ratios for optimum solubilization, which the initial experiment is unlikely to determine. The detergent-to-protein ratio, therefore, may be a critical parameter for successful solubilization, although little attention has been given to routine examination of this parameter. Since solubilization is primarily achieved by dispersion of phospholipids, and since a stoichiometric relationship appears to exist between phospholipids and detergents in the formation of mixed micelles and solubilized complexes, a single detergent concentration without knowledge of the protein concentration would not be expected to provide an optimum yield. Obviously, the value that should be presented is the optimum detergent-to-*lipid* ratio; since the concentration relationship of protein and lipid in a given membrane is relatively constant, and protein concentration is much easier to measure, detergent-to-protein ratio is a reasonable substitute.

The optimum detergent-to-protein ratio may be experimentally evaluated by solubilization at several different detergent concentrations for three different protein concentrations. The detergent concentrations should span the initial trial carried out at a protein concentration of 5 mg/ml. In addition, experiments at higher concentrations of protein, possibly 7.5 and 10 mg/ml, should be performed with detergent concentrations that correspond in multiples of 1.5 and 2 times those used for the experiment with protein at 5 mg/ml. It seems to be less useful to work at concentrations of 1 and 3 mg of protein per milliliter, since detergents with a high CMC may not be present at solubilizing concentrations for a given deter-

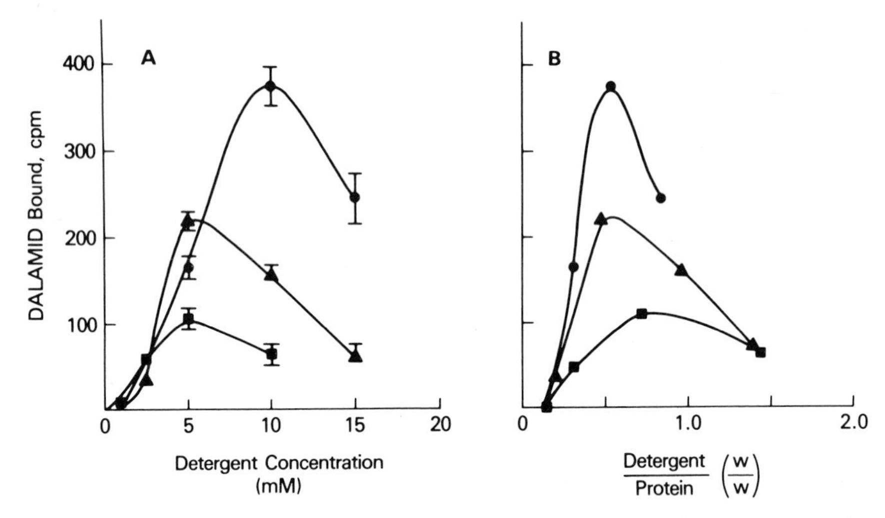

FIG. 3. Solubilization of [^{3}H]DALAMID binding activity from NG108 membranes by CHAPS at three different initial protein concentrations[20] as (A) a function of detergent concentration, and (B) a function of the ratio of detergent to protein. Protein concentrations (mg/ml) are: 11.1 (●), 6.5 (▲), and 4.3 (■).

gent-to-protein ratio when the concentration of protein is low. This observation points to the fact that protein solubilization occurs at or near the CMC for most detergents. Therefore, optimum detergent-to-protein ratios for low concentrations of protein are expected to be slightly higher than those for relatively high concentrations of protein.

Two plots of the results should be prepared. The first is a solubilization curve for three different values of initial protein concentrations. The ordinate would be solubilized activity, and the abscissa, detergent concentration. The second plot uses an abscissa showing the ratio of detergent to protein (w/w). One of two types of curves can be expected in the first plot. Either the percentage of particulate activity will increase and plateau at a given concentration for each protein concentration, or the soluble activity will rise and then fall to yield an optimum detergent concentration for each protein concentration. The second type of plot will reflect these two cases as well. If soluble activity rises and then plateaus, the selection of an appropriate detergent concentration can be based on considerations of the size of the solubilized species (see below). If, on the other hand, activity rises and falls as a function of detergent concentration, the plot of detergent-to-protein ratio will provide an optimum value of this parameter. That value will in turn allow the choice of a suitable detergent concentration for further study whenever protein concentration

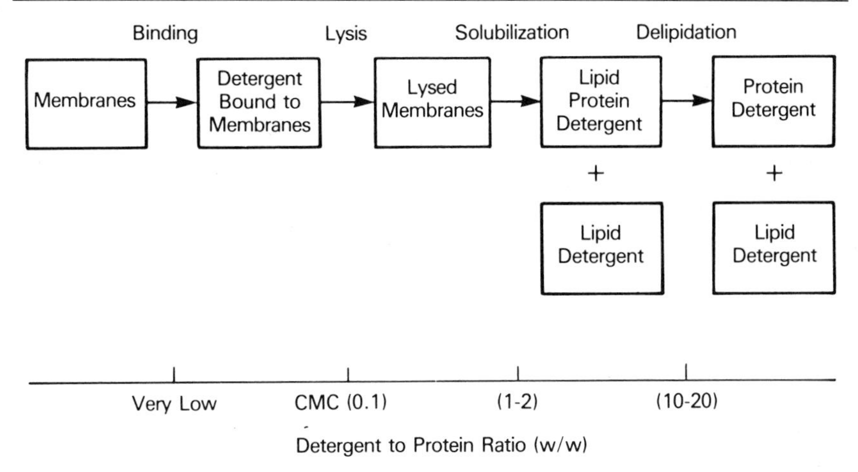

FIG. 4. Progressive solubilization of whole membranes to form protein–detergent complexes as a function of the ratio of detergent to protein. Adapted from Helenius and Simons.[1]

is known. A representative example[20] of the second case is given in Fig. 3. Panel A shows activity-detergent concentration profiles for three different protein concentrations. In panel B the same data are plotted as a function of the detergent-to-protein ratio. At the two higher protein concentrations, a reasonable optimum of the detergent-to-protein ratio is observed at approximately 0.4. At the lower initial protein concentration this value is shifted to 0.6. Presumably, this discrepancy is due to a lowering of the CHAPS concentration below the value needed for solubilization (approximately 0.15%), when at a protein concentration of 4.3 mg/ml the protein-detergent ratio is 0.4. As noted, low protein concentrations may lead to slightly higher values for detergent-to-protein ratios when detergents with high values of the CMC are used. Since all these parameters are interrelated, it is obvious that buffer composition, pH, and temperature could be optimized again at this stage. However, such repetition is probably unnecessary.

Physical Characterization of the Solubilized Species

Once solubilization has been achieved, the further effects of change in detergent concentration on the size and composition of solubilized complexes should be examined. A schematic representation of the changes in size and state of association of the species to be solubilized as a function of detergent-to-protein ratio is presented in Fig. 4. At extremely low

[20] W. A. Klee and L. M. Hjelmeland, unpublished data.

concentrations of detergent, monomers partition into the membrane without gross alterations in membrane structure. As the concentration of detergent increases, the structure of the membrane is grossly changed, leading to lysis. Finally, at detergent-to-protein ratios of about 1 : 1, the production of slowly sedimenting complexes occurs, which we define as soluble. At this point, the species being generated are usually large, heterogeneous complexes of lipids, detergent, and protein, with molecular weights on the order of 0.5 to 1 million. An increase in the detergent-to-protein ratio from 10 : 1 to 20 : 1 leads to the formation of protein–detergent complexes that are free of lipid, and to mixed micelles of lipid and detergent. Residual interactions between proteins, including those that are artifactual, may not be dissociated at this point, nor may they be capable of being dissociated by detergents that preserve biological function.

Electrostatic, and possibly hydrogen bonding, interactions also play an important role in protein interactions and must be recognized and dealt with by reagents directed at these forces. Equally, the choice of detergent may qualitatively affect the degree of protein interaction. It is now widely accepted, for example, that bile salts and their derivatives are much more efficient than the nonionic detergents in dissociating protein complexes. Within the class of bile salts and their derivatives, more subtle effects can be obtained by changing polar groups. With cholic acid as detergent, for example, the molecular weight of cytochrome *P*-450 is of the order of 500,000, whereas, with the zwitterionic cholate derivative CHAPS, the weight of this complex is only about 100,000.[21] In this respect, the combining of different detergents also appears to be useful in disaggregating complexes. Specifically, the mixing of steroidal detergents and nonionics, as has been noted, may lead to reduction in the total size of protein–detergent complexes, an effect observed for adenylate cyclase[22] and cytochrome *P*-450.[19]

Perhaps the simplest means of determining an approximate molecular weight is by gel filtration.[23] Modern gel filtration media are stable to each of the detergents listed in the table and the fundamental protocol for carrying out such a study for solubilized proteins is the same as that for a soluble protein. It is, however, important to realize that the molecular weight estimate yielded by this method includes all bound lipid and detergent. Since many well-characterized detergent protein complexes contain up to 50% of their total weight as detergent, this is a significant contribu-

[21] L. M. Hjelmeland, D. W. Nebert, and J. C. Osborne, Jr., *Anal. Biochem.* **130,** 72 (1983).
[22] A. C. Newby and A. Chrambach, *Biochem. J.* **177,** 623 (1978).
[23] C. Tanford, Y. Nozaki, J. A. Reynolds, and S. Makino, *Biochemistry* **13,** 2369 (1974).

tion. The matter is often further confused by calibration of the gel filtration column with soluble proteins in the presence of detergents. Many soluble proteins may bind very small amounts or no detergent at all; i.e., their molecular weight is unaffected by the presence of detergents. More fundamental problems exist as well. The axial ratios and partial specific volumes of membrane detergent complexes are usually quite different from those of soluble proteins, and assumptions based on the equality of such parameters between standards and unknowns are prone to systematic errors. The magnitude of the errors is unknown. It would seem appropriate that relative values produced by different detergents be assessed in order to determine a detergent that yields a functional species of minimum size.

The other general method for measuring the size of detergent–protein complexes is gel electrophoresis.[6] Although more laborious in some respects, several advantages apply to it. Several conditions can be examined at the same time in several different gel tubes and, since the physical dimension of such gels is small, a marked economy of sample is achieved. The resolving power of gel electrophoresis is also somewhat higher than that obtained with standard gel filtration columns. In principle, such studies are carried out by examining the relative mobility or the desired activity in gels of several different total concentrations of acrylamide. The results are then plotted as the logarithm of the R_f versus the percentage gel concentration, the familiar Ferguson plot.[24]

Comparison of Particulate and Solubilized Function

The final task in assessing the success of solubilization is to examine the function of the solubilized species relative to its particulate origin. Standard methods for evaluation of kinetic parameters should be used if the activity in question is catalytic, whereas binding analyses such as the Scatchard plot would evaluate the function of receptors. The need for such evaluation is emphasized by solubilization of prolactin receptors by Triton X-100, which results in changes of both the binding capacity and the affinity of the ligand.[25]

The use of detergents is associated with a number of special problems. It is clear that the binding of ligand to macromolecules, whether they are enzymes or receptors, can be affected by detergents. Detergents may bind directly at the active site, thereby serving as a competitive ligand, or free detergent may trap ligand and thereby appear as a high-capacity, low-affinity site. A final consideration is the effect of detergent on the macro-

[24] A. C. Newby, M. Rodbell, and A. Chrambach, *Arch. Biochem. Biophys.* **190,** 109 (1978).
[25] R. P. Shiu and H. G. Friesen, *J. Biol. Chem.* **249,** 7902 (1974).

molecule itself, either in dissociating subunits or in changing the conformation of the macromolecule.[26]

Comments

The solubilization of membrane proteins with maintenance of function is frequently, but not always, feasible with presently available reagents and relatively simple procedures. An outline of procedural steps aimed at designing an appropriate program for solubilization has been presented.

[26] D. S. Liscia, T. Alhadi, and B. K. Vonderhaar, *J. Biol. Chem.* **257**, 9401 (1982).

[17] Separating Detergent from Proteins

By ANNA J. FURTH, HILARY BOLTON,
JENNIFER POTTER, and JOHN D. PRIDDLE

The Need for Detergent Separation. Separating unbound detergent from hydrophobic protein may become necessary at three stages in the purification of a detergent-solubilized protein. In the initial extraction from membrane or other lipoprotein particle, the protein is integrated into a detergent micelle; here its hydrophobic surfaces, previously in contact with lipid or other hydrophobic protein, become occupied by detergent. This step needs detergent at high concentration, both to maximize solubilization and to reduce the danger of micelle sharing by unrelated protein.[1,2] Excess detergent, together with unwanted phospholipid, can then be removed in the first detergent-separation step.

The second step arises because detergents used for the initial extraction are often bulky molecules with low critical micelle concentration (CMC) values and high aggregation numbers (see the table). This means that they tend to be less suitable for later stages of the project. Low CMC values, for instance, are undesirable when the micellar protein solution is concentrated (as for enzymic or immunological work), whereas high aggregation numbers give large micelles, where detergent rather than protein is the major component and may dominate the properties of the micelle. Therefore a second detergent-separation step is often needed, so that protein can be transferred from the extraction detergent to one forming smaller micelles.

[1] A. Helenius, D. R. McCaslin, E. Fries, and C. Tanford, this series, Vol. 56, p. 734.
[2] C. Tanford and J. A. Reynolds, *Biochim. Biophys. Acta* **457**, 133 (1976).

ISBN 0-12-182004-1

Detergents with high CMC values are also preferable when working with liposomes in which purified protein is inserted into a closed phospholipid bilayer (vesicle) of known composition. In such "reconstitution" procedures, removal of detergent by dialyis or dilution is often the critical step (see also this volume [19]). Because of kinetic factors, rapid removal, and hence a high CMC value, may be crucial for determining vesicle size and stability. This subject has been reviewed elsewhere,[3] and here we consider it only as a third detergent-separation step, in which protein-bound detergent is first exchanged for phospholipid and then separated from vesicle-incorporated protein.

Purification Strategy for a Hydrophobic Protein, Incorporating Detergent-Separation Steps. The overall procedure, taking hydrophobic protein from crude extract to small micelle and possibly on to liposome, is exemplified here mainly from work in this laboratory on lactase, a membrane-bound disaccharase from the intestinal brush border. The first detergent-separation step is carried out on the crude membrane extract and involves gel chromatography in buffer of high detergent concentration (1% Triton X-100 or Emulphogen BC-720) to remove unbound detergent and phospholipid. Lactase-active fractions may be further purified at this stage, e.g., by ion-exchange chromatography. Since this does not involve detergent separation, we move here directly to the second detergent-separation step, where micellar lactase is transferred from Triton to deoxycholate. The exchange involves a second gel filtration of the partially purified membrane extract, this time in 2% deoxycholate. In this form the protein is readily concentrated, and its properties can be investigated as a concentrated micellar solution or, following a third detergent-separation step, as a vesicle-incorporated protein.

A problem not discussed here is that of removing the final traces of tightly bound detergent from a purified protein. Given that undesirable detergents from the extraction step can be removed by exchange with other detergents, and that fully detergent-free hydrophobic proteins tend to be insoluble—or at best, highly aggregated—it may be preferable to retain a belt of solubilizing monomeric detergent for physicochemical studies or, alternatively, to replace detergent with phospholipid for reconstitution studies; either procedure circumvents the problem of total removal of detergent.

Choice of Detergent for Pilot Experiments on Detergent Separation. Nonionic detergents such as Triton X-100 and Emulphogen BC-720 are often used for the initial extraction. Where the removal of such detergents needs to be monitored—as in the pilot experiments described here—we

[3] E. Racker, this series, Vol. 55, p. 699.

use Triton X-100 with its high UV absorbance ($A^{1\%}_{280} = 22$). But for routine preparations with established procedures, we use Emulphogen BC-720, a nonabsorbing detergent with similar hydrodynamic properties (see the table). This permits protein concentration to be estimated from absorbance at 280 nm. Emulphogen absorbs very little above 235 nm and has an $A^{1\%}_{230}$ value of 0.30.

I. Micellar Dimensions

Stokes Radius. Protein-free and protein-containing micelles can be separated from one another, from detergent monomers, or from phospholipid vesicles by methods based on differences in charge, hydrophobicity, density, or micellar dimensions.[4] Here we shall concentrate on gel filtration, which has the advantage of high protein recoveries and the possibility of monitoring separation clearly. Ultrafiltration is also worthy of note. Both methods exploit differences in *effective* micellar dimensions; these take into account the shape and hydration of the particle, as well as the summed molecular weights of its components. Effective particle radius or Stokes radius (a) is related to diffusion coefficient (D) by the equation $D = kT/6\pi\eta a$ (where k is Boltzmann constant, T absolute temperature, and η viscosity of water). In practice, Stokes radius is usually determined indirectly, by comparing the hydrodynamic behavior of standard and unknown particles on gel filtration. Elution volume can be related graphically[5,6] to Stokes radius through calculation of the partition or *distribution coefficient* K_D (which approximates to K_{av}, where K_{av} is $v_{sample} - v_{void}/v_{total} - v_{void}$).

The important point is that Stokes radius is an empirical value, influenced by molecular mass, hydration, and shape and measured from behavior *under specified conditions* of particles of unknown shape and hydration. Where the particle is a water-soluble protein, small changes in ionic strength or the presence of other molecules have comparatively little effect on Stokes radius and hence on gel filtration behavior. With detergent-solubilized proteins the situation is different; we have a fluid aggregate of protein and detergent in which particle dimensions may be strongly influenced by small environmental changes. Therefore to separate protein-free and protein-containing micelles successfully and reproducibly, conditions must be carefully standardized.

Size of Pure Detergent Micelles. The table gives the molecular properties of a range of detergents, using the traditional classification into *ionic*

[4] A. J. Furth, *Anal. Biochem.* **109,** 207 (1980).

[5] S. Clarke, *J. Biol. Chem.* **250,** 5459 (1975).

[6] D. Snary, P. Goodfellow, W. F. Bodmer, and M. J. Crumpton, *Nature (London)* **258,** 240 (1975).

PROPERTIES OF COMMONLY USED DETERGENTS

Common name	Description[a]	CMC (moles/liter)	Average M_r	Aggregation number	Micellar M_r	References
Ionic						
Sodium dodecyl sulfate	$CH_3(CH_2)_{11}OSO_3^-Na^+$	2.3×10^{-3}	288	84	24,200	b
Cetyltrimethylammonium bromide	$CH_3(CH_3)_3N^+(C_{16}H_{32})Br^-$	—	364	169	62,000	b
Ether deoxylysolecithin (C_{16})	—	10^{-5} to 10^{-6}	—	140	68,000	c
Sodium cholate[d]	—	3.3×10^{-3}	408	4.8	2100	c
Sodium deoxycholate[d]	—	0.91×10^{-3}	392	22	9100	c
Nonionic						
Triton X-100	*tert*-$C_8\phi E_{9.6}$	3×10^{-4}	628	140	90,000	b,c
Emulphogen BC-720[e]	$C_{12}E_8$	8.7×10^{-5}	—	120	65,000	e
Brij 35	$C_{12}E_{23}$	9.1×10^{-5}	—	40	49,000	c
Tween 80	$C_{18:1}$ sorbitan E_{20}	1.2×10^{-5}	1300	60	76,000	b,c
Dodecyl dimethylamine oxide	$C_{12}NMe_2O$	2.2×10^{-3}	229	75	17,000	c
Octyl-β-D-glucoside	—	2.5×10^{-2}	292	—	—	c

[a] *tert*-C_8 refers to a tertiary octyl group; $C_{18:1}$ indicates an 18-carbon fatty acid with one double bond; other formulas are shown in full in Helenius *et al.*[1] and Furth.[4]

[b] From Furth.[4]

[c] From Helenius *et al.*[1]

[d] At 25° in 0.05 *M* NaCl or KBr.

[e] From data for $C_{12}E_8$ in Table III of Helenius *et al.*[1]

and *nonionic*. In practice, this grouping may be misleading, since not all nonionic detergents have the low CMC values characteristic of this group, nor do ionic detergents invariably have low aggregation numbers. More important, the blanket term *ionic* fails to emphasize the difference between denaturing detergents (such as sodium dodecyl sulfate) and gentler detergents, like the bile salts, capable of solubilizing hydrophobic proteins in native form. It is these that feature in protein purification, as described here.[7]

Detergent properties affect micellar dimensions in several ways. The contribution of molecular mass can be expressed in terms of *aggregation number*. This is the number of monomer detergent molecules that can be accommodated within a single micelle, and it tends to be low for small polar molecules like the bile salts, and several orders of magnitude higher for bulky detergents like Triton (see the table). Aggregation number may rise sharply with ionic strength, e.g., from 2.2 to 22 for deoxycholate and from 2.8 to 4.8 for cholate, when salt concentration is increased from 10 m*M* to 150 m*M*.[8]

Data on the effect of solution variables on aggregation number are still scanty. High concentrations of counterions, for example, tend to reduce mutual repulsion between polar "heads," allowing monomers to close ranks and so reduce micellar dimensions; the molecular weight of a Triton micelle falls from 95,000 to 86,000 (as determined by sedimentation equilibrium) when Tris-HCl is replaced by a phosphate buffer of higher ionic strength.[9] Conversely, raising the temperature or adding phospholipid tends to increase Stokes radius,[8] e.g., from 41 Å in the pure detergent to 54 Å in mixed micelles, with detergent-to-phospholipid molar ratios of 3 : 1. Whatever the mechanism, it is clear that changes in pH, temperature, and ionic strength, and the presence of phospholipid impurities from the initial membrane extraction, may all affect the dimensions of detergent micelles, and hence their removal by gel filtration. For reproducible results, all these variables must be carefully defined.

The quickest way to separate unbound detergent from proteins is to dilute below the detergent CMC, so that protein-free micelles disperse into monomers, readily distinguishable from protein-detergent micelles by size. This requires information on CMC value, which is effectively the highest concentration of detergent monomer attainable. In practice it represents a narrow range of values rather than a single concentration and

[7] If bile salts also cause denaturation, as with deoxycholate,[1] a popular alternative has been diiodosalicylate.

[8] H. H. Paradies, *J. Phys. Chem.* **84,** 599 (1980).

[9] R. J. Robson and E. A. Dennis, *Biochim. Biophys. Acta* **508,** 513 (1978).

fluctuates widely with changes in ionic strength, pH, and temperature. Again, for reproducible results, conditions must be carefully defined.

It is immediately apparent from the table that detergents commonly used in the initial extraction may have CMC values so low that monomers are present in only very small amounts. To separate such detergents in monomer form requires the use of special micelle-dispersing tricks, to be described later.

Size of Protein–Detergent Micelles. To separate protein-free and protein-containing micelles by size, it would clearly be an advantage to predict the dimensions of protein-detergent micelles. These depend partly on the number of bound detergent molecules and partly on the aggregation state of the protein. Both parameters may change as protein purification proceeds. In the high detergent concentrations of the initial extraction, each protein molecule may be associated with a complete detergent micelle; at lower concentrations, only a few detergent molecules may remain, bound to discrete hydrophobic sites on the protein. Similarly, as lipid and other proteins are removed, the ratio of protein to detergent rises, again altering self-association and detergent-binding properties of the protein. This may account for some of the frustratingly irreproducible behavior of hydrophobic proteins[10] and makes it crucial to define not just detergent concentration, but the *ratio* of detergent to protein. Fortunately our chapter is restricted to the removal of protein-free micelles and does not concern those situations where small differences in protein micellar size are critical, as in the fractionation of mixtures of detergent-solubilized proteins.

II. Separating Protein and Excess Detergent by Gel Filtration of Crude Membrane Extracts

Strategy. This method applies particularly to our first detergent-separation step, the removal of excess detergent—together with phospholipid—after the initial extraction in high-detergent buffer. We exemplify it by the chromatography of Triton-solubilized brush border membranes on Sephacryl S-400, eluting with high-detergent buffer (1% Triton), and assaying fractions for the required enzyme (lactase). Detergent concentration must be kept high during chromatography, to prevent micelle-sharing of unrelated proteins in the crude extract.

Before starting, the likely separation between protein-free and protein-containing micelles may be prejudged from a preliminary calibration run, using pure detergent (2% Triton) as the starting sample and low-

[10] R. D. C. Mcnair and A. J. Kenny, *Biochem. J.* **179,** 379 (1979).

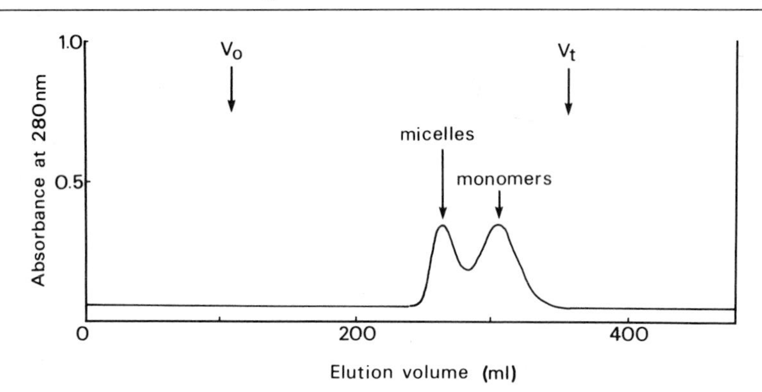

FIG. 1. Sephacryl S-400 chromatography of Triton X-100 micelles after partial dispersion into monomers by addition of deoxycholate. The starting sample contains 1% Triton and 2% deoxycholate; the elution buffer contains 0.02% Triton and 0.06% deoxycholate.

detergent buffer (0.02% Triton) for elution. Detergent concentration during elution must be kept around the CMC value to prevent trailing of the Triton micelle peak.

Method. A column of Sephacryl (approximately 2.6 by 64 cm) is eluted at 4° with constant flow rate (20 ml/hr), using an LKB Varioperpex II pump, and monitoring eluent at 278 nm. Column void volume (v_0) is determined from the elution position of high-molecular-weight Triton aggregates (prepared by incubating a 2% Triton sample at 37° for 5 min). The v_t value is determined from the elution position of sodium chloride, applied as a 0.2 *M* solution, and monitored with a conductivity meter. (With ε-DNP lysine, v_t values are anomalously high, suggesting adsorption.) The elution position of pure Triton micelles is determined from a calibration run, applying 2 ml of a 2% (w/v) solution of Triton X-100 in 10 m*M* Veronal buffer at pH 8.2. Elution buffer is 10 m*M* Veronal–0.02% Triton. Identical chromatographic arrangements are maintained for the micellar protein solution, applying 2 ml of a brush border membrane extract in 10 m*M* Veronal buffer–2% Triton X-100. Elution buffer is 10 m*M* Veronal with only 1% Triton. Elution position of protein-containing micelles of the required enzyme, here lactase, is determined by enzyme assay.[11]

Result. The elution position of pure Triton micelles is shown in Fig. 1. (the production of monomers is described in Section IV of this chapter). When the same column is used to remove excess detergent from the crude membrane extract, lactase–Triton micelles elute near the void volume, leaving a second, clearly separated UV-absorbing peak of mixed phos-

[11] M. W. Ho, S. Povey, and D. Swallow, *Am. J. Hum. Genet.* **34**, 650 (1982).

pholipid–Triton micelles. The elution position of this second peak encompasses that of pure Triton micelles, but is considerably broader, owing to the combined effect of phospholipid-induced swelling[9] and resin adsorption.[12]

In a similar procedure, Helenius and Simons[13] have used tritiated Triton to demonstrate removal of unbound detergent on Sepharose 6B. Clearly, the choice of gel filtration medium depends on protein molecular weight and aggregation state and on the detergent used. For example, monomeric detergent-solubilized lactase has an apparent molecular weight of 260,000 in Triton and of 160,000 in deoxycholate,[14] whereas phage coat protein (dimer molecular weight 10,000) is excluded from Sephadex G-200 when solubilized in Triton or Brij 96, but is well included when solubilized in sodium dodecyl sulfate.[15]

III. Transferring Proteins from Triton to Bile Salt Micelles, Using Gel Filtration

Micelle-Dispersal Techniques. Proteins are difficult to concentrate when solubilized in the bulky detergents often used for membrane extraction. Dialyzable monomers form only a small proportion of the detergent population, and the micelles, even when protein free, are too large to pass through small-pore dialysis membranes. Therefore a concentration technique such as ultrafiltration tends to produce only viscous gels, in which both protein and detergent have been simultaneously concentrated.

However, by adding such highly polar molecules as ethanediol[16] or bile salt,[17] large micelles of Triton and similar detergent may be rapidly dispersed into monomers. Figure 1 shows that gel filtration may be used to demonstrate the simultaneous presence of micelles and monomers in a partially dispersed Triton solution. Dispersal is brought about by mixing 1% Triton with 2% deoxycholate and eluting with buffer containing 0.02% Triton and 0.06% deoxycholate. Complete dispersal of micelles to monomers can be achieved by raising the ratio of bile salt to Triton, e.g., by eluting in 2% deoxycholate.

Detergent Exchange within a Protein-Containing Micelle. The same micelle-dispersing techniques can be used with protein-containing Triton micelles, giving two desirable results. First, Triton is dispersed into

[12] C. Huang, *Biochemistry* **8,** 344 (1969).

[13] A. Helenius and K. Simons, *J. Biol. Chem.* **247,** 3656 (1972).

[14] H. Bolton, A. J. Furth, M. W. Ho, and J. Potter, *Biochem. Soc. Trans.* (in press).

[15] See Fig. 2 in S. Makino, J. L. Woolford, C. Tanford, and R. E. Webster, *J. Biol. Chem.* **250,** 4327 (1975).

[16] C. E. Frasch, *Dialog,* February 1–3. Amicon, Lexington, Massachusetts.

[17] H. Bolton and A. J. Furth, unpublished results, 1982.

monomers, and can readily be removed by gel filtration or dialysis. At the same time, deoxycholate replaces Triton at hydrophobic binding sites on the protein surface, leaving protein as the major component in a small mixed micelle. For bile salt detergents of low aggregation number, such micelles contain only a few detergent molecules and allow the protein to remain soluble yet with minimum detergent presence to interfere with physicochemical or enzymic characterization. Finally, this micelle-dispersal method has the great advantage of producing a hydrophobic protein solution amendable to concentration by ultrafiltration.

The method can be demonstrated with lactase using a partially purified solution in 1% Triton, previously freed of unbound Triton chromatographically, as described in Section II. When lactase-active fractions from this first chromatography are applied to the same column, but eluted in 2% deoxycholate, the micellar M_r drops from 260,000 to 160,000. Within the resolving power of the column, this is consistent with a lactase monomer (M_r 160,000[18]) transferring from a Triton micelle to one of deoxycholate. At the same time, displaced Triton monomers can be seen eluting near the v_t position.[19]

Method. A lactase–Triton micellar solution in 10 m*M* Veronal at pH 8.2 and 1% Triton is chromatographed as described under Method in Section II.[11]

It is not necessary to add deoxycholate directly to the starting sample provided the elution buffer is high in deoxycholate (2% deoxycholate in 10 m*M* Veronal at pH 8.2). To determine the shift in micellar lactase M_r, the column may be calibrated in detergent-free buffer, using standard proteins.

IV. Separation of Phospholipid Vesicles and Detergent Micelles by Gel Filtration

Transferring Protein from Micelle to Vesicle. Mixed micelles of protein in bile salt detergent, prepared as above, are an excellent starting point for incorporating protein into phospholipid vesicles by Racker's cholate dilution method.[20] Mixed phospholipid–protein–detergent micelles are formed by incubating together concentrated solutions of detergent-solubilized phospholipid and detergent-solubilized protein. The detergent concentration is then rapidly reduced by dilution, allowing

[18] H. Skovberg, H. Sjöstrom, and O. Noren, *Eur. J. Biochem.* **114,** 653 (1981).

[19] Compared to Fig. 1 there is an increase in Triton monomer elution volume. This is observed frequently after a column has been used for several consecutive runs, and it appears to be caused by adsorption to Sephacryl. The trend could be reversed and the original elution position reproduced by unpacking the column and suspending the resin overnight in a large volume of detergent-free buffer before repacking.

[20] E. Racker, T. F. Chien, and A. Kandrach, *FEBS Lett.* **57,** 14 (1975).

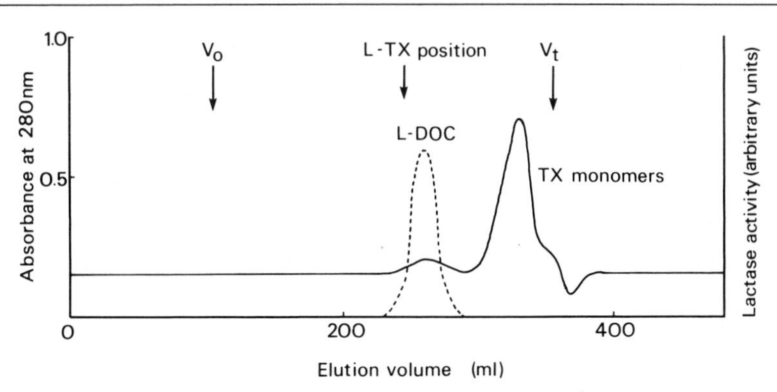

FIG. 2. Use of Sephacryl S-400 chromatography to transfer micellar lactase from Triton to deoxycholate. Elution buffer is 2% deoxycholate in 10 mM Veronal buffer, pH 8.2. The starting sample is a partially purified membrane extract, freed of unbound detergent by a previous run on the same column (eluted in 1% Triton–10 mM Veronal). The elution position of lactase–Triton micelles in a previous run is indicated by the arrow; L-DOC indicates the elution position of lactase-DOC micelles, and TX indicates that of displaced Triton monomers, both in the present run. Solid line represents absorbance at 278 nm, and dashed line represents lactase activity.

spontaneous formation of phospholipid vesicles into which protein is inserted via the hydrophobic domain. To separate detergent from vesicles and at the same time estimate Stokes radius of the vesicles, the dilution step may be followed immediately by gel filtration. On Sepharose CL-4B, for example, vesicle-incorporated lactase elutes near the void volume ($K_D = 0.02$), leaving traces of micellar lactase well included ($K_D = 0.3$).[21]

Method for Vesicle Formation.[21] A sample (40 μl) of phosphatidylcholine (type V from egg white, supplied by Sigma as a 1 mg/10 μl solution in chloroform–methanol) is pipetted into a 5-ml vial, dried in a stream of nitrogen, and twice washed by dissolving in 0.5 ml of diethyl ether. Then 0.2 ml of deoxycholate (25 mg/ml in 10 mM Tris-HCl at pH 8.5) is added. The mixture is stirred with a Vortex mixer to yield a clear yellow solution. After standing at 20° for 60 min, the solution is rapidly diluted to 4 ml with detergent-free buffer. Where dilution is to be followed by gel filtration, this buffer should be of high ionic strength, e.g., 10 mM Tris-HCl at pH 8.5 containing 0.5 M NaCl. When protein is to be incorporated, the 0.2 ml of deoxycholate solution is an aliquot of protein-containing eluate, from the detergent-exchange chromatographic step described in Section III.

V. Removing Detergent by Ultrafiltration

Detergents with high CMC value may be separated from protein-containing micelles by simple dialysis. The main requirement is that a signifi-

[21] A. J. Furth and J. D. Priddle, unpublished results, 1980.

cant proportion of detergent be present as monomer rather than micelle. However, this proportion may be artificially boosted, using either dilution or micelle dispersers. Dilution involves reducing the concentration to well below the CMC, by adding detergent-free buffer, then rapidly concentrating on ultrafiltration, using a membrane of low cutoff point. Alternatively, a suitable micelle-dispersing agent is ethylene glycol. At a concentration of 30%, ethylene glycol allows 48% of a 0.5% Triton solution to be removed through a Diaflo PM-10 membrane, using an Amicon Model 52 stirred ultrafiltration cell.[16] Lower concentrations (25%) have been used to dispel 1% Triton and thereby separate unbound detergent from peptidase-containing micelles.[22]

Given the range of ultrafiltration membranes now available, it should also be possible to separate detergents of low aggregation number as micelles rather than as monomers. However, repeated exposure to concentrated (1%) detergent solutions may cause damage to membranes. Furthermore, most detergent micelles are ellipsoidal, so that surprisingly large micellar proteins may pass through small pores "head on."

VI. Comments

Because the protein–detergent micelle is a fluid, noncovalently linked aggregate, its size and shape are much more susceptible to environmental changes than those of a simple, water-soluble protein. Therefore, to achieve reproducible results with detergent-solubilized proteins, it is imperative to standardize such parameters as ionic strength, detergent concentration, and protein concentration.

Protein detergent micelles may be separated from unbound detergent by the methods described here. Also suggested is a protocol that enables membrane proteins to be extracted (without the problem of micelle sharing) into bulky detergent micelles, then to be transferred to small micelles (with the aid of bile salts or similar detergent), and, finally, to be transferred from micelle to phospholipid vesicle. Gel filtration may be used at each of the three detergent-separation steps, both to remove unwanted detergent and to estimate the Stokes radius of the protein-containing particle.

Acknowledgments

This work was supported in part by grants to M. W. Ho from Nuffield Foundation and Medical Research Council.

[22] B. Svensson, M. Danielsen, M. Staun, L. Jeppesen, O. Noren, and H. Sjöstrom, *Eur. J. Biochem.* **90,** 489 (1978).

[18] Purification of Integral Membrane Proteins

By JOS VAN RENSWOUDE and CHRISTOPH KEMPF

The study of structure and function of biological membranes has been hampered by the lack of efficient methods for separating the functional components of the membrane, most of which are amphipathic in nature. Although this chapter focuses on some of the methods available for the isolation of integral membrane proteins, it will be obvious that no step-by-step outline can be presented.

Integral membrane proteins behave as amphiphiles[1,2]; they possess hydrophilic domains, formed by ionic and polar residues (including carbohydrates), which are exposed to an aqueous environment at the surface of the membrane, and hydrophobic domains, rich in apolar residues, which are buried within the hydrophobic core of the membrane's lipid bilayer. The topological distribution of the hydrophobic and hydrophilic domains of a protein determines its arrangement in the membrane. Some of these proteins are situated in the bilayer so that the hydrophilic portion is exposed at only one surface of the membrane, whereas others span the membrane, allowing hydrophilic domains to interact with the aqueous phase on both sides of the membrane. Endoplasmic reticulum cytochrome b_5 and erythrocyte acetylcholinesterase are probably organized in the first way. There are several examples of proteins that span the membrane: the major erythrocyte sialoglycoprotein, glycophorin; the erythrocyte anion transport protein; bacteriorhodopsin; some membrane transport proteins; and viral envelope glycoproteins. It is conceivable that a third type of membrane protein may be totally immersed within the hydrophobic phase of the bilayer and have totally hydrophobic surfaces. Some of the proteolipids from myelin and mitochondria that are soluble in organic solvents may be of this class.

It is the amphiphilic character that presents great difficulty in isolation and purification. In the intact membrane, the proteins are associated with a phospholipid bilayer and, at some stage, must be separated from the phospholipid. Once dissociated, they will exhibit the same preferential interactions that cause them to be located in a hydrophobic environment in the first place, and their amphiphilic properties tend to make them unstable in both aqueous and organic solvents. In aqueous media, intermolecular self-association at the hydrophobic surfaces of the protein is

[1] J. Lenard and S. J. Singer, *Proc. Natl. Acad. Sci. U.S.A.* **56,** 1828 (1966).

[2] D. F. H. Wallach and P. H. Zahler, *Proc. Natl. Acad. Sci. U.S.A.* **56,** 1552 (1966).

ISBN 0-12-182004-1

favored, leading to the formation of aggregates that are insoluble in most instances. In organic solvents, on the other hand, the polar domains will tend to self-associate in order to gain maximum protection by exposure of hydrophobic domains to the solvent.

In devising an approach to the isolation of a specific membrane protein, it is important to define the purpose of isolating the molecule. When the aim is structural and chemical characterization, without regard for biological activity, a relatively wide range of isolation procedures is available, including those that result in irreversible denaturation of the protein; such procedures form the bulk of those described. If, however, the biological activity of the isolated protein is to be studied, the choice is limited to techniques that do not cause irreversible denaturation; examples are few.

Since methods for the isolation of integral membrane proteins have not yet reached the degree of discrimination of those available for separating soluble proteins, it is often advantageous to "simplify" the protein source as much as possible before solubilization and fractionation. Unless highly selective methods such as affinity chromatography can be applied, it is worth making the effort to ensure that the population of cells used as starting material is as homogeneous as possible and that the membrane fractions obtained from these cells contain the membrane of interest in the highest purity achievable. A useful preliminary step in isolating membrane proteins is the removal of contaminating cytosolic and peripheral membrane proteins from the membrane preparation. A number of procedures have been used to solubilize peripheral membrane proteins. These procedures include the use of chelating agents, salt in high and low concentration, and variations of the pH from acid to basic.

Here, we outline an approach to the isolation of an integral membrane protein and note some of the available methods. It is necessary to stress that isolation of these proteins, particularly isolation with residual biological activity, is a demanding process that is not always successful.

Membrane Isolation and Purification

The (sub)cellular membranes of eukaryotic and prokaryotic cells constitute the sources of membrane proteins. The usual first step in isolation of a membrane protein is fractionation of the source tissues or cells to obtain highly purified subcellular membranes. A multitude of cell fractionation procedures are available, many of which are summarized in this series and elsewhere.[3,4] There are a few instances in which relatively pure

[3] This series, Vol. 31.

[4] G. D. Birnie (ed.). "Subcellular Components, Preparation and Fractionation." Butterworth, London, and University Park Press, Baltimore, Maryland, 1972.

membranes can be obtained simply. The erythrocyte membrane is readily available after converting red cells to red cell ghosts, and vesicles released by secretory cells and plasma membrane vesicles, obtained by forced blebbing,[5] are also convenient starting materials. Even at this first step it is important to minimize the risk of degradation of the proteins of interest by endogenous or exogenous proteolytic activity. This can be achieved by working quickly, in the cold whenever possible, and by adding protease inhibitors such as phenylmethylsulfonyl fluoride (PMSF), *N*-tosyl-L-phenylalanylchloromethyl ketone (TCPK), and leupeptin.

After the desired membrane is obtained, soluble proteins that have in some way remained associated with the membrane and peripheral membrane proteins should be removed. This is usually achieved by exposing the membranes to one or more of the following agents or conditions:

1. KCl or NaCl in relatively high concentrations (0.15–3.0 *M*). This treatment will result in greatly decreased electrostatic interaction between proteins and charged lipids. It should be noted that certain endogenous proteases are activated under conditions of high ionic strength, so that adequate precautions (inhibitors, low temperature) should be taken.
2. Washing with buffers of acid or basic pH or with sodium carbonate.[6]
3. Chelating agents such as EDTA and EGTA (up to 10 m*M*). These substances destabilize membranes by complexing Mg^{2+} and Ca^{2+}.
4. Chaotropic ions (I^-, Br^-, ClO_4^-, SCN^-), in high concentrations (2–4 *M*), act by disordering the structure of water. They disrupt hydrophobic bonds near the surface of membrane structures and promote the transfer of hydrophobic groups from an apolar environment to the aqueous phase.
5. The phenolic compound lithium 3,5-diiodosalicylate[7] presumably acts by virtue of detergent-like properties. It has relatively low chemical reactivity and is easily removed by dialysis. At concentrations above 1 *M*, membrane structure generally breaks down completely.
6. Denaturing agents such as urea and guanidine hydrochloride break noncovalent bonds if used at high concentrations (6–10 *M*). They are often used in combination with reducing agents (2-mercaptoethanol or dithiothreitol) that would cleave disulfide bonds. Use of urea and guanidine renders mixtures of denatured, dissociated polypeptide chains.
7. Protein-modifying reagents, such as *p*-chloromercuribenzoate, *p*-chloromercuribenzene sulfonate, and acid anhydrides (e.g., succinic or maleic anhydrides).

[5] D. W. Tank, E.-S. Wu, and W. W. Webb, *J. Cell Biol.* **92,** 207 (1982).
[6] Y. Fujiki, A. L. Hubbard, S. Fowler, and P. B. Lazarow, *J. Cell Biol.* **93,** 97 (1982).
[7] V. T. Marchesi and E. P. Andrews, *Science* **174,** 1247 (1971).

All of the above-mentioned reagents are able to denature protein, leading to irreversible loss of biological activity. In order to warrant selective removal of peripheral membrane proteins and membrane-associated soluble proteins, while leaving integral membrane proteins in place, the conditions in the purification procedure should be chosen such that the lipid backbone of the membrane remains intact. After application of the above-mentioned protocols, recovery of the membranes by centrifugation allows the extent of loss of lipid bilayer structure to be easily assessed as a loss of residue after centrifugation. A large decrease in turbidity of the membrane suspension upon addition of a reagent may also be taken as indicative of extensive lipid bilayer collapse. The process of membrane purification and all subsequent steps in the isolation of a membrane protein can be conveniently monitored by conventional sodium dodecyl sulfate–polyacrylamide electrophoresis. Once membranes have been "purified" using one or more of the above procedures, they may be solubilized to liberate their constituents, lipids as well as integral membrane proteins.

Solubilization

Solubilization, by definition, involves the disintegration of the lipid bilayer and can be achieved in a number of different ways, of which only a few are suitable for our purposes. Solubilization for the purpose of isolating membrane proteins is usually accomplished by treating the membranes with detergents or by extracting them with organic solvents.

Detergents

The principles of detergent solubilization are discussed in detail in this volume.[8] The detergents used in solubilization can be divided into three classes: (1) nonionic detergents, such as octylglucoside and the polyoxyethylenes (e.g., Triton), which form relatively large micelles (M_r 50,000–100,000) and have relatively low CMC values (10^{-4} to 10^{-5} M); (2) zwitterionic detergents like 3-(3′-cholamidopropyl)dimethylammonio-1-propanesulfonate (CHAPS) and sulfobetaine; and ionic detergents, such as cetylammonium bromide and sodium dodecyl sulfate (SDS), which possess strongly acidic or basic polar head groups (e.g., sulfate or a quaternary nitrogen). These detergents arrange into relatively small micelles (M_r 10,000–20,000) and have relatively high CMC values (10^{-2} to 10^{-3} M); (3) bile salts, such as cholate, taurocholate, and deoxycholate, which form small aggregates consisting of 2–8 monomers, at monomer

[8] L. M. Hjelmeland and A. Chrambach, this volume [16].

concentrations of 10^{-2} to 10^{-3} *M*. The choice of a detergent for solubilization will be determined by several factors. If, for instance, preservation of the biological activity of the membrane protein is desired, nonionic detergents such as Triton X-100 are preferable. Nonionic detergents have the added advantage of not interfering with subsequent separation procedures based on charged groups on the protein, e.g., ion-exchange chromatography. A disadvantage in their use is that they are hard to remove from the solubilization mixture because of their low CMC.[9] Removal of detergent may be desirable if a membrane protein is to be reconstituted.[10] The use of certain resins, e.g., SM-2 Biobeads, that strongly bind nonionic amphiphiles, will improve the speed and completeness of their removal.[11] In the absence of a specific interest in maintaining biological activity of the membrane protein, solubilization of the source membrane in such ionic detergents as the bile salts may be the method of choice; one limitation in their use is that they form very large, precipitating aggregates at low pH. They should be used at a pH greater than 7.8.

Extraction with Organic Solvents

Some membrane proteins (proteolipids) are soluble in organic solvents, especially if they are neutralized by suitable counterions. Much of this work has been discussed in this series.[12,13] This property can be used, in principle, for purification by allowing the proteins to partition into the organic phase of a water–organic solvent two-phase system. For any given protein, the overall hydrophobicity of the polypeptide chain and the presence of such polar side groups as sugars will determine the extent of partitioning between organic and aqueous phases. Obviously, heavily glycosylated membrane proteins will tend to partition into the aqueous phase.[14,15] A number of organic solvents have been used for this purpose,[12,13] the more common of which are noted below.

1. *n*-Butanol was originally described as a solvent for extraction of erythrocyte membrane proteins and glycoproteins. Butanol extraction is a relatively mild procedure that tends to allow preservation of antigenic properties and, sometimes, enzymic function.[12,13] Most membrane pro-

[9] A. J. Furth, H. Bolton, J. Potter, and J. D. Priddle, this volume [17].
[10] R. D. Klausner, J. van Renswoude, and B. Rivnay, this volume [19].
[11] P. W. Holloway, *Anal. Biochem.* **53,** 304 (1973).
[12] H. S. Penefsky and A. Tzagoloff, this series, Vol. 22, p. 204.
[13] A. Tzagoloff and H. S. Penefsky, this series, Vol. 22, p. 219.
[14] D. J. Anstee and M. J. A. Tanner, *Biochem. J.* **138,** 381 (1974).
[15] D. J. Anstee and M. J. A. Tanner, *Eur. J. Biochem.* **45,** 31 (1974).

teins partition into the aqueous phase of the butanol–water system, whereas lipids find the butanol phase.

2. *n*-Pentanol is less selective than butanol in that polar lipids will, together with membrane proteins, partition into the aqueous phase. Pentanol extraction may be considered if lipid–protein interactions are to be preserved, e.g., in the case of membrane enzymes that require a lipid environment for activity.

3. Aqueous phenol, 50% or greater by volume, is used at either neutral or acid (acetic acid, 25–33%) pH.[16] Lipids and a large proportion of membrane protein partitions into the organic phase, whereas glycoproteins are predominantly found in the water layer.

4. Pyridine, used in pyridine–water (2 : 1, v/v) mixtures, is reputed to be a good solubilization reagent.[17] At low temperatures, it has been successfully applied in solubilizing relatively labile membrane components with preservation of biological activity.

5. Chloroform–methanol mixtures (chloroform–methanol–water, 18 : 9 : 1) have been used extensively as lipid extraction solvent systems.[18,19] Membrane proteins generally copartition with the lipid into the organic phase, whereas most membrane glycoproteins are recoverable from the aqueous phase.

6. Both acetic and formic acid interact strongly with hydrogen bonds; both act as proton donors to the carbonyl moiety of the amide group, competing with the amide proton. Both solvents also perturb hydrophobic interactions between nonpolar residues of the protein. The result is denaturation, aggregation in aqueous solvents, and even covalent modification (formylation) of membrane proteins.[20] Acetic acid and formic acid should be avoided if biological activity of membrane proteins is to be maintained.

7. 2-Chloroethanol is a weakly protic solvent with relatively high solubilizing power, presumably due to its content of HCl (2-chloroethanol is unstable and slowly decomposes to form HCl). It does not, to any appreciable extent, affect the hydrophobic domains of membrane proteins, and it may represent one of the more suitable solvents for the study of protein–lipid interactions.[21,22] 2-Chloroethanol is generally used at a

[16] K. Takayama, D. H. McLennan, A. Tzagoloff, and C. D. Stoner, *Arch. Biochem. Biophys.* **114,** 223 (1964).

[17] O. Blumenfeld, *Biochem. Biophys. Res. Commun.* **30,** 200 (1968).

[18] E. G. Bligh and W. J. Dyer, *Can. J. Biochem. Physiol.* **37,** 911 (1959).

[19] J. Folch, M. Lees, and G. H. Sloane Stanley, *J. Biol. Chem.* **226,** 496 (1957).

[20] W. Menke and H. G. Ruppel, *Z. Naturforsch. B Anorg. Chem. Org. Chem.* **26B,** 825 (1971).

[21] P. Zahler and D. F. H. Wallach, *Biochim. Biophys. Acta* **135,** 371 (1967).

[22] P. Zahler and R. E. Weibel, *Biochim. Biophys. Acta* **219,** 320 (1970).

concentration of 90%, by volume. It allows full recovery of lipid-free membrane proteins in a soluble, presumably largely monomeric, form, although biological activity may be lost.

8. Other solvents (Methyl Cellosolve, Ethyl Cellosolve, dimethylformamide, formamide, *N*-methylpyrrolidone, hexafluoroacetone, and diethylene glycol monobutyl ether) have been used, mainly under acid conditions, in extracting membrane proteins. Their application, however, has been limited to specific examples.[23-25]

Organic solvent extraction procedures may be useful in attempts to isolate some of the integral membrane (glyco)proteins. Organic solvents, however, tend to result in at least some degree of protein denaturation; their use should, as a rule of thumb, be considered mainly in those instances in which biological activity need not be preserved.

Fractionation of Membrane Proteins

The choice of methods for fractionating a mixture of membrane proteins depends, in part, on the procedure used in solubilization. After solubilization, a membrane protein exists either in a mixed micelle with detergent, dissolved in an organic solvent, or in an aqueous environment, usually in an aggregated state. Upon solubilization with a detergent, the resultant mixture includes mixed lipid–detergent micelles, mixed protein–detergent micelles, and mixed lipid–protein–detergent micelles. Protein-containing mixed micelles can be conveniently separated from nonprotein-containing ones by density gradient centrifugation, since the buoyant density of the latter is usually lower than that of the protein-containing species.[26] Protein-containing micelles can then be subjected to a variety of fractionation methods (see below). Membrane proteins that are dissolved in organic solvents cannot, in general, be fractionated as such; the organic solvent should be removed by evaporation and the proteins subsequently taken up in a solution of detergent. Similarly, aqueous dispersions of aggregated membrane proteins should be solubilized in detergent before additional fractionation can be undertaken. All chromatographic procedures noted below, except gel filtration, can be carried out in a batch-wise manner, with often significant gain in speed.

Phase Separation. Solubilization of membrane proteins with a number of detergents leads to formation of protein–detergent micelles. Several

[23] B. Kohl and H. Sandermann, *FEBS Lett.* **80,** 408 (1977).

[24] R. L. Juliano, *Biochim. Biophys. Acta* **266,** 301 (1972).

[25] P. H. Zahler, D. F. H. Wallach, and E. F. Luescher, *Protides Biol. Fluids* **15,** 67 (1967).

[26] J. Yu, D. A. Fischman, and T. L. Steck, *J. Supramol. Struct.* **1,** 233 (1973).

properties of these mixed micelles, including size and hydrophilicity, are a property of the detergent used as well as of the solubilized protein. The common detergent Triton forms clear micellar solutions in a given range of temperature. Upon warming, the solution undergoes phase separation resulting in two clear phases, one depleted and the other enriched in detergent. The phase separation temperature is dependent on the detergent structure (Triton X-100 at approximately 64°; Triton X-114 at approximately 20°). Utilizing this property, integral membrane proteins have been separated from the bulk of the erythrocyte membrane proteins with Triton X-114.[27] The method seems to be efficient in separating integral membrane proteins from soluble and extrinsic proteins. Differences between detergent micelles and protein–detergent micelles with regard to their partitioning in a two-phase system have also been used; the light-harvesting chlorophyll *a*/*b* protein was isolated with Triton X-100 in combination with the aqueous two-phase system dextran–polyethylene glycol.[28] Phase separations of this sort have been described in detail by Albertsson,[29–31] and certain variations in the technique are described in this volume.[32]

Gel Filtration. An early step in the purification of membrane proteins will usually be gel filtration, which achieves at least partial purification. Proteins originating from a wide variety of membranes have been purified by this means.[33–35] The large size of nonionic detergent micelles usually necessitates separation on agarose rather than on dextran or acrylamide media, and the viscosity of the detergent-containing solutions tends to make the separation rather slow. A new generation of gel matrices based on a hydrophilic vinyl polymer (Fractogel, Merck) may prove to be superior.

Ion-Exchange Chromatography. Ion-exchange chromatography can be used for membrane proteins in nonionic or zwitterionic detergents. Some proteins have the unfortunate characteristic of binding to the resin in an apparently irreversible fashion. Several proteins have been purified by means of ion-exchange chromatography on DEAE-cellulose, e.g.,

[27] C. Bordier, *J. Biol. Chem.* **256,** 1604 (1981).

[28] P.-A. Albertsson and B. Andersson, *J. Chromatogr.* **215,** 131 (1981).

[29] P.-A. Albertsson, "Partition of Cell Particles and Macromolecules." Wiley, New York, 1971.

[30] P.-A. Albertsson, *Endeavour* **1,** 69 (1977).

[31] P.-A. Albertsson, *J. Chromatogr.* **159,** 111 (1978).

[32] K. C. Ingham, this volume [20]; G. Johansson, this volume [21].

[33] M. J. A. Tanner, R. G. Jenkins, D. J. Anstee, and J. R. Clamp, *Biochem. J.* **155,** 701 (1976).

[34] M. Sone, M. Yoshida, H. Hirata, and Y. Kagawa, *J. Biol. Chem.* **250,** 7917 (1975).

[35] K. Kameyama, T. Nakae, and T. Takagi, *Biochim. Biophys. Acta* **706,** 19 (1982).

5′-nucleotidase from liver plasma membranes (solubilized in sulfobetaine-14, a zwitterionic detergent),[36] and the erythrocyte glucose transport protein.[37]

Affinity Chromatography. Affinity chromatography involves the selective adsorption of a protein to a matrix bearing molecules that will specifically bind to the protein. The bound protein can then be eluted from the resin by disruption of the specific interaction.[38] Since most membrane proteins are glycosylated, lectins coupled to a resin have found application. The dissociation of the lectin–sugar linkage often can be released by simple sugars under mild conditions. One limitation of the method is that most immobilized lectins are stable only in nonionic detergents. Cationic and zwitterionic detergents, for example, inhibit concanavalin A and soybean agglutinin, whereas SDS interferes with the binding capacity of most lectins.[39] A serious complication in the use of immobilized lectins is that the glycoproteins may be heterogeneous in their glycoconjugate content and, therefore, will not bind uniformly to any given lectin; such proteins would be recovered in low yields.[40] Other affinity methods also have been successful for the purification of membrane receptors; instead of lectins, receptor-specific ligands have been coupled to a resin matrix. Examples include the insulin receptor, purified on insulin–agarose[41]; the acetylcholine receptor, purified with the help of quaternary ammonium ligands as analogs of cholinergic compounds[42]; and the β-adrenergic receptor, purified from digitonin-solubilized frog erythrocyte membranes by chromatography on a Sepharose–alprenolol column.[43] For proteins with very great affinity for the ligand, the dissociation step may prove to be too difficult to be practical. An affinity ligand should be chosen with this constraint in mind so as to achieve freely reversible binding.

The development of the monoclonal antibody technique to individual antigens of complex biological mixtures has already greatly facilitated the isolation of membrane proteins and will continue to do so. Several have been isolated by monoclonal antibody affinity chromatography, including the Ia antigens,[44] HL-A antigen,[45] and the human transferrin receptor.[46]

[36] E. M. Baiyles, A. C. Newby, K. Siddle, and J. P. Luzio, *Biochem. J.* **203,** 245 (1982).

[37] M. Kasahara and P. C. Hinkle, *J. Biol. Chem.* **252,** 7384 (1977).

[38] A detailed discussion of the method is presented in Vol. 34 of this series and is updated in this volume [1]–[4].

[39] R. Lotan, G. Beattie, W. Hubbell, and G. L. Nicolson, *Biochemistry* **16,** 1787 (1977).

[40] M. J. A. Tanner and D. J. Anstee, *Biochem. J.* **152,** 265 (1976).

[41] P. Cuatrecasas, *Proc. Natl. Acad. Sci. U.S.A.* **69,** 1277 (1972).

[42] J. O. Dolly and E. A. Barnard, *Biochemistry* **16,** 5053 (1977).

[43] R. G. L. Shorr, S. L. Heald, P. W. Jeffs, T. N. Lavin, M. W. Strohsacker, R. J. Lefkowitz, and M. C. Caron, *Proc. Natl. Acad. Sci. U.S.A.* **79,** 2778 (1982).

[44] W. R. McMaster and A. F. Williams, *Immunol. Rev.* **47,** 117 (1979).

The use of monoclonal antibodies for general protein purification is presented in this volume [24].

The choice of detergent for membrane protein solubilization in immunoaffinity chromatography is restricted to the use of nondenaturing species. The dissociation step in immunoaffinity chromatography often may require such harsh methods as low pH or chaotropic agents, all undesirable if a functional protein is to be recovered. An electrophoretic desorption method[47] for dissociating membrane proteins from immobilized high-affinity ligands such as antibodies is a relatively mild procedure through which the use of immunoabsorbents in membrane protein isolation may be extended. Immunoaffinity chromatography has also been used in combination with covalent chemical modification of membrane proteins. Rothman and Linder[48] labeled intact cells, isolated the washed membranes, and purified the chemically modified proteins with antibodies against haptens on the labeling reagents. Isolated platelet membrane proteins were labeled in the carbohydrate moiety with 2,4-dinitrophenylalanine hydrazide, and erythrocyte membrane proteins were labeled with diazodiiodoarsanilic acid (a tyrosine-specific reagent); antidinitrophenyl and antiarsanilic acid antibodies, respectively, served in the immobilized phase.

Covalent Chromatography. Covalent chromatography allows the selective absorption of proteins from a mixture to an insoluble matrix by means of covalent, reversible reactions between reactive groups on the protein and immobilized chemical reagents on the resin. Most commonly, thiol groups on the protein and the matrix are used to form a disulfide that can subsequently be cleaved from the column with a low-molecular-weight thiol. Major erythrocyte membrane proteins, glycophorin and band III, have been purified by this approach with an organic mercurial linked to agarose.[49] The selectivity of the method is not high because of the frequency of free SH groups on proteins. Nevertheless, useful purification can be achieved.

Hydrophobic Interaction Chromatography and HPLC. Some attempts have been made to separate integral membrane proteins on the basis of their variation in hydrophobicity. If proteins are applied to chromatographic media that themselves have hydrophobic characteristics,

[45] P. Parham, *J. Biol. Chem.* **254,** 8709 (1979).

[46] I. S. Trowbridge and M. B. Omary, *Proc. Natl. Acad. Sci. U.S.A.* **78,** 3039 (1981).

[47] M. R. A. Morgan, P. J. Brown, M. J. Leyland, and P. D. G. Dean, *FEBS Lett.* **87,** 239 (1978).

[48] A. Rothman and S. Linder, *Biochim. Biophys. Acta* **641,** 114 (1981).

[49] M. L. Lukacovic, M. B. Feinstein, R. I. Shaafi, and S. Perrie, *Biochemistry* **20,** 3145 (1981).

e.g., octyl- or phenylagarose,[50,51] separation on the basis of hydrophobic interactions may be possible (see this volume [3]). Mitochondrial proteins have been partially resolved on columns containing hydrophobic acrylamide derivatives,[52] and attempts have been made to purify erythrocyte membrane proteins on a column of *N*-(3-carboxypropionyl) aminodecylagarose,[53] a compound having ionic as well as hydrophobic characteristics. The expression of hydrophobicity of such a resin can be controlled by pH.

Methods for separating membrane proteins by means of reversed-phase HPLC are in an early stage of development but are covered in this volume [9]. The molecular weight of porin, a protein of the outer membrane of *Escherichia coli,* has been determined[54] by this method, although the protein was previously purified by conventional chromatographic procedures. Purification of two membrane proteins from T cells, H-2 and Lyt-2, made use of reversed-phase HPLC (Vydac RP-18 column) as well as gel permeation HPLC (BioGel TSK column) with a Nonidet P-40-containing elution buffer.[55] However, the relatively small pore size (100 Å) led to low yields in protein recovery. Much better results in fractionating T-cell membrane proteins were obtained when the membranes were solubilized in trifluoroacetic acid (0.1%) and acetonitrile (10%) and then separated on a wide-pore (300 Å) reversed-phase C_{18} column eluted with an acetonitrile gradient (V. L. Alvarez, personal communication).

Other Methods. The qualitative protein composition of membranes is most often determined by SDS–polyacrylamide electrophoresis. Proteins have been successfully fractionated on an analytical scale from erythrocyte and platelet membranes by preparative detergent electrophoresis on a 2 × 6 cm acrylamide gel.[56] The capacity of such gels is moderate; up to 60 mg of proteins was separated by the method. Inherent to this preparative method is the exposure of proteins to SDS and the high risk of denaturation. Additional limitations of the method include potential comigration of nonidentical protein species, since the separation is mainly based on the molecular weight of the proteins. This electrophoretic approach nonetheless has the advantage of speed and simplicity.

[50] J. Rosengren, S. Pahlman, M. Glad, and S. Hjertén, *Biochim. Biophys. Acta* **412,** 51 (1975).

[51] S. D. Carson and W. H. Konigsberg, *Anal. Biochem.* **116,** 398 (1981).

[52] H. Weiss and T. Buecher, *Eur. J. Biochem.* **17,** 561 (1970).

[53] R. J. Simmonds and R. J. Yon, *Biochem. J.* **157,** 153 (1976).

[54] K. Kameyama, T. Nakae, and T. Tagaki, *Biochim. Biophys. Acta* **706,** 19 (1982).

[55] V. L. Alvarez and M. Mage, *Fed. Proc. Fed. Am. Soc. Exp. Biol.* **41,** 838 (1982).

[56] W. L. Nichols, D. A. Grastineau, and K. G. Mann, *Biochim. Biophys. Acta* **554,** 293 (1979).

[19] Reconstitution of Membrane Proteins

By RICHARD D. KLAUSNER, JOS VAN RENSWOUDE, and BENJAMIN RIVNAY

Reconstitution of membrane proteins continues to be a crucial step in studying the function and structure of these molecules. The word reconstitution is poorly defined but, in general, refers to the reincorporation of a solubilized membrane protein into a natural or artificial membrane. The major virtue of this technique is realized only when functional reconstitution of activity is accomplished. The necessity for reconstitution arises because many membrane proteins express their full activity only when correctly oriented and inserted in a lipid bilayer. In order to purify a membrane protein to any degree, it is necessary first to remove it from its natural membrane. Thus, it becomes necessary to develop a reconstitution scheme in order to study the function of these purified (or partially purified) components. The last review of methods for membrane protein reconstitution to appear in this series provided a thorough overview of the field in 1979,[1] with concentration primarily on membrane enzymes. During the past several years, increasing attention to another class of membrane proteins, receptors, has provided a number of approaches to the reconstitution of integral membrane receptors.

The vast majority of reconstitution procedures can be summarized as involving (1) solubilization of the protein with a suitable detergent (this volume [16] and [18]); (2) mixing the solubilized protein with either lipid–detergent micelles or preformed lipid or natural membranes; and (3) removing the detergent (see also this volume [17]). The result, if successful, is the integral incorporation of the protein into the lipid bilayer. This summary gives little feeling for the great complexity and variety of approaches. The specific detergent and the precise conditions of the solubilization, the membrane or lipid preparation, the order of mixing, and the technique of detergent removal have been varied freely, and only recently has there begun to be a semblance of rational order to the process of successful reconstitution.

Detergent, Lipid, and the Denaturation of Proteins

The loss of function during reconstitution can be broadly assumed to be the result of denaturation, which, in turn, is the result of the detergent

[1] E. Racker, this series, Vol. 55, p. 699.

ISBN 0-12-182004-1

solubilization. The functions that can be lost range from the ability to insert into a lipid bilayer, ligand-binding activity, enzyme activity, and ion fluxes, among others. By trial and error, optimal detergents have been found that minimize the denaturation of specific membrane proteins when used for solubilization (see this volume [16]). Examples include digitonin for the β-adrenergic receptor,[2] CHAPS for the IgE receptor[3] and the opiate receptor,[4] cholate for the acetylcholine receptor,[5] and octyl glucoside for rhodopsin.[6] It has been known for some time that phospholipid, present during solubilization and reconstitution, protects against denaturation of integral membrane enzymes. This has been extended to receptors as well.[7] In an attempt to bring order to the chaos of detergent-induced denaturation, Rivnay and Metzger[3] have proposed a single parameter, ρ, which defines the relationship of lipids and detergent. This parameter is defined by Eq. (1) in which CMC_{eff} is the critical micelle concentration under the particular experimental conditions and will depend on salt and lipid concentrations.

$$\rho = [(\text{detergent}) - \text{CMC}_{\text{eff}}]/(\text{phospholipid}) \qquad (1)$$

The function ρ is similar to the parameter R_{eff}, the effective ratio, derived by Jackson *et al.*[8] [Eq. (2)].

$$R_{\text{eff}} = [(\text{detergent}) - 0.22]/(\text{phosphatidylcholine}) \qquad (2)$$

An analysis of the literature on reconstitution reveals that the likeliness of detergent-induced denaturation rises as ρ increases.[9] The ρ value is minimized by using a low detergent concentration while maintaining a high phospholipid concentration. On the other hand, if ρ is too small, the protein will not be solubilized. Experience with the acetylcholine receptor suggests that $\rho \geq 1.5$ is required for solubilization. In general, every molecule of protein is not solubilized; for the acetylcholine receptor, cholate solubilizes only 70% of the receptor. The sensitivity of a receptor to denaturation during the solubilization and reconstitution process depends on the criteria chosen to assess denaturation. This is well illustrated for the acetylcholine receptor. Agonist-induced Na^+ flux appears to be

[2] R. Neubig, E. Krodel, N. Boyd, and J. Cohen, *Proc. Natl. Acad. Sci. U.S.A.* **76,** 690 (1979).
[3] B. Rivnay and H. Metzger, *J. Biol. Chem.* **257,** 12800 (1982).
[4] W. Simonds, G. Koski, R. A. Streoty, L. Hjelmeland, and W. A. Klee, *Proc. Natl. Acad. Sci. U.S.A.* **77,** 4623 (1980).
[5] R. Anholt, J. Lindstrom, and M. Montal, *J. Biol. Chem.* **256,** 4377 (1981).
[6] G. W. Stubbs, H. G. Smith, and B. J. Litman, *Biochim. Biophys. Acta* **425,** 46 (1976).
[7] M. Epstein and E. Racker, *J. Biol. Chem.* **253,** 6660 (1978).
[8] M. L. Jackson and B. J. Litman, *Biochemistry* **21,** 5601 (1982).
[9] R. D. Klausner, J. van Renswoude, R. Blumenthal, and B. Rivnay (in press).

most sensitive to functional loss, and high ρ values, owing to either excess detergent or the absence of lipids, lead to loss of this activity.[10] Less than 10% loss was reported with cholate at $\rho = 2$, whereas no channel activity remained at a ρ of 40 or more. From several studies, the optimum ρ for this receptor appears to be $1.5 < \rho < 10$ for solubilization and $\rho = 2$ for the reconstitution. Agonist binding is less readily lost, and the ability to insert into a lipid bilayer is the function that appears most resistant to denaturation. Studies with the receptor for IgE confirm and extend these findings.[3] Solubilization of 90% of the receptor is achieved at $\rho = 2$. Again, the presence of lipids, which keeps ρ low, protects against denaturation. The situation is more complex than might be implied by the ρ value alone. Thus, the IgE receptor is relatively better protected from denaturation by inclusion of natural lipids from the cells from which the receptor is isolated than is evident when soybean lecithin is used.

Specific Examples of Receptor Reconstitution

Virtually every reconstitution system that has been reported is different. It is, therefore, difficult to assign specific and rational explanations for each of the details unique to each study. However, specific examples of successful reconstitutions are useful as menus, if not exact recipes, for the general procedures that are available. It is the purpose of this section to outline the details of specific examples of receptor reconstitution so as to allow adaptation of this group of experimental protocols to the needs of other investigators.

Acetylcholine Receptor

Anholt *et al.* reported a well-documented reconstitution of this receptor into lipid vesicles.[11] The vesicles demonstrate activation and desensitization and distinguish between agonists and antagonists. Numerous other groups have reported successful reconstitution of this receptor.[12–16]

[10] R. L. Huganir, M. A. Schell, and E. Racker, *FEBS Lett.* **108,** 105 (1979).

[11] R. Anholt, D. R. Fredkin, T. Deernick, M. Ellisman, M. Montal, and J. Lindstrom, *J. Biol. Chem.* **257,** 7127 (1982).

[12] W. Wu and M. A. Raftery, *Biochem. Biophys. Res. Commun.* **89,** 26 (1979).

[13] J. Lindstrom, R. Anholt, B. Einarson, A. Engel, M. Osame, and M. Montal, *J. Biol. Chem.* **255,** 8340 (1980).

[14] J.-P. Changeux, J. Heidmann, J.-L. Popot, and A. Sobel, *FEBS Lett.* **105,** 181 (1979).

[15] J. M. Gonzalez-Ros, A. Paraschos, and M. Marting-Carrion, *Proc. Natl. Acad. Sci. U.S.A.* **77,** 1796 (1980).

[16] R. Anholt, *Trends Biochem. Sci. (Pers. Ed.)* **6,** 288 (1981).

Membranes enriched in the acetylcholine receptor may be prepared from *Torpedo californica* electroplax organs.[17] Such membranes consist predominantly of receptor subunits but include as well an extrinsic membrane protein of $M_r = 43,000$. The latter protein, not considered part of the receptor system, can be removed by alkaline extraction.[18] The resultant preparation is then available for solubilization and reconstitution.[6]

Solubilization. Membranes are solubilized in 2% sodium cholate in the presence of crude soybean lecithin (L-α-phosphatidylcholine type II, Sigma) at a lipid concentration of 5 mg/ml in 10 m*M* sodium phosphate at pH 7.4 containing 0.1 *M* NaCl. Lipids are prepared as a stock solution containing 150 mg/ml by dispersion in distilled water with a bath sonicator. Sonication is performed under a stream of argon to reduce lipid oxidation. The cholate–membrane suspension is gently shaken at 4° for 18–24 hr. At the end of this solubilization period, the mixture is centrifuged at 165,000 *g* for 30 min and the insoluble material is discarded. The process results in 65% solubilization of receptor as shown by ^{125}I-labeled α-bungarotoxin-binding activity. Raising the cholate concentration above 3% does not enhance receptor solubilization but does lead to diminished functional activity after reconstitution.

Reconstitution. After solubilization, soybean lipids (from the 150 mg/ml dispersion) are added to the extract to a final lipid concentration of 25 mg/ml. At this point, the concentration of the receptor is 1–2 μM. Cholate is removed by dialysis for 16–18 hr against 500 volumes of 100 m*M* NaCl, 10 m*M* NaN_3, and 10 m*M* sodium phosphate at pH 7.4. This step is followed by 16–18 hr of dialysis against 500 volumes of 145 m*M* sucrose, 10 m*M* NaN_3, and 10 m*M* sodium phosphate at pH 7.4.

The vesicles formed have a mean diameter of 520 Å. Two maneuvers can increase their size. The vesicles can be frozen at −20°, followed by thawing at room temperature and resulting in vesicles with an average diameter of 620 Å. If cholesterol is incorporated during reconstitution at a ratio (cholesterol : phospholipid, w/w) of 1 : 4, the vesicles have an average diameter of 760 Å after a single cycle of freeze-thawing.

Receptor for Immunoglobulin E (IgE)

A number of variables have been evaluated as to their effect on reconstitution,[3] success being measured by the incorporation of the receptor into lipid vesicles.

[17] J. Elliott, S. G. Blanchard, W. Wu, J. Miller, C. D. Strader, P. Hartig, H.-P. Moore, J. Racs, and M. A. Raftery, *Biochem. J.* **185,** 667 (1980).

[18] R. Neubig, E. Krodel, N. Boyd, and J. Cohen, *Proc. Natl. Acad. Sci. U.S.A.* **76,** 690 (1979).

Membranes are prepared from rat basophilic leukemia cells by first washing the cells in buffer containing 120 mM NaCl, 15 mM Tris-HCl at pH 7.6, 5 mM KCl, 0.5 mM $MgCl_2$, and 0.5 mM $CaCl_2$. The cells are swollen in a 1 : 3 dilution of this buffer with water, including 1 mM phenylmethylsulfonyl fluoride (PMSF), 20 mM iodoacetamide, and 10 mg of DNase I per milliliter for 3–5 min. They are then sonicated for 1 min in a cuphorn sonicator at 0° and 60% of maximum output. The suspension is centrifuged at 31,000 g, and the pelleted membranes are suspended in full-strength buffer by brief sonication.

Lipids are prepared from the rat basophilic leukemia cells grown as tumors. Tumor (10 g) is homogenized in 25 ml of 0.16 M sodium borate at pH 8.0 containing 0.2 M NaCl and mixed with 4 volumes of chloroform–methanol (2 : 1, v/v). The upper layer and interface formed after centrifugation for 20 min at 3000 rpm are extracted with chloroform and combined with the initial lower layer. Liposomes are made by first drying the chloroform solution containing 20 mg of phospholipids under nitrogen, leaving a lipid film that is subjected to lyophilization overnight. Vesicles are formed by swelling the lipid in 1 ml of the Tris-saline buffer (described above) for 15 min at room temperature. The suspension is sonicated for 15 min with a microtip (Model W225R, Heat Systems Ultrasonics, Inc.) at an output of 50% of maximum.

Reconstitution is achieved by mixing membranes, tumor lipids as liposomes, and CHAPS. Total phospholipid is 6 mM, of which approximately 4% is derived from the receptor-containing membranes (determined by inorganic phosphate). The detergent concentration is 12.2 mM, which yields a ρ value of 1.25. The mixture can be dialyzed either immediately or after incubation at 2° for periods of up to 18 hr against 500 volumes of the Tris-saline buffer using 3 to 5 changes over 3–4 days. The result is the incorporation of about 90% of the receptor into vesicles as determined by banding on 3 to 50% sucrose gradients run for about 40 hr in an SW 50.1 rotor at g_{av} = 189,000 in the cold.

β-Adrenergic Receptor

For the β-adrenergic receptor from rat erythrocyte membranes,[2] the membranes are washed and suspended at 10 mg of protein per milliliter in 20 mM HEPES at pH 8.0 containing 2 mM EDTA. Digitonin, 70 mg/ml, is added to give a detergent : protein ratio of 3.5 : 1 (w/w). After stirring at 2° for 1 hr, the solution is diluted with an equal volume of saturated ammonium sulfate. After 30 min at 2°, the material is centrifuged at 50,000 rpm (50 Ti rotor) for 30 min, leaving the solubilized receptor in the supernatant fluid. The procedure results in 50% solubilization with a twofold purification of the receptor.

Dimyristoylphosphatidylcholine vesicles are sonicated at a lipid concentration of 150 m*M* to form small unilamellar vesicles. They are added to the membrane extract (all in 55 m*M* HEPES at pH 8.0–5.5 m*M* EDTA–10.6 m*M* $MgCl_2$–98 m*M* NaCl–0.89 *M* NH_4SO_4) to give a final lipid concentration of 5 m*M*. The mixture is chromatographed on Sephadex G-50 in 50 m*M* Tris-HCl at pH 7.5–5 m*M* $MgCl_2$–100 m*M* NaCl in the absence of detergent. The early void volume fractions are collected. These fractions are turbid and contain 50–65% of the receptors, total protein, and phospholipid, but only 10–20% of the digitonin. The detergent concentration at this point is 0.5–1 m*M*, still higher than the CMC for digitonin (0.08–0.3 m*M*). This fraction is further treated by rate zonal centrifugation on a linear sucrose density gradient. Reconstituted vesicles are found in a turbid band at 27% sucrose. This band contains 50% of the receptor activity, 6 μmol of lipid per milligram of protein, and no detectable digitonin (the limit of detection is 40 μ*M*). Electron microscopic studies reveal unilamellar vesicles, 500–900 Å in diameter. The reconstituted vesicles display ligand-binding properties similar to those found for the intact receptor in native membranes. Interestingly, the agonist, iodohydroxybenzylpindolol, which binds to the native receptor, does not bind to the digitonin-solubilized receptor. However, full binding activity is regained upon reconstitution and detergent removal.

One of the most exciting aspects of reconstitution studies with this receptor involves that of the interacting components of the adrenergic receptor–cyclase system.[19] Separate solubilization of the receptor and the nucleotide binding subunit (G) with subsequent reconstitution into phospholipid vesicles has been described.

Receptor Solubilization. Washed turkey erythrocyte membranes are treated with 50 m*M* potassium phosphate at pH 11.9 for 20 min at 4°, a procedure that does not affect the receptor, but destroys G protein activity. The membranes are centrifuged at 27,000 *g* for 15 min and suspended in 20 m*M* MOPS at pH 7.0 and 1 m*M* mercaptoethanol at a final protein concentration of 1 mg/ml. Phospholipid vesicles are prepared by sonication of 10 mg of soybean lecithin per ml of 10 m*M* Tris-HCl (pH 7.5)–2 m*M* EDTA. For each 1 ml of suspended membranes, 0.2 ml of vesicle suspension is added. The mixture is incubated at 4° in 10 m*M* $MgCl_2$. Upon addition of 20 μ*M* isoprenaline, an adrenergic ligand, the incubation temperature is raised and maintained at 30° for 10 min, and the mixture is

[19] Y. Citri and M. Schramm, *Nature (London)* **282,** 297 (1980).

centrifuged for 10 min at 18,000 *g*. The membrane pellet is suspended in 20 m*M* MOPS (pH 7.0)–0.5 *M* sucrose–0.2 m*M* EDTA–0.1 m*M* mercaptoethanol–0.01 m*M* PMSF and then solubilized by the addition of an equal volume of sodium deoxycholate (12 mg/ml). Insoluble material is removed by centrifugation for 30 min at 200,000 *g*.

G Protein Solubilization. Turkey erythrocyte membranes are washed in 20 m*M* HEPES (pH 8.0)–2 m*M* mercaptoethanol–1 m*M* $MgCl_2$–0.2 m*M* EDTA–0.02 m*M* PMSF and resuspended at a protein concentration of 10 mg/ml. An equal volume of sodium cholate (20 mg/ml) is added, and the mixture is incubated at 4° for 30 min. Insoluble material is removed by centrifugation for 30 min at 200,000 *g*. The supernatant liquid contains solubilized G protein but is without receptor activity.

Reconstitution. Phospholipid vesicles are prepared by sonication (5 min) of 10 mg of lecithin and 1 mg of stearylamine in 1 ml of 10 m*M* Tris-HCl at pH 7.5 containing 1 m*M* EDTA. The two solubilized proteins are mixed in desired proportions and treated with SM-2 resin (Bio-Rad) to remove detergent. This is accomplished by mixing 1.2 g of wet resin per milliliter of mixture, followed by shaking for 1 hr. The resin is removed by centrifugation, and the preformed vesicles are mixed with the supernatant fluid to give 10 mg of phospholipid per milliliter of supernatant liquid. The suspension is incubated briefly at 4°, and the resultant reconstituted vesicles are centrifuged at 20,000 *g* for 15 min.

Insertion of Reconstituted Vesicles into Biological Membranes. Reconstituted proteoliposomes[20] can be reinserted into biological membranes in a manner such that the inserted receptors will couple functionally with integral membrane enzymes.[21] The reconstituted vesicles that have been described are centrifuged at 18,000 *g* for 10 min.

Friend erythroleukemia cells are suspended at 5×10^6 cells per milliliter in a solution of 135 m*M* NaCl, 5 m*M* KCl, 0.8 m*M* $MgCl_2$, and 20 m*M* Tris-HCl at pH 7.4. One milliliter is layered over the vesicles, and the cells are allowed to sediment at room temperature. The buffer is removed, and the pellet of cells and vesicles is mixed. A solution of polyethylene glycol, 0.5 ml at 37°, is added and mixed. This solution contains 1040 mg of PEG 6000 per milliliter of the cell buffer, modified by containing 5 m*M* glucose, 4 m*M* $MgCl_2$, 2 m*M* ATP, and 0.1 m*M* EDTA. The modified buffer is added in progressively increasing volumes after 100 sec and every 2 min thereafter, employing the following additions: 0.2, 0.3, 0.5, 1.5, 3.5, and 7 ml. The resultant mixture is stored at 2° after centrifugation

[20] M. Schramm, *Proc. Natl. Acad. Sci. U.S.A.* **76,** 1174 (1979).

[21] S. Eimerl, G. Neufeld, M. Korner, and M. Schramm, *Proc. Natl. Acad. Sci. U.S.A.* **72,** 760 (1980).

and suspension in the buffer. The procedure is reported to allow the implantation of the receptor and G protein into the cell membrane.

Insulin Receptor

The insulin receptor may be solubilized from a preparation of turkey erythrocyte membranes by treatment of 2.9 mg per milliliter of membrane protein with 1% octyl β-glucoside in 30 m*M* NaCl, 10 m*M* sucrose, 1 m*M* EDTA, and 85 m*M* Tris-HCl at pH 7.8.[22] The mixture is stirred for 15 min at room temperature, and insoluble membrane components are removed by centrifugation for 60 min at 4° and 104,000 *g*. The solubilized receptor in the liquid phase is mixed with a 20-fold excess of phospholipid in 2% octyl glucoside. Two types of phospholipid mixtures have been used: soybean phosphatidylcholine–bovine brain phosphatidylserine (4 : 1, w/w) and dimyristoylphosphatidylcholine–bovine brain phosphatidylserine (4 : 1, w/w). Each mixture of lipid, detergent, and solubilized protein can be purified with Sephadex G-50 equilibrated in the absence of detergent at a temperature that is above the phase transition temperature of the lipid system used. Material eluting in the void volume of the eluate is dialyzed for 48 hr. The vesicles are centrifuged at 104,000 *g* for 1 hr at 4°, suspended in buffer, and centrifuged in a 2 to 30% continuous sucrose gradient until equilibrium is reached. Vesicles of either lipid composition band at 1.071 g/ml. Both types of vesicle preparations consist almost entirely of unilamellar structures about 1000 Å in diameter. These vesicles contain the insulin receptor as demonstrated by the specific binding of iodoinsulin. Although both types of lipid vesicles allow successful receptor reconstitution, the affinity of the reconstituted receptor for insulin is markedly different for each lipid system.

[22] R. J. Gould, B. H. Ginsberg, and A. A. Spector, *J. Biol. Chem.* **257,** 477 (1982).

Section IV

Other Separation Systems

[20] Protein Precipitation with Polyethylene Glycol

By KENNETH C. INGHAM

The use of nonionic water-soluble polymers, in particular polyethylene glycol (PEG), for fractional precipitation of proteins was introduced by Polson *et al.*[1] and discussed by Fried and Chun in this series.[2] The intervening years have provided an improved understanding of the molecular basis of the protein-precipitating action of PEG and additional documentation of the unique advantages of this polymer over other reagents used for this purpose. Although much of the literature on this subject deals with purification of proteins from blood plasma,[3] the approach is applicable to any complex mixture. The principles involved have been further clarified by studies with purified proteins, and the purpose of this chapter is to summarize briefly these principles with emphasis on practical information enabling the reader to assess the potential applicability of this technique to specific separation problems.

Advantages of Polyethylene Glycol

The advantages of PEG as a fractional precipitating agent stem primarily from its well-known benign chemical properties. Unlike ethanol and other organic precipitating agents, PEG has little tendency to denature or otherwise interact with proteins even when present at high concentrations and elevated temperatures. Careful experiments designed to test this principle revealed that PEG 400[4] and PEG 4000[4] at concentrations up to 30% (w/v) had no detectable effect on the circular dichroic spectrum or thermal denaturation temperature of ribonuclease.[5] The low heat of solution and the relative insensitivity of PEG-precipitation curves to minor variations in temperature eliminate the need for controlling temperature during reagent addition. Another advantage of PEG over ethanol or ammonium sulfate is the shorter time required for the precipitated proteins to equili-

[1] A. Polson, G. M. Potgieter, J. F. Largier, G. E. F. Mears, and F. J. Joubert, *Biochem. Biophys. Acta* **82,** 463 (1964).

[2] M. Fried and P. W. Chun, this series, Vol. 22, p. 238.

[3] Y. L. Hao, K. C. Ingham, and M. Wickerhauser, *in* "Methods of Protein Fractionation" (J. M. Curling, ed.), p. 57. Academic Press, New York, 1980.

[4] PEG = poly(ethylene glycol), poly(ethylene oxide), polyoxyethylene. Chemical formula: $HOCH_2CH_2(CH_2CH_2O)_nCH_2CH_2OH$. PEG 400 and PEG 4000 signify heterogeneous mixtures having nominal average molecular weights of 400 and 4000, respectively.

[5] D. H. Atha and K. C. Ingham, *J. Biol. Chem.* **256,** 12108 (1981).

ISBN 0-12-182004-1

brate and achieve a physical state suitable for large-scale centrifugation. The advantages of PEG in facilitating the growth of protein crystals is well documented.[6,7]

Mechanism of Action

Careful measurements with a variety of purified proteins indicate that their solubilities decrease exponentially with increasing concentration of PEG according to Eq. (1)

$$\log S = \log S_0 - \beta C \qquad (1)$$

where S is the solubility in the presence of PEG at concentration C (% w/v) and S_0 is the *apparent* intrinsic solubility obtained by extrapolation to zero PEG.[5] Plots of log S vs [PEG] exhibit striking linearity over a wide range of protein concentration, the slope for a given protein being relatively insensitive to pH and ionic strength but markedly dependent upon the size of the PEG up to about 6000 daltons. The slopes also tend to increase with increasing size of the protein, reinforcing the popular notion of a steric exclusion mechanism whereby proteins are concentrated in the extrapolymer space, eventually exceeding their solubility limit under the given solution conditions. Although a quantitative explanation of this behavior is yet to come, it is clear that, in the absence of specific interactions, the sequence of precipitation of several proteins in a mixture will depend primarily on the ratios of their initial concentrations relative to their respective solubilities in the absence of PEG. Thus, even though larger proteins have steeper slopes, a large protein initially present at high concentration could precipitate later than a small one present at low concentration if the intrinsic solubility of the latter is much less than that of the former. Manipulation of the solution conditions is expected to improve the separation of a given pair of proteins to the extent that their intrinsic solubilities diverge.

Which PEG to Use?

Most workers use material with a nominal average molecular weight in the 4000–6000 range. Polymers larger than this offer no advantage, since their solutions are more viscous and the precipitation curves are not much different from those obtained with PEG 6000.[1,5] Decreasing the molecular weight below 4000 spreads the precipitation of a mixture over a broader

[6] W. B. Jakoby, this series, Vol. 22, p. 248.
[7] A. McPherson, Jr., *J. Biol. Chem.* **251,** 6300 (1976).

range of PEG concentrations. The improved resolution that might be thus anticipated is partially offset by the shallower slopes obtained for individual proteins. Nevertheless, Honig and Kula[8] found the degree of purification of γ-glucosidase from yeast extract to be about twofold greater with PEG 400 than with PEG 4000 or 6000. That PEG 400 is a liquid at room temperature whose solutions are substantially less viscous than those of the higher polymers, coupled with the potentially greater ease of removing it by molecular sieve methods, indicates a need for further comparisons.

The Analytical Precipitation Curve

The following simple experiment is designed to quickly overcome ignorance about the amount of PEG required to precipitate a given protein(s) from a complex mixture. The scale of this experiment is dictated by the sensitivity of the assay employed; the availability of a radiolabeled tracer is a definite advantage. One dispenses a fixed amount (0.1–0.5 ml) of the mixture into a series of tubes (preferably in duplicate) to each of which is subsequently added an equal volume of buffer containing increasing amounts of PEG to produce a final concentration of 25–30% in the most concentrated tubes. It is important to buffer the PEG stock solutions to avoid PEG-induced changes in pH.[5,9] The increment in PEG concentration is arbitrary, but 3% (w/v) is adequate for initial screening. The vigor with which one mixes these solutions depends on the extent to which the desired protein(s) can withstand mechanical stress; gentle agitation on a vortex mixer is one approach. After 0.5–1.0 hr of incubation at room temperature or on ice, the samples are centrifuged and the percentage of the desired activity remaining in the supernatant liquid is determined. Inspection of the resulting "analytical precipitation curve" provides an estimate of the *maximum* concentration of PEG that can be added at one time without precipitating the protein of interest as well as the *minimum* concentration required to bring it out of solution, parameters that can then be more precisely defined with a second experiment that focuses on the relevant concentration range. With luck, the curve will fall either far to the left or far to the right on the PEG axis, defining a simple one-step method for removing a large portion of unwanted macromolecules and/or concentrating the desired activity prior to further processing by other methods. Otherwise, it may be necessary to obtain a "PEG cut"

[8] W. Honig and M.-R. Kula, *Anal. Biochem.* **72,** 502 (1976).

[9] G. Eichele, D. Karabelnik, R. Halonbrenner, J. N. Jansonius, and P. Christen, *J. Biol. Chem.* **253,** 5239 (1978).

via two precipitation steps utilizing in turn the *maximum* and *minimum* concentration of PEG referred to above.

It is always possible to manipulate the precipitation curve horizontally along the PEG axis by varying solution conditions. For screening purposes, it is expedient to choose a fixed concentration of PEG that causes approximately 50% precipitation of the desired protein under a given set of solution conditions in order to determine rapidly the extent to which altering those conditions might enhance or inhibit precipitation. The most gratifying result of this approach would be to identify substances or conditions that selectively influence the solubility of the desired protein. This concept is further developed in the following section.

Influence of Protein–Protein and Protein–Ligand Interactions

Studies with purified self-associating and heteroassociating proteins have shed some light on the role of protein–protein interactions on solubility in the presence of PEG.[5,10–12] Based on the above-mentioned excluded volume considerations, one predicts that conditions that foster protein association should enhance precipitation because of the larger size of the complexes, whereas those that inhibit association would have the opposite effect. This is the case with almost all systems that have been examined. Of particular relevance in the present context was the observation[10] that bovine liver glutamate dehydrogenase at 2.8 mg/ml in 0.2 *M* potassium phosphate at pH 7.0, conditions known to promote extensive self-association, was quantitatively precipitated by PEG 4000 at concentrations above 15% (w/v). Such precipitation was completely inhibited, even at higher concentration of PEG, by the combined presence of 10^{-3} *M* NADH and GTP, cofactors known to reverse the self-association. It remains to be seen whether this ligand-specific manipulation of the solubility could be exploited in a fractionation scheme. Similar effects were observed with chymotrypsin, chymotrypsinogen, and β-lactoglobulin A, in which cases self-association was manipulated by varying pH and ionic strength, parameters likely to be less selective. Nonspecific electrostatic interactions between oppositely charged proteins such as albumin and lysozyme can also have profound effects on solubility that are most pronounced at low ionic strength at a pH between the p*I* of each of the two proteins.[11] While such interactions are frequently viewed as a nuisance, to

[10] S. I. Miekka and K. C. Ingham, *Arch. Biochem. Biophys.* **191,** 525 (1978).
[11] S. I. Miekka and K. C. Ingham, *Arch. Biochem. Biophys.* **203,** 630 (1980).
[12] J. Wilf and A. P. Minton, *Biochim. Biophys. Acta* **670,** 316 (1981).

be minimized by maintaining near-physiological ionic strength, the possibility of using them to advantage in a purification scheme should be kept in mind.

A more specific type of heteroassociation of the type that might be exploited in purification is the functional interaction between human plasma fibronectin and denatured collagen, i.e., gelatin. The precipitation curve for the plasma protein in phosphate-buffered saline shifted from 11% PEG to less than 3% PEG upon addition of gelatin, which by itself was not precipitated by PEG under these conditions.[13] Since the complex between the two proteins is very stable, even at high ionic strength, it should be possible to precipitate fibronectin selectively from a complex mixture by this method. The contaminating gelatin could then be removed, e.g., by ion-exchange chromatography in the presence of urea. Although the advantage of this approach over affinity chromatography on immobilized gelatin is debatable, the example serves as an additional illustration of the application of bioaffinity principles to fractional precipitation. Any substance that interacts specifically with the desired protein has the potential to alter its solubility selectively and should thus be tested. Enzymes are ideal candidates for this approach, since they often interact with one or more effectors or cofactors, sometimes with large changes in the state of association.

Methods of Removing PEG

In many applications, PEG is used early in the purification scheme and is removed during subsequent chromatographic steps on ion-exchange or affinity columns to which PEG has no tendency to adsorb. A word of caution is in order regarding the application of PEG-containing solutions to some exclusion columns, the performance of which can be significantly altered owing to osmotic effects of the polymer.[14] Alternative approaches to removing PEG include ultrafiltration[15,16] and salt-induced phase separation[17] as reviewed.[18] The latter method is particularly useful for solutions containing relatively high concentrations of PEG and has the potential advantage that the protein may be concentrated in a low-volume, salt-

[13] K. C. Ingham, S. A. Brew, and S. I. Miekka, *Mol. Immunol.* **20,** 287 (1983).

[14] K. Hellsing, *J. Chromatogr.* **36,** 170 (1968).

[15] T. F. Busby and K. C. Ingham, *J. Biochem. Biophys. Methods* **2,** 191 (1980).

[16] K. C. Ingham, T. F. Busby, Y. Sahlestrom, and F. Castino, *in* "Ultrafiltration Membranes and Applications" (A. R. Cooper, ed.), p. 141. Plenum, New York, 1980.

[17] T. F. Busby and K. C. Ingham, *Vox Sang.* **39,** 93 (1980).

[18] K. C. Ingham and T. F. Busby, *Chem. Eng. Commun.* **7,** 315 (1980).

rich phase. For many research purposes it is probably unnecessary to remove all traces of polymer from the final product, since it is optically transparent[19] and helps prevent loss of protein by adsorption on glass.

Summary

Polyethylene glycol is a nondenaturing water-soluble polymer whose ability to precipitate protein from aqueous solution can be qualitatively understood in terms of an excluded volume mechanism. The increment in PEG concentration required to effect a given reduction in solubility is unique for a given protein–polymer pair, being insensitive to solution conditions and primarily dependent on the size of the protein and polymer. Selective manipulation of the solubility of specific proteins through control of their state of association or ligand environment can potentially remove some of the empiricism otherwise involved in fractional precipitation. Adequate methods for removing the polymer are available.

[19] The low level of UV absorbance frequently found in some PEG preparations is not inherent to the polymer but is due to a small amount of antioxidant sometimes added by the manufacturer.

[21] Affinity Partitioning

By GÖTE JOHANSSON

Partition of enzymes and other proteins between two liquid aqueous phases can be strongly influenced by specific or group-specific ligands bound to a water-soluble polymer included in the two-phase system. The two nonmiscible phases are obtained by mixing water solutions of two polymers. Several pairs of polymers can be used,[1] but the most popular system has been the one containing dextran and polyethylene glycol (PEG). To obtain two phases, the concentrations of the two polymers must exceed certain values that depend on the molecular weights of the polymers and the temperature. The composition of the phases can be found in the phase diagrams elaborated by Albertsson.[1] The partition of proteins between the phases depends on a number of factors that include

[1] P.-Å. Albertsson, "Partition of Cell Particles and Macromolecules." Wiley, New York, 1971.

METHODS IN ENZYMOLOGY, VOL. 104

ISBN 0-12-182004-1

the concentrations of polymers, the type and concentration of salt included, the pH, the nature of polymers carrying charged groups, and the temperature.[1,2] This large number of variables allows the design of a system in which most proteins are concentrated in one phase. An enzyme can then be extracted into the opposite phase by introducing a specific ligand (for the target enzyme) that is bound to the polymer concentrated in this phase, i.e., affinity partitioning. For large-scale extractions it is necessary to choose ligand-polymers that can be prepared easily in sufficient quantities and at moderate cost. Polymer derivatives of reactive triazine dyes fulfill these requirements and can be used for extraction of a number of dehydrogenases and kinases.[3-5] A number of other ligand-polymers, useful for affinity partitioning, and methods for their preparations have been summarized elsewhere.[6,7]

Synthesis of Triazine Dye–Polyethylene Glycol

Polyethylene glycol (M_r = 6000–7500), 100 g, is dissolved in 200 ml of water. The solution is heated in a water bath to 80°, and 10 g of triazine dye, e.g., Cibacron Blue F3G-A, is added followed by 2 g of LiOH. The LiOH can be replaced by NaOH, but this results in a lower yield. The mixture is maintained at 80° for 2 hr, preferably under stirring. Solid NaH_2PO_4 is added until a sample, diluted with water, reaches pH 7.0 to 7.5. Excess water is evaporated, and the residue is extracted with five 300-ml portions of hot chloroform. The pooled extracts are dried with anhydrous Na_2SO_4 and filtered through a glass-fiber paper (Whatman GF/A). The solvent is recovered by distillation, and the remaining polymer is dissolved in 2 liters of water. Washed DEAE-cellulose (Whatman DE-52) at about pH 7 is added, and the mixture is mechanically stirred until nearly all colored material is bound to the ion exchanger. This is collected by suction filtration and is washed several times with water to remove unreacted polymer. The dye-polymer is eluted by slowly passing a 2 *M* KCl solution through the exchanger bed; residual dye remains bound to the bed. The polymer is extracted from the KCl-containing eluate with chloroform. The chloroform layer is dried with Na_2SO_4 and filtered, and the solvent is removed by evaporation. The product consists of PEG covalently bound to the triazine dye in a 1 : 1 molar ratio.

[2] G. Johansson, *Mol. Cell. Biochem.* **4,** 169 (1974).
[3] M.-R. Kula, G. Johansson, and A. F. Bückmann, *Biochem. Soc. Trans.* **7,** 1 (1979).
[4] G. Kopperschläger and G. Johansson, *Anal. Biochem.* **124,** 117 (1982).
[5] G. Kopperschläger, G. Lorenz, and E. Usbeck, *J. Chromatogr.* **259,** 97 (1983).
[6] I. N. Topchieva, *Russ. Chem. Rev.* (*Engl. Transl.*) **49,** 260 (1980).
[7] A. F. Bückmann, M. Morr, and G. Johansson, *Makromol. Chem.* **182,** 1379 (1981).

Preparation of Two-Phase Systems

The two-phase systems are prepared from stock solutions of dextran, 20–30% (w/w) and PEG, 20–40% (w/w). Concentration of the dextran stock solution is checked by polarimetry because of the varying water content of solid dextran (2–5%). About 5 g of the dextran solution is weighed and diluted to 25 ml with water. Optical rotation is measured using sodium light at 20°. The concentration of dextran, C_{dx}, as a percentage (w/w) is calculated from the angle of rotation, α (in degrees), the weight of stock solution, m (in grams), and the length of the light pathway, ℓ (in dm), using Eq. (1).

$$C_{dx} = \frac{\alpha 25}{m 199 \ell} 100 \tag{1}$$

Stock solutions of ligand-polymer are used at 10% (w/w) or less because of their high viscosity. Calculated amounts of the polymer solutions are weighed, and salt solutions, buffer, and water are added, leaving sufficient volume for the protein solution. After mixing and attaining constant temperature, the protein solution at the same temperature is added. The phases are carefully mixed by inverting the vessel 10–20 times. For systems up to 100 ml in final volume, the total mixing time is 15–30 sec. The two phases separate within 15 min by gravity or in less than 1 min by centrifugation at low speed.

Analysis

Samples of known volumes (25–500 μl) are withdrawn from each phase for analysis. Enzymic activity can be determined directly, since the phase-forming polymers do not usually interfere with the assay at low concentrations. The presence of ligand-polymer may result in interference by competition with substrate binding; increasing the concentration of substrate should be tried if inhibition is noted. Several methods for determination of proteins are strongly influenced by the polymers. The method based on Coomassie Brilliant Blue G[8] has, however, been found suitable. Because of the high viscosity, particularly of the lower phase, great care must be taken in transferring samples of correct volume and mixing them thoroughly with the assay solutions. Reproducible volumes are obtained by filling and emptying the plastic tip of automatic pipettes several times with the assay solution.

Factors Determining the Partition

The partition of a substance between the two phases is described by a partition coefficient, K [Eq. (2)].

[8] M. M. Bradford, *Anal. Biochem.* **72,** 248 (1976).

$$K = \frac{\text{concentration in upper phase}}{\text{concentration in lower phase}} \quad (2)$$

For a high degree of purification, the target enzyme should have a large K value (>3) when the polymer-bound ligand is in the upper phase, whereas other proteins should have low partition coefficients ($K < 0.1$). The factors determining the partition of proteins that are without affinity for the ligand include the following.

1. Molecular weight of dextran; K increases with M_r.
2. Molecular weight of PEG; K decreases with increasing M_r.
3. Increasing polymer concentrations result in more extreme, usually decreasing, K values.
4. Salt and pH. When a negative protein (pH > pI) is partitioned, its K depends on the type of salt included in the system. K decreases in the series phosphate > sulfate > fluoride > acetate > chloride > bromide > iodide > perchlorate, as well as lithium > ammonium > sodium > potassium.[9] When pH < pI the partition is influenced in the opposite manner. The effect of salt increases with the net charge of the protein. The partition of proteins depends strongly on the concentration of salt in the range up to 50 mM and is dependent on the amount of protein. K is nearly constant at higher salt concentrations.[9]
5. Negatively charged PEG, e.g., PEG sulfonate,[10] repels negatively charged proteins toward the lower phase and attracts positive proteins into the upper phase. Positively charged PEG, e.g., trimethylamino-PEG, has the opposite effect. Charged PEG has a much stronger influence on the K value than does salt, but the effect is counteracted by salts at low concentration (5–25 mM).[11]
6. Temperature. In most instances, increasing temperature increases the partition coefficient.

Factors Determining the Affinity Partitioning Effect

Affinity partitioning is measured as the increase in logarithmic partition coefficient, $\Delta\log K$, of an enzyme caused by introducing the ligand-polymer into the system [Eq. (3)].

$$\Delta\log K = \log K \text{ (with ligand)} - \log K \text{ (without ligand)} \quad (3)$$

An enzyme can easily be extracted if $\Delta\log K \geq 2$. When triazine dyes are used as ligands bound to PEG, $\Delta\log K$ depends on the following factors.

[9] G. Johansson, *Acta Chem. Scand. Ser. B* **28,** 873 (1974).

[10] G. Johansson, A. Hartman, and P.-Å. Albertsson, *Eur. J. Biochem.* **33,** 379 (1973).

[11] G. Johansson and A. Hartman, in "Proceedings of International Solvent Extraction Conference, Lyon 1974" (G. V. Jeffreys, ed.), p. 927. The Society of Chemical Industry, London, 1974.

1. The amount of ligand-polymer in relation to ligand-binding enzyme. With increasing concentration of ligand–polymer, $\Delta\log K$ approaches a saturation value, $\Delta\log K_{max}$.

2. Increasing concentrations of dextran and PEG give larger $\Delta\log K_{max}$ (Fig. 1).

3. Salts decrease the affinity-partitioning effect. The reduction depends on salt concentration, the nature of the salt, and the specific ligand and enzyme. For phosphofructokinase extracted with Cibacron Blue F3G-A, sodium phosphate (pH 7.0) and sodium formate (pH 7.0) have only marginal influence on $\Delta\log K$ at 135 and 250 m*M* salt, respectively, compared with low concentration (10 m*M*) of phosphate buffer.[12] Sodium perchlorate and potassium phthalate (pH 7.0), on the other hand, have greater influence and $\Delta\log K$ is reduced to half its initial value at 60 m*M* salt. Similar effects have been observed in the case of glucose-6-phosphate dehydrogenase.[13]

4. $\Delta\log K$ depends strongly on pH. Working at low pH, the number of extractable enzymes increases and $\Delta\log K$ values are larger (see the table).

5. The M_r of the polymer can influence $\Delta\log K_{max}$,[12] but only limited effects have been found.

6. It has been observed that $\Delta\log K$ can decrease markedly with increasing temperature from about −2 to 60°. Partition can often be carried out at room temperature, since many enzymes are stabilized by the polymers.

7. Among the triazine dyes there are large differences in the interaction with a given enzyme. A suitable ligand can therefore be selected by screening a number of dyes,[3,5] e.g., the Procion dyes Blue MX-R, Red MX-2B, and Yellow MX-4R.

To find optimal conditions for purification by affinity partitioning, each of the described parameters should be systematically varied and analyzed by following the partition of both the desired enzyme and the mass of protein.

Strategy of Selecting a Two-Phase System

1. A 4-g system, containing 7% dextran (M_r = 500,000), 5% PEG (M_r = 6000–7500), and 25 m*M* buffer (sodium phosphate at pH 7.0, or Tris

[12] G. Johansson, G. Kopperschläger, and P.-Å. Albertsson, *Eur. J. Biochem.* **131,** 589 (1983).

[13] K. H. Kroner, A. Cordes, A. Schelper, M. Morr, A. F. Bückmann, and M.-R. Kula, *in* "Affinity Chromatography and Related Techniques" (T. C. J. Gribnau, J. Visser, and R. J. F. Nivard, eds.), p. 491. Elsevier, Amsterdam, 1982.

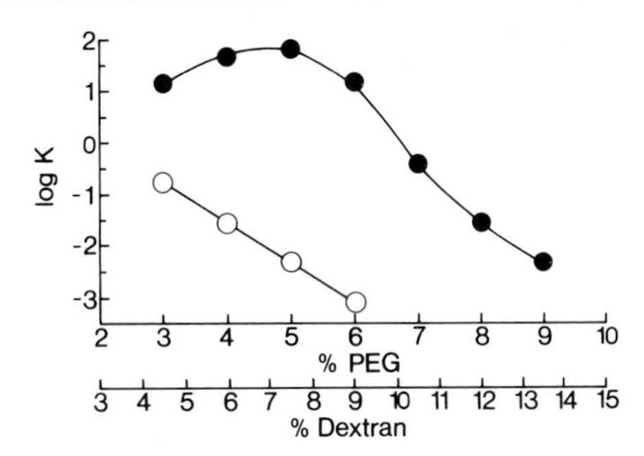

FIG. 1. Partition of phosphofructokinase when an extract of bakers' yeast is included in systems containing increasing concentrations of dextran (M_r = 70,000) and polyethylene glycol (PEG) (M_r = 38,000), with (●) and without (○) Cibacron Blue F3G-A PEG (M_r = 6500), 6.25% of total PEG. The system also contains 25 mM sodium phosphate buffer, 5 mM 2-mercaptoethanol, and 0.25 mM EDTA, pH 7.0; temperature, 0°.

acetate at pH 7.5) is prepared by mixing 1.4 g of 20% (w/w) dextran, 0.5 g of 40% (w/w) PEG, 0.75 ml of 100 mM buffer, and 0.35 ml of water in a graduated centrifuge tube. The mixture is maintained at a constant temperature (0 or 25°), and 1 ml of protein solution, containing 25 mM buffer, is added. After mixing for 15 sec, the tube is centrifuged at the same temperature for 2 min at 1000 *g* The volumes of upper and lower phases are measured, and samples are removed for analysis. Before determining

EFFECT OF pH ON ΔLOG *K* OF SOME GLYCOLYTIC ENZYMES FROM BAKERS' YEAST[a]

	ΔLog *K*		
pH	Enzyme 1[b]	Enzyme 2[b]	Enzyme 3[b]
4	2.6	2.3	1.1
5	2.5	2.6	0.5
6	1.4	0.9	0.2
7	0.7	0.1	0
8	0.3	0	0

[a] Two-phase system: 7.5% dextran (M_r = 70,000), 5% PEG (M_r = 38,000), 12.5 mM sodium phosphate at 0°. In the ligand-containing system, one-fifth of the PEG was replaced with Cibacron Blue F3G-A PEG (M_r = 6500).

[b] Enzymes 1, 2, and 3 are, respectively, glyceraldehydephosphate dehydrogenase, 3-phosphoglycerate kinase, and enolase.

protein the phases are diluted fivefold or more with water. In this two-phase system proteins are generally concentrated in the lower phase.

2. Extraction curves are prepared by replacing increasing amounts of PEG with ligand–PEG, e.g., 1/500, 1/250, 1/100, 1/50, 1/25, and 1/10. The curve of $\Delta\log K$ versus ligand–PEG in total PEG provides a guide for the amount of ligand–PEG necessary for maximal extraction effect. Straight lines are obtained when inversed plots ($1/\Delta\log K$ versus 1/{concentration of ligand}) are used,[12] which allows the determination of a saturation value, a V_{max}, from only a few measurements. A selective ligand is sought by testing several triazine dyes, e.g., the several Procion dyes produced by ICI and obtainable from Sigma Chemical Co. (St. Louis, MO) or Serva Feinbiochemia (Heidelberg).

3. The $\Delta\log K_{max}$ value can be further increased by using higher concentrations of dextran and PEG, as shown in Fig. 1, and retaining a constant ligand–PEG : PEG ratio. Since the partition coefficient of the enzyme in the presence of ligand–PEG usually goes through a maximum when the polymer concentration increases, this curve is used to choose a polymer composition that provides a reasonable recovery of the enzyme in the upper phase.

4. The selectivity of the extraction or $\Delta\log K_{max}$ may additionally be improved by changing the pH or by introducing salt to the system chosen under 3. Decreasing pH results in less specific extraction but in larger $\Delta\log K$ values; lower salt concentrations have the same influence.

5. The percentage of enzyme extracted into the upper phase can be chosen freely within practical limits by adjusting the ratio between the volumes of the two phases. The volume ratio, V (volume of upper phase : volume of lower phase), that yields the percentage, P, of the enzyme in the upper phase, is calculated by Eq. 4, where K is the partition coefficient of the enzyme.

$$V = P/[(100 - P)K] \qquad (4)$$

A study of this type has been made with phosphofructokinase from bakers' yeast,[12] and affinity partitioning has been applied to the isolation of this enzyme.[4]

Repeated Extractions

The effectiveness of purification can be enhanced by using repeated extractions that might include the following steps.

1. Preliminary extraction with an upper phase not containing ligand-PEG for removal of material with relatively high K values.

2. Several extractions with ligand-containing upper phase to improve the recovery of enzyme from a lower phase.
3. Preliminary extraction with an upper phase containing a ligand that can bind several enzymes but *not* the target enzyme.
4. Repeated washing of ligand-containing upper phase, after the affinity-partitioning step, with a fresh lower phase to remove coextracted proteins that are less strongly extracted by the ligand-PEG.

In several instances $\Delta \log K$ has been observed to increase by as much as 0.5 unit when the bulk of the protein is removed.

Preparative Extractions

Preparative affinity partitioning with systems selected by the above procedure can easily be carried out on a large scale. Two-phase systems of up to several liters can be mixed in separatory funnels; the phases will settle within 30–60 min. Very rapid extraction (≤ 5 min) is achieved by carrying out the partition in large centrifuge bottles that are used in preparative centrifuges. The phases are collected by siphoning. Larger quantities of partitioning polymers may be mixed with a motor-driven stirrer of the paddle type at low speed in a vessel with an outlet in its lower part; such separation may, however, require several hours. For rapid separation of very large volumes of the two phases, standard centrifugal separators have been used.[14]

Removal of Polymers

Enzyme can be separated from polymer by dissolving a mixture of solid potassium phosphates (yielding a pH of 7 to 9) in the recovered upper phase; 15 g or more are required per 100 g of solution. The result is another two-phase system with PEG and ligand-PEG in the viscous upper phase and the enzyme in the lower salt-containing phase.

Another possibility is to dilute the original upper phase with several volumes of buffer, e.g., 5 m*M* sodium phosphate at pH 7 to 9, and then bind the enzyme to a bed of DEAE-cellulose (or CM-cellulose for more basic proteins). Nonbound protein and PEG are washed away, and the enzyme is eluted with a salt gradient. The bound ligand-PEG may be eluted by increasing the concentrations of salt.

[14] K. H. Kroner, H. Hustedt, S. Granada, and M.-R. Kula, *Biotechnol. Bioeng.* **20,** 1967 (1978).

Comments

The amount of protein that can be included in the system is limited by the solubility in the two phases, but 50 g of protein per kilogram (final weight) of partition system can often be used. Since a solid matrix is not involved, nonspecific interactions and irreversible binding of "multiple attachment" type are eliminated. The recovered ligand–PEG may be reused, resulting in a process economical for large-scale operations.

[22] Affinity Precipitation of Dehydrogenases

By PER-OLOF LARSSON, SUSANNE FLYGARE, and KLAUS MOSBACH

Affinity precipitation of enzymes is a novel technique closely related to affinity chromatography and immunoprecipitation. The technique is applicable to oligomeric enzymes and may be used for purification,[1,2] for analysis,[1] and possibly for studies of molecular architecture and enzyme subunit arrangement.[1,2]

The first step in enzyme precipitation is the mixing of the enzyme with a bifunctional ligand composed of two ligand entities connected by a spacer. If the spacer is sufficiently long to bridge the distance between the respective binding sites of two enzyme molecules, and if the bioaffinity between the ligand and the enzyme is sufficiently strong, the two enzyme molecules will be linked to each other. If, in addition, the enzyme is oligomeric, more than one Bis-ligand can bind to each enzyme molecule, and it is conceivable that a network of enzymes and Bis-ligands will form. When this network reaches a sufficient size, it will no longer be retained in solution and will consequently precipitate.

Clearly a critical aspect of affinity precipitation is the length of the spacer connecting the two ligand entities. If the ligand binding is too short it may not be able to span the distance between two enzymes, a situation that might occur if the sites are deeply hidden beneath the contour line of the enzyme. On the other hand, a spacer that is too long may lead to intramolecular cross-linking, a situation bearing interest on questions concerning subunit arrangement, but giving no precipitation. For precipitation of dehydrogenases a Bis-NAD with a 17-Å spacer length, N_2, N_2' adipodihydrazido-bis(N^6-carbonylmethyl-NAD),[1] has functioned

[1] P.-O. Larsson and K. Mosbach, *FEBS Lett.* **98,** 333 (1979).
[2] S. Flygare, T. Griffin, P.-O. Larsson, and K. Mosbach, *Anal. Biochem.* **733,** 409 (1983).

ISBN 0-12-182004-1

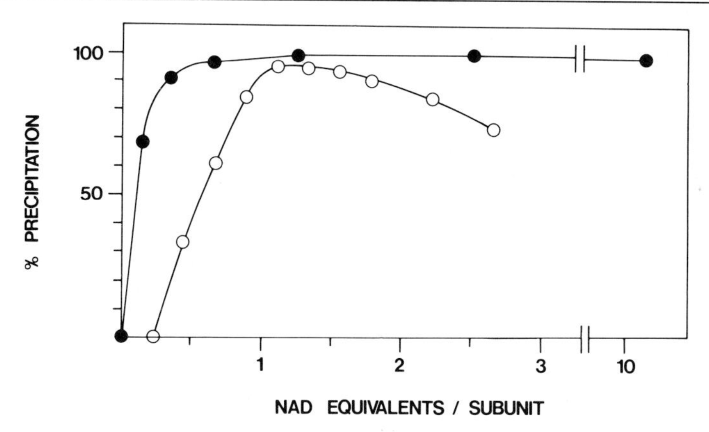

FIG. 1. Precipitation yield as a function of the ratio NAD equivalents per enzyme subunits after 16 hr. ○——○, Lactate dehydrogenase; ●——●, glutamate dehydrogenase.

well. Two other spacer lengths, 7 and 32 Å, have also been used with some systems.

In order to improve the selectivity of the affinity precipitation process and to increase the effective binding strength, ternary complex formation may be used. For example, Bis-NAD together with pyruvate or oxalate will form strong ternary complexes with lactate dehydrogenase, causing it to precipitate. In fact, in the absence of oxalate or pyruvate precipitation does not occur, an observation that is readily understood when the rather weak binary enzyme–coenzyme complex is considered (lactate dehydrogenase K_d for NAD = 3×10^{-4} *M*).[3]

A major point in affinity precipitation is the ratio between ligand entities and enzyme subunits. Maximum precipitation occurs at a ratio of unity. At lower ratios not all enzymes will be engaged in complexes, and at higher ratios only one end of the Bis-ligand may become attached to the enzyme. Moderate deviation from unity will be of less importance if the enzyme is an oligomer with a high subunit number. A comparison has been made between lactate dehydrogenase and glutamate dehydrogenase with respect to precipitation yield as a function of the ligand : subunit ratio (Fig. 1). Glutamate dehydrogenase is a hexamer, known to aggregate spontaneously when present in high concentrations, whereas lactate dehydrogenase is a tetramer without such properties. The precipitation yield for lactate dehydrogenase is satisfactory for ratios between 0.8 and 2.5 and almost quantitative at a ratio of 1.3. For glutamate dehydrogenase, on the other hand, a very wide range of ratios gives almost quantitative

[3] R. A. Stinson and J. J. Holbrook, *Biochem. J.* **131,** 719 (1973).

precipitation (0.3–10). The observation that the precipitation yield is very good, even at a ratio as low as 0.16, suggests that glutamate dehydrogenase under the prevailing conditions is an oligomer with at least 12 subunits.

In the following sections, procedures are given for affinity precipitation of lactate dehydrogenase from a crude extract and for glutamate dehydrogenase in a model experiment. Affinity precipitation within a gel is also described.

The Bis-Functional NAD-Derivative, Bis-NAD

The Bis-NAD derivative used below, N_2,N_2'-adipodihydrazido-bis-(N^6-carbonylmethyl—NAD), has a 17-Å spacer joining the two NAD entities to each other by their N^6 amino groups [NAD—$CH_2CONHNHCO(CH_2)_4CONHNHCOCH_2$—NAD]. The reagent is commercially available (Sigma Chemical Co., St. Louis, MO) or can be prepared as has been described.[1]

Affinity Precipitation of Lactate Dehydrogenase[2]

A homogenate of ox heart, subjected to centrifugation for 30 min at 20,000 *g* and dialyzed overnight at 4° against 0.05 *M* sodium phosphate at pH 7.5, serves as the source of lactate dehydrogenase for illustrating the procedure.

Optimum Bis-NAD Concentration (Pilot Affinity Precipitation)

To maximize yield in the affinity precipitation step, optimum Bis-NAD concentration is first determined in a small-scale affinity precipitation experiment. To a sample of the crude extract (2 ml), 0.25 ml of 0.1 *M* sodium oxalate and 0.25 ml of Bis-NAD are added. The concentrations of the Bis-NAD are chosen so that the nominal ratio between NAD equivalents and lactate dehydrogenase subunits are 1 : 4, 1 : 2, 1 : 1, 2 : 1, 3 : 1, and 4 : 1; the nominal enzyme concentration is estimated from activity measurements. Precipitation occurs almost immediately; after 30 min the precipitate is collected by centrifugation at 10,000 *g* for 10 min. Only the supernatant fluid need be analyzed for enzyme activity with the assumption that the tube with the lowest residual activity represents the best ratio for affinity precipitation. These conditions are then applied on a preparative scale.

Preparative Affinity Precipitation

When 200 ml of extract was treated at 4° with 25 ml of 0.1 *M* sodium oxalate and 20 ml of 130 μM Bis-NAD, as determined by pilot affinity precipitation, a precipitate formed that was collected by centrifugation after 20 hr. The precipitate was dissolved in 10 ml of sodium phosphate at pH 7.5 containing 0.6 m*M* NADH and dialyzed against the phosphate buffer. This solution represents a 50-fold increase in specific activity of lactate dehydrogenase. The recovery is about 90%, and the purity, as judged by electrophoresis, is greater than 95%.

Comments

Oxalate is the preferred third component when affinity-precipitating lactate dehydrogenase, although pyruvate can also be used.[2] A concentration of 10 m*M* oxalate was chosen since it is sufficient to promote complete precipitation under the conditions used. When the enzyme concentration is low, a higher oxalate concentration is recommended to further enhance complex formation. However, a high oxalate concentration has the drawback of interfering with activity measurements.

It is advisable to carry out a pilot affinity precipitation prior to the preparative affinity precipitation. Activity measurements of the crude extract usually give an underestimation of the lactate dehydrogenase concentration by a factor of 1.5–3, possible owing to the presence of inhibitory substances in crude extracts. Addition of Bis-NAD, based only on determination of enzyme activity, would give a low yield of precipitate as is evident from the data in Fig. 1. Although the pilot trial may seem a cumbersome extra step, it can be carried out rapidly. The precipitation yield for the best ratio is, at near maximum, 85% after only 20 min, justifying the 30 min allowed for the procedure. The actual maximum yield is reached after 2 hr. Thus, the 20 hr used for precipitation in the preparative procedure may be shortened considerably if convenient.

Model Affinity Precipitation of Glutamate Dehydrogenase[2]

Procedure

To glutamate dehydrogenase[4] (1.5 ml, 1.8 mg/ml) in 0.05 *M* sodium phosphate at pH 7.5, is added 0.25 ml of 0.8 *M* glutarate followed by 0.25 ml of 0.12 m*M* Bis-NAD. Upon gentle mixing, the solution rapidly be-

[4] Bovine liver, type I, Sigma, St Louis, MO.

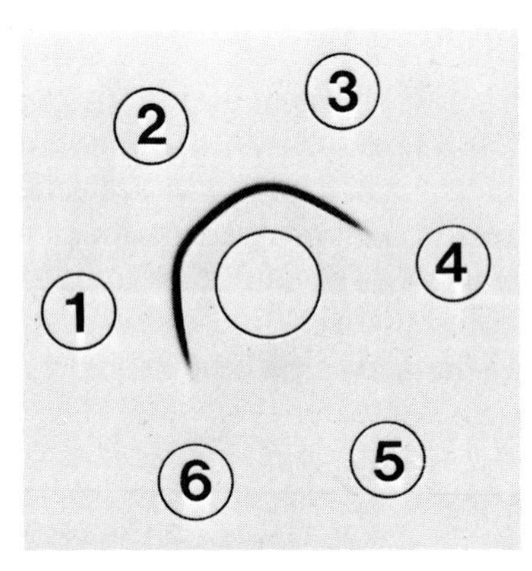

FIG. 2. Affinity precipitation in agarose gel. The center well contained 30 μg beef heart lactate dehydrogenase. For the peripheral wells 1 to 3, 1.5 nmol bis-NAD; well 5, 3 nmol NAD; and wells 4 and 6, buffer only. Reproduced, with permission, from Larsson and Mosbach.[1]

comes opaque; a precipitate is formed after 10 min; it is allowed to settle at 4° overnight. The precipitate is dissolved in 0.5 ml of 10 m*M* NADH.

Comments

Affinity precipitation of glutamate dehydrogenase is not particularly dependent on a correct ligand : subunit ratio as is the case for lactate dehydrogenase (Fig. 1). Affinity precipitates formed near a ratio of unity are extensively cross-linked, judged from the observation that a rather high NADH concentration is needed to achieve a reasonably fast dissolution. Affinity precipitation has been carried out at lower glutamate dehydrogenase concentrations (0.1 mg/ml), also with a precipitation yield of over 90%.

Affinity Precipitation in Gels

Procedure

Agarose gels containing 0.8% agarose, 0.3 *M* sodium pyruvate, and 0.05 *M* sodium phosphate at pH 7.5 are cast on microscope glass slides. Wells are punched out with a die as shown in Fig. 2. The center well is filled with 15 μl of lactate dehydrogenase solution (5–50 μg of enzyme). The peripheral wells are filled with 7 μl of Bis-NAD solution (0.5–10 nmol), with NAD solution (0.5–10 nmol), or with buffer. The slides are placed in a moist chamber at room temperature and allowed to develop for 1.5–30 hr, typically 16 hr. Bands of precipitated proteins may be developed subsequently with Amido Black.[5]

Comments

Affinity precipitation of lactate dehydrogenase in agarose gels closely resembles the corresponding immunoprecipitation technique.[5] Pyruvate has been used as the third component to enhance complex formation, and oxalate would probably work even better, judging from the experience with affinity precipitation from solution.[2] Since bands of precipitate are directly visible only when a high concentration of enzyme is used, a protein stain must usually be applied. Figure 2 shows that precipitation occurs only where lactate dehydrogenase diffuses toward Bis-NAD.

Other Systems

Yeast alcohol dehydrogenase has also been affinity precipitated.[2] For precipitation to occur (in the presence of pyrazole) 0.2 *M* sodium chloride must also be added. The precipitation was shown not to be a salting-out effect. However, the process was slow and of low yield.

Clearly, precipitation may be difficult to achieve also in other systems. An alternative approach would then be to apply the nonprecipitating mixture to a gel permeation column (e.g., a Sephacryl S-400 column) and subsequently collect the desired enzyme in the high-molecular-weight fractions.

Similar to the affinity precipitation described here is a report on the synthesis of *N,N'*-bis-3(dihydroxyboronylbenzene)adipamide and the demonstration of its ability to agglutinate red blood cells by complexing with cell surface carbohydrates.[6]

[5] J. Clausen, *in* "Laboratory Techniques in Biochemistry and Molecular Biology" (T. S. Work and E. Work, eds.), Vol. 1, p. 397. North-Holland Publ., Amsterdam, 1969.

[6] T. J. Burnett, H. C. Peebles, and J. H. Hageman, *Biochem. Biophys. Res. Commun.* **96,** 157 (1980).

[23] Protein Crystallization: The Growth of Large-Scale Single Crystals

By GARY L. GILLILAND and DAVID R. DAVIES

The establishment of the conditions necessary for growing large single crystals of a protein has always seemed to the uninitiated very much like black magic. Since these conditions can only be established empirically, and since there are many variables, each of which can be changed by small increments, it would appear at first sight that the number of possible combinations is close to infinite. We suggest that an examination of the factors (pH, temperature, precipitant) that have been used to produce large single protein crystals is useful in restricting the number of experiments necessary to find suitable growth conditions. It is also suggested that an analysis of the procedures that have been used successfully in crystallizing proteins indicates that certain methods are more practical in the laboratory than others. We shall outline the general procedures that have been used in the search for large single crystals of a size (>0.2 mm) suitable for use in an X-ray diffraction investigation and shall describe in detail the three with the most general utility for screening crystal growth conditions.

Factors Important in Crystallization

The growth of crystals can be separated into the distinct phases of nucleation, postnucleation growth, and cessation of growth.[1] Nucleation and postnucleation growth are driven by supersaturation of the protein in solution. Crystal nuclei will form only at a critical point of supersaturation in an appropriate environment, e.g., ionic strength, pH, temperature. After nucleation a crystal will continue to grow as long as a state of supersaturation exists. Growth stops when the protein environment is altered or when errors in the crystal lattice become too great to permit continued growth.

The usual procedure for growing large single crystals of a protein is to approach the point of supersaturation slowly in order to reduce the number of nucleation sites. This can be achieved by gradually altering the ionic strength, pH, temperature, or dielectric constant of a protein solu-

[1] Z. Kam, H. B. Shore, and G. Feher, *J. Mol. Biol.* **123,** 539 (1978).

METHODS IN ENZYMOLOGY, VOL. 104 ISBN 0-12-182004-1

tion. The many physical methods that have been employed for growing protein crystals have been reviewed by McPherson[2,3] and Blundell and Johnson.[4]

Protein Concentration

Proteins have been crystallized from solutions containing from one to several hundred milligrams per milliliter. However, for crystallization trials we would recommend a concentration of 10–20 mg/ml if possible. Some proteins that are relatively insoluble under the initial conditions of concentration may be rendered more soluble by changing the pH or ionic strength.

Preliminary Analysis of Precipitation Conditions

In order to obtain some indication of the quantities of precipitant that should be used, a preliminary examination should be made of the solubility of the protein as a function of precipitant concentration. This can be carried out in a depression slide with small quantities of protein solution (~10 μl) by adding the precipitant in small aliquots, sealing the depression with a coverslip, and observing the droplet for the next 15 min or so with a low-power microscope for signs of precipitation. The quantity of precipitant necessary to effect precipitation should be noted, and this concentration gradually be approached in the subsequent crystallization experiments.

This preliminary examination should be carried out over a broad range of pH, since there can be a large variation in solubility at different pH values. In many cases, a protein is least soluble at its isoelectric point, so that a prior investigation by techniques such as isoelectric focusing can be used to suggest conditions for certain types of crystallization (in particular, low ionic strength). For many enzymes it is desirable to attempt crystallization at the pH at which the enzyme is fully active.

Temperature is another important factor to consider. In general, laboratories are not equipped to allow crystallization trials to be carried out over a large range of temperatures, so that this work is usually done at room temperature or in a cold room (4–6°). The solubilities of individual proteins vary markedly with temperature.[4,5]

[2] A. McPherson, Jr., *Methods Biochem. Anal.* **23,** 249 (1976).

[3] A. McPherson, "Preparation and Analysis of Protein Crystals." Wiley, New York, 1982.

[4] T. L. Blundell and L. Johnson, "Protein Crystallography." Academic Press, New York, 1976.

[5] A. A. Green, *J. Biol. Chem.* **93,** 495 (1931).

Choice of Precipitant

Examination of the table indicates clearly that certain precipitants have been much more successful than others in inducing crystallization. Whether this reflects a genuine preferred ability of these materials to produce crystals, or merely reflects the fashions of crystallizers, is not always clear, but it would appear that the aspiring crystallizer would be well advised to begin with those precipitants that have proved to be the most successful. In particular, ammonium sulfate, polyethylene glycol (low ionic strength), and 2-methyl-2,4-pentanediol seem to be the precipitants of choice for an initial investigation.

General Methods

Direct Addition of Precipitants

Direct addition or batch procedures involve the use of salts, acids, bases, water, metal ions, or other effector molecules of the protein solution to induce crystallization. After the addition of the precipitant, if the solution is left undisturbed, nucleation and crystal growth will often occur. Nearly all of the early crystallizations were carried out by this method, but when used for exploring crystallization conditions, it does require a great deal of protein. If small volumes are used to conserve a precious supply of protein, it becomes difficult to add the precise amounts that are necessary for reproducibility of crystallization.

Nevertheless, when the crystallization conditions are known, and especially when they involve simply adjusting the pH or ionic strength of the protein solution, the batch procedure can often be reproducibly used on a microscale to produce large crystals.[6,7]

Dialysis

Traditionally, dialysis through a semipermeable membrane has been a favored technique. Dialysis is particularly applicable for proteins that crystallize at low ionic strength. The method is simple to use, and the point of supersaturation can be approached gradually. However, a dialysis bag requires relatively large amounts of material, and consequently this method has now been superseded by the Zeppezauer[8,9] procedure of

[6] M. Dobler, S. D. Dover, K. Laves, A. Binder, and H. Zuber, *J. Mol. Biol.* **71,** 785 (1972).
[7] R. R. Bott, M. A. Navia, and J. L. Smith, *J. Biol. Chem.* **257,** 9883 (1982).
[8] M. Zeppezauer, H. Eklund, and E. S. Zeppezauer, *Arch. Biochem. Biophys.* **126,** 564 (1968).
[9] M. Zeppezauer, this series, Vol. 22, p. 253.

PRECIPITATING AGENTS USED TO INDUCE CRYSTALLIZATION[a]

Precipitating agent	Crystal forms/subgroup	Crystal forms/group	Overall ranking
Salts			
Ammonium acetate	—	1	20
Ammonium citrate	—	2	19
Ammonium nitrate	—	1	20
Ammonium sulfate	296	361	1
+ Acetate[b]	2	—	—
+ Acetone	1	—	—
+ Ammonium chloride	1	—	—
+ Ammonium phosphate	1	—	—
+ Cesium chloride	12	—	—
+ Citrate[b]	4	—	—
+ Dimethylformamide	1	—	—
+ Dimethyl sulfoxide	1	—	—
+ Dioxane	3	—	—
+ Isocitrate[b]	1	—	—
+ Lithium nitrate	1	—	—
+ Phosphate[b]	3	—	—
+ Potassium phosphate	6	—	—
+ Sodium chloride	13	—	—
+ Sodium citrate	1	—	—
+ Sodium formate	1	—	—
+ Sodium phosphate	3	—	—
+ Cesium chloride	1	—	—
+ Sodium potassium phosphate	6	—	—
+ Tris[b]	3	—	—
Citrate[b]	—	4	17
Cadmium sulfate	—	4	17
Lithium chloride	—	1	20
Lithium sulfate	—	8	14
Magnesium chloride	—	1	20
Magnesium sulfate	—	9	13
Phosphate[b]	—	15	9
Potassium borate	—	1	20
Potassium nitrate	—	1	20
Potassium phosphate	—	11	11
Potassium sodium phosphate	15	16	8
+ Ammonium sulfate	1	—	—
Potassium tartrate	—	1	20
Sodium chloride	23	25	7
+ Sodium citrate	1	—	—
+ Sodium potassium phosphate	1	—	—
Sodium citrate	—	10	12
Sodium iodide	—	1	20
Sodium nitrate	4	5	16
+ Acetone	1	—	—

(continued)

TABLE (*continued*)

Precipitating agent	Crystal forms/subgroup	Crystal forms/group	Overall ranking
Sodium phosphate	—	4	17
Sodium sulfate	2	3	18
+ Acetate[b]	1	—	—
Organic precipitating agents			
Acetone	3	4	17
+ Salt	1	—	—
Dioxane	—	7	15
Dimethylsulfoxide	—	2	19
+ Ammonium sulfate	1	—	—
+ Calcium acetate	1	—	—
tert-Butanol	—	7	15
Ethanol	32	33	5
+ Ammonium sulfate	1	—	—
Glucose	—	1	20
Glutamic acid	—	1	20
Isopropanol	—	14	10
Methanol	—	7	15
2-Methyl-2,4-pentanediol	70	77	4
+ Dimethyl sulfoxide	1	—	—
+ Magnesium acetate	1	—	—
+ Magnesium chloride	2	—	—
+ Methanol	1	—	—
+ Sodium chloride	2	—	—
1,3-Propanediol	—	1	20
n-Propanol	—	5	16
Polyethylene glycol			
Polyethylene glycol (unspecified M_r)	10	11	11
+ Ammonium acetate	1	—	—
Polyethylene glycol 400 (avg M_r)	—	5	16
Polyethylene glycol 1000 (avg M_r)	—	1	20
Polyethylene glycol 2000 (avg M_r)	—	3	18
Polyethylene glycol 4000 (avg M_r)	29	30	6
+ Ammonium sulfate	1	—	—
Polyethylene glycol 6000 (or 8000) (avg M_r)	69	78	3
+ Calcium acetate	1	—	—
+ Lithium chloride	1	—	—
+ Sodium chloride	5	—	—
+ Potassium chloride	1	—	—
+ Phosphate[b]	1	—	—
Polyethylene glycol 20,000 (avg M_r)	—	3	18
Low ionic strength[c]	—	79	2

microdialysis into a capillary. Although not simple, the Zeppezauer method has several advantages. It requires very little material, and precipitated protein can be dissolved readily for another trial. However, the dialysis capillaries are not easy to assemble, and, without practice, it is difficult to load the protein solution into the capillary without introducing air bubbles. Crystals, when formed in the capillary, are difficult to observe and sometimes difficult to extract. Although there are several modifications of this procedure that use different types of dialysis cells to try to avoid the problems mentioned,[4,8–12] the Zeppezauer procedure remains the most popular. A description of the method is presented in detail.

Vapor Diffusion

Vapor diffusion is perhaps the most popular of the methods currently in use. It is effective in concentrating a protein solution containing a nonvolatile precipitant, or for diffusion of a volatile precipitant into the protein solution.

In the first case, the protein solution is prepared in a drop containing an appropriate amount of precipitant, e.g., ammonium sulfate or polyethylene glycol, with the intent of maintaining the protein below supersaturation. The drop is then equilibrated in a sealed chamber with a larger volume of a solution of the precipitant at a concentration that is just at, or slightly above, the concentration required for supersaturation. The protein solution then gradually approaches the supersaturation point as water evaporates from the drop and disperses into the reservoir.

[10] B. H. Weber and P. E. Goodkin, *Arch. Biochem. Biophys.* **141,** 489 (1970).
[11] V. Lagerkvist, L. Reyno, O. Lindquist, and E. Andersson, *J. Biol. Chem.* **247,** 3897 (1972).
[12] I. Rayment, *J. Appl. Crystallogr.* **14,** 153 (1981).

[a] This table was compiled by one of us (G. L. G.) from a data base consisting of data extracted from published crystallization procedures in which production of crystals suitable for X-ray diffraction studies was described. Included in this table are data for more than 500 macromolecules with more than 800 different crystal forms. The precipitating agents included in this table have been divided into groups according to the predominant additive. The subgroups define those cases in which additional substances have been added to induce crystallization. The subgroup elements have been tabulated if the concentration of additional salts was $>0.2\ M$ or if the concentration of organic agents was $>5\%$ (v/v).
[b] Unspecified counterion.
[c] Included in this category are many proteins whose crystallization has proved to be very sensitive to the presence of small amounts of divalent metal ions, ligands, products, or substrates or other effector molecules.

In the second case, in which a volatile precipitant is used, a similar procedure can be followed, although it is not necessary to add precipitant directly to the drop. Equilibration will bring the protein solution gradually to supersaturation through a transfer of precipitant from the larger volume to the drop.

There are two principal techniques that utilize vapor diffusion. They are the "sandwich box" technique[2-4,13-21] and the "hanging drop" method.[3,22] Despite a basic similarity, each has advantages for certain types of crystallization. In the "sandwich box" procedure, the crystallization drop is placed in a well of a depression slide; several slides may be enclosed in a sealed box with a larger volume of appropriate precipitant solution. The advantages of the procedure include the ability to use large drops (>100 μl), ease of preparation, and ease of observation. However, only a limited number of samples can be prepared per box, all must be equilibrated against the same reservoir solution, and, if one sample has to be removed from the box, the others will be disturbed.

In the "hanging drop" method, a drop of protein solution is suspended on the lower face of a microscope slide coverslip placed above equilibrating solution in the well of a 24-well tissue culture plate. Since each well can contain a different equilibrating solution, the method can be utilized to explore a variety of conditions in a compact manner. The drops can be easily observed, and the plates stack together in compact form for convenient storage. There are disadvantages; the drop size is limited to approximately 20 μl and larger drops will fall off, and it is difficult to use the procedure with many organic precipitants because the droplets spread out owing to low surface tension.

Crystallization by Concentration

Gradual concentration of a protein solution to produce crystals is a method that has been used infrequently in comparison with those meth-

[13] B. F. C. Clark, B. P. Doctor, K. C. Holmes, A. Klug, K. A. Marcker, S. J. Morris, and H. H. Paradies, *Nature (London)* **219,** 1222 (1968).

[14] S. H. Kim and A. Rich, *Science* **162,** 1381 (1968).

[15] A. Hampel, M. Labanauskas, P. G. Connors, L. Kirkegard, V. L. Rajbhandary, P. B. Sigler, and R. M. Bock, *Science* **162,** 1384 (1968).

[16] F. Cramer, F. Van Den Haar, W. Saenger, and E. Schlimme, *Angew. Chem.* **80,** 969 (1968).

[17] J. R. Fresco, R. D. Blake, and R. Langridge, *Nature (London)* **220,** 5174 (1968).

[18] C. D. Johnson, K. Adolph, J. J. Rosa, M. D. Hall, and P. B. Sigler, *Nature (London)* **226,** 1246 (1970).

[19] D. R. Davies and B. P. Doctor, "Procedures in Nucleic Acid Research," Vol. 2. Harper, New York, 1971.

[20] D. R. Davies and D. M. Segal, this series, Vol. 22, p. 266.

ods that reduce the solubility of the protein by the addition of precipitants. Nevertheless, it has been used, and in its simplest form allows evaporation to increase protein concentration.[23] Other techniques of concentration include a collodion bag apparatus[24] and dialysis against polyethylene glycol 20,000 or Lyphogel.[2,3]

Crystallization by Varying the Temperature

Another infrequently used procedure lowers the solubility by gradually raising or lowering the temperature. Baker and Dodson[25] described a method for crystallizing insulin by raising the temperature of the protein solution to 55° and then allowing it to cool slowly in a Dewar flask. The addition of insulation about the flask can further slow down the cooling process.[4]

Jakoby[26] has presented a general method of growing small crystals (about 1–10 μm) that depends on the differential solubility with respect to temperature of protein solutions in the presence of ammonium sulfate. Protein, precipitated with ammonium sulfate, is extracted with solutions of decreasing concentrations of ammonium sulfate at or near 0°. The resultant extracts are allowed to warm to room temperature when crystallization usually results. The crystals, although very small, could be used as seeds for growing larger crystals.

The Use of Detergents in Crystallizing Membrane Proteins

Spectacular advances have been made in crystallizing integral membrane proteins.[27-29] The methods use small amounts of nonionic or zwitterionic detergents such as octyl β-glucopyranoside and *N*,*N*-dodecyldimethylamine *N*-oxide in order to inhibit micelle formation and permit crystallization. Three different proteins, bacteriorhodopsin from *Halobacterium halobium*,[27] the photosynthetic reaction center from *Rhodopseudomonas viridis*,[28] and porin from *Escherichia coli*,[29] have been crystallized using vapor diffusion methods with salts or polyethylene glycol as

[21] S.-H. Kim and G. H. Quigley, this series, Vol. 59, p. 3.
[22] A. Wlodawer and K. O. Hodgson, *Proc. Natl. Acad. Sci. U.S.A.* **72,** 398 (1975).
[23] N. Camerman, T. Hofmann, S. Jones, and S. C. Nyburg, *J. Mol. Biol.* **44,** 569 (1969).
[24] D. C. Richardson, J. C. Bier, and J. S. Richardson, *J. Biol. Chem.* **247,** 6368 (1972).
[25] E. N. Baker and G. Dodson, *J. Mol. Biol.* **54,** 605 (1970).
[26] W. B. Jakoby, this series, Vol. 22, p. 248.
[27] H. Michel and D. Oesterhelt, *Proc. Natl. Acad. Sci. U.S.A.* **77,** 1283 (1980).
[28] H. Michel, *J. Mol. Biol.* **158,** 567 (1982).
[29] R. M. Garavito, J. Jenkins, J. N. Jansonius, R. Karlsson, and J. P. Rosenbusch, *J. Mol. Biol.* **164,** 313 (1983).

the precipitant with protein that had been solubilized with 0.5–1.0% detergent.

Crystallization Procedures

The Hanging Drop Procedure[3,22]

Equipment

A tissue culture tray with 24 wells (Linbro, Catalog No. 76-033-05)

Plastic or siliconized glass, 22 × 22 mm, microscope coverslips. Round coverslips, 22 mm in diameter, may also be used. Plastic coverslips are cleaned by rinsing in ethanol followed with distilled water

Stopcock grease

Procedure

1. Apply the stopcock grease around the rim of those wells in the tissue culture tray that will be used for crystallization attempts.
2. Prepare the precipitant solution and add approximately 1 ml at the appropriate concentration to each well.
3. Mix some precipitant with the protein solution and centrifuge in a small-volume bench-top centrifuge to remove any precipitate or debris. It is common to mix equal volumes of the well solution with the protein solution. Where precipitation is not a problem this can be done on the coverslip.
4. Place 5–20 μl of the protein–precipitant mixture on the coverslip, invert the coverslip, and seal it over the well chamber.

The Sandwich Box Procedure[2-4,3-21]

Equipment

The "sandwich box," a plastic or glass container (approximately 11 × 11 × 3 cm) that can easily be sealed

Glass microculture slides containing one or more depressions. These are usually siliconized to reduce spreading of the droplet

Stopcock grease

Procedure

1. Add a suitable volume (5–25 ml) of precipitant solution to the bottom of the box. Alternatively, the solution may be placed in a beaker or other suitable vessel.

2. Place the slides in the sandwich box on top of a support such as an inverted petri dish to raise it above the level of the surrounding precipitant solution.
3. Mix some precipitant with the protein solution and centrifuge to remove any precipitate or debris. As in the hanging drop method, it is common to mix equal volumes of precipitant in the box with the protein solutions.
4. Pipette 10–100 μl of the protein–precipitant mixture into the depression slide. Seal the sandwich box.

The Zeppezauer Microdialysis Procedure[8,9]

Equipment

Glass capillary, usually about 30 mm in length, with outer diameter of ~8 mm and an inner diameter of 0.5–2.0 mm. The ends of the capillary should be flattened and rounded to remove sharp edges that might damage the dialysis membrane. It is recommended that the glass capillaries be siliconized

Dialysis membrane, 2–3 cm^2 square. The membrane should have the appropriate molecular weight cutoff and be prepared according to the manufacturer's or one's own specifications

Tygon or PVC tubing is cut into ~1.5-cm pieces with notches of ~0.5 cm cut into one end. The tubing should have an inner diameter slightly less (~7 mm) than the diameter of the capillary to ensure a snug fit

Parafilm

Scintillation vial or other appropriate vessel that might be sealed and will allow observation inside the capillary

Procedure

1. Assemble the dialysis cell by covering one end of the glass capillary with the semipermeable membrane, attaching it securely with the Tygon or PVC tubing.
2. Transfer the protein solution into the dialysis cell using a syringe or small plastic tubing. Care must be taken to ensure that the membrane is not ruptured and that air bubbles are not trapped between the protein solution and the dialysis membrane.
3. Seal the open end of the dialysis cell with Parafilm.
4. Place the dialysis cell into the scintillation vial, which contains about 5 ml of the equilibration solution. Ensure that there are no bubbles between the outside of the dialysis membrane and the equilibration solution. Seal the vial.

Seeding

If crystals can be grown but remain too small for X-ray analysis in spite of attempts to make them larger by increasing the volume of the drop or increasing the protein concentration, seeding may be helpful. This may be done microscopically or macroscopically.

Microseeding[4] involves crushing a crystal or crystals followed by serial dilution of the suspension in a liquid that will not dissolve the crystals. Serial solutions are generally by factors of 10, and each is tested by adding a small amount, about 1 μl, to separate crystallizing droplets until the fewest crystals are generated. For this method, the protein solution is usually equilibrated to a point at or just below normal crystallization conditions.

In macroseeding, on the other hand, an intact single seed crystal is transferred to a fresh protein solution that has been brought very close to the original crystallization conditions.[30,31] The seed crystal is first washed several times in a solution that will dissolve potential nucleation sites and generate "fresh" surfaces on the crystal that will not inhibit further growth. It has been possible in this way to obtain large single crystals from proteins that, in the absence of seeding, were of inadequate size.

Final Considerations

As a general strategy for attacking the problem of crystallizing a specific protein, we recommend that the initial step be the testing of these precipitating agents most commonly used to grow protein crystals (see the table) at several pH values and at both room temperature and 4°. If crystallization does not occur in the presence of buffer and precipitating agent alone, occasionally the addition to the protein solution of small quantities of monovalent or divalent cations, ligands, substrates or products, substrate or product analogs, alcohols, and other organic molecules may aid in inducing crystallization. For compounds that bind to the protein, concentrations may range from stoichiometric amounts to 10 m*M*. For monovalent and divalent cation additives, the concentrations can range from 1 to 10 m*M*. For alcohols and other organic molecules the concentrations may range from only trace amounts to 5–10% (v/v). Mixtures of polyethylene glycol and salt, and even of polyethylene glycol and alcohols (>10%), have been reported to be effective.[32,33]

[30] C. Thaller, L. H. Weaver, G. Eichele, E. Wilson, R. Karlsson, and J. N. Jansonius, *J. Mol. Biol.* **147,** 465 (1981).

[31] J. D. G. Smit and K. H. Winterhalter, *J. Mol. Biol.* **146,** 641 (1981).

[32] R. J. Collier, E. M. Westbrook, D. B. McKay, and D. Eisenberg, *J. Biol. Chem.* **257,** 5283 (1982).

[33] R. J. Collier and D. B. McKay, *J. Mol. Biol.* **157,** 413 (1982).

Although the total number of possible experiments is infinite, crystals are often obtained after relatively few attempts. If a protein proves to be difficult to crystallize, however, the number of trials necessary may become too large to be practically feasible. One remedy to this situation has been proposed by Carter and Carter,[34] who have devised a statistically effective sampling method using an incomplete factorial procedure potentially to reduce the number of trials. Despite the existence of stubborn proteins that fail to crystallize, other proteins occasionally will crystallize over a wide range of conditions, with crystals appearing in the first few attempts. The usual case lies between the extremes.

[34] C. W. Carter, Jr. and C. W. Carter, *J. Biol. Chem.* **254,** 12219 (1979).

[24] Immunosorbent Separations

By GARY J. CALTON

Immunosorbent separation is based on the principle of molecular recognition, defined as the formation of a complex between two specific molecules and exemplified by the immunological complex between antigen and antibody. The application of chromatographic methods to immunosorbent separation can lead to spectacular one-step purification schemes for proteins from such complex mixtures as urine, cell extracts, and microbial fermentation broths. The reasons for the rising popularity of this process include simplicity, a high degree of specificity, selective affinities, rapid, single-step isolation, and minimal contamination by nonspecific proteins.

Preparation of Affinity Reagents

Immunosorbent chromatography reagents are prepared by immobilizing a specific antibody by means of an immobilization reagent to an appropriate chromatography matrix. The use of monoclonal antibodies for affinity reagents is superior in all cases to polyclonal antibodies from animal sera. The preparation of monoclonal antibodies cannot be covered here and the reader is referred to the excellent reviews that are available.[1–3]

[1] J. W. Goding, *J. Immunol. Methods* **39,** 285 (1980).
[2] R. H. Kennet, T. J. McKearn, and K. B. Bechtol, "Monoclonal Antibody Hybridomas: A New Dimension in Biological Analysis." Plenum, New York, 1980.
[3] J. J. Langone and H. Van Vunakis, this series, Vol. 92.

ISBN 0-12-182004-1

Antigen Preparation and Testing

In order to obtain a monoclonal antibody it is best to have a supply of homogeneous antigen, which can be obtained by use of alternate methods of isolation, e.g., by chromatography or electrophoresis. The purified antigen is then used to immunize an appropriate strain of animal in which the monoclonal antibodies are to be produced. Purified antigen is also best for immunization if animal antiserum, i.e., polyclonal antibody, is to be used. Similarly, homogeneous antigen is useful as a reagent for an enzyme-linked immunosorbent assay (ELISA)[4] or radioimmunoassay (RIA)[5] for identification of hybridomas producing monoclonal antibodies that react with the antigen. The quantities of purified antigen needed for immunization are quite small; 10 μg with an appropriate adjuvant may be sufficient. Tissue culture methods can be used for immunization of spleen cells *in vitro* and require even less antigen (10–100 ng).[6]

It is possible to use crude preparations of the antigen in order to obtain monoclonal antibodies and for the detection of the specific hybridomas that produce monoclonal antibodies to the antigen of interest. One scheme for using crude mixtures as the starting point for monoclonal antibodies[7] is the following:

Separation of the crude material is attempted on an ion-exchange resin and a purified fraction is obtained. The course of purification is determined, in this instance, by a pharmacological assay of each of the eluted fractions. A second and completely different method of ion exchange for separation of the crude mixture is also carried out and the chromatographic fractions obtained are again assayed pharmacologically. The chromatographic fractions from each of the two separations having the highest specific activity in the pharmacological assay are then used to develop an enzyme-linked immunosorbent assay. The protein of interest, which is enriched in each of these fractions, is used as the initial coat for an ELISA plate. Antibodies produced by murine hybridomas may then be incubated with each of the enriched chromatographic fractions, after which enzyme-labeled rabbit antimouse antibody is incubated in the ELISA. Following workup of the ELISA, the data are surveyed with the intent of determining which of the antibodies reacted with each of the chromatographic fractions having highest specific activity. The likelihood

[4] A. Voller, D. E. Biddwell, and A. Bartlett, "The Enzyme Linked Immunosorbent Assay (ELISA)." Dynatech Laboratories, Inc., Alexandria, Virginia, 1979.

[5] H. Van Vunakis and J. J. Langone, this series, Vol. 70, p. 201.

[6] R. A. Luben and M. A. Mohler, *Mol. Immunol.* **17,** 635 (1980).

[7] P. K. Gaur, R. L. Anthony, T. S. Cody, G. J. Calton, and J. W. Burnett, *Proc. Soc. Exp. Biol. Med.* **167,** 374 (1981).

of having the same ratio of contaminating proteins present in peak fractions from two diverse methods of chromatography is relatively low. Those wells having cells producing antibody to the active fractions are then cloned and the testing procedure repeated in order to determine which clones are producing specific antibody. The monoclonal antibodies that are obtained are tested for neutralization of the pharmacological activity as a check for the antibody actually bearing the desired activity.

This scheme provides a rational method for selection of hybridomas producing antibody to specific substances, even if the substance has not been purified. The method is valid, of course, only when pharmacological or enzyme activity can be tested. Direct neutralization of such activity may also be used to determine the presence of monoclonal antibodies against a specific protein in a mixture, although such assays are often more time consuming and difficult than the above procedure. Alternatively, analytical procedures such as electrophoresis or isoelectric focusing may be used; the resulting bands are identified by tagging the monoclonal antibody of interest with a radioactive isotope and following the reactivity with the proteins in the gel by means of a nitrocellulose blot.[8] Once a group of antibodies has been isolated, evaluation of their usefulness for immunosorbent chromatography can begin.

Specificity

A high degree of specificity is one of the major advantages in using monoclonal as compared to polyclonal antibodies. For example, monoclonal antibodies may be selected for a specific site on a protein molecule even if that site is not the most antigenic one. This differentiation can be fine tuned to allow recognition of peptides that differ by as little as one amino acid. This degree of selectivity also allows the isolation of individual subunits of a protein, thereby obtaining undegraded or partially degraded material. For instance, urokinase, which consists of two chains connected by a disulfide bridge, has been isolated as the preferred high-molecular-weight species[9] by use of a monoclonal antibody. This route was used because the light chain of urokinase is readily susceptible to proteolytic degradation. In urine and tissue culture fluids, both proteolytically degraded and nondegraded urokinases exist. The separation of the two species is difficult by alternate methods because the light chain is degraded most rapidly by proteases. It was possible, by selection of a monoclonal antibody for the light chain, to isolate only high-molecular-

[8] H. Towbin, T. Staehelin, and J. Gordon, *Proc. Natl. Acad. Sci. U.S.A.* **76,** 4350 (1979). See also this volume [33].

[9] D. A. Vetterlein and G. J. Calton, *Thromb. Haemostasis* **49,** 24 (1983).

weight urokinase consisting of both chains. This method provides a means of obtaining homogeneous, high-molecular-weight urokinase in a single step from such diverse, relatively complex starting materials as urine or tissue culture fluids. It also allows the isolation of materials that differ in pharmacological activity *in vitro* and which are extremely difficult to separate by alternate means.

Stability

Stability of the monoclonal antibody must be assessed if immunosorbent chromatography is to be used for separating mixtures that may contain proteolytic enzymes, as with crude protein extracts. Monoclonal antibodies from the same fusion may have widely differing stabilities to proteolytic degradation. If the stability of the antibody is not determined prior to immobilization, the number of times that an immunosorbent chromatography column can be reused may be severely reduced. Stability to proteolytic degradation may be determined by subjecting each of the monoclonal antibodies to each of the proteases present in the mixture to be separated, if they are known. If the identity of the protease is unknown, pepsin and papain should be used to examine the stability of the monoclonal antibody to proteolysis. Gel electrophoresis of the mixture after proteolysis will easily distinguish between stable and unstable antibodies by examination of the molecular weights of the fragments obtained in relation to the number of hours that the mixture was incubated with the protease.

Affinity

Of greatest importance for any chromatographic scheme is the affinity of the antibody for the protein of interest. It is possible to identify monoclonal antibodies with specific affinities that will dissociate from their antibody–antigen complex under conditions appropriate to maximum stability for the protein to be isolated. Most of the work with immunosorbent chromatography has used extremes of pH or chaotropic agents for dissociation of the antibody–antigen complex.[10] Such extreme conditions are unnecessary, because antibodies can be obtained that dissociate under specified circumstances. To accomplish this end, a previously developed ELISA can be used for antigen determination. The ELISA may be conducted under standard conditions by coating a 96-well plate with antigen followed by application of an excess of monoclonal antibody. A series of dissociation buffers, in which the antigen is stable, are added to the wells

[10] M. Wilchek, this volume [1].

containing the antigen–monoclonal antibody complex. Sufficient time for dissociation of the complex is allowed, from 1 to 2 hr, and the degree of dissociation is evaluated by addition of enzyme-labeled antiserum. The data obtained from this procedure will indicate the degree of dissociation of the antibody–antigen complex. Thus, one can search each of the clones for specific dissociation conditions. In this manner, it has been possible to select monoclonal antibodies that dissociate from their antibody–antigen complex under conditions as mild as 20 m*M* potassium phosphate and 750 m*M* sodium chloride at pH 7.[11] Under very mild conditions, a matrix for immobilization must be chosen that retains only slight quantities of nonspecifically adsorbed protein, because washing of the column must be limited, rapid, and gentle.

Polyclonal vs Monoclonal Antibodies

As noted, monoclonal antibodies are superior to polyclonal antibodies for immunosorbent chromatography. However, polyclonal antibodies can be used for this procedure with excellent results.[12–14] Basic differences in the use of monoclonal and polyclonal antibodies involve specificity, affinity, and capacity. In each area monoclonal antibodies possess significant advantages over the use of polyclonal antibodies. On the other hand, polyclonal antibodies are simpler for most investigators to develop because the procedures are clearly defined[15] whereas production of monoclonal antibodies requires a significant investment in equipment, time, and skills.

The specificity of monoclonal antibodies is much higher than that of polyclonal antibodies because a binding protein recognizing only a single site is produced by an appropriately cloned hybridoma. This is in contrast with the situation with polyclonal antibodies, in which each of the spleen cells in the animal is capable of producing an antibody, thereby providing a vast array of antibodies directed against many sites on the protein to which the animal has been exposed. Cross-reactivity with such animal sera is recognized.

Within an animal serum, both high- and low-affinity antibodies exist. In immunosorbent chromatography, antibodies with very low affinity tend to allow the antigen to bleed, i.e., to wash slowly from the column

[11] C. S. Cobbs, P. K. Gaur, A. J. Russo, J. E. Warnick, G. J. Calton, and J. W. Burnett, *Toxicon,* **21,** 385 (1983).

[12] J. W. Eveleigh and D. E. Levy, *J. Solid Phase Biochem.* **2,** 45 (1977).

[13] D. Vetterlein, T. E. Bell, P. L. Young, and R. Roblin, *J. Biol. Chem.* **225,** 3665 (1980).

[14] See this series, Vol. 34, [90–94].

[15] P. H. Maurer and H. J. Callahan, this series, Vol. 70, p. 49.

regardless of eluant, whereas antibodies with very high affinity are virtually impossible to dissociate from their antibody–antigen complex without resorting to denaturing agents. A similar situation holds with monoclonals, but with them it is possible to *select* the degree of affinity that is desired for chromatographic purposes.

The degree of affinity of the antibody also affects capacity of the immunosorbent chromatography reagent. With monoclonal antibodies, a defined moderate-level affinity can be obtained and recognized (by the buffer/elution experiments suggested above) and used for immunosorbent chromatography. However, because animal antisera contain antibodies with a large range of binding affinities, the antibody–antigen complex at these sites is normally not dissociable, with capacity decreasing severely upon initial use of the reagent. With polyclonal antibodies, 5 to 10 runs are required to obtain a level at which antigen is released under desirable conditions. Low-affinity antibodies, present in animal antisera, preclude isolation of the antigen of interest in high yields because the washing step required to remove extraneously adsorbed protein on the matrix results in bleeding of the antigen of interest due to dissociation of the weakly bound antigen.

Immobilization

Once appropriate monoclonal antibodies have been obtained, a number of methods of immobilization are available to the investigator.[16] Because the monoclonal antibodies differ from one another, immobilization conditions must be tailored to the specific monoclonal antibody. For optimum results in immobilization, it is recommended that a series of monoclonal antibodies be used, all of which react on ELISA with the antigen of interest. The antibodies may be from one or more fusions, but it is not unusual to find that not more than 1 in 10 monoclonal antibodies is suitable for use as an immunosorbent reagent. A major reason for this concerns the conformational constraints that may occur when the monoclonal antibody is immobilized. An erroneous impression of suitability for immunosorbent chromatography may be obtained from the ELISA. If the monoclonal antibody is complexed with the antigen that has been immobilized on a polystyrene ELISA plate, absolute freedom of conformation of the antibody is possible. When the monoclonal antibody is immobilized on a polymer matrix, its structure may be deformed, active sites may be bound, or the Fab portion may be in a position in which it is not possible to form an antibody–antigen complex. These observations explain the

[16] K. Mosbach, this series, Vol. 44.

discrepancy that is often found between reactivity in an ELISA and results with a monoclonal antibody in immunosorbent chromatography.

Chromatography

The chromatographic step itself is carried out by the techniques of standard affinity chromatography.[17] The antigen should be charged onto the column in a buffer system adequate for protein stability, while at the same time being sufficiently dilute to avoid dissociation of the antibody–antigen complex. Such conditions are determined by a study of the protein itself and from the ELISA/buffer dissociation experiments that have been described. The concentration of the antigen in the solution is of little importance because the concentration of the antigen is normally dilute when compared to the quantity of antibody immobilized on the column (2–10 mg/ml of matrix). Because the antibody will be in large excess as the chromatographic front advances, the equilibrium will favor formation of the antibody–antigen complex. Antibody–antigen complexes are rapidly formed under chromatographic conditions and flow rates as high as 20 bed volumes per hour have been used. The normal limitation imposed upon the flow rate is due to the polymer matrix on which the monoclonal antibody has been immobilized, rather than to the formation of the antibody–antigen complex. For laboratory isolations this is not a hindrance, but in large-scale purifications the flow rate becomes a limiting factor.

Washing conditions are based upon the previously described ELISA, which determines the dissociation conditions to be used. Such conditions should be sufficiently mild that dissociation of the antibody–antigen complex will not occur, but sufficiently strong that the column will be freed of nonspecifically adsorbed proteins. Once extraneous proteins have been desorbed, as determined by adsorbance or other suitable analytical methods, dissociation buffers are used as indicated by the ELISA buffer assay. It is not unusual that either stronger or weaker buffers than those selected by ELISA be required for dissociation of the antibody–antigen complex because immobilization of the antibody may alter the binding affinity of the antibody.

The method of immunosorbent chromatography, particularly with the use of monoclonal antibodies, is rapid, specific, and gentle, and the affinity reagents can be used repeatedly.

[17] W. B. Jakoby and M. Wilchek, this series, Vol. 34.

Section V

Related Techniques

[25] Preparation of High-Purity Laboratory Water

By GARY C. GANZI

High-purity water is critically important for separations, analyses, and biological preparations.

Water Impurities

There is ample evidence that research results may be affected significantly by levels of water impurities that are minute in comparison to levels found in typical potable supplies. Contaminants may occur naturally, be added by potable treatment processes, or be leached from distribution systems. Levels of contaminants in source waters may vary seasonally and, at any given location in the distribution system, may vary daily. Such impurities are commonly categorized as follows.

Organic and inorganic *particulates* are found in all natural waters. They are also added by sloughage from distribution system components and can be formed by the interaction and precipitation of dissolved constituents. Particles may clog precision equipment and shield bacteria from disinfection. Colloidal materials (particles smaller than about 0.1 μm in diameter), can interact with biological systems as a result of their inherent electrostatic charge. Particulate concentrations in potable water are commonly in the part per million range.

Concentrations of *dissolved organic compounds* (expressed as total organic carbon) are in the part per million range in many potable water supplies. In most surface water supplies, the primary process of synthesis of dissolved organics is the natural biodecomposition of living matter. For example, tannins, lignins, and humic acids are formed by the decomposition of vegetation. Other sources are agricultural runoff and compounds added during treatment of municipal water. Pesticides, herbicides, detergents, and polyelectrolytes are commonly found in the part per billion range of concentrations. In addition, a wide variety of organic compounds are often synthesized in water supplies by the interaction of organic and inorganic impurities with oxidizing additives used for microbial control.

Dissolved inorganics are primarily ionic in nature and are found in relatively large concentrations in most water supplies. Typical concentrations of major constituents are shown in Table I. Trace inorganic constituents, i.e., less than about one part per million, commonly include metals and chlorine-containing sanitizing agents, which even at parts per billion

METHODS IN ENZYMOLOGY, VOL. 104

ISBN 0-12-182004-1

TABLE I
MAJOR INORGANIC CONSTITUENT CONCENTRATIONS OF TYPICAL POTABLE WATER SUPPLIES

Constituent	Concentration of inorganic constituents (parts per million)								
	New York City[a]	Los Angeles[b]	Chicago[c]	Boston[d]	Palo Alto[e]	Dallas[f]	St. Louis[g]	Suffern, N.J.[h]	San Juan, P.R.[i]
Calcium	17	81	36	9	68	26	—	43	38
Magnesium	6	16	10	5	19	4	45	11	6
Sodium	10	22	5	11	69	17	44	61	14
Potassium	2	5	1	1	11	—	—	2	6
Bicarbonate	59	190	64	8	429	61	26	137	73
Sulfate	16	83	25	5	7	38	152	35	50
Chloride	24	56	11	11	35	18	24	100	30
Nitrate	1	4	0.2	0.1	2	1	5	3.5	6
Silica	5	22	3	—	106	—	—	15	30
Oxygen	11	—	12	10	—	—	—	—	—

[a] Groton Supply from representative sites: average values for 1978.
[b] Municipal Supply, February, 1982.
[c] City of Chicago, South Water District, February, 1976.
[d] Weston Reservoir, 1980 average.
[e] Municipal Supply, December, 1981.
[f] Dallas Utilities East Side Plant, June, 1979.
[g] Municipal Supply, May, 1972.
[h] Municipal Well Source, January, 1982.
[i] Municipal Supply, November, 1981.

concentrations, e.g., the heavy metals, may affect both biological and biochemical systems.

Microorganisms are found in most water sources and tend to proliferate in distribution systems. In addition to being a direct source of contamination, enzymes produced by them may interact with biochemical compounds and reduce product purity. Some endotoxins from gram-negative bacteria are known to affect biological systems in concentrations in the part per trillion range.

Water Treatment Systems

A wide variety of water treatment systems are available to the investigator that provide water suitable for laboratory use. Each system has its advantages and limitations and, since no individual unit process is capable of removing all contaminants, water treatment equipment is often installed in sequential arrangements to provide adequate purity. Some of

TABLE II
RETENTION EFFICIENCIES OF CARTRIDGE-TYPE DEPTH FILTERS[a]

Cartridge description	Weight percentage retention of AC fine dust[b]
Porous cellulose (5-μm rating)	57
Spun polyethylene (5-μm rating)	60
Polypropylene wound fibers (10-μm rating)	72
Polypropylene wound fibers (5-μm rating)	79
Cotton wound fibers (10-μm rating)	81
Cotton wound fibers (5-μm rating)	85

[a] Data obtained by F. Badmington, Millipore Corp., Ashby Road, Bedford, MA.

[b] Purchased from the AC Spark Plug Division of General Motors Corp. The cartridges were challenged with 200 ppm dust in water suspension. Dust particle size distribution as specified by the vendor was as follows:

Particle diameter range (μ*m*)	*Weight percentage*
0–5	39 ± 2
5–10	18 ± 3
10–20	16 ± 3
20–40	18 ± 3
40–80	9 ± 3

the more common laboratory water treatment processes are discussed here.

Depth Filtration. Depth filtration is primarily effected by passing water through beds of granular material such as sand or through cartridges constructed of closely wound polymeric fibers or packed porous matrices. Depth filters primarily remove particles larger than 1 μm in diameter. Removal is more efficient as particle size increases. Typical removal efficiencies of common cartridge-type depth filters are shown in Table II. The predominant mechanism of particle removal is physical exclusion. Surface modification of the filter matrix can often result in higher removal efficiencies by adsorption mechanisms. In some cases, depth filtration is preceded by chemical reagent addition. This can cause agglomeration of colloids or precipitation of dissolved materials. The particles formed are then large enough to be removed by depth filters. Most depth filters are discarded and replaced when plugging of the filter matrix causes excessive pressure losses in the purification system. However, beds of sand

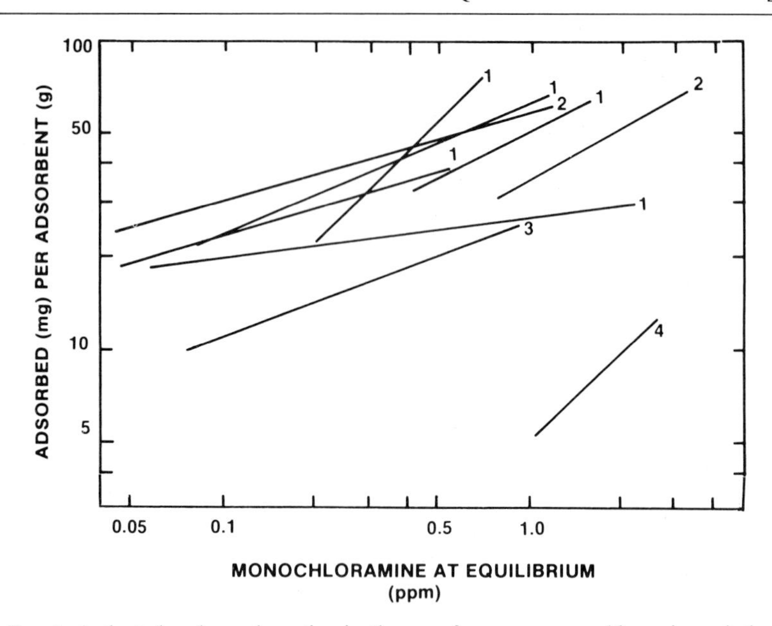

FIG. 1. Activated carbon adsorption isotherms of aqueous monochloramine solutions at 25° and pH 7.5. Curves: 1, bituminous coal based; 2, coconut shell based; 3, petroleum based; 4, synthetic polymer based. Data obtained by K. Siu, Continental Water Systems Corp., Ashby Road, Bedford, MA.

and similar media can be flushed clean of particles by periodic high velocity backwash cycles.

Activated Carbon Treatment. Activated carbons are treated carbonaceous materials with extremely high porosity and surface areas. The structure of activated carbon allows physical adsorption and catalytic decomposition of some organic contaminants and chlorine-based sanitizing agents. Granular or powdered carbons are available in a wide variety of pore size distributions. As a result, their ability to adsorb specific contaminants varies widely. Adsorption isotherms for monochloramine, a common potable water-sanitizing agent, are shown in Fig. 1. In general, a carbon with a predominance of large pores performs more efficiently on tannins, lignins, humic acids, and colloidal particles. A high-surface-area carbon with a predominance of small pores removes lower molecular weight compounds more efficiently. For this reason, many high-performance water purification systems utilize a large-pore carbon bed for primary organic removal, followed by a small-pore polishing bed to remove low-molecular-weight compounds. Dissolved organic carbon levels usually can be reduced to concentrations below 50 parts per billion in this way. Carbon filters are discarded and replaced when their effectiveness diminishes, although commercial regeneration services are available.

Ultrafiltration. Ultrafilters are available primarily as polymeric membranes in flat sheet or tubule form. When contaminated water is driven through them under pressure, their extremely small pores efficiently, but not entirely, prevent passage of particles, colloids, microorganisms, and bacterial endotoxins. Typical retention efficiencies are shown in Fig. 2. Ultrafilters range in nominal pore size. For water treatment, they are commonly specified to exclude 90–95% of contaminants of 10,000 daltons. Despite their small pore size ratings, water flux through ultrafilters is usually high at relatively low transmembrane pressure differentials (170–340 kPa). In order to reduce membrane fouling and loss of flux, ultrafiltration equipment is designed to optimize the rate of diffusion of excluded impurities away from the membrane surface. When membrane fouling does occur, ultrafiltration modules can be chemically cleaned either in place or by a commercially available service.

Ion Exchange. Ion exchangers contain chemically bound ionic groups

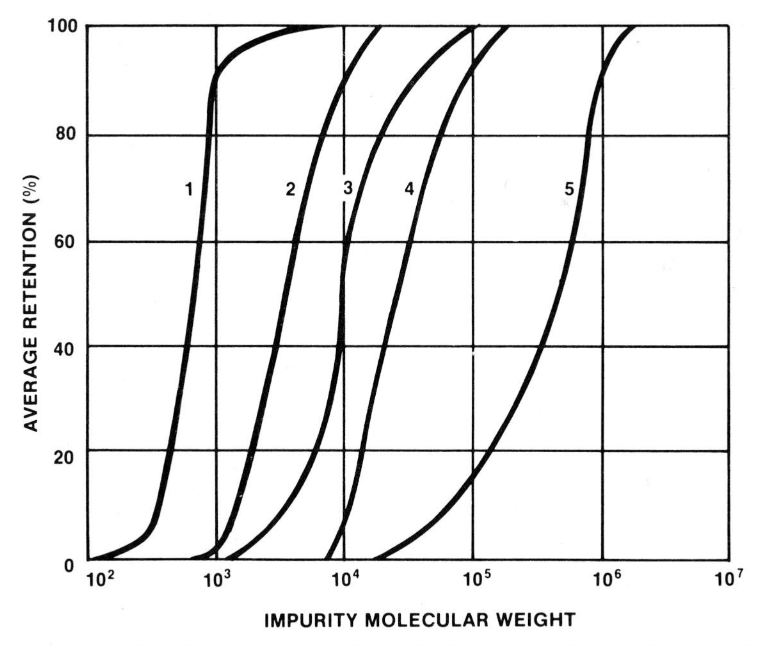

FIG. 2. Retention efficiencies for ultrafilters of different pore size as a function of impurity molecular weight. Curves: 1, 1×10^3 dalton rating; 2, 1×10^4 dalton rating; 3, 2.5×10^4 dalton rating; 4, 1×10^5 dalton rating; 5, 1×10^6 dalton rating. Data were obtained from Millipore Corp., Ashby Road, Bedford, MA.

TABLE III
TYPICAL METAL PURITY LEVELS ATTAINABLE BY USE OF ION EXCHANGE AND DISTILLATION[a]

Metal	Two-stage still	Two-stage quartz still	Single-pass ion exchange	Ion exchange followed by distillation	Milli-Q[b]
Al	10	0.5	—	0.1	<0.05
Ag	1	—	—	—	0.01
B	0.01	—	—	—	3
Ca	50	0.07	0.2	0.03	<0.05
Cr	—	—	0.02	—	<0.1
Cu	50	—	—	—	<0.05
Fe	0.1	—	0.02	—	<0.01
Mg	80	0.05	2	0.01	0.03
Mn	0.01	—	0.02	—	<0.01
Na	1	—	—	—	0.07
Ni	—	—	0.002	—	<0.01
Pb	50	—	0.02	—	<0.05
Si	50	5	—	1	1
Sn	50	—	—	—	<0.1
Zn	10	—	0.06	—	0.03

[a] A. R. Knott, *At. Absorpt. Newsl.* **14,** No. 5, 126 (1975). Data are given in parts per billion.

[b] Trademark of Millipore Corp. The Milli-Q contains activated carbon and ion-exchange beds followed by a microporous filter.

in equilibrium with free replaceable ions of opposite charge. They are primarily available in the form of polymeric beads. The most common laboratory configuration are beds consisting of a mixture of cation- and anion-exchange resins. Water is deionized as it passes through these beds by the exchange of ionized contaminants with equal equivalents of hydrogen and hydroxide ions. Since ion-exchange beds are extremely efficient deionizers, purification levels approaching the theoretical electrical resistivity of water (18.3 Ω-cm $\times$ 10^6) are routinely obtained. Typical purity levels attainable are shown in Table III.

Ion-exchange resins are available with variations in charge density, porosity, and chemical structure and, therefore, have a wide range of properties and purification capabilities. Strongly ionized resins are utilized for high degrees of deionization and the removal of dissolved silica and carbon dioxide. They are most effective in the absence of compounds that may irreversibly foul them, e.g., dissolved iron or fulvic acid. Macroporous resins are utilized to deionize in the presence of colloids. Weakly ionized resins are available for pretreatment and partial deionization of highly fouling feeds.

Ion-exchange resins can also be used as adsorbents for removal of colloids and dissolved organics. In addition, tailored resins are available for the preferential reaction and reduction of concentrations of specific ions, particularly heavy metals, to extremely low levels.

Since ion exchangers remove ionic impurities by chemical substitution, their ion removal capacity is limited. For applications involving high ionic loads, where resin disposal is not economical, resins are regenerated by use of special equipment on site or by a commercially available regeneration resin exchange service. For applications involving low ionic loads, exhausted resins are often discarded.

The proper use of ion exchangers depends on a thorough understanding of the multicomponent chemical equilibria involved. The reversible nature of ion-exchange reactions can result in leakage of trace quantities of ions (e.g., sodium, silicate, and chloride) by elution with ions (e.g., calcium and sulfate) in the feed water that associate more strongly with the resins.

For applications that indicate the use of disposable resins, leakage of ions is minimized with resins of maximum acidity and basicity (such anion resins are called type I resins). To minimize residual ions from resin synthesis, a multistep chemical conditioning process is performed using large excesses of acid and caustic. Cation and type I anion resins conditioned in this manner are called nuclear grade resins.

In cases where resins are regenerated, the extent of ionic leakage also depends on the service history of the resin, the thoroughness of regeneration, and the ease of contaminant removal from the resin. The advantage of the high basicity of type I anion resin becomes a disadvantage during regeneration because of its high affinity for contaminants. For example, relatively small concentrations of chloride ion in the regeneration caustic can cause significant chloride leakage upon subsequent deionization with type I resins.[1] Higher water purity can often be obtained with resins having improved regeneration characteristics, such as the dimethylethanolamine-based type II anion resins.[2] These resins have a higher affinity for hydroxide ion than the trimethylamine-based type I resins. Con-

[1] F. X. McGarvey, S. M. Ziarkowsky, E. W. Hauser, and M. C. Gottlieb, "Effect of Caustic Quality on the Performance of Strong Base Anion Exchangers." Sybron Chemical Division, Birmingham, NJ, IWC-81-31.

[2] M. C. Gottlieb and G. P. Simon, "Type II Strongly Basic Anion Exchange Resins for Two Bed Demineralizers." Presented at the 13th Liberty Bell Corrosion Conference, September 18, 1975, Philadelphia, PA.

TABLE IV
TYPICAL SEPARATION EFFICIENCIES OF CELLULOSE ACETATE REVERSE OSMOSIS MEMBRANES[a]

Solute	Percentage separation
LiF	98.5
LiCl	94.2
LiBr	93.2
$LiNO_3$	88.9
NaCl	94.4
KF	94.8
KCl	94.5
KBr	93.7
$LClO_3$	91.2
KNO_3	86.5
$KClO_4$	86.3
RbCl	94.7
RbBr	94.5
CsCl	95.0
CsBr	94.5
$MgCl_2$	98.0
$MgBr_2$	96.9
$Mg(NO_3)_2$	96.1
$CaCl_2$	96.4
$SrCl_2$	95.8
$BaCl_2$	96.3

[a] T. Matsuura, L. Pageau, and S. Sourirajan, *J. Appl. Polym. Sci.* **19,** 179 (1975).

taminants are more readily eluted from type II resins, and they are less prone to recontamination by reagents and rinse water during regeneration.

Distillation. Equipment for distillation purifies by vaporizing feed water and condensing the vapor as product. Impurity removal efficiencies generally increase as the impurity vapor pressure decreases below that of water. Therefore, distillation efficiently removes particles, microorganisms, macromolecules, and ionic contaminants. Product purity generally can be improved by increasing the number of consecutive times the product is distilled, by increasing the ratio of feed to product water, and by using stills with an inert construction material, such as quartz. Typical product quality from stills is listed in Table III.

Reverse Osmosis. Reverse osmosis equipment makes use of membranes, usually in the form of spirally wrapped flat sheets or hollow fibers, that are highly permeable to water and exclude most impurities (primarily on the basis of molecular size and charge). Such systems efficiently remove particles, microorganisms, macromolecules, and most dissolved organic materials. Ion rejections generally vary from 90 to 99%. Since product flow per unit membrane area is a function of the transmembrane pressure differential, most reverse osmosis equipment utilizes a high-pressure feed pump to increase productivity. Typical removal efficiencies are presented in Table IV.

Ultraviolet Irradiation. Irradiation of water can destroy microorganisms and, depending on intensity and frequency, decompose a number of organic materials. Typical capabilities are shown in Table V. Irradiation equipment is designed to maximize energy input density and to prevent coating of the ultraviolet lamps with water impurities.

TABLE V
INCIDENT ULTRAVIOLET ENERGIES AT 2537 Å NECESSARY FOR MICROORGANISM INHIBITION[a]

	Energy required for inhibition (microwatt – sec/cm²)	
Organism	90%	100%
Bacteria		
Bacillus subtilis	5,800	11,000
B. subtilis (spores)	11,600	22,000
Escherichia coli	3,000	6,600
Micrococcus candidus	6,050	12,300
Pseudomonas aeruginosa	5,500	10,500
Pseudomonas fluorescens	3,500	6,600
Staphylococcus albus	1,840	5,720
Staphylococcus aureus	2,600	6,600
Streptococcus lactis	6,150	8,800
Yeast		
Saccharomyces ellipsoideus	6,000	13,200
Saccharomyces cerevisiae	6,000	13,200
Mold spores		
Aspergillus glaucus	44,000	88,000
Aspergillus flavus	60,000	99,000
Aspergillus niger	132,000	330,000

[a] R. Nagy, *Am. Ind. Hyg. Assoc. J.* **25,** 274 (1964).

TABLE VI
TYPICAL REMOVAL CAPABILITIES AND LIMITATIONS OF PURIFICATION PROCESSES

Treatment step	Primary impurities removed	Primary impurities added	Physical and chemical limitations
Depth filtration	Particles	Ions, dissolved organics, microorganisms, endotoxin	May release particles at high pressure differentials
Activated carbon	Dissolved organics, chlorine	Ions, carbon fines, microorganisms, endotoxin	Particulate fouling may reduce removal efficiency
Ultrafiltration	Particles, microorganisms, endotoxin	—	Particulate fouling may reduce throughput
Ion exchange	Ions	Microorganisms, endotoxin	Particulate or chemical precipitation may reduce removal efficiency; chlorine sensitive
Distillation	Particles, microorganisms, endotoxin, ions	Components of construction materials	Particulate fouling or chemical precipitation may reduce throughput and removal efficiency
Reverse osmosis	Particles, microorganisms, endotoxin, dissolved organics, ions	—	Particulate fouling or chemical precipitation may reduce throughput and removal efficiency; Certain membranes are pH or chlorine sensitive
UV irradiation	Microorganisms	Carbon dioxide, decomposed organics	Particulate fouling may reduce inhibition efficiency
Microporous membrane filtration	Particles, microorganisms	—	Particulate fouling may reduce throughput

Microporous Membrane Filtration. Microporous membrane filters are polymeric filters of extremely well-defined pore size that quantitatively screen particles and microorganisms larger than their rated size. Microporous membranes also remove particles smaller than their rating by a depth filtration mechanism (although removal for these smaller particles is not absolute). Membrane filters are available primarily in the form of flat sheets or pleated cartridges in disposable configurations, designed to maximize membrane surface area per unit volume. Since microporous filters tend irreversibly to plug in the presence of high particulate loads, their predominant use in water purification is as a final polishing and microbiological barrier.

Process Considerations

Choosing the appropriate combination of purification steps for production of laboratory-grade water depends not only on the desired product water quality and quantity, but also on the physical and chemical limitations of the purification equipment. For example, several of the processes remove certain impurities efficiently, but may themselves add trace quantities of other impurities. For this reason, the order of the purification steps must be carefully selected. In addition, the active agent in the equipment can sometimes be chemically degraded, fouled, or plugged by specific water impurities, necessitating the use of pretreatment steps to maintain equipment performance. Typical removal capabilities and common limitations of each purification process are presented in Table VI. The schematics in Fig. 3 show some common unit operation sequences for the production of laboratory-grade water.

Practical Considerations

Practical considerations in the choice of water purification equipment should include not only initial costs, operating and maintenance costs, and laboratory space required but such other factors as may influence water quality. Equipment that is unreliable, unsuited to feed water conditions, or requires frequent cleaning may produce a product of poor quality. The capability of the purification equipment in providing the needed quantity of water on demand is a distinct advantage, since ultrapure water tends to become contaminated when stored. Clearly, the ability of the equipment manufacturer to provide consistent, high-quality, and properly treated membranes, resins, and wetted materials of construction and to provide the user with the information required to maintain the equipment is a critical factor in the ability to produce high-quality water.

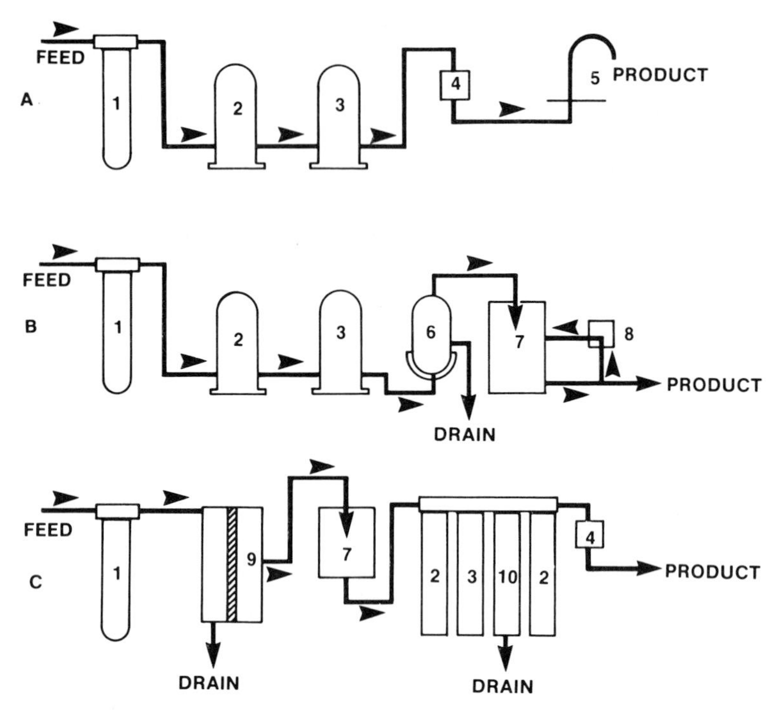

FIG. 3. Schematics of common unit operation sequences for the production of laboratory-grade water: 1, depth filter; 2, activated carbon bed; 3, ion exchanger; 4, microporous filter; 5, faucet; 6, still; 7, storage tank; 8, ultraviolet sterilizer; 9, reverse osmosis system; 10, ultrafilter. (A) High-purity water using service ion exchange. (B) High-purity water production using distillation. (C) Ultrapure water production using reverse osmosis, two-stage carbon adsorption, and ultrafiltration. Depending on feed water quality, the systems shown may require pretreatment by processes such as depth filtration, ultrafiltration, softening, and pH adjustment.

Ideally, the investigator has been able to identify the water contaminants that may affect experimental results and can therefore specify equipment that provides water of the required purity. Since this is usually not the case, the investigator can review published water quality standards and determine which equipment manufacturers meet or exceed them. Table VII summarizes the purity specifications established by the American Society for Testing and Materials (ASTM), the College of American Pathologists (CAP), and the National Committee for Clinical Laboratory Standards (NCCLS).

Standard methods for the examination of water and waste water also include a "distilled water suitability test" that determines whether puri-

TABLE VII
LABORATORY WATER PURITY STANDARDS[a]

Standard	Society	Type I	Type II	Type III	Type IV
Electrical resistivity, minimum,	ASTM	16.66	1.0	1.0	0.2
Ω-cm $\times$ 10^6 at 25°	CAP	10.00	2.0	0.1	—
	NCCLS	10.00	2.0	0.1	—
pH at 25°	ASTM	—	—	6.2–7.5	5.0–8.0
	CAP	—	—	5.0–8.0	—
	NCCLS	—	—	5.0–8.0	—
Color-retention time of potassium	ASTM	60	60	10	10
permanganate (min)	CAP	60	60	60	—
	NCCLS	60	10	—	—
Total bacteria (colonies/ml)	ASTM	—	—	—	—
	CAP	10	10^4	—	—
	NCCLS	10	10^3	—	—
Particulate matter, maximum size	ASTM	—	—	—	—
(larger than 0.2 μm)	CAP	None	—	—	—
	NCCLS	None	—	—	—
Silica (parts per million, SiO_2,	ASTM	—	—	—	—
maximum)	CAP	0.05	0.1	1.0	—
	NCCLS	0.05	0.1	1.0	—

[a] Type I, reagent grade water; type II, analytical grade water; types III and IV, general laboratory grade water.

fied water inhibits or promotes the growth of bacteria.[3] Other evaluations report the growth of cells in tissue culture and evaluations of cell yields as a function of water quality.

The technological advances in biology and biochemistry and improved sensitivity of analytical methods have increased the need for water of high purity. The requirement of highly purified water should be recognized as a necessary component of a successful research program.

[3] B. L. Green and W. Litsky, *J. Food Prot.* **8,** 654 (1979).

[26] Buffers for Enzymes

By JOHN S. BLANCHARD

The desirability of maintaining a stable pH during an enzyme-catalyzed reaction has long been realized.[1] Since 1900, when phosphate was first used as a means of maintaining a stable pH in enzyme studies,[2] a large number of inorganic compounds have served as buffers, including phosphate (pK_2 7.2), cacodylate (pK 6.3), borate (pK 9.2), and bicarbonate (pK 6.4 and 10.0). The use of weak organic acids and bases extended the range of pH beyond that which could be obtained with inorganic buffers. Unfortunately, many buffers have intrinsic handicaps associated with their use in biochemical systems, e.g., interactions with substrates or enzymes, resulting in reduced activity. Because buffers are generally present in much higher concentration than any other component in reaction mixtures, interactions of any sort can seriously influence the interpretation of enzymological data. In this context, the pioneering development by Good and his colleagues[3] in preparing a series of N-substituted taurine and glycine buffers with pK values in the region of most interest to biochemists has allowed the systematic evaluation of buffers and specific buffer effects. Since the first report by Good, a number of other zwitterionic, amino-containing, sulfonic acid buffers with pK values ranging from 6.1 to 10.4 have been developed[4,5] and are commercially available. A comprehensive discussion of buffers for biological research is available,[6] and should be consulted for more complete information.

Buffer Selection

Purity and availability are of major practical importance in the choice of a buffer. Most, if not all, of the buffers listed in Table I meet these standards and are, as well, reasonably inexpensive. Many are routinely analyzed commercially for their heavy-metal content.

[1] This series, Vol. 22, p. 3, and Vol. 24, p. 53.

[2] A. Fernbach and L. Hubert, *C. R. Hebd. Seances Acad. Sci.* **131,** 293 (1900).

[3] N. E. Good, G. D. Winget, W. Winter, T. N. Connolly, S. Izawa, and R. M. M. Singh, *Biochemistry* **5,** 467 (1966).

[4] M. A. Jermyn, *Aust. J. Chem.* **20,** 183 (1967).

[5] W. J. Ferguson, K. I. Braunschweiger, W. R. Braunschweiger, J. R. Smith, J. J. McCormick, C. C. Wasmann, N. P. Jarvis, D. H. Bell, and N. E. Good, *Anal. Biochem.* **104,** 300 (1980).

[6] D. D. Perrin and B. Dempsey, "Buffers for pH and Metal Ion Control." Chapman & Hall, London, 1974.

ISBN 0-12-182004-1

TABLE I
SELECTED BUFFERS AND THEIR pK VALUES AT 25°

Trivial name	Buffer name	pK_a[a]	dpK_a/dt[b]
Phosphate (pK_1)	—	2.15	0.0044
Malate (pK_1)	—	3.40	—
Formate	—	3.75	0.0
Succinate (pK_1)	—	4.21	−0.0018
Citrate (pK_2)	—	4.76	−0.0016
Acetate	—	4.76	0.0002
Malate	—	5.13	—
Pyridine	—	5.23	−0.014
Succinate (pK_2)	—	5.64	0.0
MES	2-(*N*-Morpholino)ethanesulfonic acid	6.10	−0.011
Cacodylate	Dimethylarsinic acid	6.27	—
Dimethylglutarate	3,3-Dimethylglutarate (pK_2)	6.34	0.0060
Carbonate (pK_1)	—	6.35	−0.0055
Citrate (pK_3)	—	6.40	0.0
BIS-Tris	[Bis-(2-hydroxyethyl)imino]tris(hydroxymethyl)methane	6.46	0.0
ADA	*N*-2-Acetamidoiminodiacetic acid	6.59	−0.011
Pyrophosphate	—	6.60	—
EDPS (pK_1)	*N*,*N*′-Bis(3-sulfopropyl)ethylenediamine	6.65	—
Bis-Tris propane	1,3-Bis[tris(hydroxymethyl)methylamino]propane	6.80	—
PIPES	Piperazine-*N*,*N*′-bis(2-ethanesulfonic acid)	6.76	−0.0085
ACES	*N*-2-Acetamido-2-aminoethanesulfonic acid	6.78	−0.020
MOPSO	3-(*N*-Morpholino)-2-hydroxypropanesulfonic acid	6.95	−0.015
Imidazole	—	6.95	−0.020
BES	*N*,*N*-Bis-(2-hydroxyethyl)2-aminoethanesulfonic acid	7.09	−0.016
MOPS	3-(*N*-Morpholino)propanesulfonic acid	7.20	0.015
Phosphate (pK_2)	—	7.20	−0.0028
EMTA	3,6-Endomethylene-1,2,3,6-tetrahydrophthalic acid	7.23	—
TES	2-[Tris(hydroxymethyl)methylamino]ethanesulfonic acid	7.40	−0.020
HEPES	*N*-2-Hydroxyethylpiperazine-*N*′-2-ethanesulfonic acid	7.48	−0.014
DIPSO	3-[*N*-Bis(hydroxyethyl)amino]-2-hydroxypropanesulfonic acid	7.60	−0.015
TEA	Triethanolamine	7.76	−0.020
POPSO	Piperazine-*N*,*N*′-bis(2-hydroxypropanesulfonic acid)	7.85	−0.013
EPPS, HEPPS	*N*-2-Hydroxyethylpiperazine-*N*′-3-propanesulfonic acid	8.00	—
Tris	Tris(hydroxymethyl)aminomethane	8.06	−0.028
Tricine	*N*-[Tris(hydroxymethyl)methyl]glycine	8.05	−0.021

(*continued*)

TABLE I (*continued*)

Trivial name	Buffer name	pK_a[a]	dpK_a/dt[b]
Glycinamide	—	8.06	−0.029
PIPPS	1,4-Bis(3-sulfopropyl)piperazine	8.10	—
Glycylglycine	—	8.25	−0.025
Bicine	*N*,*N*-Bis(2-hydroxyethyl)glycine	8.26	−0.018
TAPS	3-{[Tris(hydroxymethyl)methyl]amino} propanesulfonic acid	8.40	0.018
Morpholine	—	8.49	—
PIPBS	1,4-Bis(4-sulfobutyl)piperazine	8.60	—
AES	2-Aminoethylsulfonic acid, taurine	9.06	−0.022
Borate	—	9.23	−0.008
Ammonia	—	9.25	−0.031
Ethanolamine	—	9.50	−0.029
CHES	Cyclohexylaminoethanesulfonic acid	9.55	0.029
Glycine (pK_2)	—	9.78	−0.025
EDPS	*N*,*N*′-Bis(3-sulfopropyl)ethylenediamine	9.80	—
APS	3-Aminopropanesulfonic acid	9.89	—
Carbonate (pK_2)	—	10.33	−0.009
CAPS	3-(Cyclohexylamino)propanesulfonic acid	10.40	0.032
Piperidine	—	11.12	—
Phosphate (pK_3)	—	12.33	−0.026

[a] These data are compiled from references 3–6.
[b] See footnote 10.

Because buffering capacity is maximal at the p*K*, the compound chosen should have a p*K* in the range of the desired pH. If the pH optimum of an enzymic reaction is not known, it is desirable to use a related series of buffers and measure reaction rates at several pH values; pH 6.0 (MES), pH 7.0 (PIPES), pH 8.0 (EPPS), and pH 9.0 (TAPS or CHES) provide a useful range. Once the pH optimum is approximated experimentally, structurally unrelated buffers with a similar p*K* should be tested. Thus, the use at pH 7.5 of TES (p*K* 7.5), phosphate (p*K* 7.2), and triethanolamine (p*K* 7.7) would serve to alert the investigator to interactions between buffer and reaction components.

The problem of specific buffer interaction with other reaction components is most prevalent when inorganic buffers are used. Phosphate, for example, inhibits many kinases and dehydrogenases as well as enzymes with phosphate esters as substrates. Phosphate also inhibits carboxypeptidase, fumarase, and urease. Borate forms covalent complexes with *gem*-diols, including such important biological molecules as mono- and oligosaccharides, the ribose moieties of nucleic acids and pyridine nucleotides, and such polyols as glycerol among other metabolic intermediates. Bicarbonate, because it is in equilibrium with CO_2, requires closed systems, and since equilibration of CO_2 and bicarbonate is slow, carbonic anhydrase is often added to such buffers. Buffers containing such primary amines as Tris form Schiff bases with aldehydes and ketones.

Perhaps the most serious problems encountered when using inorganic, primary amine or carboxylic acid buffers are their propensity for forming coordination complexes with di- and trivalent metal ions.[7-9] Complexation of a metal ion by the buffer results in proton release with consequent decrease in pH. This is not as serious a problem as the formation of insoluble precipitates upon complexation, or the chelation of a metal required for enzymic activity (e.g., Mg^{2+} for kinases, Cu^{2+} or Fe^{2+} for hydroxylases). The Good buffers generally have low or known metal-binding capabilities[3] and are favored for use in studies of metal-requiring enzymes.

Once a noninteracting buffer of appropriate pK is found, a decision as to the concentration to be used must be made. In general, the lowest possible concentration of buffer should be used in order to avoid nonspecific ionic strength effects. The most straightforward way to accomplish this is to set up a reaction mixture with a low (10 mM) concentration of buffer. The pH is checked before the addition of enzyme, and a suitable period of time after addition of the enzyme, possibly the standard reaction time or after equilibrium is reached. If a significant change in pH has occurred (± 0.05 pH), a higher concentration of buffer must be used, perhaps 20 or 50 mM. pH stability is particularly important in the measurement of systems in which protons are generated or consumed stoichiometrically with substrate utilization.

Stock buffer solutions should be prepared in glass containers (plastic containers may leach UV-absorbing plasticizers) at a temperature close to the working temperature. The pK, and therefore pH, of buffers, particularly amine buffers, are sensitive to temperature (see Table I[10]). Tris, for example, when titrated to pH 8.06 at 25° will have a pH of 8.85 at 0°. The Good buffers generally have only a small inverse dependence of pK on temperature, whereas the pK of carboxylic acid buffers are even less temperature sensitive. Since the dilution of a more concentrated stock buffer solution will change the pH, as will the addition of salts, the pH of a

[7] R. M. Smith and A. E. Martell, "Critical Stability Constants," Vol. 2. Plenum, New York, 1975.

[8] L. G. Sillen and A. E. Martell, "Stability Constants of Metal-Ion Complexes," *Spec. Publ.—Chem. Soc.* n17 (1964).

[9] L. G. Sillen and A. E. Martell, "Stability Constants of Metal-Ion Complexes. Supplement" *Spec. Publ.—Chem. Soc.* n25 (1971).

[10] The change in pK per degree centigrade. These values can be used to correct the pKs in Table I to higher or lower temperatures.

reaction should be checked, with all components added, by direct measurement of the pH with a micro-combination electrode. By this means guesswork is removed in the reporting of experimental conditions. Titration of the buffer at the approximate working concentration and temperature is an important consideration.

Since many enzyme rate studies rely on spectroscopic measurement of substrates utilized or products formed, ideally the buffer used should be transparent in the spectral region of interest. Although all the buffers in Table I are nonabsorbing in the visible region, several have appreciable absorption in the UV. ADA and ACES absorb strongly below 260 nm, but most of the other Good buffers can be used at and above 240 nm. A series of buffers that are transparent in the visible range and down to 240 nm have been developed[11] for spectrophotometric determination of pK_a.

Factors in Buffer Preparation

After selection of the buffer, it may be prepared by titrating the crystalline-free acid to the desired pH. The pH meter used should be calibrated with two or more commercially available pH standards that bracket the desired pH region. If the effects of monovalent cations are being investigated, mineral countercations can be avoided by titrating the buffer with tetramethylammonium hydroxide. It may be desirable to filter the titrated buffer through a sterile ultrafiltration device to prevent bacterial or fungal growth, especially with solutions in the pH 6–8 range; this is particularly important if large quantities are prepared for extended storage. It may also be desirable to add a small amount of EDTA (10 μM) to chelate heavy-metal ions which may be present.

The purification of proteins or enzymes requires large amounts of buffered solutions for homogenization, chromatographic separations, and dialysis. Cost, then, becomes a significant concern in buffer selection, and the relatively inexpensive inorganic buffers and Tris are advantageous. Tris, in particular, has been used extensively in enzyme purification, although it has disadvantages: it is not a good buffer below pH 7.5, has a highly temperature-dependent pK, and, since it is a primary amine, interferes with the Bradford dye-binding protein assay.[12] Glycine and other primary amines interfere similarly in that assay whereas Bicine, HEPES, and EPPS all give false-positive colors with the Lowry assay.

The choice of suitable buffers for use in protein purification will depend on the pH required for maximal stability and resolution of the pro-

[11] D. D. Perrin, *Aust. J. Chem.* **16,** 572 (1963).

[12] M. M. Bradford, *Anal. Biochem.* **22,** 248 (1976).

tein of interest, and the particular method employed. Although any buffer is suitable for gel permeation chromatography, absorption and elution of proteins from hydroxyapatite is generally performed in phosphate solutions. Cationic buffers, e.g., Tris, are used in anion-exchange chromatography, whereas cation-exchange chromatography requires the use of anionic buffers such as phosphates. Buffer exchange, after chromatography, is readily accomplished by passage of the protein solution through a Sephadex G-10 or G-25 column equilibrated with the desired buffer.

Some buffers contain volatile components, thereby allowing their removal by lyophilization. A number of such volatile buffer systems have been described,[6] and are used extensively for preparative ion-exchange chromatography.

Metal Ions

Specific reaction mixtures may require that mono-, di-, or trivalent metal ions be completely absent from the reaction; in other work, a saturating or intermediate concentration of such ions may be required. If a specific monovalent cation requirement is being investigated, the buffer should be titrated with tetramethylammonium hydroxide. The K_m for activation of the enzyme can be determined by subsequent addition of the appropriate mineral cation salt to the assay system. If other enzymic parameters of a mineral cation-stimulated enzyme are being investigated, then a relatively high concentration of buffer (50–100 mM) containing the desired cation as counterion, e.g., K^+ with pyruvate kinase, would be used. This method generally ensures saturation with the monovalent cation and prevents the need for adding additional salts to the reaction mixture with concomitant increase in ionic strength.

A number of assays require the control of di- and trivalent cations. If such metals must be absent from the reaction, effective "removal" is generally accomplished by addition of a chelating agent (0.1–5.0 mM). A number of polydentate chelators serve this purpose, including citric acid, tripolyphosphoric acid, nitrilotriacetic acid (NTA), ethylene glycol bis(β-aminoethyl ether)-N,N,N',N'-tetraacetic acid (EGTA), ethylenediaminetetraacetic acid (EDTA), and diethylenetriaminepentaacetic acid (DTPA). The stability constants of these chelators with numerous metal ions have been reported[9,10] and are shown in Table II for several of biochemical interest.

The maintenance of fixed concentrations of necessary metal ions is more difficult. The above-mentioned polydentate chelators have been used as metal ion buffers to ensure constant, but low, levels of free metal ions in nutrient media. Metal buffering is particularly important when salt

TABLE II
STABILITY CONSTANTS[a] OF METAL ION–CHELATE COMPLEXES

	Chelator					
Metal ion	Citrate	PPP_i	NTA	EGTA	EDTA	DTPA
Ca^{2+}	3.6	5.2	6.5	11.0	10.6	10.6
Cd^{2+}	3.8	6.6	10.0	16.1	16.6	19.1
Co^{2+}	5.0	6.9	10.8	12.4	16.5	19.0
Cu^{2+}	5.9	8.3	13.1	17.7	18.9	21.1
Fe^{3+}	11.4	—	15.9	20.5	25.1	28.6
Mg^{2+}	3.4	5.8	5.5	5.2	8.7	9.3
Mn^{2+}	3.7	7.2	7.4	12.2	14.1	15.1
Ni^{2+}	5.4	6.8	11.5	12.7	18.7	20.2
Zn^{2+}	5.0	7.5	10.4	12.7	16.7	18.7

[a] Expressed as log K, where $K = (ML_n)/(M)(L)^n$ for the reaction $M + nL \rightarrow ML_n$ where $n = 1$. Data from Smith and Martell[7] and Sillen and Martell.[8]

solubility is low or when the free hydrated metal ion is susceptible to hydrolysis, e.g., the reaction, $M^{n+} + H_2O \rightarrow M(OH)^{(n-1)+} + H^+$.[13]

Broad-Range Buffers

Attempts have been made to formulate buffers that can span pH regions of 5 or more pH units. One method employs a mixture of buffers to achieve a buffering capacity that is approximately constant across the pH region of interest. A second, more common, approach already noted is to use a series of structurally related individual buffers, e.g., the Good buffers, with pK values evenly spaced across the pH range. Each approach has limitations of which the investigator should be aware. First, each component of the buffer mixture will be buffering optimally over a relatively small pH region (pH = p$K \pm 1$), outside of which it will play no role. A specific ionized form of such a nonparticipatory buffer may be inhibitory across the pH region in which that form is present. Furthermore, the presence of extraneous nonbuffering components will increase the ionic strength of the solution with possible undesirable effects. A thorough and detailed account of buffer mixtures that provide a wide

[13] For an extensive list of pK values of hydrolysis, see Perrin and Dempsey.[6] For a thorough discussion of metal-ion control in the study of metal-activated enzymes, in particular the phosphotranferases, see J. M. Morrison, this series, Vol. 63, p. 257.

range of buffering capacity with constant ionic strength is available[14] and has been used successfully.[15]

When using a series of individual buffers, it is important to overlap the working pH with buffers of different pK values to ensure that specific buffer interactions are absent. A hypothetical pH profile from pH 6–10 could be determined, for example, with the following series of buffers at the indicated pH values: MES (pH 6.0–6.7), PIPES (pH 6.5–7.4), HEPES (pH 7.2–8.2), TAPS (pH 8.0–9.0), CHES (pH 8.9–9.9), and CAPS (pH 9.8–10.0). This procedure has been found to be useful and results in smooth pH profiles of kinetic parameters[16,17] without breaks indicative of specific buffer effects.

Effect of Organic Solvents

Apparent pH values have been determined[18,19] for a large number of buffers in various nonaqueous solvent mixtures. These studies have resulted in the formulation of a pH* scale, formally equivalent to the pH scale, for nonaqueous solvent mixtures. Simon[20] reported pH* values of over 1000 organic compounds in a number of organic solvents, including Methyl Cellosolve, acetone, acetonitrile, dimethylformamide, and dioxane. For acid–base equilibria, pK^* values for charge-generating ionizations (neutral acids) were found to be higher than the corresponding pK_{H_2O} values, whereas those in which a charge was not generated (cationic acids) had pK^* values lower than pK_{H_2O}.

Similar effects of added organic solvents are expected for the ionization behavior of functional groups on the enzymes that participate in catalysis and binding. Thus, the addition of nonaqueous solvents should elevate the pK of a neutral acid (carboxyl, sulfhydryl, tyrosyl, or metal-coordinated water), but have little effect on the pK of a cationic acid (amino or imidazole). The experimental protocol and the choice of suitable buffers for use in such experiments has been outlined,[21] and has been successfully applied to the identification of groups involved in cataly-

[14] K. J. Ellis and J. F. Morrison, this series, Vol. 87, p. 405.
[15] J. W. Williams and J. F. Morrison, *Biochemistry* **20,** 6024 (1981).
[16] P. F. Cook, G. L. Kenyon, and W. W. Cleland, *Biochemistry* **20,** 1204 (1981).
[17] R. E. Viola and W. W. Cleland, *Biochemistry* **17,** 4111 (1978).
[18] R. G. Bates, M. Paabo, and R. A. Robinson, *J. Phys. Chem.* **67,** 1833 (1963).
[19] C. L. deLigny, P. F. M. Luykx, M. Renbach, and A. A. Wiereke, *Recl. Trav. Chim. Pays-Bas* **79,** 699 (1960).
[20] W. Simon, *Angew. Chem. Int. Ed. Engl.* **3,** 661 (1964).
[21] W. W. Cleland, *Adv. Enzymol. Relat. Areas Mol. Biol.* **45,** 273 (1977).

sis.[16,17] This technique, in conjunction with the measurement of the temperature dependence of the pK of a functional group, should allow identification of enzymic groups to be made.

Buffers at Low Temperature

Buffers suitable for the study of enzyme-catalyzed reactions at subzero temperatures in mixed, organic cryosolvents have been investigated.[22,23] The pH dependence of nine buffers, spanning a range of pH 3–10 as a function of cryosolvent mixture and temperature, has been examined for which a linear dependence of pH* on $1/T$ (°K) was found.[22] By this means, it has been possible to predict the pH* of a buffered solution in a specific cryosolvent at a specific subzero temperature. The practical problems of low buffer solubility at low temperature and high percentages of organic solvent remain to be solved.[24]

Use of Deuterated Buffer

Deuterated buffer solutions are increasingly being used in the investigation of deuterium isotope effects and proton NMR behavior of enzyme-catalyzed reactions. The preparation of deuterated buffers requires the titration of the free acid, dissolved in D_2O, with KOH(D) to the required pD. *Note that the pD of the buffer is 0.4 pH unit higher than the reading on the pH meter.* If highly enriched deuterated buffers (>95 atom % D) are required, it is advisable to exchange the protons in both the buffer and the base prior to titration by dissolving each in a small amount of D_2O, and removing the solvent by rotary evaporation or lyophilization. Since D_2O solutions are hydroscopic, they should be stored in a tightly stoppered container, preferably in a desiccator (without desiccant).

Deuterated buffer solutions have been used for the enzymic synthesis of stereospecifically deuterated compounds and in studies of either primary deuterium kinetic isotope effects or solvent isotope effects. In the former case, D_2O is usually a substrate in the reaction[25,26] and solvent-derived deuterium is thus incorporated into products to give isotope effects, as shown for a number of other enzymes.[27] Deuterated buffer solutions used for the study of isotope effects should be measured at

[22] G. H. B. Hoa, P. Douzou, and A. M. Michelson, *Biochim. Biophys. Acta* **182,** 334 (1969).
[23] G. H. B. Hoa and P. Douzou, *J. Biol. Chem.* **248,** 4649 (1973).
[24] A. L. Fink and M. A. Geeves, this series, Vol. 63, p. 336.
[25] J. S. Blanchard and W. W. Cleland, *Biochemistry* **19,** 4506 (1980).
[26] T. Y. S. Shen and E. W. Westhead, *Biochemistry* **12,** 3333 (1973).
[27] T. B. Dougherty, V. R. Williams, and E. S. Younathan, *Biochemistry* **11,** 2493 (1972).

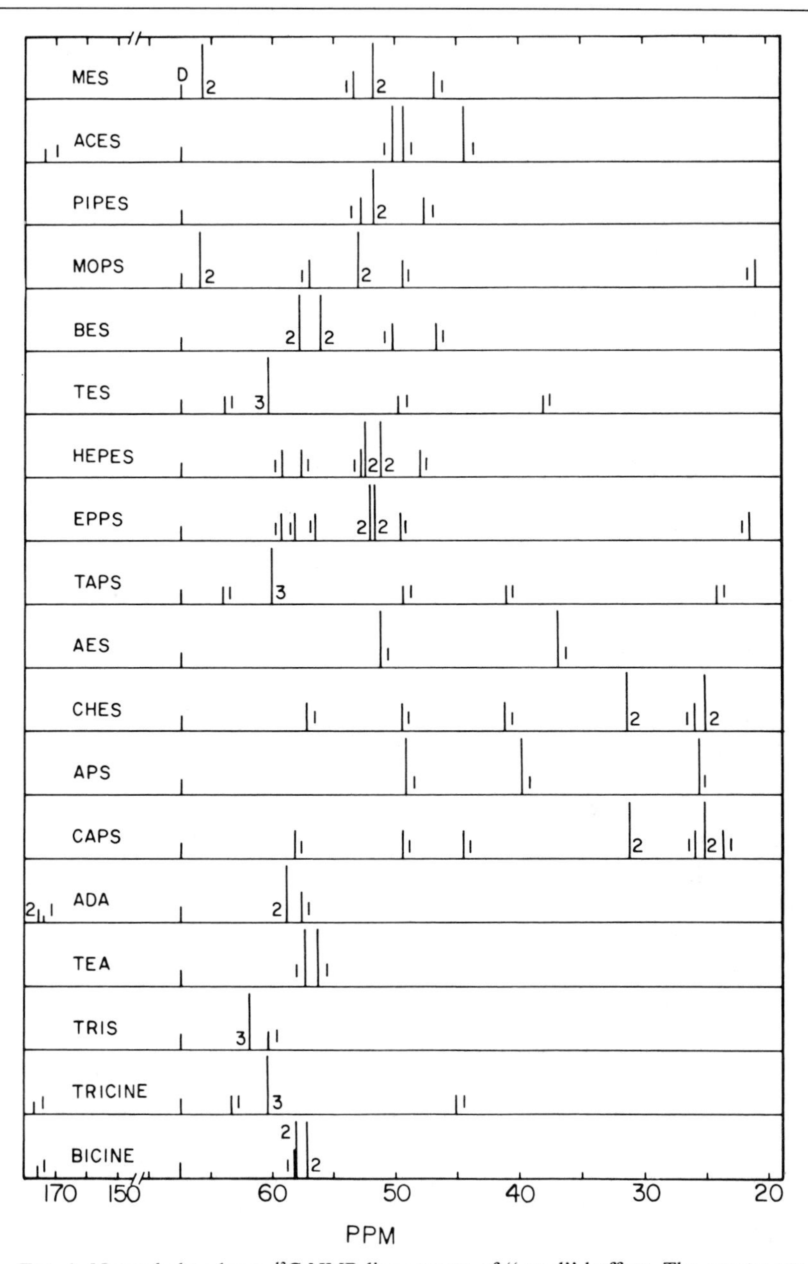

FIG. 1. Natural abundance ^{13}C NMR line spectra of "good" buffers. The spectra were obtained at 50 MHz, and contained 300 m*M* buffer (titrated to their respective p*K*s) with 50 m*M* dioxane (D, 67.4 ppm) as internal standard. Integrated intensities are shown next to the upfield ^{13}C lines; downfield carbonyl ^{13}C resonances show reduced intensities due to longer relaxation times.

several different mole fractions of deuterated solvents to ensure linearity with deuterium content. Solvent isotope effects,[28] on the other hand, are exhibited when deuterium from the solvent is not incorporated into substrates or products, e.g., with the pyridine nucleotide dehydrogenases.[29,30] Such effects may be small and may be due to true rate effects, different solvation effects in D_2O, or perturbations in enzymic pK values when in deuterated solvents.

Buffers for NMR Studies

There has been increasing interest in NMR approaches to the study of enzyme-catalyzed reactions, both *in vitro* and *in vivo*. The choice of a specific buffer will depend on the experimental design and the nucleus being used as probe, as well as factors considered earlier. Proton NMR investigations are conducted in D_2O and interference by buffer resonances can be avoided by using inorganic buffers or those with only exchangeable protons. However, such buffers are the same as those discussed above as being associated with a number of specific buffer effects, and they are generally of limited usefulness. For NMR investigations of ^{19}F and ^{31}P, any of the buffers in Table I are suitable, since they do not contain such nuclei.

^{13}C NMR investigations are the most broadly applicable to enzyme-catalyzed reactions and to studies of metabolism,[31] as they do not rely on the presence of phosphorus or fluorine in the compound, or require the use of highly enriched, deuterated solvent. Since ^{13}C NMR is so broadly applicable to mechanistic and metabolic investigations, and since the "Good" buffers are generally superior for enzyme work, the analysis of the ^{13}C NMR chemical shifts of 18 "Good" buffers have been determined, and are presented in Fig. 1. These data may be valuable to investigators planning to use ^{13}C NMR as a tool in the study of enzyme-catalyzed reactions by allowing buffers to be chosen whose ^{13}C resonances will not interfere with the resonances of substrates, products, or the enzyme itself.

[28] R. L. Schowen, *Prog. Phys. Org. Chem.* **9,** 275 (1972).

[29] J. Schmidt, J. Chen, M. DeTraglia, D. Minkel, and J. T. McFarland, *J. Am. Chem. Soc.* **101,** 3634 (1979).

[30] K. M. Welsh, D. J. Creighton, and J. P. Klinman, *Biochemistry* **19,** 2005 (1980).

[31] "Biological Applications of Magnetic Resonance" (R. G. Shulman, ed.). Academic Press, New York, 1979.

[27] Protein Assay for Dilute Solutions

By ENRICO CABIB and ITZHACK POLACHECK

Protein solutions of concentration lower than 25–50 μg/ml require a concentration step before assay by Lowry's method.[1] In the procedure described below,[2] trichloroacetic acid is used to concentrate proteins by quantitative precipitation in the presence of yeast-soluble ribonucleic acid as a carrier. Many substances that interfere with the colorimetric assay[3] remain in solution and are thereby eliminated.

Assay Method

Principle. Protein is precipitated with 10% trichloroacetic acid, in the presence of ribonucleic acid as a carrier. The precipitated protein is dissolved in sodium hydroxide and measured by a slight modification of Lowry's procedure.

Reagents

- Yeast-soluble ribonucleic acid, 5 mg/ml. Each batch of ribonucleic acid should be checked, and those producing a blank above 0.15, as measured against water, are rejected. The product of Calbiochem-Behring has been satisfactory
- Trichloroacetic acid, 100 g per 100 ml of solution
- Sodium hydroxide, 0.1 *M*
- Sodium carbonate, 3% (w/v) in 0.1 *M* sodium hydroxide
- Sodium tartrate, 4% (w/v)
- Cupric sulfate ($CuSO_4 \cdot 5\ H_2O$), 2% (w/v)
- Lowry reagent: Mix 9.6 ml of sodium carbonate–sodium hydroxide reagent with 0.2 ml each of sodium tartrate and copper sulfate
- Folin reagent (phenol reagent solution 2 *N*; Fisher Scientific Company)
- Bovine serum albumin, 1 mg/ml

Procedure. To 1 ml of sample, containing between 5 and 25 μg of protein, in a 10 × 75 mm Pyrex tube (DuPont, No. 00100), 25 μl of soluble RNA is added, followed by 0.11 ml of trichloroacetic acid. The tube

[1] O. H. Lowry, N. J. Rosebrough, A. L. Farr, and R. J. Randall, *J. Biol. Chem.* **193,** 265 (1951).

[2] I. Polacheck and E. Cabib, *Anal. Biochem.* **117,** 311 (1981).

[3] G. L. Peterson, *Anal. Biochem.* **100,** 201 (1979).

ISBN 0-12-182004-1

contents are mixed by vortexing. After standing in ice for 45 min, the tubes are centrifuged for 10 min at 27,000 *g* in a Sorvall RC2-B centrifuge with an SS34 rotor. The supernatant fluid is aspirated, and the pellet is suspended, by vortexing, in 0.5 ml of 0.1 *M* sodium hydroxide. If dissolution is not complete the tubes are placed in a boiling water bath for 5 min and then cooled in tap water. After addition of 0.5 ml of Lowry reagent, the tubes are incubated for 10 min at 37°. Folin reagent, 50 μl, is added with immediate vortexing of each tube. After an additional 15 min at 37°, the absorbance at 750 nm is measured. A series of standards containing between 5 and 25 μg of bovine serum albumin is processed and measured at the same time.

Comments. The standards yield a straight line up to at least 25 μg of protein, compared to the convex curve observed in this range with the direct Lowry assay. The presence of several substances that interfere with the Lowry procedure[2] [examples: 0.5% digitonin, 0.5% Zwittergent 3-08 or 3-16 (Calbiochem), 1% sodium dodecyl sulfate, or 0.05 *M* Tris-chloride] does not alter the results. When Tris is present, however, it is necessary to wash the trichloroacetic acid precipitate once with 1 ml of cold 10% trichloroacetic acid. When 0.1% Triton X-100 is present in the sample, absorbance values are about 55% of the normal.

Similar procedures have been published[4,5] in which deoxycholate is used as a coprecipitating carrier. Deoxycholate, however, does not yield a precipitate with trichloroacetic acid in the presence of certain detergents, such as sodium dodecyl sulfate, digitonin, and sulfobetaines.

[4] A. Bensadoun and D. Weinstein, *Anal. Biochem.* **70,** 241 (1976).
[5] G. L. Peterson, *Anal. Biochem.* **83,** 346 (1977).

[28] Enzyme Localization in Gels

By MARY J. HEEB and OTHMAR GABRIEL

This chapter is an update of an earlier one in this series[1] and provides general information for the location of functionally active enzymes after separation by electrophoresis or electrofocusing in various anticonvection media such as agarose, starch, or polyacrylamide gels. The scope of this chapter is limited to illustrating recent developments in the detection

[1] O. Gabriel, this series, Vol. 22, p. 578.

METHODS IN ENZYMOLOGY, VOL. 104

ISBN 0-12-182004-1

of enzymes *in situ* as perceived by the authors. Although no attempt is made to provide a comprehensive review of the literature, many different enzymes are covered in detail and references to others are presented. Several publications contain valuable information on the localization of enzymes, including reviews that provide extensive lists of reagents and methods for enzyme detection.[2,3] A list of references for a wide variety of enzymes[4] and a general review[5] have been published.

Some of the more recent trends in localizing enzymes on gels include the use of renaturing techniques, as applied to enzymes that have been subjected to electrophoresis in detergent-containing gels.[6] A number of potentially carcinogenic reagents, such as benzidine and *o*-dianisidine, have been replaced in some assay systems with eugenol or tetrabase.[7,8] Fluorophores attached to substrates (e.g., umbelliferyl or dansyl derivatives) have been used extensively, providing sensitive methods whereby the progress of reaction can be observed directly. Ultrathin agar overlays on polyester sheets have been used for more sensitive sandwich-type incubations[9] and are particularly useful when applied after ultrathin-layer isoelectric focusing. A procedure has been described by which aliquots of a large number of column fractions may be subjected to electrophoresis at the same time and then stained for several different enzymes.[10]

Retention of Enzymic Activity

For any of the enzyme detection methods that will be discussed, a major consideration should be the preservation of the maximum level of enzymic activity throughout the process. Special attention must be paid to the possible detrimental effect of reagents used in preparation of the gel: oxidizing agents (ammonium persulfate), incomplete reaction of polymerizing agents, buffer ions, and a pH incompatible with the enzyme. Appropriate precautions should be taken to limit inactivation of the en-

[2] M. J. Siciliano and C. R. Shaw, *in* "Chromatographic and Electrophoretic Techniques" (I. Smith, ed.), Vol. 2, p. 185. Heinemann, London, 1976.
[3] H. Harris and D. A. Hopkinson, "Handbook of Enzyme Electrophoresis in Human Genetics." North-Holland Publ., Amsterdam, 1976.
[4] B. D. Hames and D. Rickwood, "Gel Electrophoresis of Proteins: A Practical Approach," IRL Press, London and Washington D.C., 1981.
[5] W. Ostrowski, *J. Chromatogr. Libr.* **18B,** 287 (1983).
[6] R. E. Manrow and R. P. Dottin, *Anal. Biochem.* **120,** 181 (1982).
[7] E. H. Liu and D. M. Gibson, *Anal. Biochem.* **79,** 597 (1977).
[8] B. Lomholt, *Anal. Biochem.* **65,** 569 (1975).
[9] M. Hofelmann, R. Kittsteiner-Eberle, and P. Schrier, *Anal. Biochem.* **128,** 217 (1983).
[10] P. H. Odense, C. Annand, and J. Barlow, *Anal. Biochem.* **108,** 257 (1980).

zyme. For example, prior electrophoresis of the gel alone,[11] followed by tests for the absence of persulfate ions,[12] and inclusion of sulfhydryl group-containing components into buffer systems as protection against oxidation[13] were reported in many instances to result in retention of otherwise labile activity. Direct addition of sulfhydryl reagents to polymerization mixtures, however, is not useful, since polymerization will be prevented or delayed.

With isoelectric focusing gels, it may be necessary to remove the ampholytes at the end of the procedure by soaking in an appropriate buffer prior to application of enzyme detection methods.[14] If the ampholytes or the pH of the gel do not interfere with enzyme detection, omission of this step will reduce diffusion of bands.

A number of monomeric enzymes and a few enzymes containing identical subunits have been renatured after electrophoresis in sodium dodecyl sulfate (SDS)-containing gels[15]; the extent of renaturation often depends on the commercial source[16] of the SDS.[17] In some cases it has been sufficient to soak gels in the appropriate buffer, but for enzymes that are tolerant of the procedure, SDS can be removed more quickly, with greater recovery of enzymic activity, by incubation in buffered 25% isopropanol[18] (see example 3 in the last section of this chapter). A combination of nonionic detergent, high salt, glycerol, substrate, and a sulfhydryl reducing agent has been used successfully to renature a number of enzymes in SDS gels or O'Farrell gels[6] (see example 4 in the last section of this chapter). Another approach is to subject an SDS gel to isoelectric focusing in urea and nonionic detergent in a second dimension; since SDS is removed in the second dimension, many proteins can be renatured from urea.[19,20] Alternatively, for some separations the isoelectric focusing step might be sufficient in a single dimension. An advantage to renaturing enzymes within gels is that part of the activity may be physically bound to the gel.[15]

[11] O. Gabriel, this series, Vol. 22, p. 565.
[12] A. Bennick, *Anal. Biochem.* **26,** 453 (1968).
[13] J. M. Brewer, *Science (Washington, D.C.)* **156,** 256 (1967).
[14] H. Mukasa, A. Shimamura, and H. Tsumori, *Anal. Biochem.* **123,** 276 (1982).
[15] S. A. Lacks and S. S. Springhorn, *J. Biol. Chem.* **255,** 7467 (1980).
[16] Inhibition of renaturation is correlated with the amount of hexadexyl sulfate and tetradecyl sulfate present as contaminants.
[17] S. A. Lacks, S. S. Springhorn, A. L. Rosenthal, *Anal. Biochem.* **100,** 357 (1979).
[18] A. Blank, R. H. Sugiyama, and C. A. Dekker, *Anal. Biochem.* **120,** 267 (1982).
[19] R. E. Manrow and R. P. Dottin, *Proc. Natl. Acad. Sci. U.S.A.* **77,** 730 (1980).
[20] R. P. Dottin, R. E. Manrow, B. R. Fishel, S. L. Aukerman, and J. L. Culleton, this series, Vol. 68, p. 513.

Localization of Enzymes

It is essential to obtain as much information as possible about the properties of the enzyme before attempts are made to locate it in gels. The localization must follow separation as quickly as possible, and several approaches can be tried.

Elution of Enzyme from the Gel. Cutting of the gel into segments followed by elution with appropriate buffer and assay of the eluted enzyme by conventional methods is tedious and time consuming. Nevertheless, much valuable information is gained concerning the properties of the enzyme and its chances of remaining enzymically active during the separation and elution procedure. Elution of active enzyme will depend not only on prevention of protein denaturation, but also on recovery from the gel. Recovery depends in part on the molecular weight of the enzyme protein and the extent of cross-linkage used for the electrophoretic separation. It should be kept in mind that individual bands of separated components contain only microgram quantities of protein so that stabilization of such minute amounts, once eluted, may be necessary. Addition of carrier proteins or of polyhydroxyalcohols, such as glycerol, can be effective.

Staining for Protein prior to Elution. Most of the commonly used protein stains result in irreversible denaturation of enzymes, and it is not wise to attempt to detect enzyme activity after using a protein stain. However, the detection of proteins under relatively nondenaturing conditions such as UV-scanning, limited tannic acid treatment,[21] or binding of fluorophore,[1] followed by elution and localization of enzymic activity, represents a compromise for detection of protein as well as enzymic activity. The number of gel slices that must be eluted and assayed can be greatly reduced by this approach.

Staining for Enzymic Activity in Situ after Electrophoretic Separation. The general principles for the localization of enzymes *in situ* can be divided into several major groups according to the methods and type of reagents involved. Most, if not all, of these techniques were originally described by histochemists.

1. Simultaneous capture. This method is widely used and is applied to enzymes that remain active in the presence of the staining reagents. The substrate is converted by the enzyme to a product that, in turn, couples with a reagent present in the incubation mixture to form an insoluble colored product.

[21] K. Aoki, S. Kajiwara, R. Shinke, and H. Nishira, *Anal. Biochem.* **95,** 575 (1979).

2. Postincubation coupling. Incubation of substrate with enzyme results in a product, and this step is followed by addition of reagent to yield colored material. Diffusion of product during the first incubation period results in broadening of bands.

3. Autochromic methods. Direct visualization of the progress of reaction is made possible when there are changes in the optical properties of either substrate or product. An advantage of the method is direct observation of the progress of the reaction as well as allowing an estimate of the extent of diffusion during the incubation period.

4. Sandwich-type incubation. The method uses auxiliary indicator enzymes or high-molecular-weight substrates that require the overlay of the gel with another gel, paper, or other suitable matrix containing the necessary components. Incubation of the separating gel with the "indicator matrix" permits localization of enzymic activity.

5. Copolymerization of substrate in the gel. This method is possible and desirable for certain high-molecular-weight substrates, such as nucleic acid, starch, gelatin, or pectin. It is often necessary to prevent the enzyme from acting on the substrate during electrophoresis by deletion of a cofactor, inclusion of an inhibitor or chelating agent, use of a less than optimal pH for enzymic activity, or other means. Conditions can be restored to near optimal during the subsequent incubation period in order to detect bands of enzyme activity.

For most of the general methods just described, it is advisable to mark the position of the enzyme with a small piece of fine wire, or other means, at the earliest time at which bands are detectable and the least diffuse. The enzyme detection step can then be followed with a general protein stain.

Limitations and Problems Encountered with in Situ Localization of Enzymes

Generally, diffusion of both the enzyme and its reaction products will cause reuniting and broadening of separated components. It is imperative to minimize these factors by variation of incubation and reaction conditions to optimize staining while minimizing diffusion. Considering the complex kinetic events in a gel during staining, it requires extensive experimental evidence to establish the rate-limiting factors in a specific staining procedure. For most systems, it is unrealistic to expect that quantitative conclusions can be derived from the staining intensity, and investigators should use extreme caution in interpreting densitometric data from enzyme activity stains in a quantitative manner.

A common problem encountered with *in situ* staining techniques concerns the issue of specificity of the procedure. It is important to establish the same pattern of substrate specificity for the enzyme in a gel as that established using conventional solution methods. A number of control incubations (absence of substrate, presence of specific inhibitors) will eliminate artifactual staining, such as the "nothing dehydrogenases."[1,22,23]

Locating Specific Enzymes

A number of enzymes are presented in tabular form with appropriate references to guide the reader to recent literature that provides detailed description for localizing enzymes *in situ*. A division into major groups of enzymes, such as oxidoreductases, transferases, and the like, was thought to be useful since analogous functional enzyme properties will lend themselves to the use of similar techniques for detection. The methods selected for inclusion are primarily those published in readily available journals since 1974 and are ones that are generally reproducible under standard laboratory conditions. A similar compilation in Vol. 22 of this series[1] should be consulted for methods applied to a number of other enzymes. The methods are selected to demonstrate the principles for detecting specific enzymes but can, with slight modifications, be applied to the localization of similar enzymes. The included methods are based on the detection of enzyme activity; immunological techniques are excluded. Many valuable contributions are not included in the chapter, and there is no intent to be totally comprehensive.

Abbreviations used in Tables I–V are phenazine methosulfate, PMS; nitro blue tetrazolium, NBT; 3-(4,5-dimethylthiazoyl-2)-2,5-diphenyltetrazolium bromide, MTT; p-iodonitrotetrazolium violet, INT; 2,3,5-triphenyltetrazolium chloride, TTC; and 4-methylumbelliferyl, 4-MU.

Oxidoreductases (Table I)

Many enzymes in this group use NAD or NADP as a cosubstrate. Oxidation of one substrate, then, results in formation of a reduced pyridine nucleotide. The transfer of reducing power to various tetrazolium dyes can be mediated, in turn, by phenazine methosulfate to yield a deeply colored, insoluble formazan. The advantage of the technique is that only a single incubation containing all components is required with

[22] J. L. O'Conner, D. P. Edwards, and E. D. Bransome, *Anal. Biochem.* **78,** 205 (1977).
[23] P. H. Springell and T. A. Lynch, *Anal. Biochem.* **74,** 251 (1976).

TABLE I
OXIDOREDUCTASES (EC 1)

EC No.	Enzyme	Substrate	Principle of detection	Method[a]	References
1.1.1	Hydroxysteroid dehydrogenase	Pregnenolone	Formation of NADH, PMS, NBT	i	*b*
1.1.1.8	Glycerol-3-phosphate dehydrogenase	α-Glycerophosphate	Formation of NADH, PMS, NBT	i	*c*
1.1.1.27	Lactate dehydrogenase	Lactate	Formation of NADH, PMS, NBT	i	*d–f*
1.1.1.40	Malate dehydrogenase (decarboxylating)	Malate	Formation of NADPH, PMS, MTT	i	*g, h*
1.1.1.42	Isocitrate dehydrogenase	Isocitrate	Precipitation of CO_2 with Ca^{2+}	i	*i*
1.2.1.12	Glyceraldehyde-3-phosphate dehydrogenase	3-Phosphoglycerate	Precipitation of phosphate with Ca^{2+}	i	*i*
1.4.1.3	Glutamate dehydrogenase	α-Ketoglutarate	Quenching of fluorescence of NAD(P)H	i	*j*
1.4.3.1	D-Aspartate oxidase	D-Aspartate	Enzyme[s] coupling; quenching of fluorescence of NAD(P)H on paper overlay	i, iv	*k*
1.4.3.3	D-Amino-acid oxidase	D-Alanine	Enzyme[s] coupling; quenching of fluorescence of NAD(P)H on paper overlay	i, iv	*k, l*
1.4.3.4	Amine oxidase	Benzylamine	Densitometry for product	iii	*m*
1.11.1.6	Catalase	H_2O_2	Couple with peroxidase, diaminobenzidine	i	*n*
1.11.1.7	Peroxidase	H_2O_2	Eugenol fluorescence	i	*o, p*

1.13.11.12	Lipoxygenase	Linoleic acid	Hydroperoxide formation, 3,3′-dimethoxybenzidine	ii	*q*
1.14.18.1	Monophenol monooxygenase (tyrosinase)	Catechin	Colored product on overlay	iii, iv	*r*

[a] Mode of detection: i, simultaneous capture method; ii, postincubation capture reaction; iii, autochromic method; iv, sandwich indicator (overlay); v, copolymerization of substrate in gel. The numerals refer to methods described in the text under the section *Staining for Enzymic Activity in Situ after Electrophoretic Separation*.

[b] J. L. O'Conner, *Anal. Biochem.* **78,** 205 (1977).

[c] F. Leibenguth, *Biochem. Genet.* **13,** 263 (1975).

[d] P. Carda-Abella, S. Perez-Cuadrada, S. Lara-Baruque, L. Gil-Grande, and A. Nunez-Puertas, *Cancer* **49,** 80 (1982).

[e] K. Simon, E. R. Chaplin, and I. Diamond, *Anal. Biochem.* **79,** 571 (1977).

[f] P. H. Springell and T. A. Lynch, *Anal. Biochem.* **74,** 251 (1976).

[g] S. Povey, D. E. Wilson, H. Harris, I. P. Gormley, P. Perry, and K. E. Buckton, *Ann. Hum. Genet.* **39,** 203 (1975).

[h] A. C. Peterson, *Nature (London)* **248,** 561 (1974).

[i] H. G. Nimmo and G. A. Nimmo, *Anal. Biochem.* **121,** 17 (1982).

[j] R. L. Nelson, M. S. Povey, D. A. Hopkinson, and H. Harris, *Biochem. Genet.* **15,** 87 (1977).

[k] R. L. Nelson, S. Povey, D. A. Hopkinson, and H. Harris, *Biochem. Genet.* **15,** 1023 (1977).

[l] R. F. Barker and D. A. Hopkinson, *Ann. Hum. Genet.* **41,** 27 (1977).

[m] A. W.-S. Ma Lin and D. O. Castell, *Anal. Biochem.* **69,** 637 (1975).

[n] E. M. Gregory and I. Fridovich, *Anal. Biochem.* **58,** 57 (1974).

[o] E. H. Liu and D. M. Gibson, *Anal. Biochem.* **79,** 597 (1977).

[p] C. A. Cullis and K. Kolodynska, *Biochem. Genet.* **13,** 687 (1975).

[q] B. O. DeLumen and S. J. Kazeniac, *Anal. Biochem.* **72,** 428 (1976).

[r] P. Thomas, H. Delincee, and J. F. Diehl, *Anal. Biochem.* **88,** 138 (1978).

[s] The enzyme used here is L-glutamate dehydrogenase.

the formation of an insoluble product that is not subject to diffusion. The disadvantage of the method is sensitivity to light and oxygen[24] as well as the necessity of carefully controlled reaction conditions and of appropriate controls (such as omission of substrate) to avoid erroneous interpretation.[1,22,23]

Transferases (Table II)

Transferases catalyze the transfer of a group (such as methyl or glycosyl) from one compound to another. These enzymes are frequently detected by coupling the primary reaction product with auxiliary enzyme(s) which, in turn, can be coupled to tetrazolium dyes or other indicators. Consideration must be given to the slow penetration of the coupling enzymes into the gel. For this reason, many of the indicator systems described are applied peripherally in agar or paper overlays in a sandwich type of incubation. A device for recovering expensive coupling enzymes has been described.[25] Some of the more recent methods for the detection of transferases use different principles. For example, 1-chloro-2,4-dinitrobenzene is used as an acceptor for glutathione with glutathione *S*-transferase[26]; phosphorylase *b* is copolymerized in the separating gel as a coupling enzyme for the detection of oligo-1,4 $\xrightarrow{\alpha}$ 1,6-glucosyltransferase (branching enzyme)[27]; and precipitation of pyrophosphate with manganese ion is used for detection of hypoxanthine-guanine phosphoribosyltransferase.[28]

Hydrolases (Table III)

Hydrolases catalyze the addition of water to various bonds, resulting in cleavage. They compose a large group of enzymes, many of which are very stable and, therefore, amenable to a wide variety of detection methods. Several techniques use substrates such as umbelliferyl,[29,30] *p*-nitrophenyl,[31] or other chromogenic derivatives that change their optical properties as a consequence of the enzymic reaction. Enzymes that release inorganic phosphate, pyrophosphate, or carbon dioxide as products can be visualized by precipitation with Ca^{2+} ions.[32] Enzymes that are

[24] W. Worsfold, M. J. Marshall, and E. B. Ellis, *Anal. Biochem.* **79,** 152 (1977).
[25] T. Yamashita and H. Utoh, *Anal. Biochem.* **84,** 304 (1978).
[26] P. G. Board, *Anal. Biochem.* **105,** 147 (1980).
[27] K. Sato and K. Sato, *Anal. Biochem.* **108,** 16 (1980).
[28] B. Vasquez and A. L. Bieber, *Anal. Biochem.* **84,** 504 (1978).
[29] P. L. Chang, S. R. Ballantyne, and R. G. Davidson, *Anal. Biochem.* **97,** 36 (1979).
[30] P. M. Coates, M. A. Mestriner, and D. A. Hopkinson, *Ann. Hum. Genet.* **39,** 1 (1975).
[31] M. E. Hodes, M. Crisp, and E. Gelb, *Anal. Biochem.* **80,** 239 (1977).
[32] H. G. Nimmo and G. A. Nimmo, *Anal. Biochem.* **121,** 17 (1982).

TABLE II
TRANSFERASES (EC 2)

EC No.	Enzyme	Substrate	Principle of detection: Coupling enzyme(s)	Principle of detection: Detection procedure	Method[a]	References
2.1.3.	Aspartate or ornithine carbamoyltransferase	Asp or ornithine, carbamoyl phosphate	—	Precipitation of phosphate with Ca^{2+} or Pb^{2+}, Alizarin Red	i, ii	*b* *c*
2.3.2.2	γ-Glutamyltransferase	*N*-γ-L-Glutamyl naphthylamide	—	Coupling with fast garnet GBC	ii	*d*
2.4.1	Fructosyltransferase	Raffinose	—	Opalescent product (fructan)	iii	*e*
2.4.1.18	α-1,4-Glucan branching enzyme	Glucose-1-phosphate	Phosphorylase *b* in gel	Scan for turbidity (glycogen)	i	*f*
2.4.1.22	Lactose synthase and other galactosyltransferases	UDPGal, Glucose-1-phosphate, GlcNAc	UDPGphosphorylase, UDPGdehydrogenase, nucleoside diphosphate kinase	NADH fluorescence	i, iv	*g*
2.4.2.7	Adenine phosphoribosyltransferase	5-Ribosyl pyrophosphate	—	Adsorption of [^{14}C]AMP to DEAE paper overlay	i, iv	*h*
2.4.2.8	Hypoxanthine-guanine phosphoribosyltransferase	Phosphoribosyl pyrophosphate	—	Precipitation of PP_i with Mn^{2+}	i	*i*
2.5.1.18	Glutathione *S*-transferase	Glutathione chlorodinitrobenzene	—	Scan for absorbance of product	iii	*j, k*
2.6.1.1	Aspartate aminotransferase	Cysteine sulfinate	—	Formation of SO_3^{2-}, NBT	i	*l, m*
2.6.2	Transamidases	Casein, dansyl cadaverine	—	Fluorescence	iii	*n, o*

(continued)

TABLE II (*continued*)

EC No.	Enzyme	Substrate	Principle of detection: Coupling enzyme(s)	Principle of detection: Detection procedure	Method[a]	References
2.7	Glucosyltransferase	Sucrose, dextran	—	Periodate–Schiff stain (for reducing sugar)	ii	*p*
2.7.1.37	Protein kinase	Histone or casein, $[^{32}P]$ATP	—	TCA precipitation of $[^{32}P]$phosphohistone or casein	v, i	*q, r*
2.7.1.40	Pyruvate kinase	PEP, ADP, fructose-1-6-diphosphate, glucose	Hexokinase, glucose-6-phosphate dehydrogenase	Formation of NADPH, PMS, NBT; quenching of fluorescence of NADH	i	*s, t*
		PEP, fructose 1,6-diphosphate			i	*s, t*
2.7.3.2	Creatine kinase	Creatine phosphate	Hexokinase, glucose-6-phosphate dehydrogenase	Formation of NADPH, fluorescence or NBT, overlay	i, iv	*u, v* *w*
2.7.4.3	Adenylate kinase	AMP, GTP, PEP	Pyruvate kinase, hexokinase, glucose-6-phosphate dehydrogenase	Formation of NADPH, PMS, MTT	i	*x*
2.7.7.6	RNA polymerase	Nucleotide triphosphates	—	Precipitation of PP_i with Ca^{2+}	i	*y*
2.7.7.9	Glucose-1-phosphate uridylyltransferase	UDPG, glucose-1,6-diphosphate	Phosphoglucomutase, glucose-6-phosphate dehydrogenase	Formation of NADH, PMS, INT; precipitation of pyrophosphate with Ca^{2+}	i	*z*
		UTP, glucose-1-phosphate			i	*aa, bb* *y*

[a] See footnote *a* in Table I.
[b] J. E. Grayson, R. J. Yon, and P. J. Butterworth, *Biochem. J.* **183,** 239 (1979).
[c] D. N. Baron and J. E. Buttery, *J. Clin. Pathol.* **25,** 415 (1972).
[d] M. Izumi and K. Taketa, *in* "Electrophoresis '81" (R. C. Allen and P. Arnaud, eds.), p. 709. de Gruyter, Berlin, 1981.
[e] R. R. B. Russell, *Anal. Biochem.* **97,** 173 (1979).
[f] K. Satoh and K. Sato, *Anal. Biochem.* **108,** 16 (1980).
[g] M. Pierce, R. D. Cummings, and S. Roth, *Anal. Biochem.* **102,** 441 (1980).
[h] S. Mowbray, B. Watson, and H. Harris, *Ann. Hum. Genet.* **36,** 153 (1972).
[i] B. Vasquez and A. L. Bieber, *Anal. Biochem.* **84,** 504 (1978).
[j] P. G. Board, *Anal. Biochem.* **105,** 147 (1980).
[k] W. C. Kenney and T. D. Boyer, *Anal. Biochem.* **116,** 344 (1981).
[l] J. G. Scandalios, J. C. Sorenson, and L. A. Ott, *Biochem. Genet.* **13,** 759 (1975).
[m] T. Yagi, H. Kagamiyama, and M. Nozaki, *Anal. Biochem.* **110,** 146 (1981).
[n] P. Stenberg and J. Stenflo, *Anal. Biochem.* **93,** 445 (1979).
[o] L. Lorand, G. E. Siefring, Y. S. Tong, J. Bruner-Lorand, and A. J. Gray, *Anal. Biochem.* **93,** 453 (1979).
[p] H. Mukasa, A. Shimamura, and H. Tsumori, *Anal. Biochem.* **123,** 276 (1982).
[q] F. Phan-Dinh-Tuy, A. Weber, J. Henry, D. Cottreau, and A. Kahn, *Anal. Biochem.* **127,** 73 (1982).
[r] M. Gagelmann, W. Pyerin, D. Kubler, and V. Kinzel, *Anal. Biochem.* **93,** 52 (1979).
[s] E. Melendez-Hevia, J. Corzo, and J. Perez, *in* "Electrophoresis '81" (R. C. Allen and P. Arnaud, eds.), p. 693. de Gruyter, Berlin, 1981.
[t] M. J. Siciliano and C. R. Shaw, *in* "Chromatographic and Electrophoretic Techniques" (I. Smith, ed.), Vol. 2, p. 207. Heinemann, London, 1976.
[u] G. M. Graeber, M. J. Reardon, A. W. Fleming, H. D. Head, R. Zajtchuk, W. H. Brott, and J. H. Foster, *Ann. Thor. Surg.* **32,** 230 (1981).
[v] T. Yamashita and H. Utoh, *Anal. Biochem.* **84,** 304 (1978).
[w] N. Hall and M. DeLuca, *Anal. Biochem.* **76,** 561 (1976).
[x] D. E. Wilson, S. Povey, and H. Harris, *Ann. Hum. Genet.* **39,** 305 (1976).
[y] H. G. Nimmo and G. A. Nimmo, *Anal. Biochem.* **121,** 17 (1982).
[z] R. E. Manrow and R. P. Dottin, *Proc. Natl. Acad. Sci. U.S.A.* **77,** 730 (1980).
[aa] R. E. Manrow and R. P. Dottin, *Anal. Biochem.* **120,** 181 (1982).
[bb] R. P. Dottin, R. E. Manrow, B. R. Fishel, S. L. Aukerman, and J. L. Culleton, this series, Vol. 68, p. 513.

TABLE III
HYDROLASES (EC 3)

EC No.	Enzyme	Substrate	Principle of detection[a]	Method[a]	References
3.1	Esterases	4-MU-fatty ester/ Naphthyl ester/ Indoxyl ester	Fluorescence/ Coupling with fast garnet/ Blue product on overlay	iii/ i/ iii	*b, c* *d, e* *f*
3.1	Lipases	Trioleine	Rhodamine, fluorescence	i, iv	*g*
3.1	Phosphodiesterases	Thymidine-3′-nitrophenyl phosphate/ 4-MU-phosphate/ Dinucleotides/ 4-MU-thymidylate	Alkaline development, paper overlay/ Fluorescence/ Coupling with nucleotide phosphorylase, xanthine oxidase, adenine deaminase, PMS, MTT/ Fluorescence	ii/ iii/ i/ iii, iv	*h, i* *j, k*
3.1	DNases	DNA	Methyl green or ethidium bromide, overlay	ii, iv	*h, l–p*
3.1.1.4	Phospholipase A_2	Lecithin in gel	Stain for free unsaturated fatty acid	v, ii	*q*
3.1.3.2	Acid phosphatase	α-Naphthyl acid phosphate	Coupling with fast garnet	i	*c, r*
3.1.3.11	Fructose-1,6-bisphosphatase	Fructose-1,6-diphosphate	Precipitation of PP_i with Ca^{2+}	i	*r*
3.1.4.3	Phospholipase C	Egg yolk	Opaque zones in agar overlay	iii, iv	*s*
3.1.27	Ribonuclease	RNA/ Dinucleotides/ RNA in gel/[^{32}P]RNA in gel	Pyronin B/ Coupling with adenosine deaminase, xanthine oxidase, nucleoside phosphorylase, PMS, MTT/ Autoradiography	ii/ i/ ii	*n* *t* *o*
3.1.6.1	Arylsulfatase	Hydroxynitrophenyl sulfate/ 4-Methyl umbelliferyl sulfate/ Nitrocatechol sulfate	Coupling with cupric ferricyanide and diaminobenzidine/ Fluorescence/ Colored product	i/ iii/ iii	*u* *v, w* *x*
3.2	Glycosidases	Sucrose/ Starch/ Dextran/ Pectin in gel	Reducing sugar with TTC/ Ruthenium red	i/ ii	*y* *z*
3.2.1.1	α-Amylase	Lyosine red/ Soluble starch/ Starch in gel	Colored product/ Reaction of undigested starch with I_2/ TTC for reducing sugar	iii/ ii/i	*aa, g* *bb, y, cc*
3.2.1.11	Dextranase	Dextran 10	Reducing sugar with TTC	i	*y*
3.2.1.20	α-Glucosidase	4-MU- or Naphthyl-α-D-glucopyranoside	Fluorescence on paper overlay/ Coupling with fast blue B	iii, iv/ or i	*dd*

3.2.1.21	β-Glucosidase	4-MU-β-D-glucopyranoside	Fluorescence	iii	*ee*
3.2.1.26	β-D-Fructofuranosidase	Sucrose	Coupling with peroxidase and glucose oxidase, diaminobenzidine/ TTC for reducing sugar/ Coupling with glucose dehydrogenase, mutarotase, PMS, INT	i/ i/ i	*ff* *y* *gg*
3.2.1.28	Trehalase	Trehalose, glucose	Coupling with peroxidase, glucose oxidase, eugenol	i	*hh, ii*
3.4	Proteinases	Casein/ Azocoll/ Gelatin in gel/ Fluorescein-casein	TCA precipitation of undigested protein; scan at 520 nm/ Salt precipitation of undigested protein/ Fluorescence	ii, iv/ ii, v/ iii, v	*bb, g, jj, kk, aa*
3.4.11.4	Tripeptide aminopeptidase	L-Leu-Gly-Gly	Coupling with amino acid oxidase, PMS, TTC	i	*ii*
3.4.13.11	Dipeptidase	L-Leu-L-Leu	Coupling with amino acid oxidase, PMS, TTC	i	*ii*
3.4.17	Carboxypeptidase	Carbonaphthoxy-DL-phenylalanine	Coupling with diazo blue B	i	*ll*
3.4.21.4	Trypsin	Cytochrome *c*/ Azocoll/ Benzoyl-L-Arg-L-Arg-*p*-nitroanilide	TCA precipitation of undigested protein/ Scan/ Detect under UV	ii/ iii/ iii	*mm* *nn* *jj*
3.4.21.10	Acrosin	Arg-β-naphthylamine	Coupling with fast black	i	*rr*
3.4.21.11	Elastase	*N*-Acetyl-DL-Ala-α-naphthyl ester/ Elastin-orcein	Scan at 550 nm/ Clear zone in overlay	iii/ iii, iv	*oo, pp* *qq*
3.4.21.14	Subtilisin	Benzoxyl-Gly-Gly-L-Leu-*p*-nitroanilide	Product visible under UV light	iii	*hh*
3.4.21.31	Urokinase, plasminogen activators	Plasminogen, fibrin/ Plasminogen, gelatin in gel	Clear zones in opaque overlay/ Stain for undigested protein with Amido Black	i, iv/ ii, v	*ss,* *tt*
3.4.22.1	Cathepsin B	Carboxybenzoyl-L-Arg-L-Arg-methoxy-naphthylamide/ Azocoll	Coupling with fast garnet GBC/ Scan	i iii	*uu,* *jj*
3.4.24.4	Thermolysin	Cytochrome *c*	TCA precipitation of undigested protein	ii	*pp*
3.5.1.2	Glutaminase	L-Glutamine	Coupling with enzyme,* PMS, NBT	i	*vv*
3.5.1.5	Urease	Urea	Cresol red, lead acetate	i/	*ww*

(*continued*)

TABLE III (*continued*)

EC No.	Enzyme	Substrate	Principle of detection[a]	Method[a]	References
3.5.3.1	Arginase	Arginine/ Arginine, α-ketoglutarate	Coupling with urease, dithiothreitol, NBT/ Coupling with enzyme* and urease; quenching of fluorescence of NADPH	i i	*xx*, *yy*
3.5.4.4	Adenosine deaminase	Adenosine, α-keto glutarate	Coupling with enzyme;* quenching of fluorescence of NADPH	i	*yy*
3.5.4.5	Cytidine deaminase	Cytidine, α-keto glutarate/ Cytidine	Coupling with enzyme;* Quenching of fluorescence of NADPH/ Dithiothreitol, MTT	i/ i	*yy*, *zz*, *zz*
3.6.1	Phosphatases and pyrophosphatases	β-Naphthyl phosphate/ *p*-Nitrophenyl phosphate or other phosphate	Ammonium molybdate, tetrabase in overlay/ Precipitation of phosphate or PP_i with Ca^{2+}	i/ i, iv	*aaa* *r*

* The enzyme used is L-glutamate dehydrogenase.

[a] The first letter of each principle is capitalized; a slant line denotes a new procedure. For i–v under Method, see Table I, footnote *a*.

[b] P. M. Coates, M. A. Mestriner, D. A. Hopkinson, *Ann. Hum. Genet.* **39,** 1 (1975).

[c] C. A. Cullis and K. Kolodynska, *Biochem. Genet.* **13,** 687 (1975).

[d] J. Yourno and W. Mastropaolo, *Blood* **58,** 939 (1981).

[e] M. Rosenberg, V. Roegner, and F. F. Becker, *Anal. Biochem.* **66,** 206 (1975).

[f] J. K. Herd and J. Tschida, *Anal. Biochem.* **68,** 218 (1975).

[g] M. Hofelmann, R. Kittsteiner-Eberle, and P. Schreier, *Anal. Biochem.* **128,** 217 (1983).

[h] M. E. Hodes, M. Crisp, and E. Gelb, *Anal. Biochem.* **80,** 239 (1977).

[i] D. M. Hawley, K. C. Tsou, and M. E. Hodes, *Anal. Biochem.* **117,** 18 (1981).

[j] M. E. Hodes and J. E. Retz, *Anal. Biochem.* **110,** 150 (1981).

[k] K. W. Lo, S. Aoyagi, and K. C. Tsou, *Anal. Biochem.* **117,** 24 (1981).

[l] A. C. G. Porter, *Anal. Biochem.* **117,** 28 (1981).

[m] H. S. Kim and T.-H. Liao, *Anal. Biochem.* **119,** 96 (1982).

[n] A. Blank, R. H. Suigiyama, and C. A. Dekker, *Anal. Biochem.* **120,** 267 (1982).

[o] J. Huet, A. Sentenac, and P. Fromageot, *FEBS Lett.* **94,** 28 (1978).

[p] T. Karpetsky, G. E. Brown, E. McFarland, A. Rahman, K. Rictro, W. Roth, M. B. Haroth, A. Ansher, P. Duffey, and C. Levy, *in* "Electrophoresis '81" (R. C. Allen and P. Arnaud, eds.), p. 674. de Gruyter, Berlin, 1981.

[q] W. T. Shier and J. T. Trotter, *Anal. Biochem.* **87,** 604 (1978).

[r] H. G. Nimmo and G. A. Nimmo, *Anal. Biochem.* **121,** 17 (1982).

[s] C. J. Smyth and T. Wadstrom, *Anal. Biochem.* **65,** 137 (1975).
[t] R. C. Karn, M. Crisp, E. A. Yount, and M. E. Hodes, *Anal. Biochem.* **96,** 464 (1979).
[u] E. A. Nichols, V. M. Chapman, and F. H. Ruddle, *Biochem. Genet.* **8,** 47 (1973).
[v] P. Mannowitz, L. Goldstein, and F. Bellomo, *Anal. Biochem.* **89,** 423 (1978).
[w] P. L. Chang, S. R. Ballantyne, and R. G. Davidson, *Anal. Biochem.* **97,** 36 (1979).
[x] E. Shapira, R. R. DeGregorio, R. Matalon, and H. L. Nadler, *Anal. Biochem.* **62,** 448 (1975).
[y] H. Mukasa, A. Shimamura, and H. Tsumori, *Anal. Biochem.* **123,** 276 (1982).
[z] R. H. Cruickshank and G. C. Wade, *Anal. Biochem.* **107,** 177 (1980).
[aa] P. E. Burdett, A. E. Kipps, and P. H. Whitehead, *Anal. Biochem.* **72,** 315 (1976).
[bb] S. A. Lacks and S. S. Springhorn, *J. Biol. Chem.* **255,** 7467 (1980).
[cc] T. L. Brown, M.-G. Yet, and F. Wold, *Anal. Biochem.* **122,** 164 (1982).
[dd] D. M. Swallow, G. Corney, H. Harris, and R. Hirschhorn, *Ann. Hum. Genet.* **38,** 391 (1975).
[ee] M. J. Sciliano and C. R. Shaw, *in* "Chromatographic and Electrophoretic Techniques" (I. Smith, ed.), Vol. 2, p. 199. Heinemann, London, 1976.
[ff] L. Faye, *Anal. Biochem.* **112,** 90 (1981).
[gg] P. Babczinski, *Anal. Biochem.* **105,** 328 (1980).
[hh] K. A. Killick and L.-W. Wang, *Anal. Biochem.* **106,** 367 (1980).
[ii] M. Suguira, Y. Ito, K. Hirano, and S. Sawaki, *Anal. Biochem.* **81,** 481 (1977).
[jj] K. R. Lynn and N. A. Clevette-Radford, *Anal. Biochem.* **117,** 280 (1981).
[kk] D. Every, *Anal. Biochem.* **116,** 519 (1981).
[ll] M. Pacaud and J. Uriel, *Eur. J. Biochem.* **23,** 435 (1971).
[mm] C. W. Ward, *Anal. Biochem.* **74,** 242 (1976).
[nn] A. Gertler, Y. Tencer, and G. Tinman, *Anal. Biochem.* **54,** 270 (1973).
[oo] G. Feinstein and A. Janoff, *Anal. Biochem.* **71,** 358 (1976).
[pp] J. Dijkhof and C. Poort, *Anal. Biochem.* **83,** 315 (1977).
[qq] J. L. Westergaard and R. C. Roberts, *in* "Electrophoresis '81" (R. C. Allen and P. Arnaud, eds.), p. 677. de Gruyter, Berlin, 1981.
[rr] D. L. Garner, *Anal. Biochem.* **67,** 688 (1975).
[ss] A. Granelli-Piperano and E. Reich, *J. Exp. Med.* **148,** 223 (1978).
[tt] C. Heussen and E. B. Dowdle, *Anal. Biochem.* **102,** 196 (1980).
[uu] J. S. Mort and M. Leduc, *Anal. Biochem.* **119,** 148 (1982).
[vv] J. N. Davis and S. Prusiner, *Anal. Biochem.* **54,** 272 (1973).
[ww] M. B. Shaik-M, A. L. Guy, and S. K. Pancholy, *Anal. Biochem.* **103,** 140 (1980).
[xx] F. Farron, *Anal. Biochem.* **53,** 264 (1973).
[yy] R. L. Nelson, S. Povey, D. A. Hopkinson, and H. Harris, *Biochem. Genet.* **15,** 1023 (1977).
[zz] Y.-S. Teng, J. E. Anderson, and E. R. Giblett, *Am. J. Hum. Genet.* **27,** 492 (1975).
[aaa] B. Lomholt, *Anal. Biochem.* **65,** 569 (1975).

TABLE IV
LYASES (EC 4)

EC No.	Enzyme	Substrate	Principle of detection	Method[a]	References
4.1.1.1	Pyruvate decarboxylase	Pyruvate	Precipitation of CO_2 with phosphate, Alizarin Red	i, ii	*b*
4.1.1.28	Aromatic-amino-acid decarboxylase	[^{14}C]Tyrosine	^{14}C product binds to ion-exchange paper overlay	ii, iv	*c, d*
4.1.1.31	Phosphoenolpyruvate carboxylase	PEP, bicarbonate	Coupling with fast violet B	i	*e*
4.1.2	Aldolases	2-Keto-4-hydroxyglutarate/ Fructose-1,6-bisphosphate	Coupling with *o*-aminobenzaldehyde, glycine/ Porphyrindin reduction (negative stain)	i/ ii, iv	*f, g*
4.1.3.1	Isocitrate lyase	Isocitrate	Coupling with Schiff reagent	i	*h*
4.1.3.2	Malate synthase	Glyoxalate, acetyl-CoA	Reduction of potassium ferricyanide	ii	*i*
4.1.3.7	Citrate synthase	Oxaloacetic acid, acetyl-CoA	Reduction of 2,4-dichlorophenolindophenol, MTT	i	*j*
4.1.3.27	Anthranilate synthase	Chorismate and NH_4^+ or glutamine	Scan for fluorescent product	iii	*k*
4.2.1.2	Fumarate hydratase	Fumarate	Coupling with malate dehydrogenase, MTT	i	*l*
4.2.1.3	Aconitate hydrolase (aconitase)	*cis*-Aconitate	Formation of NADPH, PMS, MTT	i	*m*
4.2.1.11	Enolase	2-Phosphoglycerate	Phenylhydrazine, scan for phenylhydrazone	ii	*n*

4.2.1.13	Serine dehydratase	Serine	Coupling with 1,2-diamino-4-nitrobenzene	ii	*o*
4.3.2.1	Argininosuccinate lyase	Argininosuccinate	Coupling with arginase, urease, GLUD, defluorescence of NAD(P)H	i	*p*
4.4.1.5	Lactoyl-glutathione lyase (glyoxalase)	Methyl glyoxal, glutathione	Development of starch gel with I_2	ii	*q*

[a] See footnote *a* in Table I.
[b] H. G. Nimmo and G. A. Nimmo, *Anal. Biochem.* **121,** 17 (1982).
[c] M. Landon, *Anal. Biochem.* **77,** 293 (1977).
[d] L. L. Cavalli-Sforza, S. A. Santachiara, L. Wang, E. Erdelyi, and J. Barchas, *J. Neurochem.* **23,** 629 (1974).
[e] M. C. Scrutton and F. Fatebene, *Anal. Biochem.* **69,** 247 (1975).
[f] N. D. Lewinski and E. E. Dekker, *Anal. Biochem.* **87,** 56 (1978).
[g] P. Christen and A. Gasser, *Anal. Biochem.* **109,** 270 (1980).
[h] H. C. Reeves and M. J. Volk, *Anal. Biochem.* **48,** 437 (1972).
[i] M. J. Volk, R. N. Trelease, and H. Reeves, *Anal. Biochem.* **58,** 315 (1974).
[j] I. Craig, *Biochem. Genet.* **9,** 351 (1973).
[k] T. H. Grove and H. R. Levy, *Anal. Biochem.* **65,** 458 (1975).
[l] E. Tolley and I. Craig, *Biochem. Genet.* **13,** 867 (1975).
[m] C. A. Slaughter, D. A. Hopkinson, and H. Harris, *Ann. Hum. Genet.* **39,** 193 (1975).
[n] H. K. Sharma and M. Rothstein, *Anal. Biochem.* **98,** 226 (1979).
[o] F. Gannon and K. M. Jones, *Anal. Biochem.* **79,** 594 (1977).
[p] R. L. Nelson, S. Povey, D. A. Hopkinson, and H. Harris, *Biochem. Genet.* **15,** 1023 (1977).
[q] L. W. Parr, I. A. Bagster, and S. G. Welch, *Biochem. Genet.* **15,** 109 (1977).

involved in ammonia metabolism can be coupled with glutamate dehydrogenase, a pyridine nucleotide-requiring reaction.[33] Hydrolases such as nucleases,[17,34] proteases,[15] lipase and amylase,[15] have been detected after renaturation in detergent-containing gels.

Lyases (Table IV)

Lyases catalyze the cleavage of bonds by means other than hydrolysis. In most cases they are detected on gels by reason of the chemical properties of the product of the enzymic reaction that is coupled with a chromogenic reagent.

Isomerases and Ligases (Table V)

Since only a few examples of detection methods for these classes are available, both groups of enzymes are presented in the same table. Isomerases catalyze reactions resulting in a molecular rearrangement. They are usually detected on gels by coupling with an enzyme that requires a pyridine nucleotide. Ligases catalyze the joining of two molecules with concomitant hydrolysis of a nucleoside triphosphate.

Selected Examples of Detailed Detection Methods

For many detection methods, economy of reagents may be desirable. At least 50 ml of reagent is generally needed to immerse a slab gel, but it is sufficient to wet the cut surface with as little as 6 ml of reagent. Tube gels can be immersed in reagent in test tubes of narrow diameter. For sandwich-type incubations, ultrathin layer techniques have been described.[9]

Example 1. α-Glycerophosphate Dehydrogenase (EC 1.1.1.8, Glycerol-3-phosphate Dehydrogenase). Simultaneous Capture Technique (i). The method[35] illustrates the use of tetrazolium salts for the detection of enzymes that produce reduced pyridine nucleotides. Similar techniques can be used for other dehydrogenases or for enzymes that can be coupled to dehydrogenases. The reaction must be carried out in the dark, with careful control of pH and temperature. Reagents are prepared just before use. As previously noted, control incubations without substrate are essential. If the stained gel is to be preserved, excess reagent must be removed to decrease background discoloration.

For a horizontal starch gel, 50 ml of staining reagent contains 25 mg of NAD, 15 mg of nitro blue tetrazolium (NBT), 1 mg of phenazine metho-

[33] R. L. Nelson, S. Povey, D. A. Hopkinson, and H. Harris, *Biochem. Genet.* **15,** 1023 (1977).

[34] J. Huet, A. Sentenac, and P. Fromageot, *FEBS Lett.* **94,** 28 (1978).

[35] F. Leibenguth, *Biochem. Genet.* **13,** 263 (1974).

TABLE V
ISOMERASES AND LIGASES (EC 5 and 6)

EC No.	Enzyme	Substrate	Principle of detection	Method[a]	References
5.3.1.1	Triosephosphate isomerase	Glycerol-3-phosphate, pyruvate	Couple with glyceraldehyde-3-phosphate dehydrogenase, lactate dehydrogenase, PMS, NBT	i	*b*
5.3.1.8	Mannosephosphate isomerase	Mannose-6-phosphate	Couple with glucose-6-phosphate dehydrogenase, glucosephosphate isomerase, PMS, NBT	i	*c*
5.3.1.9	Glucosephosphate isomerase	Fructose-6-phosphate	Couple with glucose-6-phosphate dehydrogenase, PMS, MTT	i	*b*
5.4.2.1	Phosphoglycerate phosphomutase	2-Phosphoglycerate, ATP	Couple with phosphoglycerokinase, glyceraldehyde-3-phosphate dehydrogenase, defluorescence of NADH	i	*b*
6.4.1.1	Pyruvate carboxylase	Pyruvate, bicarbonate	Coupling with fast violet	ii	*d*

[a] See footnote *a* in Table I.
[b] M. J. Siciliano and C. R. Shaw, *in* "Chromatographic and Electrophoretic Techniques" (I. Smith, ed.), p. 198. Heinemann, London, 1976.
[c] E. A. Nichols, V. M. Chapman, and F. H. Ruddle, *Biochem. Genet.* **8,** 47 (1973).
[d] M. C. Scrutton and F. Fatebene, *Anal. Biochem.* **69,** 247 (1975).

sulfate, 5 ml of 1 *M* sodium α-glycerophosphate at pH 7, 10 ml of 0.2 *M* Tris-HCl at pH 8.0, and 35 ml of water. Pour the solution over the cut surface of the gel and incubate in the dark at 37° until dark blue bands appear. Rinse in water and fix in ethanol–acetic acid–glycerol–water (5:2:1:4).

Example 2. Argininosuccinase (EC 4.3.2.1, Argininosuccinate Lyase). Simultaneous Capture Technique (i). The method[33] illustrates the use of coupling enzymes that require pyridine nucleotides as cosubstrates and also the use of a negative stain. As coupling enzyme, glutamate dehydrogenase is used, an enzyme that is equally useful for the detection of other enzymes that catalyze the production of ammonia, or that can be coupled to such enzymes, e.g., cytidine deaminase, adenosine deaminase, adenosine monophosphate deaminase, arginase, D-amino acid oxidase, D-aspartate oxidase, and urease.

The stain solution consists of 50 mg of barium arginosuccinate, 1 mg (40 units) of arginase (Sigma), 2 mg (20 units) of urease (Sigma type VI), 25 mg of α-ketoglutarate, 10 mg of NADH, and 50 μl of glutamate dehydrogenase (Sigma, 500 units/ml) in 5 ml of 0.1 *M* Tris-HCl at pH 7.6. The solution is applied to a Whatman 3 MM paper overlay on the cut surface of a starch slab gel, incubated at 37°, and is examined for nonfluorescent, i.e., quenched, zones under an UV lamp. The paper is photographed before the zones become diffuse. Glutamate dehydrogenase and lactate dehydrogenase bands may appear in some electrophoresed extracts. These bands are seen on the gel before they appear on the paper, in contrast to arginosuccinase bands. Bands from glutamate dehydrogenase can also be located by spraying ammonia on the paper just after the primary stain has developed. The paper will fade rapidly, but bands of this enzyme will be visible on the gel within 5 min.

Example 3. Nucleolytic Enzymes (EC 3.1). Postincubation Capture (ii) and Copolymerization of Substrate in the Running Gel (v). One type of renaturing technique is described in the following example.[18] Separating SDS slab gels are prepared by copolymerizing either rRNA (about 7 A_{260} units/ml or 0.3 mg/ml) or DNA (0.3 mg/ml) in the gels. Samples are treated by heating for 2 min at 100° in 2% SDS, 10% glycerol, 0.005% bromophenol blue, and 0.0625 *M* Tris-HCl at pH 6.8. For detection of RNase after electrophoresis, the gels are first incubated in 0.25 liter of 0.01 *M* Tris-HCl at pH 7.4, containing 25% isopropanol, for 30 min, with one change after 15 min, followed by incubation in buffer alone for 60 min with two changes at 20-min intervals. For digestion of RNA, the gel is incubated for 90 min at 37° in 0.1 *M* Tris-HCl at pH 7.4. For staining, the gel is incubated for 10 min in 0.01 *M* Tris-HCl, followed by another 10 min in 150 ml of 0.2% toluidine blue in the same buffer. The gel is washed for 60 min in the Tris buffer, with changes at 10 and 30 min, and is then ready for photography. All incubations are performed on a gently rotating shaker. For detection of DNase, the time of the first two incubations is increased to 60 min, with changes each 20 min. When the purpose is digestion, incubation is increased to 14 hr, and 10 m*M* $MgCl_2$ and 5 m*M* $CaCl_2$ are included in the buffer.

The use of isopropanol in the renaturing buffer provides a more effective means for removal of SDS and allows greater recovery of nucleolytic enzyme activity. Variations in recovery of activity due to the use of different SDS preparations are minimized in many instances.[16] It is recommended that the optimal isopropanol concentration and exposure time for a particular enzyme, with a given source of SDS, be individually adjusted.

Example 4. UDPGpyrophosphorylase (EC 2.7.7.9, Glucose-1-phosphate Uridylyltransferase). Simultaneous Capture (i). An alternative approach to renaturation is described in the following technique,[6] which was applied to both SDS and O'Farrell gels (isoelectric focusing in urea in the first direction followed by SDS gels in a second dimension). The success of the method relies on a renaturing buffer containing nonionic detergent, high salt, glycerol, substrate, and in many cases, sulfhydryl reducing agents. When sulfhydryl reagents are used, they are omitted from the final buffer changes to avoid interference with enzyme staining reagents. The method has been applied with variations to adenylate kinase and to the homomeric enzymes alkaline phosphatase and creatine kinase. It was not possible to renature lactate dehydrogenase in SDS gels by this technique, although this enzyme, as well as alcohol dehydrogenase, was successfully renatured after treatment with SDS followed by isoelectric focusing in urea gel.[20]

After electrophoresis, SDS gels or O'Farrell gels containing UDPG pyrophosphorylase are incubated in 25 m*M* Tricine-HCl at pH 7.5, containing 10% glycerol, 200 m*M* NaCl, 1 m*M* uridine, and 2.5% Nonidet P-40. Tube gels are incubated at room temperature for 4 hr with 4 changes of 10 volumes each; slab gels are incubated for 2.5 hr with 5 changes of 5 volumes each. The gels are then stained in a solution containing 0.08 unit/ml phosphoglucomutase, 0.35 unit/ml glucose-6-phosphate dehydrogenase, 1 m*M* UDPG, 2 m*M* sodium pyrophosphate, 1.6 m*M* NADP, 10 μM glucose 1,6-diphosphate, 1 m*M* EDTA, 4 m*M* $MgCl_2$, 85 m*M* Tricine-HCl at pH 7.6, 400 μg/ml INT, and 40 μg/ml phenazine ethosulfate. The reaction mixture is incubated in the dark at room temperature until blue bands appear, generally a matter of several hours.

Example 5. Aspartate Transcarbamylase (EC 2.1.3.2, Aspartate Carbamoyltransferase). Simultaneous Capture (i) and Postincubation Cap-

ture (ii). This method[36] illustrates the detection of enzymes that release phosphate, pyrophosphate, or CO_2 by use of calcium ion as precipitant, with additional enhancement of visualization by Alizarin Red. Other enzymes that may be detected by these means are ornithine transcarbamylase, isocitrate dehydrogenase, UDPGpyrophorylase, xanthine–guanine phosphoribosyltransferase, ribonuclease, and a variety of decarboxylases, synthases, phosphatases, and pyrophosphatases. Appropriate variations in assay conditions for a number of these enzymes have been presented.[32]

Gels are incubated in 0.1 *M* glycine–NaOH, pH 10, containing 1 m*M* aspartate, 0.2 m*M* carbamoyl phosphate, and 2 m*M* $CaCl_2$. An opalescent precipitate appears after approximately 20 min of incubation at room temperature and can be viewed or photographed against a dark background. If desired, the calcium phosphate thus formed can be better visualized, or scanned, after staining with 0.001% Alizarin Red S and 0.0005% $AlCl_3$ in 0.1 *M* potassium phosphate at pH 11 for 12–24 hr.

Example 6. Proteases or Lipases (EC 3.1). Postincubation Capture (ii) or Simultaneous Capture (i) in a Sandwich-Type Indicator Gel (iv). The method[9] involves the use of ultrathin-layer agar gels as overlays for sandwich-type incubations. The ultrathin technique is particularly suitable when applied to ultrathin-layer isoelectrofocused gels, thereby reducing the problem of diffusion. For many other applications, a thicker block of agar or acrylamide, prepared in a similar manner to the electrophoresis gel, may be substituted.[37]

For the detection of lipase, 0.4 g of agar is dissolved with heating and stirring in 20 ml of 0.1 *M* sodium succinate at pH 6. The mixture is cooled to 60–65°, and 4 mg of rhodamine B and 0.5 g of trioleine are added and emulsified for 1 min with an Ultra-Turrax homogenizer.

For detection of protease, 1 g of agar is dissolved in 25 ml of 0.3 *M* Tris-HCl at pH 7, and 0.5 g of casein is dissolved in 25 ml of the Tris buffer, each with heating and stirring. The two warm solutions are then mixed. The agar containing substrate (a measured volume, calculated to be 30% in excess) is poured onto Mylar (Technoplast, Cologne) polyester film of 100 μm thickness (300 × 125 mm), which is supported on a warmed glass base plate (40°). Greased polyester spacers of 150 μm are used along both edges of the film. The surface of the agar is covered with a polyester film that has been pretreated with 6 *N* NaOH for 15 min, rinsed, and dried. The film is pressed with a loaded glass plate. After the agar

[36] J. E. Grayson, R. J. Yon, and P. J. Butterworth, *Biochem. J.* **183,** 239 (1979).
[37] D. Every, *Anal. Biochem.* **116,** 519 (1981).

solidifies, the glass plate is removed and the film is applied to the electrofocused gel. The sandwich that is formed is incubated at 40° for 2–5 min.

For detection of lipase, light pink bands (fluorescent at 366 nm) appear on a deep red gel. For detection of proteases, the agar gel is immersed in 10% TCA; clear bands appear on an opalescent gel.

Example 7. α-Glucosidase (EC 3.2.1.20). Autochromic Method (iii) with Sandwich-Type Incubation (iv). The autochromic method, by which spectral changes in either the substrate or product can be directly detected, is obviously preferred when applicable. It is basically simple and allows rapid detection, thereby reducing the problem of diffusion. Tables I through V include a number of examples in which chromogenic derivatives are used as substrates. Umbelliferyl derivatives, in particular, have been used to detect both α- and β-glucosidases,[2,38] arylsulfatase,[29,39] esterases,[22] and phosphatases.[40]

Whatman No. 3 filter paper is soaked in 0.1 *M* sodium citrate at pH 4, containing 4-methyl umbelliferyl α-D-glucopyranoside (0.5 mg/ml). After electrophoresis of the enzyme, the paper is smoothed onto the cut surface of a starch gel and covered with Saran Wrap. The gel is viewed under UV light during incubation at room temperature for 15–60 min. Once bands appear, fluorescence may be enhanced by spraying with ammonia.

[38] D. M. Swallow, G. Corney, H. Harris, and R. Hirschhorn, *Ann. Hum. Genet.* **38,** 391 (1975).
[39] P. Manowitz, L. Goldstein, and F. Bellomo, *Anal. Biochem.* **89,** 423 (1978).
[40] D. M. Hawley, K. C. Tsou, and M. E. Hodes, *Anal. Biochem.* **117,** 18 (1981).

[29] Gel Protein Stains: A Rapid Procedure[1]

By A. H. REISNER

The triphenylmethane textile dye Coomassie Brilliant Blue R250 (Acid Blue 83) was first used by Fazekas de St. Groth *et al.*[2] to stain proteins on cellulose acetate. In a solvent system of methanol–acetic acid–water (5 : 1 : 5), proteins in polyacrylamide gels (PAG) were stained with it,[3] but the procedure caused marked shrinkage of the gel and a high background that required destaining and thereby possible loss of some bands. The use

[1] This chapter is dedicated to the memory of P. Nemes.
[2] Fazekas de St. Groth, R. G. Webster, and A. Datyner, *Biochim. Biophys. Acta* **71,** 377 (1963).
[3] T. S. Meyer and B. L. Lambert, *Biochim. Biophys. Acta* **107,** 144 (1965).

ISBN 0-12-182004-1

of 12.5% trichloroacetic acid (TCA)[4,5] as a solvent reduced these drawbacks, but background staining remained a problem.

The relatively low solubility of the methyl-substituted G form of the dye (Coomassie Brilliant Blue G250, Acid Blue 90, Xylene Brilliant Cyanin G) in 12.5% TCA led to its substitution for the R form,[6] and a concomitant reduction in background was obtained because the stain did not penetrate the gel. If, on the other hand 6% (w/v) perchloric acid replaced TCA as the solvent for the G form, the dye was highly soluble but rendered into its leuco form (orange brown).[7] Furthermore, when bound to proteins in this environment the dye was transformed to its intense blue form. Thus, the dye fully penetrated the gel, producing a pale orange background in which protein zones appeared blue. Finally, the Ampholines (LKB) used in isoelectric focusing tended *not* to bind the dye, making the system particularly useful for staining proteins after isoelectric focusing.

The method per se is not suitable for gels in which sodium dodecyl sulfate is a component.

Reagent

Coomassie Brilliant Blue G250, $HClO_4$ solution. Dilute 100 ml of 70% $HClO_4$ (density 1.70 g/ml) to 2000 ml with water and add 0.8 g of the dye. Stir at room temperature for 1 hr and filter through Whatman No. 1 paper. Follow by filtration through a 0.45 μm Millipore membrane. The solution is stable indefinitely at room temperature. Some batches of the dye cause a deeper orange background coloration than others. If the background is darker than desired, dilute the stock solution with 6% (w/v) $HClO_4$.

Procedure

Gels to be stained are placed into a volume of the stain such that the ratio of water in the gel to volume of stain is approximately 3 : 5. The gels can be developed between 20 and 37°. Dense protein bands are observed within 10 sec at room temperature. After 10 min, most bands can be seen and all the bands that are going to be observed are evident within 90 min. At 37°, the required time is about half that at room temperature. About 8–10 hr at 37° (overnight at room temperature) is required for the dye to

[4] A. Chrambach, R. A. Reisfeld, M. Wyckoff, and J. Zaccari, *Anal. Biochem.* **20,** 150 (1967).
[5] D. Rodbard and A. Chrambach, *Anal. Biochem.* **40,** 95 (1967).
[6] W. Diezel, G. Kopperschlager, and E. Hoffman, *Anal. Biochem.* **48,** 617 (1972).
[7] A. H. Reisner, P. Nemes, and C. Bucholtz, *Anal. Biochem.* **64,** 509 (1975).

penetrate completely a 4-mm-thick gel. It has been shown[8] that the sensitivity can be increased about threefold when the gels, after staining, are placed in 5% (v/v) acetic acid; under these conditions the background color changes to pale blue.

When dealing with large numbers of gel slabs, it may be convenient to stain them in polythene bags. Gels may be placed into the bags using an appropriately shaped spatula, after which stain is added and the bag is sealed by heat. After 45 min at 37°, the gel should be transferred to a new polythene bag and about 0.5 ml of 6% (w/v) $HClO_4$ containing 0.005% (w/v) Coomassie Brilliant Blue G250 added prior to sealing the bag. The gel can be stored for many months and photographed if air spaces are eliminated between the gel–polythene interface. To photograph the gels it is useful, though not essential, to use a medium red filter such as a Wratten Series A together with a fine-grained panchromatic film such as Ilford Pan F or Kodak Panatomic X.

A more sensitive but relatively elaborate procedure using Coomassie Brilliant Blue G250 to stain isoelectric focused gels has been described by Vesterberg *et al.*[9]

[8] I. B. Holbrook and A. G. Laver, *Anal. Biochem.* **75,** 634 (1976).
[9] O. Vesterberg, L. Hansen, and A. Sjosten, *Biochim. Biophys. Acta* **491,** 160 (1977).

[30] Gel Protein Stains: Silver Stain

By CARL R. MERRIL, DAVID GOLDMAN, and MARGARET L. VAN KEUREN

Silver Staining

Applications of silver-based histological[1–6] and photographic techniques[7–9] to the detection of proteins and other biopolymers, separated on

[1] L. Keranyi and R. Gallyas, *Clin. Chem. Acta* **38,** 465 (1972).
[2] P. Verheechi, *J. Neurol.* **209,** 59 (1975).
[3] D. Karcher, A. Lwenthal, and G. Van Soon, *Acta Neurol. Belg.* **79,** 335 (1979).
[4] R. C. Switzer, C. R. Merril, and S. Shifrin, *Anal. Biochem.* **98,** 231 (1979).
[5] H. R. Hubell, L. I. Rothblum, and T. C. Hsu, *Cell Biol. Int. Rep.* **3,** 615 (1979).
[6] C. R. Merril, R. C. Switzer, and M. L. Van Keuren, *Proc. Natl. Acad. Sci. U.S.A.* **76,** 4335 (1979).
[7] C. R. Merril, M. L. Dunau, and D. Goldman, *Anal. Biochem.* **110,** 201 (1981).
[8] C. R. Merril, D. Goldman, S. A. Sedman, and M. H. Ebert, *Science* **211,** 1437 (1981).
[9] C. R. Merril, D. Goldman, and M. L. Van Keuren, *Electrophoresis* **3,** 17 (1982).

METHODS IN ENZYMOLOGY, VOL. 104 ISBN 0-12-182004-1

gels, have resulted in highly sensitive protein stains. These silver stains are 100 times more sensitive for protein than the commonly used Coomassie Blue.[4,6–9]

In photography, photoreduced metallic silver on the "light-exposed" crystals acts as a catalyst, resulting in a differential reduction of silver halide crystals during chemical image development. Chemical image development usually relies on the use of organic reducing agents in alkaline solutions.

Histological silver stain development closely paralleled the evolution of photographic methods. However, the chemical basis of the selective reduction of ionic to metallic silver in most histological stains remains unknown. In some cases it may be catalyzed by reducing agents in the tissue. Reduction of acidic solutions of silver nitrate has been observed in tissue known to contain significant amounts of ascorbic acid.[10] In other tissues, differences achieved with silver staining appear to be based on physical interface phenomena.[10] Although aldehydes were thought to play a general role, neither aldehyde-creating nor aldehyde-blocking reagents appreciably affect silver staining in some tissues.[10]

A Negative-Image Silver Stain

Detection of proteins that have been separated by electrophoresis in gels with silver stains is probably due primarily to physical effects. The ability to generate a negative image in a gel containing separated proteins strongly supports this suggestion.

Procedure. All steps prior to the final washing (step 5) must be conducted in the dark. All solutions used should be in a quantity that is 10-fold larger than the volume of the gel.

1. Fix the gel in 50% methanol–12% acetic acid (v/v) for 30 min and wash for 10 min with deionized water.
2. Soak the washed gel in 0.2 *M* silver nitrate solution in the dark for 20 min, and briefly rinse with deionized water.
3. After rinsing, soak the gel in photographic developer (Kodak D76 in 1 : 5 dilution) for 10 min.
4. Rinse for 5 min with photographic fixer (Kodak general-purpose fixer, used without dilution).
5. Wash with water for three 20-min periods.

The negative image obtained shows clear regions in those portions of the gel containing protein whereas the background region is a brownish

[10] H. S. W. Thompson and R. D. Hunt, *in* "Selected Histochemical and Histopathological Methods," p. 800. Thomas, Springfield, Illinois, 1966.

gray. The production of such a negative image indicates that the protein has affected the reducibility of the silver in the region of the gel occupied by the protein. This image may be reversed by exposure to light during development or by the assistance of chemical reversal procedures.[7–9] The negative silver stain is 10-fold less sensitive then the positive silver stains.

A Positive-Image Silver Stain

As positive-image stains are easier to analyze than negative-image stains, a highly sensitive, simply performed reversal stain was developed by utilizing photochemical techniques.[7–9] In the most sensitive of these stains, image reversal is facilitated by the use of potassium dichromate.[8,9]

Procedure. All solutions used should be in a quantity that is 10-fold larger than the volume of the gel.

1. Gels may be fixed in either 20% (w/v) trichloroacetic acid or 50% methanol–12% acetic acid (v/v) for 30 min (gels thinner than 0.5 mm should be fixed only in trichloroacetic acid).
2. Wash gels twice for 15 min each with 10% ethanol or methanol and 5% acetic acid in water. This step allows the gel to swell to normal size and assists in removing contaminating buffers and ions that might otherwise reduce the sensitivity of the stain. Gels thicker than 1 mm require additional washing.
3. Soak gels for 15 min in 3.4 m*M* potassium dichromate containing 3.2 m*M* nitric acid. The amount of nitric acid added should be just sufficient to retain the potassium dichromate in solution. Addition of excess nitric acid reduces the sensitivity of the stain.
4. Soak for 20 min in 12 m*M* silver nitrate.
5. The silver nitrate is discarded and a solution of 0.28 *M* sodium carbonate containing 0.5 ml of formaldehyde (a commercially available solution of 37% formaldehyde) per liter is added. A precipitate of silver salts will form rapidly; to prevent the precipitate from adsorbing to the surface of the gel, the carbonate–formaldehyde solution should be changed at least twice. The sodium carbonate in this step is used to make the gel alkaline so that the formaldehyde can reduce ionic silver to the metallic form. Formic acid formed in the reaction is also buffered by the sodium carbonate. For maximum sensitivity, image development is allowed to continue until a yellowish background appears: a period of 15–20 min is required with a 1-mm-thick gel; longer periods are necessary with thicker gels.[9] When the image is sufficiently developed, the process is stopped by placing the gel in 3% (v/v) acetic acid for 5 min. Gels should be washed at least twice, 20 min each time, with water before storage.

6. Gels may be stored indefinitely in water or, at any time thereafter, soaked in 3% glycerol for 5 min and dried between dialysis membranes under reduced pressure at 80–82° for 3 hr. This results in a transparency that is relatively permanent and easy to store. If the gel has been properly washed, the image is intensified upon drying. For autoradiography or fluorography, gels are soaked in 3% glycerol for 5 min and dried onto Whatman 3 MM filter paper under reduced pressure at 80–82° for 3 hr.

A Modified Positive-Image Silver Stain

In some applications, it is desirable to combine the potassium dichromate and silver nitrate in a single solution, particularly when a 7% or lower concentration of acrylamide gel is used, or with applications of agarose gels. The first two steps are performed as specified (note that agarose is more efficiently fixed with trichloroacetic acid). In step 3, the gel is placed in a solution that is 19.5 m*M* silver nitrate, 1.34 m*M* potassium dichromate, and 13.5 m*M* sulfuric acid for 20 min. The gel image is then developed normally as described in steps 5 and 6.

Recycling for Increased Sensitivity

Recycling is accomplished by staining as described in the positive silver stain procedure followed by two additional 15-min rinses in 10 gel volumes of 3% (v/v) acetic acid and repeating or recycling through steps 4 and 5. This procedure results in an intensification of the stain because the image intensity in the positive image stain, i.e., without recycling, is limited by diffusion of silver from the gel during step 5. By recycling the gel through steps 4 and 5, silver ions can be replaced in the gel and additional staining intensity can be achieved.[9] Acidification with acetic acid is required in the recycling procedure to prevent the nonspecific reduction of silver that would occur if an alkaline gel, containing formaldehyde, were placed in a silver nitrate solution in step 4. Recycling can be repeated several times to intensify minor spots; however, the background also darkens and may become a problem.

Other Image Intensification and Destaining Procedures

Since the gel silver image is similar to a photograph, photographic image intensification and destaining procedures may be adapted from photographic formulas.[11] Modification of chemical concentrations and timing of the procedure are usually necessary, since most polyacrylamide

[11] E. J. Wall, F. I. Jordan, and J. S. Carrol, *in* "Photographic Facts and Formulas," p. 168. American Photographic Book Publishing Co., New York, 1976.

gels are thicker and have diffusion properties different from those of photographic emulsions. For *destaining* silver images, we have found the following method most useful.

1. Dissolve 37 g of sodium chloride and 37 g of cupric sulfate in 850 ml of deionized water. Add concentrated ammonium hydroxide until all the precipitate is dissolved and a deep blue solution is achieved before adjusting the volume to 1 liter.
2. Dissolve sodium thiosulfate, 436 g, in 1 liter of water.
3. Just prior to use, equal volumes of the reagents prepared in steps 1 and 2 are mixed and used directly if total destaining is required. They may be diluted (1 : 10 or 1 : 100) with water if light silver deposits are to be removed or if the image is to be lightened only slightly. It is advisable to photograph the gels prior to and during destaining to preserve a transient of the image, since it is difficult to stop destaining at a precise point. Gels may be restained after destaining by washing the gel three times in deionized water, 10 min each time, and then repeating steps 2 through 5 of the positive-image staining procedure. If the gel is insufficiently washed prior to restaining, silver will be reduced within the gel in step 4.

Sensitivity, Quantitation, and Protein Detection

Proteins have been detected with silver stains at concentrations as low as 20 pg/mm^2.[9] The positive-image silver stain procedure described above has been shown, with 8 purified proteins, to be linear over a 40-fold range in concentration, between 50 pg/mm^2 to 2 ng/mm^2. At concentrations greater than 2 ng/mm^2, the stain becomes nonlinear as spot densities reach saturation.[9] The dynamic range may be extended by recording the image during development. It should be noted that the relationship between density of silver staining and the concentration of protein is characteristic for each protein.[9] Quantitative use of the silver stain is possible if constitutive or marker proteins are present on each gel, so that densities can be normalized. Care must be taken to work within the linear range of the stain. There are some proteins that will stain with Coomassie Blue, but will not stain with the positive-image silver stain unless the recycling procedure is employed.[9]

Some of the histological silver stains that were developed for subcellular organelles or cellular structures have proved to be useful for staining specific proteins separated on polyacrylamide gels. One of these stains neurofilament protein,[12] and another primarily stains nucleolar proteins.[13]

[12] P. Gambetti, L. Autilio-Gambetti, and S. C. Papasozonenos, *Science* **213,** 1521 (1981).

[13] H. R. Hubbell, L. I. Rothblum, and T. C. Hsu, *Cell Biol. Int. Rep.* **3,** 615 (1979).

Colored Protein Images with Silver Stains

Most silver stains produce some colored bands or spots. Such coloration has also been observed in silver-based photographic processes. Color produced by this means was dependent on three variables: silver grain size, the refractive index of the gel or emulsion, and the distribution of silver grains in the gel. In general, the smaller grains transmit reddish or yellow-red light. Larger grains give bluish colors, and the very large grains produce black images. In a study of human cerebrospinal fluid proteins, utilizing a histochemical silver stain to stain proteins separated by two-dimensional electrophoresis, some lipoproteins stained blue whereas some glycoproteins appeared as yellowish-brown and red spots.[14] By modifying silver stain procedures, color effects can be enhanced[15] although saturation and negative staining effects can usually be accentuated by such modifications, making quantitation more difficult.

Comments

Although silver stains for the detection of protein are highly sensitive and fairly easy to perform, they are not without problems. The major loss of sensitivity in silver staining is due to inadequate water purity. Deionized water with a conductivity of less than 1 μmho is required in all reagents, including wash and fixing solutions. The second major cause of difficulty is usually inadequate fixation of proteins prior to staining. Polyacrylamide gels thinner than 0.5 mm, and all agarose gels, require fixation in 20% trichloroacetic acid. Gels thicker than 1 mm require additional washing prior to staining. Occasional surface artifacts, caused by silver carbonate adsorption, will mar a gel. They can be minimized by rapidly changing the sodium carbonate–formaldehyde solution during initial image development and by handling gels carefully. Pressure, fingerprints, and surface drying also cause surface artifacts. When sodium dodecyl sulfate gels (containing samples denatured in a solution containing mercaptoethanol) are stained to the point that the background is a yellowish brown, two horizontal lines will be observed at 60,000 and 67,000 daltons. These artifacts are caused by the mercaptoethanol and can be eliminated by reducing the amount employed.

Quenching may be observed if gels stained with silver are to be used for autoradiography or fluorography. The histological silver stain[4,16] al-

[14] D. Goldman, C. R. Merril, and M. H. Ebert, *Clin. Chem.* **26,** 1317 (1980).
[15] W. Sammons, L. D. Adams, and E. E. Nishizawa, *Electrophoresis* **2,** 135 (1981).
[16] M. L. Van Keuren, D. Goldman, and C. R. Merril, *Anal. Biochem.* **116,** 248 (1981).

most completely quenches the detection of ^3H-labeled compounds; however, the photochemical positive-image silver stain described here causes less quenching with ^3H-labeled proteins; quenching with ^{14}C-labeled proteins is barely perceptible.[16] Fluorographic detection of ^3H-labeled proteins can be restored almost completely by destaining the gel prior to fluorography.[16]

[31] Gel Protein Stains: Glycoproteins

By JOHN E. GANDER

Procedures for identifying glycoproteins on gels are available[1,2] and have been improved recently. The methods currently available are divided into the following categories: (1) thymol–H_2SO_4 method, (2) periodic acid–Schiff base method, and (3) fluorescein isothiocyanate-labeled lectin or -antibody method.

Thymol–H_2SO_4 Method

This procedure is useful for the location of glycoproteins containing at least 50 ng of carbohydrate.[2] Glycoproteins bearing hexosyl, hexuronosyl, or pentosyl residues react with H_2SO_4 to form furfural derivatives, which, in turn, react with thymol to form a chromogen. The chromogen is stable for only a few hours at ambient temperature. Furfural derivatives are not formed when 2-deoxy- or 2-acetamido-2-deoxyhexosaminyl residues are allowed to react with H_2SO_4. Gels must be washed free of low-molecular-weight contaminants before they are treated with the acid. Since protein zones become purple upon treatment with concentrated H_2SO_4 in the presence of glycine, this amino acid must be removed if Tris-glycine is used as a buffer system; the chromogen is presumably derived from the reaction of protein-bound tryptophan with the glyoxylic acid formed when H_2SO_4 interacts with glycine.[2]

The method cannot be used with gels containing glycosyl residues, e.g., agarose; such residues would be dehydrated to furfural derivatives by H_2SO_4.

Procedure. Fractionate protein(s) on 5–10% polyacrylamide gels in tubes 5 mm in diameter using any of the usual buffer systems including

[1] K. Burridge, this series, Vol. 50, p. 54.
[2] D. Rauchsen, *Anal. Biochem.* **99,** 474 (1979).

ISBN 0-12-182004-1

Tris–glycine, ampholytes, or sodium dodecyl sulfate (SDS).[2] After electrophoresis, wash the gels twice for at least 2 hr in borosilicate tubes with isopropanol–acetic acid–H_2O (25 : 10 : 65) to fix the proteins and to remove low-molecular-weight substances. Additional washes may be necessary if the protein samples contain large concentrations of sucrose or other soluble carbohydrates that can react with H_2SO_4 to form a furfural derivative. A final wash for 2 hr in the same solvent containing 0.2% thymol (w/v) results in formation of a stable gel. After washing with thymol, decant the liquid and allow the gels to drain. Add a solution of concentrated H_2SO_4–absolute ethanol (80 : 20) at ambient temperature to each gel and cap the tubes. At least 10 ml of reagent per milliliter of gel should be used. Gently shake the tubes held in a horizontal position at 35° for 2.5 hr or until the opalescent appearance of the gels just disappears. Zones containing glycoproteins stain red whereas the background is yellow. Other proteins do not form visible zones when treated in this manner.

Periodic Acid–Schiff Base Method

This procedure is another generally useful technique for locating glycoproteins on gels. Periodic acid oxidizes and cleaves *gem* secondary alcohols of glycosyl residues to dialdehydes; the aldehydes are then allowed to react with fuchsin,[3,4] Alcian blue,[5] or dansyl hydrazine[6] to form a Schiff base. A lower limit of 2–3 μg of carbohydrate can be detected using fuchsin or Alcian blue[7] compared with 40 ng using dansyl hydrazine.[6] Neutral glycosyl residues that are substituted at C-3 and 2-deoxyglycosyl or 2-acetamido-2-deoxyglycosaminyl residues substituted at either C-3 or C-4, or both positions, will not be oxidized by periodic acid.

The method is not as sensitive as the H_2SO_4–thymol method when fuchsin or Alcian blue are used as the base but is comparable when dansyl hydrazine is used. However, with dansyl hydrazine, it is necessary to modify the Eckhardt *et al.* procedure[6] if glycoproteins are separated on agarose gels.[8]

Procedure with Dansyl Hydrazine as the Base. Polymerization of acrylamide to polyacrylamide must be catalyzed by a nonfluorescent reagent such as ammonium persulfate.

[3] R. M. Zaccharias, T. E. Zell, J. H. Morrison, and J. J. Woodlock, *Anal. Biochem.* **30,** 148 (1969).

[4] G. Fairbanks, T. L. Steck, and D. F. H. Wallach, *Biochemistry* **10,** 2606 (1971).

[5] A. H. Wardi and G. A. Michos, *Anal. Biochem.* **49,** 607 (1972).

[6] A. E. Eckhardt, C. E. Hayes, and I. E. Goldstein, *Anal. Biochem.* **73,** 192 (1976).

[7] R. A. Kapitany and E. J. Zebrowski, *Anal. Biochem.* **56,** 361 (1973).

[8] M. Furlan, B. A. Perret, and E. A. Beck, *Anal. Biochem.* **96,** 208 (1979).

Dissolve proteins in 10% glycerol (v/v) and 0.005% methylene green (w/v), layer on the stacking gel, and electrophorese in 7.5% polyacrylamide gels. Fix the gels overnight in ethanol–glacial acetic acid–water (40 : 5 : 55). Treat the gels with 20 μl of 0.7% paraperiodic acid (w/v) in acetic acid–water (95 : 5) for 2 hr. Rinse gels with water, treat with 0.5% sodium metabisulfite in 5% acetic acid until the gels are colorless (1–1.5 hr), and rinse repeatedly with water. Transfer the gels to tubes of appropriate size and add equal volumes of acidic dimethyl sulfoxide (0.6 ml of 12 *N* HCl per liter of dimethyl sulfoxide) and a freshly prepared solution containing 2 mg of dansyl hydrazine per milliliter of dimethyl sulfoxide. Stopper tubes, mix the solutions, and maintain at 60° for 2 hr. Decant the solution and add sufficient $NaBH_4$ in dimethyl sulfoxide (0.2 mg of $NaBH_4$ per milliliter of dimethyl sulfoxide) to cover the gels. After 30 min at 25°, decant the liquid and rinse the gels with water. Treat the gels with acetic acid–water (99 : 1) repeatedly until the background is colorless when illuminated with ultraviolet light (λ_{max} 366 nm). If a 4 × 5 press-type camera equipped with a Polaroid 545 film adapter and Polaroid type 57 (ASA 3200) film is used, an exposure time of 2.5 sec at f 5.6 with a Kodak Wratten 16 filter should provide adequate photographic record of the gel. Controls in which the periodate oxidation step is omitted will disclose proteins with natural fluorescence or those that noncovalently bind dansyl hydrazine.

Fluorescein Isothiocyanate-Labeled Lectin Method

This procedure has specificity toward the carbohydrate region of glycoproteins to a degree dictated by the specific lectin. Fluorescein-labeled concanavalin A reacts with numerous glycoproteins because many contain residues, such as D-mannopyranosyl, for which concanavalin A has an affinity. In contrast, fluorescein-labeled monoclonal antibody, with its antigenic determinants directed toward a unique functional group on a specific glycoprotein, would provide a very sensitive and specific probe for that glycoprotein.

Furlan *et al.* have described a method for staining glycoproteins, fractionated on either polyacrylamide or agarose gels, by using fluorescein isothiocyanate conjugated to *Ricinus communis* agglutinin, to concanavalin A, or to immunoglobulins.[8] Less than 100 ng of hexosyl residues bound to protein is detectable.

Procedure. Conjugate lectin or antibody (5 mg/ml) by allowing to react with fluorescein isothiocyanate (100 μg/ml final concentration) in 0.15 *M* Na_2HPO_4 (pH 9.5) at 25° for 20 hr.[9] Dialyze the reaction mixture against

[9] T. H. The and T. E. Feltkamp, *Immunology* **18,** 865 (1970).

several changes of 0.1 *M* NaCl–0.05 *M* Tris-HCl–1 m*M* $CaCl_2$ at pH 7.0. The $A_{280}:A_{485}$ ratio should be about 0.9 depending on the molar extinction coefficient at 280 nm of the protein.

Glycoprotein separations have been carried out on polyacrylamide gels (5% acrylamide containing 5% bisacrylamide) made in 0.2% SDS–0.1 *M* Tris-HCl–6 *M* urea at pH 7.4[8]; or 2% polyacrylamide–0.5% agarose buffered with 0.075 *M* sodium 5,5′-diethylbarbiturate–0.01 *M* ethylenediaminetetraacetic acid at pH 8.6 and containing 6 *M* urea and 0.2% SDS[10]; or on 1% agarose gels in a buffer containing 0.075 *M* sodium 5,5′-diethylbarbiturate–0.01 *M* ethylenediaminetetraacetate–0.2% SDS at pH 8.6.[11] Remove the SDS by washing the gels for 20 hr in methanol–acetic acid–water (10 : 3 : 27) and in 5% acetic acid. Wash the fixed gel for 4 hr during each of four changes of 0.1 *M* NaCl–0.05 *M* Tris-HCl (pH 7.0) containing 1 m*M* $CaCl_2$ and 1 m*M* $MnCl_2$. Dialyze the fluorescein isothiocyanate–lectin or fluorescein isothiocyanate–antibody, conjugate against the NaCl–Tris-$CaCl_2$–$MnCl_2$ buffer described above, and adjust the concentration of the fluorescein-labeled protein to 1 mg/ml. Treat the gels with fluorescein-labeled protein solution for 12 hr at 25°. Follow by destaining in the same buffer for 2 days. The stained bands remaining are noted as being visible for up to 1 month at either 25 or 4°.[8] The staining solution can be used at least three times without loss of significant staining intensity.

Intense background staining of agarose gels was obtained with galactose-specific *Ricinus* lectins and the fluorescent substances could not be removed by washing with the Tris–NaCl–$CaCl_2$ buffer. However, buffer containing 0.1 *M* D-galactose did remove the background fluorescence. No difficulty should be encountered in destaining polyacrylamide or agarose gels treated with fluorescein-labeled concanavalin A.

Comments

Staining techniques that couple the action of enzymes to the formation of chromogenic or fluorogenic products have been used to locate trehalase,[12] invertases,[13] and glycoproteins that react with sialic acid-containing lectins[14] or concanavalin A.[15] The latter method depends on the forma-

[10] B. A. Perret, M. Furlan, and E. A. Beck, *Biochim. Biophys. Acta* **578,** 164 (1979).
[11] E. A. Beck, L. Tranqui-Pouit, A. Chapel, B. A. Perret, M. Furlan, G. Hudry-Clergeon, and M. Suscillon, *Biochim. Biophys. Acta* **578,** 155 (1979).
[12] K. A. Killick and L.-W. Wang, *Anal. Biochem.* **106,** 367 (1980).
[13] P. Babczinski, *Anal. Biochem.* **105,** 328 (1980).
[14] K. Yamada and S. Shimizu, *Histochem. J.* **11,** 457 (1979).
[15] K. Yamada and S. Shimizu, *Histochem.* **47,** 159 (1976).

tion of a stain when horseradish peroxidase, which is covalently attached to a lectin or an antibody, catalyzes the oxidation of diaminobenzidine.[13]

Most of the noted techniques are less than completely satisfactory. For instance, as sensitivity is increased, interference by background substances in gels also increases. Adequate controls must be included for each staining series because any one preparation of glycoproteins under investigation may contain substances that lead to artifacts.

[32] Gel Protein Stains: Phosphoproteins

By JOHN A. CUTTING

Phosphoproteins may be detected on polyacrylamide gel electropherograms by use of either the radioactivity of intrinsically incorporated $^{32}P_i$, or a stain that has some specificity for the phosphate moiety. The use of ^{32}P-labeled samples has the advantage of greater sensitivity over direct staining. However, only *de novo* incorporated phosphate is thereby detected. There are several reasons why phosphoproteins of interest may not become labeled during incubation of a biosystem with a ^{32}P-containing substrate: the protein is not being synthesized or presynthesized molecules do not have available sites for additional phosphorylation; the protein may be subject to avid phosphatases; or a necessary, specific kinase may be unavailable. Each situation precludes detection of the protein. In contrast, all phosphate linked to a particular amino acid in a phosphoprotein is available for detection by a suitable dye method.

Method I. The Entrapment of Liberated Phosphate (ELP)[1]

This is the most specific of the available phosphoprotein staining methods. Moreover, it is easy to perform a test for the authenticity of a stained band (see Controls). A modification permits detection of enzymes that release inorganic phosphate from their substrates (see below). The method depends on entrapment within the gel of insoluble calcium phosphate formed during the alkaline hydrolysis of susceptible protein phosphoester bonds in the presence of calcium chloride. Subsequently, entrapped phosphate is allowed to react with a modified Fiske–SubbaRow reagent, and detection of the resulting insoluble blue complex is enhanced

[1] J. A. Cutting and T. F. Roth, *Anal. Biochem.* **54,** 386 (1973).

ISBN 0-12-182004-1

by staining with methyl green. The location of a phosphoprotein on the gel results in a bright green band. Sensitivity is 1 nmol of phosphate.

Reagents

A. 10% (w/v) sulfosalicylic acid (SSA) in deionized water
B. 10% (w/v) SSA in 25% (v/v) 2-propanol in deionized water
C. 0.5 *M* $CaCl_2$ in 10% (w/v) SSA
D. 0.5 *N* NaOH
E. 1% (w/v) ammonium molybdate in deionized water
F. 1% (w/v) ammonium molybdate in 1 *N* HNO_3
G. 0.5% (w/v) methyl green (CI No. 42590) in 7% (v/v) aqueous acetic acid
H. 7% (v/v) aqueous acetic acid

All reagents must be free of phosphate, and glassware should be washed with a phosphate-free detergent. Suspect reagents and glassware can be checked by the addition of the Fiske–SubbaRow reagent (4.2% ammonium molybdate in 4.5 *N* HCl), which will present as a blue color in the presence of phosphate. Acceptable glassware should then be thoroughly rinsed with deionized water.

Procedure. For optimal results, the volumes of reagent solutions employed at each step should be 15 times the volume of the gel being stained.

1. If the sample or gel system contains a phosphate buffer or other low-molecular-weight phosphates, the gel must be fixed in reagent A. Gels containing SDS must be fixed in reagent B. In both instances gels should be treated for 12 hr, with several changes of the reagent, to permit diffusion from the gel of these interfering substances.
2. All gels: Fix in reagent C for 1 hr.
3. Rinse the gel thoroughly in deionized water to avoid carryover of surface $CaCl_2$, which will otherwise cause excessive background staining.
4. Transfer the gel to reagent D at 60° for 30 min.
5. Rinse the gel twice for 10 min each in reagent E.
6. Place the gel in reagent F for 30 min.
7. Stain the gel with reagent G for 30 min.
8. Destain with reagent A. This process is hastened at 60°.[2]
9. Store gels in reagent H.

Controls. Protein itself may be stained, e.g., with Coomassie Blue, in a parallel-run gel. Separate gels or lanes should contain a known phospho-

[2] D. H. Ohlendorf, G. R. Barbarash, A. Trout, C. Kent, and L. T. Banaszak, *J. Biol. Chem.* **252,** 7992 (1977).

protein (chicken phosvitin or α-casein) as a positive control, and be stained for phosphoprotein simultaneously with the unknown. Authenticity of a stained band may be verified by the following procedure.[1] Two sets of gels (a and b) are prepared, each containing the unknown and a positive control in separate lanes. After appropriate fixation, as detailed in step 1 above, treat the gel sets as follows:

A1. Fix set (a) in reagent A and set (b) in reagent B for 1 hr.

A2. Carry out step 3 above.

A3. Transfer both sets to a solution that will solubilize the protein sample and permit it to diffuse from the gel [0.2 *M* NaCl, 4 *M* urea, or 0.1% sodium dodecyl sulfate (SDS) solution]. This may require a day and several changes of the solution. Residual protein can be monitored in test gels with Coomassie Blue stain. If either phosphate buffer or SDS is used in this step it must subsequently be removed as in step 1[3] above.

A4. Use steps 4 through 9, as above.

By this strategem, phosphate liberated during hydrolysis is not trapped in the gels of set (a) since $CaCl_2$ is absent from reagent A, but is trapped at the phosphoprotein locus by the standard procedure to which the gels of set (b) were subject. Step A3 results in diffusion from the gel of all protein and nontrapped phosphate so that, following step A4, the presence of a green band in the gels of set (b) [and the absence of such a band from the gels of set (a)] is the consequence of a staining reaction with trapped calcium phosphate and is not an artifact.

Entrapment of Enzymically Released Orthophosphate.[4] The ELP procedure is adaptable for the detection on polyacrylamide gels of enzymes that release phosphate from their substrates. Pyruvate-uridine diphospho-*N*-acetylglucosamine transferase has been located on gels after incubation with its substrates, 1 m*M* phosphoenolpyruvate and 1 m*M* uridine diphospho-*N*-acetylglucosamine, in the presence of 50 m*M* $CaCl_2$. After steps 5 through 9, the enzyme was detected as a green band.

Method II. The Use of Stains-All[4]

The cationic carbocyanine dye, Stains-all {1-ethyl-2-(3-(1-ethylnaphtho[1,2-*d*]thiazolin-2-ylidene)-2-methylpropenyl)naphthol[1,2-*d*] thiazolium bromide}, stains many biopolymers varying shades of blue or red. Unconjugated proteins are stained red, whereas phosphoproteins are

[3] R. I. Zemell and R. A. Anwar, *J. Biol. Chem.* **250,** 3185 (1975).

[4] M. R. Green, J. V. Pastewka, and A. C. Peacock, *Anal. Biochem.* **56,** 43 (1973).

in shades of blue. However, glycoproteins are also stained blue,[5] as are DNA and RNA. Although the staining procedure is easily performed, the inherent lack of specificity should limit the use of this stain for monitoring phosphoproteins, the phosphate content of which is verifiable by more specific criteria. Sensitivity of the stain is at 0.3 nmol of phosphate.

Reagents

A. 25% (v/v) 2-propanol in deionized water
B. (Stock solution) 0.1% Stains-all in formamide
C. Working solution, prepared immediately before use: 10 ml of reagent B, 10 ml of formamide, 50 ml of 2-propanol, 1 ml of 3.0 *M* Tris-HCl, pH 8.8. Prepare to 200 ml with deionized water

Procedure. Since direct light causes the gel background to become opaque and stained bands to decolorize, both staining and destaining are done in the dark.

1. Agitate alkaline gels or those containing SDS in reagent A for 15 min.
2. Stain gels overnight in reagent C.
3. Destain in deionized water.

Method III. Trivalent Metal Chelation[6]

This method uses a trivalent metal ion as a mordant that will allow staining of acidic phosphoproteins (phosvitins) with the protein dye Coomassie Blue. Usually phosvitins do not stain permanently with this dye. The method depends on differential staining: a blue band, corresponding to phosvitin, is evident only when the trivalent metal ion Al^{3+} is included in the staining reagent. Since Al^{3+} appears to permit staining by reducing the net negative charge on the protein,[6] this method is of low specificity. Sensitivity is at 0.13 nmol of phosphate.

Reagents

A. 0.05% Coomassie Brilliant Blue R250 in 25% (v/v) 2-propanol, 10% (v/v) acetic acid, 1% (v/v) Triton X-100, prepared in deionized water
B. 0.1 *M* aluminum nitrate prepared in reagent A
C. 7% (v/v) aqueous acetic acid

[5] L. E. King, Jr. and M. Morrison, *Anal. Biochem.* **71,** 223 (1976).
[6] J. Hagenauer, L. Ripley, and G. Nace, *Anal. Biochem.* **78,** 308 (1977).

Procedure

1. Prepare two parallel-run gels (a and b), and stain gel (a) in reagent A, and gel (b) in reagent B.
2. Destain both gels in reagent C.

A permanently stained blue band on gel (b) that is absent from gel (a) suggests the presence of an acidic phosphoprotein.

[33] Western Blots

By JAIME RENART and IGNACIO V. SANDOVAL

The blotting[1] of electrophoretically separated components onto solid supports allows their identification, assists in the study of their binding to other molecules, and sometimes leads to the isolation of such ligands on a small scale. Because the proteins, extracted from gel, are bound to the surface of a blotting support, blotting is particularly suitable when the probes used for scrutinizing specific proteins either do not permeate the matrix of the gel or do so only so slowly that their binding to protein cannot be detected. In each instance, it is important to ensure that both the gel electrophoresis system and the blotting support are adequate for the type of experiment and for the protein probes that are used.

Technically, the simplest problem is the identification of proteins with probes that recognize denatured proteins directly coupled to a blotting support. For this situation, the following protocol can be used.

The protein mixture must be resolved into its components by electrophoresis in one or two dimensions. The sodium dodecyl sulfate (SDS)–polyacrylamide gel systems of Laemmli[2] or Neville[3] (see also this volume [12]) can be used to separate proteins in one dimension. The system of Neville[3] has the advantage of using Tris-borate as running buffer, thereby avoiding the lengthy washes required to remove glycine from gels in which Tris-glycine is used.[2] However, complex mixtures of proteins are incompletely resolved by this means, and separation in two dimensions may be necessary. The O'Farrell system[4] does so by using electrofocusing in the first dimension and electrophoresis on SDS–polyacrylamide gels in the second.

[1] J. Renart, J. Reiser, and G. R. Stark, *Proc. Natl. Acad. Sci. U.S.A.* **76,** 3116 (1979).
[2] U. K. Laemmli, *Nature* (*London*) **227,** 680 (1970).
[3] D. M. Neville, *J. Biol. Chem.* **256,** 6328 (1971).
[4] P. H. O'Farrell, *J. Biol. Chem.* **250,** 4007 (1975).

METHODS IN ENZYMOLOGY, VOL. 104

ISBN 0-12-182004-1

Proteins separated by electrophoresis remain embedded in acrylamide gel and must be blotted onto paper. For this purpose, either diazophenylthio (DPT) paper[5] or diazobenzyloxymethyl (DBM) paper[1] is available. These two papers are the best studied supports with respect to efficiency of transfer, capacity for binding protein, reactivity conditions, and stability.[1,4–6] DPT paper is easier to prepare than DBM paper,[7] and the following protocol, described by Seed,[5] is effective.

Materials

1,4-Butanediol diglycidyl ether (Aldrich 12,419-2)
2-Aminothiophenol (Aldrich 12,313-7)
Whatman 50 or 540 paper, or Schleicher & Schuell 589WH paper

Methods

All operations must be carried out in a fume hood. Gloves should be used, since both 1,4-butanediol diglycidyl ether and 2-aminothiophenol are toxic.

1. Cut the paper to the desired size (usually 16–18 cm) and place about 20 g of paper in a heat-sealable polyester bag (Sears Seal-N-Save or polyethylene bags). First add 70 ml of 0.5 *M* NaOH, which is to be followed by 30 ml of butanediol diglycidyl ether. The bag is sealed, leaving sufficient room for good mixing and the possibility for opening and resealing the bag. Rotate the bag end-over-end for 12–16 hr. After the coupling period, remove reagents by pouring them into 1 *M* NH_4OH and allow 24 hr for inactivation before disposal down the drain.

2. Mix 10 ml of 2-aminothiophenol with 40 ml of ethanol, and add this mixture to the bag. Reseal and rotate for an additional 10 hr.

3. After step 2, papers are washed by sequential immersion and agitation in ethanol, followed by washing in 0.1 *M* HCl for a total of three cycles of about 15 min each. Papers are rinsed in distilled water for 1–2 hr, immersed in ethanol, and finally dried in air. The paper should be stored at −20° in a desiccator.

4. Aminophenylthio (APT) paper is activated to form the diazophenylthio (DPT) form by diazotization immediately prior to blotting. Diazotization is performed by immersing the paper in a tray containing 100 ml of 1.2 *M* HCl at 4° to which 32 mg of $NaNO_2$ dissolved in 1.2 *M*

[5] B. Seed, *Genet. Eng.* **4,** 91 (1983).

[6] J. Reiser and J. Wardale, *Eur. J. Biochem.* **114,** 569 (1981).

[7] J. C. Alwine, D. J. Kemp, B. A. Parker, J. Reiser, J. Renart, G. R. Stark, and G. M. Wahl, this series, Vol. 68, p. 220.

HCl is added. Diazotization is allowed to occur for 30 min with occasional shaking. The paper is briefly washed subsequently with several changes of distilled water and immediately thereafter placed in contact with the gel to be blotted.

Blotting of Proteins onto Paper

The transfer of proteins from gels onto paper, i.e., blotting, can be performed either by passive diffusion[1] or by electrophoresis.[6] The rate and yield of protein transfer are both related inversely to the size of the protein and to the extent of cross-linking of the gel. Transfer of proteins by electrophoresis is faster and more efficient and minimizes diffusion of proteins over gel and paper surfaces that occurs during transfer by passive diffusion.

In both methods, the blotting paper is placed over a scouring foam pad supported by a rigid plastic grid. The gel, containing the protein, is washed three times for 10 min each in 250 ml of water and once for 10 min in 50 m*M* sodium borate. The borate is at pH 8 for blotting by passive diffusion and at pH 9.2 for electrophoresis. The gel is laid on the top of the paper and successively covered with a foam pad and a rigid plastic grid. Bubbles between gel and paper are carefully removed by rolling a pipette over the gel surface. Finally, the gel is fixed in close contact with the paper by fastening the plastic grids with rubber bands (Fig. 1). When blotting is performed by passive diffusion, the paper should be beneath the gel and on top of a dry foam pad so as to absorb the borate buffer impregnating both the gel and the foam pad that lies on top of it. Most of the protein diffusing from the gel within 4 hr reacts and becomes attached to the paper. The rate of transfer may be increased by replacing the pad underneath the blotting paper as many times as needed.

The blotting of protein by electrophoresis can be carried out by placing the assembly of gel and paper, with the paper facing the anode, in an electrophoresis tank filled with 10 m*M* sodium borate at pH 9.2. Transfer is performed at 400 V (300–400 mA) to produce a voltage gradient of

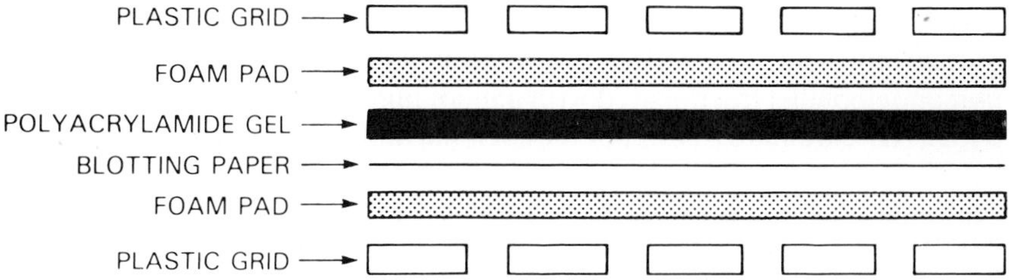

PLASTIC GRID →
FOAM PAD →
POLYACRYLAMIDE GEL →
BLOTTING PAPER →
FOAM PAD →
PLASTIC GRID →

FIG. 1. Schematic of the protein-blotting sandwich used in electrotransfer experiments.

32 V/cm for 1 hr at room temperature.[6] Lowering the pH of the buffer to between 8 and 8.5 decreases the current and allows the voltage gradient to be obtained with standard power supplies.

Other Transfer Media

In addition to DPT and DBM paper, cyanogen bromide-activated paper[8] and cyanuric chloride paper[9] have been used to form stable covalent bonds with the blotted proteins.

Whenever the formation of covalent bonds between protein and blotting support is undesirable, nitrocellulose sheets,[10] DEAE paper,[11] or papers with attached affinity ligands for the protein under scrutiny may be used. Although other buffers may be required for blotting onto these supports, the method is essentially similar to that described for DPT and DBM papers. Blotting of protein onto nitrocellulose can be performed electrophoretically by immersing the assembly of gel and paper in 0.7% acetic acid and using a voltage gradient of 6 V/cm for 1 hr at room temperature.[10]

Protein Identification

Finally, the specific protein under scrutiny must be singled out from among all others that were blotted by using specific probes. For this purpose, the blotting supports must first be treated in order to inactivate remaining diazonium groups in DPT and DBM papers. The paper is soaked for 15 min at room temperature with gentle rocking[1] in 0.1 *M* Tris-HCl at pH 9.0 containing 10% ethanolamine and 0.25% gelatin (w/v). Unreacted sites on nitrocellulose may be saturated with gelatin by soaking the membrane in 10 m*M* Tris-HCl at pH 7.4, containing 0.9% NaCl and 3% gelatin, for 1 hr at 40° with continuous rocking.[10] Bovine serum albumin may be substituted for gelatin, but, owing to its frequent contamination with IgG, its use should be avoided whenever it is anticipated that protein A will be applied.

The most commonly used probes for protein are antibodies, since they usually bind with high specificity and affinity to the denatured antigenic protein covalently ligated to the blotting support. Whole sera, various immunoglobulins fractionated with ammonium sulfate and further purified by DEAE-chromatography, or monoclonal antibodies can be used as pro-

[8] R. A. Hitzeman, L. Clarke, and J. Carbon, *J. Biol. Chem.* **255,** 12073 (1980).

[9] H. D. Hunger, H. Grutzman, and C. Cotelle, *Biochim. Biophys. Acta* **653,** 344 (1981).

[10] H. Towbin, T. Staehelin, and J. Gordon, *Proc. Natl. Acad. Sci. U.S.A.* **76,** 4350 (1979).

[11] T. McLellan and J. A. M. Ramshaw, *Biochem. Genet.* **19,** 647 (1981).

tein probes. The amount of antibody used will vary with the titer. Antibody is added to the blotting support bathed in a saline buffer (10 m*M* Tris-HCl, 0.9% NaCl, 0.1% gelatin, and 0.04% sodium azide at pH 7.4). Routinely, a paper of the size of a standard slab gel is soaked for 2 to 24 hr at room temperature in 30 ml of the saline buffer (50 to 100 μl buffer/cm^2) containing either 30 to 100 μl of whole immune serum, 0.1–0.2 mg of purified IgG, or 5 to 10 μg of a high-affinity monoclonal antibody. Unbound antibody is washed thereafter with six changes of 200 ml of the saline buffer for 1 hr each, at room temperature before labeling the antibody bound to the protein with a second labeled probe. The antibody is most frequently treated with protein A from *Staphylococcus aureus* A[12] or with anti-antibodies bearing iodine-125.[10,13] The reaction between the antibody and the labeled probe is performed by soaking the blot in 30 ml of saline buffer containing 2 to 5 $\times$ 10^6 cpm ^{125}I-labeled (100 Ci/mol) of either protein A or anti-antibody for 4 hr at room temperature.

Anti-antibodies labeled with a photofluore such as rhodamine,[10] or conjugated to horseradish peroxidase,[10] can be used with proteins blotted onto nitrocellulose. After labeling, the unbound labeled probe is removed by washing with the saline buffer (eight changes of 200 ml for 6 hr each) and dried. Blots labeled with ^{125}I are exposed to Kodak X-Omat film. Nitrocellulose blots labeled with photofluores are inspected under longwave UV light through a yellow filter.

Protein blotted onto DPT or DBM paper or onto nitrocellulose sheets can be tested repeatedly with the same or new probes after removing the old probe from the blotting support.[1] To remove either protein A or antibodies bound to blotted proteins, the papers are incubated with 10 m*M* Tris-HCl at pH 7.5 containing 0.1 *M* 2-mercaptoethanol and 10 *M* urea, for 30 min at 60°.[1] Alternatively, the paper is washed with 10 m*M* Tris-HCl at pH 7.5, containing 3 *M* potassium thiocyanate, for 30 min at 37° and then with 10 m*M* HCl for an additional 30 min at room temperature.[6] An equally effective method of removing an old probe is by soaking the paper in 0.25 *M* glycine-HCl at pH 2.8 for 10 min at room temperature. After removal of the probes by either method, papers are neutralized with 50 ml of 1 *M* Tris-HCl at pH 7.5 for 30 min and washed with saline buffer so that they may be probed again. Nitrocellulose blots can be cleaned from old probes by incubation with 8 *M* urea, 0.1 *M* 2-mercaptoethanol, and 5 mg bovine serum albumin per milliliter at 60° for 1 hr. Before incubation with a new probe, the blots must always be reequilibrated with the saline buffer.

[12] S. W. Kessler, *J. Immunol.* **115,** 1617 (1975).
[13] W. N. Burnette, *Anal. Biochem.* **112,** 195 (1981).

The use of probes that bind only to native protein, rather than to a denatured form as presented after SDS treatment, requires skill. Few general rules can be given. The proteins must be resolved by electrophoresis under nondenaturing conditions (i.e., omitting both SDS and urea) or, if resolved under denaturing conditions, require renaturation. Partial renaturation of some proteins subsequent to SDS–polyacrylamide gel electrophoresis has been achieved by soaking the gels in buffers known to stabilize them; generally 20% glycerol and 200 m*M* NaCl are included.[14]

The use of covalent attachment of proteins to DPT and DBM paper may hinder the binding of the probe to protein but need not do so. In case of interference, protein may be blotted onto nitrocellulose sheets,[10] DEAE paper,[11] or papers having attached affinity ligands for the protein.[15]

Comments

The blotting of proteins resolved by gel electrophoresis onto paper has been used for sorting antibodies interacting with them from whole serum.[16] This allows the immunization of animals with complex mixtures of proteins containing one or more antigens of interest without lengthy purification of antigen. Similarly, proteolytic digests of proteins resolved by gel electrophoresis could be blotted onto solid supports and the antigenic determinants of the protein characterized.

[14] R. E. Manrow and R. P. Dottin, *Anal. Biochem.* **120,** 181 (1982).
[15] H. A. Ehrlich, S. N. Cohen, and H. O. McDevitt, *Cell* **13,** 681 (1978).
[16] J. B. Olmstead, *J. Biol. Chem.* **256,** 11955 (1981).

[34] Fluorography for the Detection of Radioactivity in Gels

By WILLIAM M. BONNER

Fluorographic techniques can greatly increase the efficiency of radioactive isotope detection. In these techniques, scintillants or fluors are placed in the path of the radioactive disintegration. Light emitted by the fluors in response to a radioactive disintegration can then be detected on film. Because the wavelength of the emitted light depends on the fluor, not on the energy of the radioactive disintegration, the same film can be used to detect the small range of photoenergies from isotopes of widely varying

METHODS IN ENZYMOLOGY, VOL. 104 ISBN 0-12-182004-1

energies more efficiently than would be possible if the film had to interact directly with those isotopes.

Weak β-Emitters in Polyacrylamide Gels

For ^{3}H, ^{14}C, ^{33}P, and ^{35}S, the fluor must be placed inside the gel since most of their radiation never emerges from the gel to expose the film. However, most fluors are hydrophobic and not soluble in the hydrophilic compounds used for keeping polyacrylamide gels solvated. Dimethyl sulfoxide (DMSO) was found to have the necessary properties; 2,5-diphenyloxazole (PPO) dissolved in DMSO can be easily and efficiently introduced into polyacrylamide gels. For ^{14}C, ^{33}P, and ^{35}S, the increase in sensitivity over autoradiography is approximately 15-fold. For ^{3}H, the enhancement in sensitivity is in excess of 1000-fold. The detection of ^{3}H in polyacrylamide gels is a major achievement of fluorography.

Original PPO–DMSO[1] and Related Procedures. Although DMSO and other compounds used in fluorography have no known mutagenicity or carcinogenicity, they are not without potential hazard, and appropriate precautions should be used. DMSO readily penetrates the skin and may facilitate the entry of dissolved substances such as PPO. Mixtures based on glacial acetic acid must also be used with caution. The procedures described here should be carried out in a hood and with gloves.

After electrophoresis, the gel may be either stained or processed directly for fluorography. Formaldehyde may be used to fix small proteins and peptides in the gel.[2] All soaking steps should be performed with gentle shaking in closed containers sufficiently large to contain the gel without folding. Plastic freezer trays with tight-fitting lids can readily be obtained for this purpose.

The following procedure is for 0.8-mm-thick gels (increase by a factor of 3 for 1.5-mm-thick gels[1]). The gels are soaked (1) twice for 10 min each in DMSO to remove water; (2) once in four volumes of a 22% (w/v) solution of PPO in DMSO for 1 hr; (3) twice for 10 min each in water to precipitate PPO within the gels. Gels are dried for about 1 hr under vacuum at 80°. Commercial vacuum gel dryers are available for this purpose. For exposure overnight using unflashed film, sensitivities are approximately 20 dpm/mm^2 for ^{14}C, ^{35}S, and ^{33}P and 300 dpm/mm^2 for ^{3}H.

Glacial acetic acid may be substituted for DMSO as the solvent; under such conditions, the gels may be immersed directly in 22% PPO (w/v) in glacial acetic acid, eliminating step 1. The staining pattern of the gel is

[1] W. M. Bonner and R. A. Laskey, *Eur. J. Biochem.* **46,** 83 (1974).

[2] G. Steck, P. Leuthard, and R. R. Bürk, *Anal. Biochem.* **107,** 21 (1980).

more faithfully retained when acetic acid is used instead of DMSO. PPO can be recovered from either solvent for reuse by adding water and collecting the precipitated PPO on a filter.[3]

Some time after publication of the original procedure using PPO dissolved in DMSO, a commercial preparation, Enhance (New England Nuclear), became available. Since Enhance is based on acetic acid, gels can be soaked directly in it for 1 hr, then in two 10-min washes of water. Enhance may be considerably less expensive than PPO-containing solutions if one does not recover the PPO for reuse.

Aqueous Procedures. Water-soluble fluorography preparations are becoming available as water-soluble fluors are found or synthesized. The original water-soluble fluorographic procedure uses sodium salicylate.[4] This method is simpler than, and may be as sensitive as, the methods just described, but its resolution is significantly less than that obtained with PPO–DMSO. It is useful primarily when a few widely spaced radioactive bands are to be detected.

The following procedure is used for a 1.5-mm-thick gel. Gels at neutral or alkaline pH may be added directly to 1 *M* (16% w/v) sodium salicylate for 30 min, but stained gels need to be presoaked for 30 min in water to remove acetic acid. The gel is removed from the sodium salicylate and dried at 80° directly on paper under an acetate sheet. The usual Mylar sheets should not be used, since they become glued to the dried gel.

Commercial water-soluble fluorography solutions have appeared. The procedures are basically identical to that used with sodium salicylate, but the resolution is better and appears to be as good as that obtained with PPO–DMSO. Two preparations presently available are Autofluor (National Diagnostics) and Enlightning (New England Nuclear). None of the above methods is obviously the best for every situation; investigators must choose according to their requirements.

Weak β-Emitters in Agarose Gels

After they have been dehydrated in several changes of absolute ethanol, agarose gels can be soaked in a 3% (w/v) solution of PPO in absolute ethanol for 3 hr.[5] The PPO is precipitated thereafter by immersing the gel in water. During drying, the gel should not be heated or heated only gently so that it does not melt.

[3] R. A. Laskey, A. D. Mills, and J. S. Knowland, Appendix in R. A. Laskey and A. D. Mills, *Eur. J. Biochem.* **56,** 335 (1975).

[4] J. P. Chamberlain, *Anal. Biochem.* **98,** 132 (1979).

[5] R. A. Laskey, A. D. Mills, and N. R. Morris, *Cell* **10,** 237 (1977).

Weak β-Emitters on Solid Porous Supports

Efficient fluorography of ^{3}H and ^{14}C on solid supports such as thin-layer plates, paper chromatograms, and nitrocellulose filters can be obtained by dipping the completely dried support into a slightly warm solution (about 37°) of 0.4% PPO in 2-methylnaphthalene (Aldrich Chemical Co.).[6] The plate or sheet is allowed to stand or hang vertically so that the excess solution can run off. The solution solidifies as it cools to ambient temperature. The support can then be exposed to flashed or unflashed film at −70°. One caution is that plates should not be left uncovered overnight, since the 2-methylnaphthalene is volatile. The levels of detection are similar to those for polyacrylamide gels.

Strong β- and γ-Emitters in Polyacrylamide Gels or on Solid Supports

For strong β-emitters (^{32}P) and γ-emitters (^{125}I), most of the radiation passes through the film without exposing it. By means of an intensifying screen on the other side of the film, some of this radiation is converted into light, which then passes back into the film, exposing it.[7] Film cassettes can be purchased with intensifying screens permanently attached to one or both faces. In general, calcium tungstate intensifying screens are most suitable. Dried gels should be exposed at −70°; wet gels are better exposed at 0°, because freezing may crack them. In some cases, flashed film and intensifying screens should not be combined, since some screens contain enough endogenous radioactivity to blacken the film during a long exposure. Approximately 1 dpm/mm^2 of ^{125}I and 0.5 dpm/mm^2 of ^{32}P can be detected after exposure overnight.

Film Exposure Conditions

For fluorography the dried gel is placed in a cassette with the correct film and exposed at −70°.[8,9] The appropriate type of film is essential for good results since the sensitivities of different types can vary by factors of 10.[1] Kodak X-Omat AR or the equivalent from other manufacturers should be used. Film images can be quantitated by microdensitometric techniques, but, for small numbers of spots, another useful method is merely to cut out the spots from the dried fluorographed gel. If the gel is dried on thin paper and spotted around the edges with radioactive ink

[6] W. M. Bonner and J. D. Stedman, *Anal. Biochem.* **89,** 247 (1978).
[7] R. A. Laskey and A. D. Mills, *FEBS Lett.* **82,** 314 (1977).
[8] K. Randerath, *Anal. Biochem.* **34,** 188 (1970).
[9] U. Lüthi and P. G. Waser, *Nature (London)* **205,** 1190 (1965).

(about 1 μCi of ^{14}C per milliliter) the fluorograph can easily be placed in register over the dried gel and the desired spots marked with pencil on the paper backing while holding the film–gel sandwich in front of a strong light. Spots are excised and placed in scintillation vials. Pieces of dried gel infused with fluor will be recorded in a scintillation counter, but the geometry must be kept constant for accurate quantitation. It is generally more accurate to digest the pieces of dried gel overnight at 37° in 1 ml of a fresh solution of 95 parts of 30% H_2O_2 and 5 parts of concentrated NH_4OH in capped scintillation vials. Plastic vials should be used, since glass occasionally explodes. The next morning, vials are cooled and a water-miscible scintillation fluid is added. Since this is a homogeneous system, internal standards can be added to calculate counting efficiency and, hence, the absolute amount of radioactivity in the area. This method also works well for areas in which different isotopes are present, i.e., ^{3}H and ^{14}C.

Flashing Film to Linearize Its Response.[10] The response of film to light is nonlinear. At low light intensities, the density of the resultant image is approximately proportional to the square of the radioactive density. The response of the film can be made to approach linearity by exposing it to a hypersensitizing light flash. A single flash of less than 1 msec duration from an electronic flash unit is suitable, but it is usually necessary to decrease its intensity with a Kodak Wratten 21 or 22 (orange) filter and several layers of exposed film. In addition to the hypersensitized grains, some of the grains in the film are also exposed by the flash, resulting in an increased background fogging. When the fogging density is about 0.15 Å, the film response is nearly linear. The distance between the film and flash unit should be varied until that density is obtained. The density can easily be measured by developing a piece of flashed film, cutting out a small piece, and placing it in the cuvette holder of a spectrophotometer. Since exposed film acts as a neutral density filter, any visible wavelength, such as 650 nm, can be used. Once these conditions are ascertained, the film is placed on a yellow background and flashed; its flashed face is then placed against the gel.

Flashed film is useful for accurate microdensitometric quantitation of spots as well as for two situations in which accurate quantitation may not be required. When the areas of interest are near the minimum level of detection, flashing the film results in a two- to threefold increase in relative density for them.[6,10] This is because the nonlinear response of unflashed film underrepresents the faintest spots. When storage at −70° is unavailable, ^{3}H and ^{14}C fluorography is possible with flashed film but not unflashed film at ambient temperature; however, the sensitivity is about half that at −70°.

[10] R. A. Laskey and A. D. Mills, *Eur. J. Biochem.* **56,** 335 (1975).

Gel Preparation for Easier Fluorography

Fluorography is facilitated if two factors are considered even before acrylamide gels are prepared. First, gels should be elastic for ease in handling and resistance to cracking during drying. Blattler *et al.*[11] found that, if the final acrylamide percentage (A) times the final bisacrylamide percentage (B) satisfies the equation $AB = 1.3$, the gels will have satisfactory elasticity throughout the range from 4% acrylamide to 40%. This laboratory has routinely used the formula $AB = 1.5$ for gels from 5 to 60% acrylamide and found these gels to have good elasticity and resistance to cracking. An SDS gel with 15% acrylamide and 0.1% bisacrylamide follows this formula and separates proteins greater than 10,000 daltons. Stacking gels that are removed before fluorography do not need to follow this formula.

A second factor to be considered is gel thickness. Since the time required for a substance to diffuse a given distance is proportional to the square of that distance (Fick's second law), the use of thinner gels would greatly decrease the time necessary for infusion with fluors and solvents. In practice, gels of 0.8 mm satisfying the formula $AB = 1.5$ can be handled easily and can be infused with scintillant in less than 1 hr instead of the 3 hr required for 1.5-mm gels. Spacers and slot combs for such gels can be cut easily with scissors or a razor blade from skived Teflon tape (0.8 mm = 0.031 in.), available from most plastic supply houses.

[11] D. P. Blattler, F. Garner, K. Van Slyke, and A. Bradley, *J. Chromatogr.* **64,** 147 (1972).

Author Index

Numbers in parentheses are footnote reference numbers and indicate that an author's work is referred to although the name is not cited in the text.

A

B

C

E

F

G

H

I

J

K

L

M

N

O

Q

R

S

T

U

V

W

X

Y

Z

Subject Index

A

B

C

H

M

N

O

P

S

T

U

V

W

Y

Z